Springer Series in Computational Physics

Editors: C. A. J. Fletcher   R. Glowinski   W. Hillebrandt   M. Holt   P. Hut
H. B. Keller   J. Killeen   S. A. Orszag   V. V. Rusanov

Springer Series in Computational Physics

Editors: C. A. J. Fletcher   R. Glowinski   W. Hillebrandt   M. Holt   P. Hut
H. B. Keller   J. Killeen   S. A. Orszag   V. V. Rusanov

A Computational Method in Plasma Physics
F. Bauer, O. Betancourt, P. Garabedian

Implementation of Finite Element Methods for Navier-Stokes Equations
F. Thomasset

Finite-Difference Techniques for Vectorized Fluid Dynamics Calculations
Edited by D. Book

Unsteady Viscous Flows.   D. P. Telionis

Computational Methods for Fluid Flow.   R. Peyret, T. D. Taylor

Computational Methods in Bifurcation Theory and Dissipative Structures
M. Kubicek, M. Marek

Optimal Shape Design for Elliptic Systems.   O. Pironneau

The Method of Differential Approximation.   Yu. I. Shokin

Computational Galerkin Methods.   C. A. J. Fletcher

Numerical Methods for Nonlinear Variational Problems
R. Glowinski

Numerical Methods in Fluid Dynamics.   Second Edition
M. Holt

Computer Studies of Phase Transitions and Critical Phenomena
O. G. Mouritsen

Finite Element Methods in Linear Ideal Magnetohydrodynamics
R. Gruber, J. Rappaz

Numerical Simulation of Plasmas.   Y. N. Dnestrovskii, D. P. Kostomarov

Computational Methods for Kinetic Models of Magnetically Confined Plasmas
J. Killeen, G. D. Kerbel, M. C. McCoy, A. A. Mirin

Spectral Methods in Fluid Dynamics. Second Edition
C. Canuto, M. Y. Hussaini, A. Quarteroni, T. A. Zang

Computational Techniques for Fluid Dynamics 1.   Second Edition
Fundamental and General Techniques.   C. A. J. Fletcher

Computational Techniques for Fluid Dynamics 2.   Second Edition
Specific Techniques for Different Flow Categories.   C. A. J. Fletcher

**Methods for the Localization of Singularities in Numerical Solutions
of Gas Dynamics Problems.**   E. V. Vorozhtsov, N. N. Yanenko

Classical Orthogonal Polynomials of a Discrete Variable
A. F. Nikiforov, S. K. Suslov, V. B. Uvarov

**Flux Coordinates and Magnetic Field Structure:
A Guide to a Fundamental Tool of Plasma Theory**
W. D. D'haeseleer, W. N. G. Hitchon, J. D. Callen, J. L. Shohet

Monte Carlo Methods in Boundary Value Problems
K. K. Sabelfeld

A.F. Nikiforov   S.K. Suslov
V.B. Uvarov

Classical Orthogonal Polynomials of a Discrete Variable

With 26 Figures

Springer-Verlag

Berlin Heidelberg New York
London Paris Tokyo
Hong Kong Barcelona
Budapest

Professor Dr. Arnold F. Nikiforov
Professor Dr. Vasilii B. Uvarov

M.V. Keldysh Institute of Applied Mathematics, Academy of Sciences
of the USSR, Miusskaya pl. 4, SU-125047 Moscow, USSR

Sergei K. Suslov

Kurchatov Institute of Atomic Energy, SU-123182 Moscow, USSR

Editors

C. A. J. Fletcher

Department of Mechanical Engineering
The University of Sydney
New South Wales, 2006
Australia

R. Glowinski

Institut de Recherche d'Informatique
et d'Automatique (INRIA)
Domaine de Voluceau
Rocquencourt, B. P. 105
F-78150 Le Chesnay, France

W. Hillebrandt

Max-Planck-Institut für Astrophysik
Karl-Schwarzschild-Straße 1
W-8046 Garching, Fed. Rep. of Germany

M. Holt

College of Engineering and
Mechanical Engineering
University of California
Berkeley, CA 94720, USA

P. Hut

The Institute for Advanced Study
School of Natural Sciences
Princeton, NJ 08540, USA

H. B. Keller

Applied Mathematics 101-50
Firestone Laboratory
California Institute of Technology
Pasadena, CA 91125, USA

J. Killeen

Lawrence Livermore Laboratory
P. O. Box 808
Livermore, CA 94551, USA

S. A. Orszag

Department of Mechanical and
Aerospace Engineering
Princeton University
Princeton, NJ 08544, USA

V. V. Rusanov

M. V. Keldysh Institute
of Applied Mathematics
Miusskaya pl. 4
SU-125047 Moscow, USSR

ISBN-13: 978-3-642-74750-2   e-ISBN-13: 978-3-642-74748-9
DOI: 10.1007/978-3-642-74748-9

Library of Congress Cataloging-in-Publication Data. Nikiforov, A. F. [Klassicheskie ortogonal 'nye polinomy diskretnoĭ peremennoĭ. English]. Classical orthogonal polynomials of a discrete variable / A. F.
Nikiforov, S. K. Suslov, V. B. Uvarov. p. cm. — (Springer series in computational physics). Translation
of: Klassicheskie ortogonal 'nye polinomy diskretnoĭ peremennoĭ. Includes bibliographical references
and index. ISBN 0-387-51123-7 (U.S. : alk. paper). 1. Orthogonal polynomials. 2. Functions, Special.
3. Multivariate analysis. 4. Mathematical physics. I. Suslov, S. K. (Sergeĭ Konstantinovich). II. Uvarov,
V. B. (Vasiliĭ Borisovich). III. Title. IV. Series. QC20.7.075N5513   1990   515'.55—dc20   90-9793

Typesetting: Springer TEX inhouse system
57/3140-543210 – Printed on acid-free paper

Preface

Mathematical modelling of many physical processes involves rather complex differential, integral, and integro-differential equations which can be solved directly only in a number of cases. Therefore, as a first step, an original problem has to be considerably simplified in order to get a preliminary knowledge of the most important qualitative features of the process under investigation and to estimate the effect of various factors. Sometimes a solution of the simplified problem can be obtained in the analytical form convenient for further investigation. At this stage of the mathematical modelling it is useful to apply various special functions.

Many model problems of atomic, molecular, and nuclear physics, electrodynamics, and acoustics may be reduced to equations of hypergeometric type,

$$\sigma(x)y'' + \tau(x)y' + \lambda y = 0 \quad , \tag{0.1}$$

where $\sigma(x)$ and $\tau(x)$ are polynomials of at most the second and first degree respectively and λ is a constant [E7, A1, N18]. Some solutions of (0.1) are functions extensively used in mathematical physics such as classical orthogonal polynomials (the Jacobi, Laguerre, and Hermite polynomials) and hypergeometric and confluent hypergeometric functions.

On the other hand, along with the special functions which are the solutions of (0.1), in various fields of physics and mathematics wide use is made of quantities that are determined on a discrete set of argument values. Examples are the Clebsch-Gordan and Racah coefficients (the $3j$- and $6j$-symbols), which have long been used in quantum mechanics and in the theory of group representations. Moreover, in probability theory (specifically, in problems of queuing theory [F1a] and birth and death processes [F1a, K6–9]), in coding theory [L6], etc. classical orthogonal polynomials of a discrete variable (the Hahn, Meixner, Kravchuk, and Charlier polynomials) are extensively used. It has been proved that all these functions may be described by means of a unified treatment in terms of polynomials which are solutions of the difference equation of hypergeometric type, proposed in [N17]. This equation may be obtained by approximating the differential equation (0.1) on lattices of certain classes. It has been found that the polynomial solutions of the above difference equation include the polynomials earlier introduced from various considerations by Markov [M5], Chebyshev [T3], Rogers [R21–R23], Stieltjes and Wigert [S24, W6], Pollaczek [P10], and Karlin and MacGregor [K6], and also the Askey-Wilson polynomials [A27, W8, A29], introduced by means of the basic hypergeometric series.

A general theory of polynomial solutions for the difference equation of hypergeometric type was first constructed in preprint [N17] by generalizing the method applied earlier to the differential equation (0.1) [N16, N16a]. The present monograph represents an appreciably revised and extended version of the book [N14] earlier published in Russian (see also [N18]).

In the monograph the reader will find a systematic, concise presentation of the theory of polynomial solutions of the hypergeometric-type difference equation. The book contains methods for solving a wide class of difference equations and recursion relations as well as applications of the classical orthogonal polynomials of a discrete variable in computational mathematics, probability theory, and coding theory, for information compression and storage. It is shown that the basic quantities of the representation theory of the rotation group – generalized spherical harmonics and the Clebsch-Gordan and Racah coefficients – can be expressed in terms of the classical orthogonal polynomials of a discrete variable.

Moreover, a general method for obtaining particular solutions of arbitrary difference equations of hypergeometric type in the form of generalized q-hypergeometric series has been constructed.

The book has been written according to the following scheme. The first three chapters form the basis of the book. Chapter 1 gives a concise review of the theory of classical orthogonal polynomials – the Jacobi, Laguerre, and Hermite polynomials which satisfy the differential equation of hypergeometric type (0.1). Polynomial solutions of Eq. (0.1) are studied, using the fact that derivatives of the solutions of (0.1) also satisfy the equation of type (0.1) [N5, N18]. This enables us to obtain readily explicit expressions for classical orthogonal polynomials in the form of the Rodrigues formula. Then the orthogonality property is proved and differentiation formulas, recursion relations, and some other properties are deduced.

Chapter 2 considers the difference equation of hypergeometric type

$$
\begin{aligned}
&\frac{\sigma(x)}{h}\left[\frac{y(x+h)-y(x)}{h}-\frac{y(x)-y(x-h)}{h}\right] \\
&\quad +\frac{\tau(x)}{2}\left[\frac{y(x+h)-y(x)}{h}+\frac{y(x)-y(x-h)}{h}\right]+\lambda y(x)=0 \ ,
\end{aligned}
\tag{0.2}
$$

which approximates the differential second-order equation (0.1) on a lattice with constant mesh h up to the second order of accuracy [S1, G17][1]. For some values of $\lambda = \lambda_n$ ($n = 0, 1, \ldots$) the particular solutions of Eq. (0.2) are classical orthogonal polynomials of a discrete variable – the Hahn, Meixner, Kravchuk, and Charlier polynomials[2]. The theory of these polynomials is developed following

---

[1] The difference operator L_h approximates the differential operator L at point x to the order of accuracy m with respect to mesh h if $L_h y(x) - L y(x) = O(h^m)$ when $h \to 0$.

[2] The origin of the term "polynomials of a discrete variable" may be explained as follows: for the Hahn, Meixner, Kravchuk, and Charlier polynomials the orthogonality property is written in the form of a sum with a certain weight over a discrete set of lattice points (instead of the integral form for the Jacobi, Laguerre, and Hermite polynomials).

the same logical scheme as in the theory of the Jacobi, Laguerre, and Hermite polynomials, but the derivatives have to be replaced by appropriate difference quotients. For the Hahn, Meixner, Kravchuk, and Charlier polynomials an analog of the Rodrigues formula is obtained, the orthogonality property is established, and the "difference differentiation" formula, asymptotic representations, etc. are derived.[3]

Chapter 3, which is fundamental to the whole book, gives a generalization of the difference equation of hypergeometric type (0.2) to the case of a lattice with a varying mesh. In order to obtain the respective difference equation for nonuniform lattices it is convenient to pass from the variable x to the variable s assuming $x = x(s)$ and to use the values of $x(s)$ on an s-uniform lattice $s = s_i$ ($i = 0, 1, \ldots$), where $\Delta s = s_{i+1} - s_i = h$. Then the mesh $\Delta x(s_i) = x\big(s_{i+1}\big) - x\big(s_i\big)$ will be variable. After the replacement of x by $x(s)$ we obtain the difference equation

$$\frac{\sigma[x(s)]}{x(s + h/2) - x(s - h/2)} \left[\frac{y(s + h) - y(s)}{x(s + h) - x(s)} - \frac{y(s) - y(s - h)}{x(s) - x(s - h)} \right] \\ + \frac{\tau[x(s)]}{2} \left[\frac{y(s + h) - y(s)}{x(s + h) - x(s)} + \frac{y(s) - y(s - h)}{x(s) - x(s - h)} \right] + \lambda y(s) = 0 , \tag{0.3}$$

which corresponds to Eq. (0.2). In this equation $\sigma(x)$ and $\tau(x)$ are arbitrary polynomials in x of at most the second and first degree respectively and λ is a constant.

Equation (0.3) approximates the original differential equation (0.1) up to the second order of accuracy when $h \to 0$ on a nonuniform lattice $x = x(s)$. It is shown that, for a certain class of nonuniform lattices, Eq. (0.3) allows the keeping of a property similar to the fundamental property of the differential equation (0.1) and the difference equation (0.2): the function

$$v(s) = \frac{y(s + h/2) - y(s - h/2)}{x(s + h/2) - x(s - h/2)} ,$$

which is approximately equal to the derivative dy/dx at the point $x = x(s)$, satisfies an equation of the same type as the function $y(s)$.

For the above class of functions $x(s)$ this property lets us reserve all the basic points of argument used in Chap. 2 and, by applying elementary mathematical tools, obtain the basic properties of polynomial solutions of Eq. (0.3) — a discrete analog of the Rodrigues formula, the orthogonality property, the recursion relations, asymptotic representations, etc.

The class of functions $x(s)$ under consideration has the form

$$x(s) = C_1 q^s + C_2 q^{-s} + C_3 , \tag{0.4}$$

where C_1, C_2, and C_3 are arbitrary constants and q is a parameter. This class also includes linear and quadratic lattices because constants C_1, C_2, and C_3 may,

---

[3] Particular solutions of Eq. (0.2) with arbitrary complex λ are obtained in [N11] (see also [N18].

in general, depend on q, so that we may choose the constants $C_i = C_i(q)$ such that the expression (0.4) transforms into $x(s) = C_1 s^2 + C_2 s + C_3$ when $q \to 1$.

For the functions $x(s)$ of the form (0.4) Eq. (0.3) is called *the difference equation of hypergeometric type*. The polynomial solutions of this equation that have the orthogonality property in the form of a sum over a discrete set of lattice points are called *classical orthogonal polynomials of a discrete variable* (according to terminology accepted in [E7]).

The Racah polynomials and the dual Hahn polynomials for the quadratic lattice $x(s) = s(s+1)$ that are important in applications are studied in particular detail. For all nonuniform lattices at $q \neq 1$ the systems of polynomials that in the limit $q \to 1$ take the form of polynomials orthogonal on linear and quadratic lattices are constructed.

In Sect. 3.10 the polynomial solutions of the difference equation of hypergeometric type (0.3) are considered for arbitrary complex values of the equation coefficients. It is shown that under certain conditions these polynomials have the continuous orthogonality property in the form of an integral in the complex plane of variable s over a contour C (in particular, the Pollaczek polynomials).

In Sect. 3.11 the explicit expression of polynomial solutions in terms of generalized q-hypergeometric series is obtained from the Rodrigues formula for the most general case [N22]. These series are introduced by replacing in the generalized hypergeometric series the values $(a)_k = \Gamma(a + k)/\Gamma(a)$ by the values $(a|q)_k = \tilde{\Gamma}_q(a + k)/\tilde{\Gamma}_q(a)$, where $\tilde{\Gamma}_q(s) = q^{-(s-1)(s-2)/4}\Gamma_q(s)$ is a generalization of Euler's gamma function $\Gamma(s)$ [J1, N18].

From the expression of polynomials in terms of generalized q-hypergeometric series that was obtained for the most general case, the formulas for all particular cases are derived by an appropriate choice of parameters and by taking various limits. The consideration of these particular cases gives us the classification of corresponding q-polynomials. All the earlier introduced polynomials are included in our scheme.

We use *generalized q-hypergeometric series* instead of *basic hypergeometric series* [A27, A29, W8, G7a] because they have more symmetry (for example, these series do not change after replacing q by $1/q$) and for $q \to 1$ they transform into generalized hypergeometric series, since $\lim_{q\to 1} \tilde{\Gamma}_q(s) = \Gamma(s)$.

The remaining chapters (4–6), which form the second part of the book, deal with applications. Chapter 4 discusses applications of classical orthogonal polynomials of a discrete variable in computational mathematics and the theory of difference schemes, in information compression and storage, for the function approximation in a rectangle and on a sphere, in the theory of probability and coding theory, in the genetic Moran model, and in some problems of queueing theory. Here the difference analogs of spherical harmonics orthogonal on a discrete set of sphere points are constructed by using the Hahn polynomials and the q-analogs of the Racah polynomials on a cosinusoidal lattice[4].

---

[4] For other important applications see [N2a, V1].

In Chap. 5 the basic quantities of the theory of representations of the three-dimensional rotation group and the quantum theory of angular momentum – generalized spherical harmonics, the Clebsch-Gordan coefficients and the $6j$-symbols of Wigner – are expressed through the Kravchuk, Hahn, and Racah polynomials, respectively, which allows the representation of properties of these quantities in a simple form. Since the Hahn polynomials are difference analogs of the Jacobi polynomials, the relation between the Clebsch-Gordan coefficients and the Hahn polynomials easily explains an analogy between these coefficients and the Jacobic polynomials, noted by I.M. Gel'fand [G13]. Chapter 5 discusses also close connections of the Hahn polynomials with the group representations of four-dimensional rotations SO(4) and the Lorenz group SO(3.1), as well as the Racah polynomials with the representations of group SU(3). It is shown that the $9j$-symbols form up to normalization a new system of orthogonal polynomials of two discrete variables. The main properties of these polynomials are established by building on the quantum theory of angular momentum.

Chapter 6 considers the method of trees – a simple graphical technique of solving a multidimensional Laplace equation – proposed by N.Ya. Vilenkin, G.I. Kuznetsov, and Ya.A. Smorodinsky in [V8, V10]. Coefficients of transformation between solutions of the Laplace equation in different systems of spherical coordinates (the T-coefficients [K10]) are expressed through the Racah, Hahn, and Kravchuk polynomials.

The first four chapters of the book were written by A.F. Nikiforov and V.B. Uvarov except for Sects. 3.10.3–5, some material from [N18] being used in the first part on foundations of the theory; Chaps. 5 and 6 were written by S.K. Suslov.

The book is aimed at a wide range of specialists in theoretical and mathematical physics and computational mathematics. Most of the material is sufficiently elementary that it is possible to use it as a textbook for undergraduate and graduate students of physical and mathematical disciplines, those studying quantum mechanics, and also those who lecture on mathematics and physics.

<table>
<tr><td>Moscow</td><td align="right">*A. F. Nikiforov*</td></tr>
<tr><td>April 1991</td><td align="right">*V. B. Uvarov*</td></tr>
<tr><td></td><td align="right">*S. K. Suslov*</td></tr>
</table>

Foreword to the Russian Edition

Classical orthogonal polynomials of a discrete variable are an important class of special functions arising in various problems of mathematics, theoretical physics, computational mathematics and engineering; this field is now under extensive development. It should be noted that there is a deep analogy between classical orthogonal polynomials of continuous and discrete arguments, and the theory of group representation is one of its basic manifestations. This analogy was noted by I.M. Gel'fand in the mid-fifties [G13] in connection with the study of representations of the rotation group playing an important role in theoretical physics.

Studies of classical orthogonal polynomials of a discrete variable were initiated by P.L. Chebyshev in the middle of the last century and continued by many prominent scientists. However, there are no books where a theory of these polynomials is consistently developed. Up until recently it was not even clear what polynomials introduced by different authors from various viewpoints belong to the above class of special functions.

In this book the reader will find for the first time a systematic, compact presentation of both the theory of classical orthogonal polynomials of a discrete variable and its main applications. The authors have made a significant contribution to this field. They have developed a simple approach to the construction of the theory of classical orthogonal polynomials of a discrete variable as solutions of a difference equation of hypergeometric type.

Also of interest is a nonstandard approach to investigating the representations of the three-dimensional space rotation group by using the theory of classical orthogonal polynomials of a discrete variable.

This comprehensive book will be useful for both mathematicians and physicists.

Moscow   *M.I.Graev*
February 1984   (editor of the Russian edition)

Acknowledgements

The authors are grateful to Academicians A.N. Tikhonov, A.A. Samarsky, and I.M. Gel'fand and the participants of their seminars at Moscow State University for the discussion of the basic contents of the book. They also express their gratitude to Academicians S.M. Nikolsky and A.A. Gonchar, and P.I. Lizorkin, who took part in the discussions of the authors' report as well as the report of R. Askey at the V.A. Steklov Mathematical Institute of the USSR Academy of Sciences. The authors thank the editor of the book, M.I. Graev, for his close attention and valuable advice during preparation of the manuscript for the Russian edition. Also, the remarks made by A.A. Abramov, Yu.A. Danilov, D.F. Grechukhin, G.I. Kuznetsov, V.P. Pokrovsky, V.V. Rusanov, V.S. Ryabenky, A.N. Shiryaev, Yu.F. Smirnov, Ya.A. Smorodinsky, and I.M. Sobol were of great use. Thanks are also due to A. Ronveaux, A. Magnus, and J. Dehesa for discussions and suggestions for improvements in preliminary versions of this book. We would like to thank Springer-Verlag for their cooperation and patience during the preparation of the manuscript. In addition, we wish to express our sincere thanks to N.A. Sokolova for help in translating Chaps. 1–4.

Contents

Part I

Foundations of the Theory

1. Classical Orthogonal Polynomials

Classical orthogonal Polynomials – the Jacobi, Laguerre and Hermite polynomials – form the simplest class of special functions. At the same time, the theory of these polynomials admits wide generalizations. By using the Rodrigues formula for the Jacobi, Laguerre and Hermite polynomials we can come to integral representations for other special functions of mathematical physics, for example, hypergeometric functions and Bessel functions [N16, N18]. On the other hand, a construction scheme for the theory of these polynomials can naturally be generalized to classical orthogonal polynomials of a discrete variable. In view of this in Chap. 1 we shall give in a coherent way a brief description of some basic facts of the theory of classical orthogonal polynomials.

1.1 An Equation of Hypergeometric Type

Many problems of applied mathematics, and theoretical and mathematical physics lead to equations of the form

$$\sigma(x)y'' + \tau(x)y' + \lambda y = 0 , \tag{1.1.1}$$

where $\sigma(x)$ and $\tau(x)$ are polynomials of at most second and first degree, respectively, and λ is a constant. We shall refer to (1.1.1) as *an equation of hypergeometric type*, and its solutions as *functions of hypergeometric type*[1].

For any solution of (1.1.1) the following fundamental property is satisfied: *all derivatives of the functions of hypergeometric type are also functions of hypergeometric type.*

To prove this we differentiate (1.1.1). As a result we obtain that the function $v_1(x) = y'(x)$ satisfies the equation

$$\sigma(x)v_1'' + \tau_1(x)v_1' + \mu_1 v_1 = 0 , \tag{1.1.2}$$

where $\tau_1(x) = \tau(x) + \sigma'(x)$, $\mu_1 = \lambda + \tau'(x)$. Since $\tau_1(x)$ is a polynomial of at most degree 1, and μ_1 does not depend on x, Eq. (1.1.2) is an equation of hypergeometric type.

The converse is also true: every solution of (1.1.2) with $\lambda = \mu_1 - \tau_1' + \sigma'' \neq 0$ is the derivative of a solution of (1.1.1).

---

[1] Equation (1.1.1) is so called because its particular solutions are hypergeometric functions for $\sigma(x) = x(1 - x)$ and confluent hypergeometric functions for $\sigma(x) = x$.

Let $v_1(x)$ be a solution of Eq. (1.1.2). If the function $v_1(x)$ was the derivative of a solution $y(x)$ of (1.1.1), then according to this equation the functions $y(x)$ and $v_1(x)$ would be related in the following way:

$$y(x) = -\frac{1}{\lambda}\left[\sigma(x)v_1' + \tau(x)v_1\right] .$$

We can show that the function $y(x)$ defined by this formula really satisfies (1.1.1), and that its derivative is $v_1(x)$. We have

$$\lambda y' = -\left[\sigma(x)v_1'' + \tau_1(x)v_1' + \tau'(x)v_1\right] = \lambda v_1 ,$$

i.e. $y' = v_1(x)$. Substituting $v_1 = y'(x)$ in the original expression for $y(x)$ we obtain (1.1.1) for $y(x)$.

In a similar way, by differentiating (1.1.1) n times we can obtain an equation of hypergeometric type for the function $v_n(x) = y^{(n)}(x)$:

$$\sigma(x)v_n'' + \tau_n(x)v_n' + \mu_n v_n = 0 , \tag{1.1.3}$$

where

$$\tau_n(x) = \tau(x) + n\sigma'(x) , \tag{1.1.4}$$

$$\mu_n = \lambda + n\tau' + \tfrac{1}{2}n(n-1)\sigma'' . \tag{1.1.5}$$

Here every solution of (1.1.3) for $\mu_k \neq 0$ $(k = 0, 1, \ldots, n-1)$ can be represented in the form $v_n(x) = y^{(n)}(x)$, where $y(x)$ is a solution of (1.1.1).

1.2 Polynomials of Hypergeometric Type and Their Derivatives. The Rodrigues Formula

1.2.1. The property of Eq. (1.1.1) considered above lets us construct a family of particular solutions of (1.1.1) corresponding to a given λ. In fact, when $\mu_n = 0$ Eq. (1.1.3) has the particular solution $v_n(x) = \text{const}$. Since $v_n(x) = y^{(n)}(x)$, this means that when

$$\lambda = \lambda_n = -n\tau' - \tfrac{1}{2}n(n-1)\sigma''$$

the equation of hypergeometric type has a particular solution of the form $y(x) = y_n(x)$ which is a polynomial of degree n. We shall call such solutions *polynomials of hypergeometric type*. The polynomials $y_n(x)$ are the simplest solutions of (1.1.1).

1.2.2. To find the polynomials of hypergeometric type explicitly we multiply (1.1.1) and (1.1.3) by functions $\varrho(x)$ and $\varrho_n(x)$, thus reducing them to self-adjoint form:

$$(\sigma\varrho y')' + \lambda\varrho y = 0 , \tag{1.2.1}$$

$$\left(\sigma\varrho_n v_n'\right) + \mu_n\varrho_n v_n = 0 \ . \tag{1.2.2}$$

Here $\varrho(x)$ and $\varrho_n(x)$ satisfy the differential equations

$$(\sigma\varrho)' = \tau\varrho \ , \tag{1.2.3}$$

$$\left(\sigma\varrho_n\right)' = \tau_n\varrho_n \ . \tag{1.2.4}$$

Now using (1.1.4) for $\tau_n(x)$ we can easily establish the connection between $\varrho_n(x)$ and $\varrho_0(x) \equiv \varrho(x)$. We have

$$\frac{\sigma\varrho_n'}{\varrho_n} = \tau + n\sigma' = \frac{(\sigma\varrho)'}{\varrho} + n\sigma'$$

whence

$$\frac{\varrho_n'}{\varrho_n} = \frac{\varrho'}{\varrho} + \frac{n\sigma'}{\sigma} \ ,$$

and consequently

$$\varrho_n(x) = \sigma^n(x)\varrho(x) \quad (n = 0, 1, \ldots) \ . \tag{1.2.5}$$

Since $\sigma(x)\varrho_n(x) = \varrho_{n+1}(x)$ and $v_n'(x) = v_{n+1}(x)$ we can rewrite (1.2.2) in the form of the recurrence relation

$$\varrho_n v_n = -\frac{1}{\mu_n}\left(\varrho_{n+1} v_{n+1}\right)' \ . \tag{1.2.6}$$

Hence we obtain successively

$$\varrho y \equiv \varrho_0 v_0 = -\frac{1}{\mu_0}\left(\varrho_1 v_1\right)'$$

$$= \left(-\frac{1}{\mu_0}\right)\left(-\frac{1}{\mu_1}\right)\left(\varrho_2 v_2\right)'' = \cdots = \frac{1}{A_n}\left(\varrho_n v_n\right)^{(n)} \ ,$$

where

$$A_n = (-1)^n \prod_{k=0}^{n-1} \mu_k \ , \qquad A_0 = 1 \ . \tag{1.2.7}$$

We now proceed to obtain an explicit form for polynomials of hypergeometric type. If the function $y(x)$ is a polynomial of degree n, i.e. $y = y_n(x)$, then $v_n(x) = y_n^{(n)}(x) = \text{const}$, and we obtain the following expression for $y_n(x)$:

$$y_n(x) = \frac{B_n}{\varrho(x)}\left[\sigma^n(x)\varrho(x)\right]^{(n)} \ , \quad n = 0, 1, \ldots, \tag{1.2.8}$$

where $B_n = A_n^{-1}y_n^{(n)}(x)$ is a normalizing constant, and A_n is defined by Eq. (1.2.7) with

$$\mu_k = \lambda + k\tau' + \tfrac{1}{2}k(k-1)\sigma'' \ , \quad \lambda = \lambda_n = -n\tau' - \tfrac{1}{2}n(n-1)\sigma'' \ .$$

Thus the polynomial solutions of Eq. (1.1.1) are uniquely defined by (1.2.8) up to an arbitrary constant. These solutions correspond to the values $\mu_n = 0$, i.e.

$$\lambda = \lambda_n = -n\tau' - \tfrac{1}{2}n(n-1)\sigma'' , \quad n = 0, 1, \dots . \tag{1.2.9}$$

We call relation (1.2.8) the *Rodrigues formula* [E7].

1.2.3. Since the derivatives $y_n^{(m)}(x) \equiv v_{mn}(x)$ are polynomials of degree $n - m$ and satisfy the equation

$$\sigma(x)v''_{mn} + \tau_m(x)v'_{mn} + \mu_m v_{mn} = 0 , \tag{1.2.10}$$

they are also polynomials of hypergeometric type. The Rodrigues formula for $y_n^{(m)}(x)$ can be derived from (1.2.6) in the same way as Eq. (1.2.8). As a result we obtain

$$y_n^{(m)}(x) = \frac{A_{mn} B_n}{\sigma^m(x)\varrho(x)} \frac{d^{n-m}}{dx^{n-m}} \left[\sigma^n(x)\varrho(x) \right] , \tag{1.2.11}$$

where

$$A_{mn} = A_m(\lambda)|_{\lambda=\lambda_n} = (-1)^m \prod_{k=0}^{m-1} \mu_{kn} , \quad A_{0n} = 1; \quad \mu_{kn} = \mu_k(\lambda_n) .$$

Since

$$\mu_{kn} = \lambda_n - \lambda_k = -(n-k)\left(\tau' + \frac{n+k-1}{2}\sigma'' \right) ,$$

we have

$$A_{mn} = \frac{n!}{(n-m)!} \prod_{k=0}^{m-1} \left(\tau' + \frac{n+k-1}{2}\sigma'' \right) .$$

As it should be expected, Eq. (1.2.11) for $y_n^{(m)}(x)$ can be obtained from the Rodrigues formula (1.2.8) up to the normalizing factor if we replace n by $n - m$ and $\varrho(x)$ by $\varrho_m(x) = \sigma^m(x)\varrho(x)$.

1.2.4. Let us consider some *consequences of the Rodrigues formula* (1.2.11).

1) Letting $m = 1$ in (1.2.11) we obtain

$$\begin{aligned}
y_n'(x) &= \frac{A_{1n} B_n}{\sigma(x)\varrho(x)} \frac{d^{n-1}}{dx^{n-1}} \left[\sigma^n(x)\varrho(x) \right] \\
&= -\frac{\lambda_n B_n}{\varrho_1(x)} \frac{d^{n-1}}{dx^{n-1}} \left[\sigma^{n-1}(x)\varrho_1(x) \right] \\
&= -\lambda_n \frac{B_n}{\bar{B}_{n-1}} \frac{\bar{B}_{n-1}}{\varrho_1(x)} \frac{d^{n-1}}{dx^{n-1}} \left[\sigma^{n-1}(x)\varrho_1(x) \right] \\
&= -\lambda_n \frac{B_n}{\bar{B}_{n-1}} \bar{y}_{n-1}(x) .
\end{aligned} \tag{1.2.12}$$

Here $\bar{y}_n(x)$ is a polynomial arrived by replacing $\varrho(x)$ by $\varrho_1(x)$ in the expression for $y_n(x)$; $\bar{B}_n$ is a normalizing constant in the Rodrigues formula for $\bar{y}_n(x)$.

2) By using the Rodrigues formula we can express the derivatives $y'_n(x)$ in terms of the $y_n(x)$ themselves. In fact, since $\sigma^{n+1}(x)\varrho(x) = \sigma(x)\varrho_n(x)$ and $(\sigma\varrho_n)' = \tau_n\varrho_n$, then according to (1.2.8) and (1.2.11) we have

$$y_{n+1}(x) = \frac{B_{n+1}}{\varrho(x)}\left\{\tau_n(x)\frac{d^n}{dx^n}\left[\sigma^n(x)\varrho(x)\right] + n\tau'_n\frac{d^{n-1}}{dx^{n-1}}\left[\sigma^n(x)\varrho(x)\right]\right\}$$

$$= \frac{B_{n+1}}{B_n}\left[\tau_n(x)y_n(x) - \frac{n}{\lambda_n}\tau'_n\sigma(x)y'_n(x)\right] .$$

Hence

$$\sigma(x)y'_n(x) = \frac{\lambda_n}{n\tau'_n}\left[\tau_n(x)y_n(x) - \frac{B_n}{B_{n+1}}y_{n+1}(x)\right] . \tag{1.2.13}$$

3) From (1.2.11) for $m = n - 1$ it is easy to find the leading coefficients a_n and b_n in the expansion $y_n(x) = a_n x^n + b_n x^{n-1} + \dots$. Since

$$y_n^{(n-1)}(x) = n!a_n x + (n-1)!b_n ,$$

$$\frac{d}{dx}\left[\sigma^n(x)\varrho(x)\right] = \frac{d}{dx}\left[\sigma(x)\varrho_{n-1}(x)\right] = \tau_{n-1}(x)\varrho_{n-1}(x) ,$$

we have

$$n!a_n x + (n-1)!b_n = A_{n-1,n}B_n\tau_{n-1}(x) .$$

Hence

$$a_n = \frac{A_{n-1,n}B_n}{n!}\tau'_{n-1} = B_n\prod_{k=0}^{n-1}\left(\tau' + \frac{n+k-1}{2}\sigma''\right) , \quad a_0 = B_0 ;$$

$$b_n/a_n = n\tau_{n-1}(0)/\tau'_{n-1} . \tag{1.2.14}$$

1.3 The Orthogonality Property

1.3.1. The polynomial solutions of (1.1.1) have the orthogonality property. To obtain this property we write equations for $y_n(x)$ and $y_m(x)$ in the self-adjoint form

$$\left[\sigma(x)\varrho(x)y'_n\right]' + \lambda_n\varrho(x)y_n = 0 , \quad \left[\sigma(x)\varrho(x)y'_m\right]' + \lambda_m\varrho(x)y_m = 0 .$$

We multiply the first equation by $y_m(x)$ and the second by $y_n(x)$. Then we subtract the second equality from the first and integrate the result over x from a to b. Since

$$y_m(x)\left[\sigma(x)\varrho(x)y'_n\right]' - y_n(x)\left[\sigma(x)\varrho(x)y'_m\right]'$$

$$= \frac{d}{dx}\left\{\sigma(x)\varrho(x)W\left[y_m(x), y_n(x)\right]\right\} ,$$

where

$$W(u,v) = \begin{vmatrix} u & v \\ u' & v' \end{vmatrix}$$

is the Wronskian, we obtain the following relation

$$(\lambda_m - \lambda_n) \int_a^b y_m(x)y_n(x)\varrho(x)dx = \sigma(x)\varrho(x)W\left[y_m(x),\, y_n(x)\right]_a^b .$$

Let the function $\varrho(x)$ satisfy the conditions

$$\sigma(x)\varrho(x)x^k\big|_{x=a,b} = 0 , \quad k = 0, 1, \ldots \tag{1.3.1}$$

for some a and b. Then the right-hand side of the equliaty obtained is zero because the Wronskian $W\left[y_m(x),\, y_n(x)\right]$ is a polynomial in x. Therefore when $\lambda_m \neq \lambda_n$ we have

$$\int_a^b y_m(x)y_n(x)\varrho(x)dx = 0 . \tag{1.3.2}$$

Note that the condition $\lambda_m \neq \lambda_n$ in (1.3.2) can be replaced by the condition $m \neq n$ if

$$\tau' + \tfrac{1}{2}(n + m - 1)\sigma'' \neq 0 . \tag{1.3.3}$$

In order to fulfil the conditions (1.3.1) for the finite values of a and b it is sufficient to require that the function $\varrho(x)$ must satisfy the following boundary conditions:

$$\sigma(x)\varrho(x)\big|_{x=a} = 0 , \quad \sigma(x)\varrho(x)\big|_{x=b} = 0 .$$

But if, for example, a is a finite value, and $b = +\infty$ then the conditions (1.3.1) are equivalent to the conditions:

$$\sigma(x)\varrho(x)\big|_{x=a} = 0 , \quad \lim_{x\to+\infty} \sigma(x)\varrho(x)x^k = 0 , \quad k = 0, 1, \ldots .$$

The other possible cases can be considered in an analogous way.

The polynomials of hypergeometric type $y_n(x)$ for which the function $\varrho(x)$ satisfies the condition (1.3.1) are known as the *classical orthogonal polynomials*. These polynomials are usually considered under the auxiliary requirements that $\varrho(x) > 0$ and $\sigma(x) > 0$ on the interval (a,b).

Let us note in conclusion that *the system of classical orthogonal polynomials* $\{y_n(x)\}$ *is closed* on the interval (a,b) for the continuous functions $f(x)$ that satisfy the condition of square-integrability

$$\int_a^b f^2(x)\varrho(x)dx < \infty ,$$

i.e. from the equalities

$$\int_a^b f(x)y_n(x)\varrho(x)dx = 0 , \quad n = 0, 1, \ldots$$

it follows that $f(x) = 0$ for $x \in (a, b)$. For details, see [S38, E7, G15, N18].

1.3.2. From the properties of the derivatives of polynomials of hypergeometric type it follows that *the derivatives of the classical orthogonal polynomials $y_n^{(k)}(x)$ are also classical polynomials, orthogonal with weight $\varrho_k(x) = \sigma^k(x)\varrho(x)$ on the interval (a, b):*

$$\int_a^b y_m^{(k)}(x)y_n^{(k)}(x)\varrho_k(x)dx = \delta_{mn}d_{kn}^2 . \tag{1.3.4}$$

The squared norms d_{kn}^2 of polynomials $y_n^{(k)}$ can be expressed in terms of the squared norm $d_n^2 = d_{0n}^2$ of polynomial $y_n(x)$ if we use (1.2.2) for $y_n^{(k)}(x)$:

$$\frac{d}{dx}\left[\varrho_{k+1}(x)y_n^{(k+1)}(x)\right] + \mu_{kn}\varrho_k(x)y_n^{(k)}(x) = 0 . \tag{1.3.5}$$

We multiply this equation by $y_n^{(k)}(x)$ and integrate over the interval (a, b). After integrating by parts we obtain:

$$\varrho_{k+1}(x)y_n^{(k+1)}(x)y_n^k(x)\big|_a^b - \int_a^b \left[y_n^{(k+1)}(x)\right]^2 \varrho_{k+1}(x)dx$$

$$+ \mu_{kn}\int_a^b \left[y_n^{(k)}(x)\right]^2 \varrho_k(x)dx = 0 .$$

Owing to the condition (1.3.1) the integrated terms are zero, and therefore $d_{k+1,n}^2 = \mu_{kn}d_{kn}^2$. Hence, by induction, we obtain

$$d_{mn}^2 = d_{nn}^2 \Big/ \prod_{k=m}^{n-1} \mu_{kn} . \tag{1.3.6}$$

Since

$$d_{nn}^2 = \int_a^b \left[y_n^{(n)}(x)\right]^2 \sigma^n(x)\varrho(x)dx = \left(a_n n!\right)^2 \int_a^b \sigma^n(x)\varrho(x)dx , \tag{1.3.7}$$

the calculation of d_{mn}^2, $m = 0, 1, \ldots, n - 1$, can be reduced to the calculation of the integral $\int_a^b \sigma^n(x)\varrho(x)dx$. For $d_n^2 = d_{0n}^2$ we finally obtain

$$d_n^2 = (-1)^n A_{nn} B_n^2 \int_a^b \sigma^n(x)\varrho(x)dx . \tag{1.3.8}$$

1.4 The Jacobi, Laguerre, and Hermite Polynomials

For investigating properties of classical orthogonal polynomials and determining the weight functions $\varrho(x)$ it is convenient to use the fact that as a result of a linear change of independent variable x equations (1.1.1) and (1.2.3) are transformed to the ones of the same type. Meanwhile the polynomials $y_n(x)$ remain polynomials of the same degree and can still be defined by the Rodrigues formula. This enables us to carry out classification of classical orthogonal polynomials.

1.4.1 Classification of Polynomials

By using the linear change of an independent variable x, the expressions for $\sigma(x)$ and $\varrho(x)$ obtained after solving Eqs. (1.2.3) can be reduced (up to constant multipliers) to the following canonical forms:

$$\varrho(x) = \begin{cases} (1 - x)^\alpha (1 + x)^\beta & \text{for} \quad \sigma(x) = 1 - x^2 \,, \\ x^\alpha \, e^{-x} & \text{for} \quad \sigma(x) = x \,, \\ e^{-x^2} & \text{for} \quad \sigma(x) = 1 \,. \end{cases}$$

According to the form of the function $\sigma(x)$ we obtain the following systems of polynomials:

1.4.1.1. Let $\sigma(x) = 1 - x^2$, $\varrho(x) = (1 - x)^\alpha (1 + x)^\beta$. Then Eq. (1.2.3) yields $\tau(x) = -(\alpha + \beta + 2)x + \beta - \alpha$. The corresponding polynomials $y_n(x)$ with $B_n = (-1)^n / (2^n n!)$ are called the *Jacobi polynomials*[2] and are designated by $P_n^{(\alpha,\beta)}(x)$. By the Rodrigues formula

$$P_n^{(\alpha,\beta)}(x) = \frac{(-1)^n}{2^n n!} (1 - x)^{-\alpha} (1 + x)^{-\beta} \frac{d^n}{dx^n} \left[(1 - x)^{n+\alpha} (1 + x)^{n+\beta} \right] \,.$$

Hence it is seen that

$$P_n^{(\alpha,\beta)}(-x) = (-1)^n P_n^{(\beta,\alpha)}(x) \,. \tag{1.4.1}$$

For the Jacobi polynomials and their derivatives the boundary conditions (1.3.1) and the orthogonality relations (1.3.2) and (1.3.4) will be satisfied if $a = -1$, $b = 1$, $\alpha > -1$, $\beta > -1$.

The important special cases of the Jacobi polynomials are:

a) the *Legendre polynomials* $P_n(x) = P_n^{(0,0)}(x)$;
b) the *Chebyshev polynomials of the first and second kinds*:

$$T_n(x) = \cos n\varphi \,, \quad U_n(x) = \frac{T'_{n+1}(x)}{n + 1} = \frac{\sin[(n + 1)\varphi]}{\sin \varphi} \,, \tag{1.4.2}$$

where $\varphi = \arccos x$. Below it will be shown that

$$T_n(x) = \frac{n!}{(1/2)_n} P_n^{(-1/2,\,-1/2)}(x) \,, \tag{1.4.3}$$

$$U_n(x) = \frac{(n + 1)!}{(3/2)_n} P_n^{(1/2,\,1/2)}(x) \,. \tag{1.4.4}$$

---

[2] The constants B_n agree with the normalization in [E7], but in a general case they could be chosen arbitrarily.

c) The *Gegenbauer polynomials* (also known as the *ultraspherical polynomials*):

$$C_n^\lambda(x) = \frac{(2\lambda)_n}{(\lambda + 1/2)_n} P_n^{(\lambda-1/2,\, \lambda-1/2)}(x) \ .$$

We have used the notation

$$(\alpha)_n = \alpha(\alpha + 1) \ldots (\alpha + n - 1) = \frac{\Gamma(\alpha + n)}{\Gamma(\alpha)} \ ,$$

where $\Gamma(z)$ is the gamma-function.

1.4.1.2. Let $\sigma(x) = x$, $\varrho(x) = x^\alpha \, e^{-x}$. Then $\tau(x) = -x + \alpha + 1$. The polynomials $y_n(x)$ with $B_n = 1/n!$ are called the *Laguerre polynomials* and are designated by $L_n^\alpha(x)$:

$$L_n^\alpha(x) = \frac{1}{n!} e^x x^{-\alpha} \frac{d^n}{dx^n} \left(x^{n+\alpha} \, e^{-x} \right) \ .$$

The Laguerre polynomials and their derivatives will satisfy the orthogonality relations (1.3.2) and (1.3.4) when $a = 0$, $b = \infty$, $\alpha > -1$.

1.4.1.3 Let $\sigma(x) = 1$, $\varrho(x) = e^{-x^2}$. Then $\tau(x) = -2x$. The polynomials $y_n(x)$ with $B_n = (-1)^n$ are called the *Hermite polynomials* and are designated by $H_n(x)$:

$$H_n(x) = (-1)^n \, e^{x^2} \frac{d^n}{dx^n} \left(e^{-x^2} \right) \ .$$

Hermite polynomials and their derivatives are orthogonal on the interval $(-\infty, \infty)$.

Let us note that in all the cases considered above the condition $\lambda_m \neq \lambda_n$ in (1.3.2) and (1.3.4) can be replaced by the equivalent condition $m \neq n$ [see (1.3.3)].

From (1.2.12) the next differentiation formulas can be derived for the Jacobi, Laguerre and Hermite polynomials:

$$\frac{d}{dx} P_n^{(\alpha,\beta)}(x) = \tfrac{1}{2}(n + \alpha + \beta + 1) P_{n-1}^{(\alpha+1,\beta+1)}(x) \ ,$$

$$\frac{d}{dx} L_n^\alpha(x) = -L_{n-1}^{\alpha+1}(x) \ , \tag{1.4.5}$$

$$\frac{d}{dx} H_n(x) = 2n H_{n-1}(x) \ .$$

The basic information about the Jacobi, Laguerre, and Hermite polynomials is given in Table 1.1.

Table 1.1. Data for the Jacobi, Laguerre and Hermite polynomials

$y_n(x)$	$P_n^{(\alpha,\beta)}(x)\,(\alpha>-1,\beta>-1)$	$L_n^\alpha(x)\,(\alpha>-1)$	$H_n(x)$
(a,b)	$(-1,1)$	$(0,\infty)$	$(-\infty,\infty)$
$\varrho(x)$	$(1-x)^\alpha(1+x)^\beta$	$x^\alpha e^{-x}$	e^{-x^2}
$\sigma(x)$	$1-x^2$	x	1
$\tau(x)$	$\beta-\alpha-(\alpha+\beta+2)x$	$1+\alpha-x$	$-2x$
λ_n	$n(n+\alpha+\beta+1)$	n	$2n$
B_n	$\dfrac{(-1)^n}{2^n n!}$	$\dfrac{1}{n!}$	$(-1)^n$
a_n	$\dfrac{\Gamma(2n+\alpha+\beta+1)}{2^n n!\,\Gamma(n+\alpha+\beta+1)}$	$\dfrac{(-1)^n}{n!}$	2^n
b_n	$\dfrac{(\alpha-\beta)\Gamma(2n+\alpha+\beta)}{2^n(n-1)!\,\Gamma(n+\alpha+\beta+1)}$	$(-1)^{n-1}\dfrac{n+\alpha}{(n-1)!}$	0
d_n^2	$\dfrac{2^{\alpha+\beta+1}\Gamma(n+\alpha+1)\Gamma(n+\beta+1)}{n!(2n+\alpha+\beta+1)\Gamma(n+\alpha+\beta+1)}$	$\dfrac{\Gamma(n+\alpha+1)}{n!}$	$2^n n!\sqrt{\pi}$
α_n	$\dfrac{2(n+1)(n+\alpha+\beta+1)}{(2n+\alpha+\beta+1)(2n+\alpha+\beta+2)}$	$-(n+1)$	$\dfrac{1}{2}$
β_n	$\dfrac{\beta^2-\alpha^2}{(2n+\alpha+\beta)(2n+\alpha+\beta+2)}$	$2n+\alpha+1$	0
γ_n	$\dfrac{2(n+\alpha)(n+\beta)}{(2n+\alpha+\beta)(2n+\alpha+\beta+1)}$	$-(n+\alpha)$	n

1.4.2 General Properties of Orthogonal Polynomials

Some properties of the Jacobi, Laguerre and Hermite polynomials follow directly from their orthogonality. In this connection let us consider some *general properties* of polynomials $p_n(x)$ that are orthogonal on an interval (a,b) with the weight function $\varrho(x)>0$. We shall suppose that the leading coefficient of $p_n(x)$ is real and different from zero. When considering the polynomial properties we shall use the fact that every polynomial $q_n(x)$ of degree n can be represented as a linear combination of the orthogonal polynomials $p_k(x)$, $k=0,1,\ldots,n$, i.e.

$$q_n(x) = \sum_{k=0}^{n} c_{kn}p_k(x)\ . \tag{1.4.6}$$

The coefficients c_{kn} are easily determined by the orthogonality property

$$\int_a^b p_m(x)p_n(x)\varrho(x)dx = 0\ , \quad m\neq n\ , \tag{1.4.7}$$

which leads to the formula

$$c_{kn} = \frac{1}{d_k^2} \int_a^b q_n(x) p_k(x) \varrho(x) dx \, , \qquad (1.4.8)$$

where

$$d_k^2 = \int_a^b p_k^2(x) \varrho(x) dx \, .$$

1.4.2.1. Let us show that the orthogonality relation (1.4.7) is equivalent to

$$\int_a^b p_n(x) x^m \varrho(x) dx = 0 \quad (m < n) \, . \qquad (1.4.9)$$

In fact, if we insert the expansion of $p_m(x)$ in powers of x into the integral (1.4.7) with $m < n$, then (1.4.7) follows from (1.4.9). Similarly, if we expand x^m in terms of the $p_k(x)$, we obtain (1.4.9) from (1.4.7).

It follows from the relation (1.4.9) that $p_n(x)$ *is orthogonal to every polynomial of lower degree.*

1.4.2.2. We can show that the *interval (a, b) and the weight $\varrho(x)$ determine the polynomials $p_n(x)$ that satisfy the orthogonality property, up to a normalizing factor.*

Suppose that there are two polynomials $p_n(x)$ and $\tilde{p}_n(x)$ that satisfy (1.4.9). We have

$$\tilde{p}_n(x) = \sum_{k=0}^{n} c_{kn} p_k(x) \, .$$

By (1.4.8) and (1.4.9) $c_{kn} = 0$ for $k < n$, so that $p_n(x)$ and $\tilde{p}_n(x)$ must be proportional.

If we use complex-conjugate values in (1.4.9), then it can easily be seen that $p_n(x)$ and $p_n^*(x)$ satisfy the same orthogonality relations. Since the leading coefficients of these polynomials coincide with each other, we have $p_n(x) = p_n^*(x)$, i.e. $p_n(x)$ is a polynomial with real coefficients.

Example. Let us establish the connection between the Chebyshev polynomials $T_n(x) = \cos(n \arccos x)$ and the Jacobi polynomials $P_n^{(\alpha,\beta)}(x)$. By using the relation

$$\cos(n+1)\varphi + \cos(n-1)\varphi = 2 \cos \varphi \cos n\varphi \, , \quad x = \cos \varphi$$

and mathematical induction, it is easy to show that the function $T_n(x)$ is the polynomial of degree n with the leading coefficient $a_n = 2^{n-1}$ ($a_0 = 1$). The orthogonality property of $T_n(x)$ is a consequence of the orthogonality property of the functions $\cos n\varphi$:

$$\int_0^\pi \cos m\varphi \cos n\varphi \, d\varphi = 0 \, , \quad m \neq n \, .$$

After the substitution $\cos \varphi = x$ we obtain

$$\int_{-1}^{1} T_m(x)T_n(x)\frac{dx}{\sqrt{1-x^2}} = 0 , \quad m \neq n .$$

Since $T_n(x)$ and $P_n^{(-1/2, -1/2)}(x)$ are orthogonal on an interval $(-1, 1)$ with the same weight $\varrho(x) = (1 - x^2)^{-1/2}$, we have $T_n(x) = C_n P_n^{(-1/2, -1/2)}(x)$, where C_n is a constant. Comparing the coefficients at x^n yields $C_n = n!/\left(\frac{1}{2}\right)_n$.

If we use the differentiation formula (1.4.5) for the Jacobi polynomials, then the Chebyshev polynomials of the second kind

$$U_n(x) = \frac{1}{n+1}T'_{n+1}(x) \quad , \quad x = \cos\varphi$$

may be expressed in terms of the polynomials $P_n^{(1/2, 1/2)}(x)$ as

$$U_n(x) = \frac{(n+1)!}{(3/2)_n}P_n^{(1/2, 1/2)}(x) .$$

1.4.2.3. If $p_n(x)$ are orthogonal on $(-a, a)$ with the weight $\varrho(x)$, which is an even function, then the polynomials $p_n(x)$ and $p_n(-x)$ satisfy the same orthogonality relations. Hence $p_n(-x) = (-1)^n p_n(x)$, i.e.

$$p_{2n}(x) = s_n(x^2) , \quad p_{2n+1}(x) = xt_n(x^2) ,$$

where $s_n(x^2)$ and $t_n(x^2)$ are polynomials of degree n in x^2. Hence with $m \neq n$ we have

$$\int_{-a}^{a} p_{2m}(x)p_{2n}(x)\varrho(x)dx = \int_{-a}^{a} s_m(x^2)s_n(x^2)\varrho(x)dx$$

$$= 2\int_{0}^{a} s_m(x^2)s_n(x^2)\varrho(x)dx = \int_{0}^{a^2} s_m(\xi)s_n(\xi)\frac{\varrho(\sqrt{\xi})}{\sqrt{\xi}}d\xi = 0 .$$

Thus the polynomials $s_n(x) = p_{2n}(\sqrt{x})$ are orthogonal on the interval $(0, a^2)$ with the weight $\varrho_1(x) = \varrho(\sqrt{x})/\sqrt{x}$. In a quite similar way we obtain that the polynomials $t_n(x) = (1/\sqrt{x})p_{2n+1}(\sqrt{x})$ are orthogonal on the interval $(0, a^2)$ with the weight $\varrho_2(x) = \sqrt{x}\varrho(\sqrt{x})$.

Therefore in a special case $\varrho(x) = e^{-x^2}$ we have

$$H_{2n}(x) = C_n L_n^{-1/2}(x^2) , \quad H_{2n+1}(x) = A_n x L_n^{1/2}(x^2) .$$

The constants C_n and A_n can be determined by comparing the coefficients of the highest terms, then

$$H_{2n}(x) = (-1)^n 2^{2n} n! L_n^{-1/2}(x^2) , \tag{1.4.10}$$

$$H_{2n+1}(x) = (-1)^n 2^{2n+1} n! x L_n^{1/2}(x^2) . \tag{1.4.11}$$

By using the parity of the weight $\varrho(x)$ for $P_n^{(\alpha,\alpha)}(x)$ we obtain that the polynomial $s_n(x) = P_{2n}^{(\alpha,\alpha)}(\sqrt{x})$ is orthogonal on the interval $(0,1)$ with the weight $x^{-1/2}(1-x)^{\alpha}$. Therefore it coincides with the polynomial $P_n^{(\alpha,\,-1/2)}(2x-1)$ up to a multiplier. Hence

$$P_{2n}^{(\alpha,\alpha)}(x) = \frac{n!\,\Gamma(\alpha+2n+1)}{(2n)!\,\Gamma(\alpha+n+1)}\,P_n^{(\alpha,\,-1/2)}\left(2x^2-1\right) . \tag{1.4.12}$$

Similarly, we shall have

$$P_{2n+1}^{(\alpha,\alpha)}(x) = \frac{n!\,\Gamma(\alpha+2n+2)}{(2n+1)!\,\Gamma(\alpha+n+1)}\,x\,P_n^{(\alpha,\,1/2)}\left(2x^2-1\right) . \tag{1.4.13}$$

1.4.2.4. *All orthogonal polynomials satisfy a three term recurrence relation*

$$xp_n(x) = \alpha_n p_{n+1}(x) + \beta_n p_n(x) + \gamma_n p_{n-1}(x) , \tag{1.4.14}$$

where α_n, β_n and γ_n are constants.

For the proof we use the expansion

$$xp_n(x) = \sum_{k=0}^{n+1} c_{kn} p_k(x) , \tag{1.4.15}$$

$$c_{kn} = \frac{1}{d_k^2} \int_a^b p_k(x) x p_n(x) \varrho(x) dx . \tag{1.4.16}$$

Since $xp_k(x)$ is a polynomial of degree $k+1$, by the orthogonality property of $p_n(x)$ the coefficients $c_{kn} = 0$ when $k+1 < n$. Hence (1.4.15) can in fact be written in the form of (1.4.14) for $\alpha_n = c_{n+1,n}$, $\beta_n = c_{nn}$, $\gamma_n = c_{n-1,n}$.

The coefficients α_n, β_n, γ_n can be expressed in terms of the squared norm d_n^2 and the leading coefficients a_n, b_n in $p_n(x)$. From (1.4.16) it is seen that $d_k^2 c_{kn} = d_n^2 c_{nk}$. Since $\alpha_{n-1} = c_{n,n-1}$, $\gamma_n = c_{n-1,n}$, if we put $k = n-1$, we obtain

$$\gamma_n = \alpha_{n-1}\frac{d_n^2}{d_{n-1}^2} .$$

On the other hand, comparing the coefficients of the highest terms on the left-hand and right-hand sides of (1.4.15), we have $a_n = \alpha_n a_{n+1}$, $b_n = \alpha_n b_{n+1} + \beta_n a_n$. Hence

$$\alpha_n = \frac{a_n}{a_{n+1}} , \quad \beta_n = \frac{b_n}{a_n} - \frac{b_{n+1}}{a_{n+1}} , \quad \gamma_n = \frac{a_{n-1}}{a_n}\frac{d_n^2}{d_{n-1}^2} . \tag{1.4.17}$$

The coefficients α_n, β_n and γ_n for the Jacobi, Laguerre and Hermite polynomials are given in Table 1.1.

From the recurrence relation (1.4.14) there immediately follows the Darboux-Christoffel formula (see, for example, [S38, N18]):

$$\sum_{k=0}^{n} \frac{p_k(x)p_k(y)}{d_k^2} = \frac{1}{d_n^2}\frac{a_n}{a_{n+1}}\frac{p_{n+1}(x)p_n(y) - p_n(x)p_{n+1}(y)}{x-y} . \tag{1.4.18}$$

1.4.2.5. *All zeros x_i of $p_n(x)$ are simple and lie in the interval (a, b).*

For the proof we shall suppose that $p_n(x)$ on the interval (a, b) changes its sign at the points $x_1, x_2, \ldots, x_k$ $(0 \le k \le n)$ at growing x. The property in question will certainly be valid if we show that $k = n$. We put

$$q_k(x) = \begin{cases} 1 & \text{for} \quad k = 0 \\ \prod_{j=1}^{k}(x - x_j) & \text{for} \quad 0 < k \le n \, . \end{cases}$$

The product $p_n(x)q_k(x)$ evidently does not change sign for $x \in (a, b)$. Therefore

$$\int_a^b p_n(x)q_k(x)\varrho(x)dx \ne 0 \, .$$

It follows that $k = n$, since if $k < n$

$$\int_a^b p_n(x)q_k(x)\varrho(x)dx = 0$$

by (1.4.9).

1.5 Classical Orthogonal Polynomials as Eigenfunctions of Some Eigenvalue Problems

As shown above, the classical orthogonal polynomials $y_n(x)$ are the simplest solutions of the equation of hypergeometric type (1.1.1) with

$$\lambda = \lambda_n = -n\tau' - \tfrac{1}{2}n(n - 1)\sigma'' \, , \quad n = 0, 1, \ldots \, .$$

It turns out that the polynomials $y_n(x)$ are distinguished among all the solutions of (1.1.1) corresponding to various values of λ not only by their simplicity but also by their being the only nontrivial solutions of (1.1.1) for which the function $y(x)\sqrt{\varrho(x)}$ is both bounded and square integrable on (a, b). This property is extensively used in theoretical and mathematical physics, especially in quantum mechanics problems [N18].

Theorem. *Let $y = y(x)$ be a solution of the equation of hypergeometric type (1.1.1) and let $\varrho(x)$, a solution of $(\sigma \varrho)' = \tau \varrho$, be bounded on an interval (a, b) and satisfy the conditions imposed on $\varrho(x)$ for classical orthogonal polynomials. Then nontrivial solutions of (1.1.1) for which $y(x)\sqrt{\varrho(x)}$ is bounded and square integrable on (a, b) exist only when*

$$\lambda = \lambda_n = -n\tau' - \tfrac{1}{2}n(n - 1)\sigma'' \, , \quad n = 0, 1, \ldots , \tag{1.5.1}$$

and they have the form

$$y(x, \lambda_n) = y_n(x) = \frac{B_n}{\varrho(x)} \frac{d^n}{dx^n}\left[\sigma^n(x)\varrho(x)\right] \, , \tag{1.5.2}$$

*i.e. they are the classical polynomials that are orthogonal with weight $\varrho(x)$ on
(a, b) (if a and b are finite, the condition of quadratic integrability can be omitted).*

Proof. That the classical orthogonal polynomials $y_n(x)$ with $\lambda = \lambda_n$ are nontrivial
solutions can be verified immmediately.

Let us show that the problem has no other solutions. Suppose the contrary,
i.e. that for some λ there is a nontrivial solution $y = y(x, \lambda)$ which is not a
classical orthogonal polynomial. From the equations for $y(x, \lambda)$ and $y_n(x)$, i.e.

$$(\sigma \varrho y')' + \lambda \varrho y = 0 , \quad \left(\sigma \varrho y_n'\right)' + \lambda_n \varrho y_n = 0 ,$$

we can easily obtain the relation

$$\left(\lambda - \lambda_n\right) \int_{x_1}^{x_2} y(x, \lambda) y_n(x) \varrho(x) dx + \sigma(x) \varrho(x) W\left(y_n, y\right)\big|_{x_1}^{x_2} = 0 , \tag{1.5.3}$$

where

$$a < x_1 < x_2 < b , \quad W\left(y_n, y\right) = y_n(x) y'(x, \lambda) - y_n'(x) y(x, \lambda) .$$

The integral $\int_a^b y(x, \lambda) y_n(x) \varrho(x) dx$ converges owing to the square-integrability
of the functions $y(x, \lambda) \sqrt{\varrho(x)}$ and $y_n(x) \sqrt{\varrho(x)}$. Then for $\lambda \neq \lambda_n$ it follows from
(1.5.3) that

$$\lim_{x \to a} \sigma(x) \varrho(x) W\left(y_n, y\right) = c_1 ,$$
$$\lim_{x \to b} \sigma(x) \varrho(x) W\left(y_n, y\right) = c_2 , \tag{1.5.4}$$

where c_1 and c_2 are constants.

When $\lambda = \lambda_n$ are relation (1.5.4) still holds since, owing to (1.5.3),

$$\sigma(x) \varrho(x) W\left(y_n, y\right) = c , \tag{1.5.5}$$

i.e. in (1.5.4) $c_1 = c_2 = c$.

The relation (1.5.4) enables us to study the behavior of $y(x, \lambda)$ for $x \to$
a $(x \to b)$ if $c_1 \neq 0$ $\left(c_2 \neq 0\right)$. Since

$$\frac{d}{dx}\left[\frac{y(x, \lambda)}{y_n(x)}\right] = \frac{1}{y_n^2(x)} W\left[y_n(x), y(x, \lambda)\right] ,$$

we have

$$y(x, \lambda) = y_n(x)\left\{\frac{y(x_0, \lambda)}{y_n(x_0)} + \int_{x_0}^{x} \frac{W[y_n(s), y(s, \lambda)] ds}{y_n^2(s)}\right\} . \tag{1.5.6}$$

In (1.5.6) we choose the point $x_0 < b$ so that it lies to the right of all the zeros of
$y_n(x)$. Investigation of the behavior of $y(x, \lambda)$ as $x \to b$ by means of (1.5.4) and
(1.5.6) shows that when $c_2 \neq 0$ the function $y(x, \lambda)$ does not satisfy the theorem
conditions, i.e. $c_2 = 0$. Similarly it can be shown that $c_1 = 0$.

Hence if $\lambda \neq \lambda_n$ $(n = 0, 1, \ldots)$, from (1.5.3) for $x_1 \to a$, $x_2 \to b$ we obtain

$$\int_a^b y(x, \lambda) y_n(x) \varrho(x) dx = 0 , \quad n = 0, 1, \ldots .$$

By virtue of closure of the systems of classical orthogonal polynomials this equation is possible only when $y(x, \lambda) = 0$ for $x \in (a, b)$. On the other hand, if $\lambda = \lambda_n$, then by using $c_1 = c_2 = 0$ we obtain $W(y_n, y) = 0$, i.e. the solutions $y_n(x)$ and $y(x, \lambda)$ are linearly dependent, which contradicts the hypothesis.

2. Classical Orthogonal Polynomials of a Discrete Variable

The basic properties of the polynomials $p_n(x)$ that satisfy the orthogonality relations

$$\int_a^b p_n(x)p_m(x)\varrho(x)dx = 0 \quad (m \neq n) \tag{2.0.1}$$

hold also for the polynomials that satisfy the orthogonality relations of a more general form, which can be expressed in terms of Stieltjes integrals

$$\int_a^b p_n(x)p_m(x)dw(x) = 0 \quad (m \neq n) , \tag{2.0.2}$$

where $w(x)$ is a monotonic nondecreasing function (usually called the distribution function). The orthogonality relation (2.0.2) is reduced to (2.0.1) in the case when the function $w(x)$ has a derivative on (a, b) and $w'(x) = \varrho(x)$. For solving many problems orthogonal polynomials are used that satisfy the orthogonality relations (2.0.2) in the case when $w(x)$ is a function of jumps, i.e. the piecewise constant function with jumps ϱ_i at the points $x = x_i$. In this case the orthogonality relation (2.0.2) can be rewritten in the form

$$\sum_i p_n(x_i)p_m(x_i)\varrho_i = 0 \quad (m \neq n) . \tag{2.0.3}$$

The polynomials $p_n(x)$ that satisfy the relations (2.0.3) are called *the orthogonal polynomials of a discrete variable*. Among them the classical orthogonal polynomials of a discrete variable – the Hahn, Meixner, Kravchuk and Charlier polynomials for which $x_{i+1} = x_i + 1$ [E7] – are most studied. In this chapter it will be shown that these polynomials satisfy the difference equations which can be obtained from the differential equations for the classical orthogonal polynomials. These difference equations have main properties similar to those of initial differential equations, which allow the construction of the theory of classical orthogonal polynomials of a discrete variable by analogy with the theory of the Jacobi, Laguerre and Hermite polynomials.

2.1 The Difference Equation of Hypergeometric Type

2.1.1. The theory considered in Chap. 1 for polynomial solutions of the differential equation of hypergeometric type,

$$\tilde{\sigma}(x)y'' + \tilde{\tau}(x)y' + \lambda y = 0 \,, \tag{2.1.1}$$

where $\tilde{\sigma}(x)$ and $\tilde{\tau}(x)$ are polynomials of at most the second and first degree[1], and λ is a constant, admits a natural generalization to the case when the differential equation is replaced by a difference equation. Let us consider the simplest case, when differential equation (2.1.1) is replaced by the difference equation of the form

$$\tilde{\sigma}(x)\frac{1}{h}\left[\frac{y(x+h)-y(x)}{h} - \frac{y(x)-y(x-h)}{h}\right]$$
$$+ \frac{\tilde{\tau}(x)}{2}\left[\frac{y(x+h)-y(x)}{h} + \frac{y(x)-y(x-h)}{h}\right] + \lambda y(x) = 0 \,, \tag{2.1.2}$$

which approximates (2.1.1) on a lattice with the constant mesh $\Delta x = h$ up to the second order in h. We say that a difference operator $\mathcal{L}_h$ approximates the differential operator $\mathcal{L}$ at the point x to order m in h if $\mathcal{L}y(x) - \mathcal{L}_h y(x) = \mathrm{O}(h^m)$, $h \to 0$.

When transferring from (2.1.1) to (2.1.2) to approximate the derivatives $y'(x)$ and $y''(x)$ we used the linear combination of the left (backward) and right (forward) difference quotients

$$\frac{1}{h}[y(x) - y(x-h)] \quad \text{and} \quad \frac{1}{h}[y(x+h) - y(x)] \,,$$

which yields the error $\mathrm{O}(h^2)$ for $h \to 0$:

$$y'(x) = \frac{1}{2}\left[\frac{y(x+h)-y(x)}{h} + \frac{y(x)-y(x-h)}{h}\right] + \mathrm{O}(h^2) \,,$$
$$y''(x) = \frac{1}{h}\left[\frac{y(x+h)-y(x)}{h} - \frac{y(x)-y(x-h)}{h}\right] + \mathrm{O}(h^2) \,.$$

By the linear change of independent variable x by hx and the change of functions $y(hx)$ by $y(x)$, $\tilde{\sigma}(hx)/h^2$ by $\tilde{\sigma}(x)$, $\tilde{\tau}(hx)/h$ by $\tilde{\tau}(x)$ Eq. (2.1.2) can be reduced to the equation of the same form at $h = 1$:

$$\tilde{\sigma}(x)\Delta\nabla y(x) + \frac{\tilde{\tau}(x)}{2}(\Delta + \nabla)y(x) + \lambda y(x) = 0 \,, \tag{2.1.3}$$

where $\Delta f(x) = f(x+1) - f(x)$, $\nabla f(x) = f(x) - f(x-1)$.

Since $\nabla f(x) = \Delta f(x) - \Delta\nabla f(x)$, Eq. (2.1.3) is equivalent to

$$\sigma(x)\Delta\nabla y(x) + \tau(x)\Delta y(x) + \lambda y(x) = 0 \,, \tag{2.1.4}$$

where

$$\sigma(x) = \tilde{\sigma}(x) - \tfrac{1}{2}\tilde{\tau}(x) \,, \quad \tau(x) = \tilde{\tau}(x) \,. \tag{2.1.5}$$

---

[1] For the sake of convenience on further argument we denote the coefficients in Eq. (2.1.1) by $\tilde{\sigma}(x)$ and $\tilde{\tau}(x)$ instead of $\sigma(x)$ and $\tau(x)$ as used in Chap. 1.

Evidently $\sigma(x)$ is a polynomial of at most the second degree. Let us note that the difference equation (2.1.4) obtained as a result of approximating (2.1.1) arises also in some other problems and has its own meaning.

2.1.2. Further on we shall use the following properties of the operators Δ and ∇:

$$\Delta f(x) = \nabla f(x+1) \, , \tag{2.1.6}$$

$$\Delta \nabla f(x) = \nabla \Delta f(x) = f(x+1) - 2f(x) + f(x-1) \, , \tag{2.1.7}$$

$$\Delta[f(x)g(x)] = f(x)\Delta g(x) + g(x+1)\Delta f(x) \, . \tag{2.1.8}$$

From (2.1.8) we obtain the formula for summation by parts:

$$\sum_{x_i=a}^{b-1} f(x_i)\Delta g(x_i) = f(x_i)g(x_i)\Big|_a^b - \sum_{x_i=a}^{b-1} g(x_{i+1})\Delta f(x_i) \, . \tag{2.1.9}$$

Here $x_{i+1} = x_i + 1$. We observe that for a polynomial $q_m(x)$ of degree m the expressions $\Delta q_m(x)$ and $\nabla q_m(x)$ are polynomials of degree $m-1$; and that $\Delta^m q_m(x) = \nabla^m q_m(x) = q_m^{(m)}(x)$.

2.1.3. We can establish a number of properties of the solutions of (2.1.4) that are analogous to those of solutions of (2.1.1). Let us show that the function $v_1(x) = \Delta y(x)$ satisfies a difference equation of the form (2.1.4). For the proof we apply the operator Δ to both sides of (2.1.4):

$$\Delta\big[\sigma(x)\nabla v_1(x)\big] + \Delta\big[\tau(x)v_1(x)\big] + \lambda v_1(x) = 0 \, .$$

By using (2.1.8) and (2.1.6) we can write this equation in the form

$$\sigma(x)\Delta\nabla v_1(x) + \tau_1(x)\Delta v_1(x) + \mu_1 v_1(x) = 0 \, , \tag{2.1.10}$$

where $\tau_1(x) = \tau(x+1) + \Delta\sigma(x)$, $\mu_1 = \lambda + \Delta\tau(x)$.

Since $\tau_1(x)$ is a polynomial of at most the first degree, and μ_1 is independent of x, Eq. (2.1.10) for $v_1(x)$ is of the same form as (2.1.4).

It is easy to verify the converse: every solution of (2.1.10) with $\lambda \neq 0$ can be represented in the form $v_1(x) = \Delta y(x)$, where $y(x)$ is a solution of (2.1.4) that can be expressed in terms of $v_1(x)$ by

$$y(x) = -\frac{1}{\lambda}\big[\sigma(x)\nabla v_1(x) + \tau(x)v_1(x)\big] \, .$$

2.1.4. In a similar way for the function $v_m(x) = \Delta^m y(x)$ we can obtain a difference equation of hypergeometric type

$$\sigma(x)\Delta\nabla v_m(x) + \tau_m(x)\Delta v_m(x) + \mu_m v_m(x) = 0 \, , \tag{2.1.11}$$

where

$$\tau_m(x) = \tau_{m-1}(x+1) + \Delta\sigma(x), \quad \tau_0(x) = \tau(x) \quad ; \tag{2.1.12}$$

$$\mu_m = \mu_{m-1} + \Delta\tau_{m-1}(x), \quad \mu_0 = \lambda . \tag{2.1.13}$$

The converse is also valid: every solution of (2.1.11) with $\mu_k \neq 0$ ($k = 0, 1, \ldots,$ $m - 1$) can be represented as $v_m(x) = \Delta^m y(x)$, where $y(x)$ is a solution of (2.1.4).

If we rewrite (2.1.12) in the form

$$\tau_m(x) + \sigma(x) = \tau_{m-1}(x+1) + \sigma(x+1) , \tag{2.1.14}$$

we easily obtain an explicit expression for $\tau_m(x)$:

$$\tau_m(x) = \tau(x+m) + \sigma(x+m) - \sigma(x) . \tag{2.1.15}$$

To obtain an explicit formula for μ_m we have only to observe that $\Delta\tau_m(x)$ and $\Delta^2\sigma(x)$ are independent of x. Therefore

$$\Delta\tau_m = \Delta\tau_{m-1} + \Delta^2\sigma = \cdots = \Delta\tau + m\Delta^2\sigma = \tau' + m\sigma''$$

and consequently $\mu_m = \mu_{m-1} + \tau' + (m - 1)\sigma''$. Hence

$$\mu_m = \mu_0 + \sum_{k=1}^{m}(\mu_k - \mu_{k-1}) = \lambda + m\tau' + \tfrac{1}{2}m(m - 1)\sigma'' . \tag{2.1.16}$$

2.1.5. Below we shall use (2.1.4) in the self-adjoint form. For this purpose we multiply both sides of (2.1.4) by a function $\varrho(x)$. If this function satisfies the equation

$$\Delta[\sigma(x)\varrho(x)] = \tau(x)\varrho(x) , \tag{2.1.17}$$

then by virtue of (2.1.8) we obtain

$$\sigma(x)\varrho(x)\Delta\nabla y(x) + \tau(x)\varrho(x)\Delta y(x)$$
$$= \sigma(x)\varrho(x)\Delta\nabla y(x) + \nabla y(x+1)\Delta[\sigma(x)\varrho(x)]$$
$$= \Delta[\sigma(x)\varrho(x)\nabla y(x)] .$$

As a result, Eq. (2.1.4) can be reduced to the form

$$\Delta[\sigma(x)\varrho(x)\nabla y(x)] + \lambda\varrho(x)y(x) = 0 . \tag{2.1.18}$$

Similarly Eq. (2.1.11) can be reduced to the form

$$\Delta\big[\sigma(x)\varrho_m(x)\nabla v_m(x)\big] + \mu_m\varrho_m(x)v_m(x) = 0 , \tag{2.1.19}$$

where the function $\varrho_m(x)$ satisfies the equation

$$\Delta\big[\sigma(x)\varrho_m(x)\big] = \tau_m(x)\varrho_m(x) . \tag{2.1.20}$$

Like Eq. (2.1.4) for $y(x)$, which approximates the differential equation (2.1.1) up to the second order of accuracy with respect to the mesh h, Eq. (2.1.17) corresponds to the second order approximation of the differential equation

$$[\tilde{\sigma}(x)\varrho(x)]' = \tilde{\tau}(x)\varrho(x) \,, \tag{2.1.21}$$

which appears when (2.1.1) is reduced to the self-adjoint form

$$[\tilde{\sigma}(x)\varrho(x)y'(x)]' + \lambda\varrho(x)y(x) = 0 \,.$$

In fact, the difference equation that approximates (2.1.21) up to the second order of accuracy in h has the form

$$\frac{1}{h}\left[\tilde{\sigma}(x+h)\varrho(x+h) - \tilde{\sigma}(x)\varrho(x)\right] = \tfrac{1}{2}\left[\tilde{\tau}(x+h)\varrho(x+h) + \tilde{\tau}(x)\varrho(x)\right] \,.$$

After the replacement

$$x \to hx \,, \quad \frac{1}{h^2}\tilde{\sigma}(hx) \to \tilde{\sigma}(x) \,, \quad \frac{1}{h}\tilde{\tau}(hx) \to \tilde{\tau}(x) \,,$$

which transforms Eq. (2.1.2) into (2.1.3), as well as the replacement $\varrho(hx) \to \varrho(x)$, we obtain the equation of the same form with $h = 1$:

$$\Delta\left[\tilde{\sigma}(x)\varrho(x)\right] = \tfrac{1}{2}\left[\tilde{\tau}(x+1)\varrho(x+1) + \tilde{\tau}(x)\varrho(x)\right] \,. \tag{2.1.22}$$

Using (2.1.5) it is easy to see that (2.1.22) coincides with (2.1.17).

By means of (2.1.17) and (2.1.20) we can obtain the connection between the functions $\varrho_m(x)$ and $\varrho(x)$. To do this we write Eq. (2.1.20) in the form

$$\frac{\sigma(x+1)\varrho_m(x+1)}{\varrho_m(x)} = \tau_m(x) + \sigma(x) \,.$$

Hence it follows that (2.1.14) is equivalent to the relation

$$\frac{\sigma(x+1)\varrho_m(x+1)}{\varrho_m(x)} = \frac{\sigma(x+2)\varrho_{m-1}(x+2)}{\varrho_{m-1}(x+1)} \,,$$

i.e.

$$\frac{\varrho_m(x+1)}{\sigma(x+2)\varrho_{m-1}(x+2)} = \frac{\varrho_m(x)}{\sigma(x+1)\varrho_{m-1}(x+1)} = c_m(x) \,,$$

where $c_m(x)$ is any function of period 1. We only need to find any solution of Eq. (2.1.20), so we can take $c_m(x) = 1$. As a result we obtain

$$\varrho_m(x) = \sigma(x+1)\varrho_{m-1}(x+1) \,. \tag{2.1.23}$$

Since $\varrho_0(x) = \varrho(x)$, we have

$$\varrho_m(x) = \varrho(x+m)\prod_{k=1}^{m}\sigma(x+k) \,. \tag{2.1.24}$$

2.2 Finite Difference Analogs of Polynomials of Hypergeometric Type and of Their Derivatives. The Rodrigues Type Formula

2.2.1. The property of the difference derivatives $\Delta^m y(x)$ established in Sect. 2.1 allows us to construct a theory of classical orthogonal polynomials of a discrete variable along the same lines as the discussion in Chap. 1. Putting in (2.1.11) $m = n$ we obtain that equation

$$\sigma(x)\Delta\nabla v_n(x) + \tau_n(x)\Delta v_n(x) + \mu_n v_n(x) = 0 \tag{2.2.1}$$

has a particular solution $v_n(x) = \text{const}$ if $\mu_n = 0$. Since $v_n(x) = \Delta^n y(x)$, this means that if

$$\lambda = \lambda_n = -n\tau' - \tfrac{1}{2}n(n-1)\sigma'' \tag{2.2.2}$$

there is a particular solution $y = y_n(x)$ of (2.2.1) which is a polynomial of degree n, provided that $\mu_m \neq 0$ for $m = 0, 1, \ldots, n-1$. In fact, the Eq. (2.1.11) for $v_m(x)$ can be rewritten in the form

$$v_m(x) = -\frac{1}{\mu_m}\left[\sigma(x)\nabla v_{m+1}(x) + \tau_m(x)v_{m+1}(x)\right] .$$

It is clear from this that if $v_{m+1}(x)$ is a polynomial, then $v_m(x)$ is also a polynomial if $\mu_m \neq 0$.

To obtain an explicit expression for $y_n(x)$ we use (2.1.23) for writing (2.1.19) in the form of a simple relation between $v_m(x)$ and $v_{m+1}(x)$. In fact

$$\varrho_m(x)v_m(x) = -\frac{1}{\mu_m}\Delta\left[\sigma(x)\varrho_m(x)\nabla v_m(x)\right]$$

$$= -\frac{1}{\mu_m}\nabla\left[\sigma(x+1)\varrho_m(x+1)\Delta v_m(x)\right] ,$$

i.e.

$$\varrho_m(x)v_m(x) = -\frac{1}{\mu_m}\nabla\left[\varrho_{m+1}(x)v_{m+1}(x)\right] .$$

For $m < n$ we now obtain successively

$$\varrho_m v_m = -\frac{1}{\mu_m}\nabla\left(\varrho_{m+1}v_{m+1}\right)$$

$$= \left(-\frac{1}{\mu_m}\right)\left(-\frac{1}{\mu_{m+1}}\right)\nabla^2\left(\varrho_{m+2}v_{m+2}\right) = \cdots \tag{2.2.3}$$

$$= \frac{A_m}{A_n}\nabla^{n-m}\left(\varrho_n v_n\right) ,$$

where

$$A_m = (-1)^m \prod_{k=0}^{m-1} \mu_k , \quad A_0 = 1 . \tag{2.2.4}$$

If $y = y_n(x)$ we have $v_n(x) = \text{const}$, whence

$$v_{mn}(x) = \Delta^m y_n(x) = \frac{A_{mn} B_n}{\varrho_m(x)} \nabla^{n-m} \left[\varrho_n(x) \right] , \tag{2.2.5}$$

where

$$A_{mn} = A_m(\lambda)\big|_{\lambda=\lambda_n} = \frac{n!}{(n-m)!} \prod_{k=0}^{m-1} \left(\tau' + \frac{n+k-1}{2} \sigma'' \right) ,$$

$$A_{0n} = 1 , \quad m \le n \ ; \tag{2.2.6}$$

$$B_n = \frac{\Delta^n y_n(x)}{A_{nn}} = \frac{1}{A_{nn}} y_n^{(n)}(x) .$$

From (2.2.5) with $m = 0$ we obtain an explicit expression for $y_n(x)$:

$$y_n(x) = \frac{B_n}{\varrho(x)} \nabla^n \left[\varrho_n(x) \right] . \tag{2.2.7}$$

Thus the polynomial solutions of (2.1.4) are determined by (2.2.7) up to the normalizing factor B_n. These solutions correspond to the values $\lambda = \lambda_n$ from (2.2.2). By using (2.1.6) and (2.1.24) we can also write (2.2.7) in the form

$$y_n(x) = \frac{B_n}{\varrho(x)} \Delta^n \left[\varrho_n(x - n) \right] = \frac{B_n}{\varrho(x)} \Delta^n \left[\varrho(x) \prod_{k=0}^{n-1} \sigma(x - k) \right] . \tag{2.2.8}$$

Equation (2.2.5) is the finite-difference analog of the Rodriques formula (1.2.11) for the classical orthogonal polynomials and their derivatives.

2.2.2. Let us consider some consequences of the Rodriques formula (2.2.5).

2.2.2.1. The Rodrigues formula for the polynomials $y_n(x)$ and their differences $\Delta y_n(x)$ leads to a relation between $\Delta y_n(x)$ and the polynomials themselves. To find it, it is enough to observe that if $m = 1$ in (2.2.5) we have $A_{1n} = -\lambda_n$ and according to (2.1.24) we have $\left[\varrho_1(x) \right]_{n-1} = \varrho_n(x)$. In fact

$$\varrho_1(x) = \sigma(x + 1)\varrho(x + 1) ,$$

$$\left[\varrho_1(x) \right]_{n-1} = \varrho_1(x + n - 1) \prod_{k=1}^{n-1} \sigma(x + k)$$

$$= \varrho(x + n) \prod_{k=1}^{n} \sigma(x + k) = \varrho_n(x) .$$

Hence

$$\Delta y_n(x) = -\lambda_n \frac{B_n}{\varrho_1(x)} \nabla^{n-1} \left[\varrho_n(x) \right] = -\lambda_n \frac{B_n}{\bar{B}_{n-1}} \frac{\bar{B}_{n-1}}{\varrho_1(x)} \nabla^{n-1} \left\{ \left[\varrho_1(x) \right]_{n-1} \right\}$$

$$= -\lambda_n \frac{B_n}{\bar{B}_{n-1}} \bar{y}_{n-1}(x) . \tag{2.2.9}$$

Here $\bar{y}_n(x)$ is the polynomial obtained by replacing $\varrho(x)$ by $\varrho_1(x)$ in the formula for $y_n(x)$, and $\bar{B}_n$ is the normalizing constant in the Rodrigues formula for $\bar{y}_n(x)$.

2.2.2.2. By using the Rodrigues formula it is easy to derive a linear relation that connects the difference $\nabla y_n(x)$ with $y_n(x)$ and $y_{n+1}(x)$. We have

$$y_{n+1}(x) = \frac{B_{n+1}}{\varrho(x)} \nabla^{n+1} \big[\varrho_{n+1}(x)\big] = \frac{B_{n+1}}{\varrho(x)} \nabla^n \big[\Delta\varrho_{n+1}(x-1)\big] \ .$$

By using the equality

$$\Delta\varrho_{n+1}(x-1) = \Delta\big[\sigma(x)\varrho_n(x)\big] = \tau_n(x)\varrho_n(x)$$

and applying formulas (2.1.8) and (2.1.6) successively we obtain

$$y_{n+1}(x) = \frac{B_{n+1}}{\varrho(x)} \nabla^n \big[\tau_n(x)\varrho_n(x)\big]$$

$$= \frac{B_{n+1}}{\varrho(x)} \Big\{ \tau_n(x)\nabla^n\big[\varrho_n(x)\big] + n\tau_n'\nabla^{n-1}\big[\varrho_n(x-1)\big] \Big\} \ .$$

Since

$$\nabla y_n(x) = v_{1n}(x-1) = -\frac{\lambda_n B_n}{\sigma(x)\varrho(x)} \nabla^{n-1}\big[\varrho_n(x-1)\big] \ ,$$

we obtain the formula

$$\sigma(x)\nabla y_n(x) = \frac{\lambda_n}{n\tau_n'} \left[\tau_n(x)y_n(x) - \frac{B_n}{B_{n+1}}y_{n+1}(x) \right] \ , \tag{2.2.10}$$

which is analogous to (1.2.13).

2.2.2.3. By using (2.2.5) with $m = n - 1$ we can easily calculate the leading coefficients a_n and b_n in the expansion

$$y_n(x) = a_n x^n + b_n x^{n-1} + \dots \ .$$

For this purpose we first calculate the $(n-1)$th difference $\Delta^{n-1}(x^n)$, which is a polynomial of the first degree. We have

$$\Delta^{n-1}\big(x^n\big) = \alpha_n\big(x + \beta_n\big) \ ,$$

where α_n and β_n are constants. To determine α_n and β_n we observe that

$$\Delta^n\big(x^n\big) = \Delta\big[\alpha_n\big(x + \beta_n\big)\big] = \alpha_n \ .$$

Hence

$$\alpha_{n+1}\big(x + \beta_{n+1}\big) = \Delta^n\big(x^{n+1}\big) = \Delta^{n-1}\big(\Delta x^{n+1}\big)$$

$$= \Delta^{n-1}\big[(x+1)^{n+1} - x^{n+1}\big]$$

$$= \Delta^{n-1}\left[(n+1)x^n + \frac{(n+1)n}{2}x^{n-1} + \dots\right]$$

$$= (n+1)\alpha_n(x+\beta_n) + \frac{(n+1)n}{2}\alpha_{n-1} \ .$$

Comparing the coefficients of the powers of x on the two sides of this equation, we obtain

$$\alpha_{n+1} = (n+1)\alpha_n \ , \quad \alpha_{n+1}\beta_{n+1} = (n+1)\alpha_n\beta_n + \frac{(n+1)n}{2}\alpha_{n-1} \ .$$

Since $\alpha_1 = 1$ and $\beta_1 = 0$, the first equation yields $\alpha_n = n!$, whence $\beta_{n+1} = \beta_n + \frac{1}{2}$, i.e. $\beta_n = (n-1)/2$. Therefore

$$\Delta^{n-1}(x^n) = n!\left(x + \frac{n-1}{2}\right) \ .$$

Consequently

$$\Delta^{n-1}y_n(x) = \Delta^{n-1}\left(a_n x^n + b_n x^{n-1} + \dots\right)$$

$$= a_n\alpha_n(x+\beta_n) + b_n\alpha_{n-1}$$

$$= n!a_n\left(x + \frac{n-1}{2}\right) + (n-1)!b_n \ .$$

On the other hand,

$$\nabla\varrho_n(x) = \Delta\varrho_n(x-1) = \Delta\big[\sigma(x)\varrho_{n-1}(x)\big] = \tau_{n-1}(x)\varrho_{n-1}(x) \ .$$

Consequently if we take $m = n-1$ in (2.2.5) we obtain

$$n!a_n\left(x + \frac{n-1}{2}\right) + (n-1)!b_n = A_{n-1,n}B_n\tau_{n-1}(x) \ ,$$

whence

$$a_n = \frac{A_{n-1,n}B_n}{n!}\tau'_{n-1} = B_n\prod_{k=0}^{n-1}\left(\tau' + \frac{n+k-1}{2}\sigma''\right) \ , \quad a_0 = B_0 \ ; \quad (2.2.11)$$

$$\frac{b_n}{a_n} = n\frac{\tau_{n-1}(0)}{\tau'_{n-1}} - \frac{1}{2}n(n-1)$$

$$= n\frac{\tau(0) + (n-1)[\sigma'(0) + \tau'/2]}{\tau' + (n-1)\sigma''} \ . \qquad (2.2.12)$$

2.3 The Orthogonality Property

2.3.1. The polynomial solutions $y_n(x)$ have the orthogonality property under certain restrictions on coefficients of Eq. (2.1.4). To derive this property we use the equations for $y_n(x)$ and $y_m(x)$ in self-adjoint form (by analogy with the

Sturm-Liouville problem)

$$\Delta\big[\sigma(x)\varrho(x)\nabla y_n(x)\big] + \lambda_n\varrho(x)y_n(x) = 0 \;,$$

$$\Delta\big[\sigma(x)\varrho(x)\nabla y_m(x)\big] + \lambda_m\varrho(x)y_m(x) = 0 \;.$$

Multiply the first equation by $y_m(x)$ and the second by $y_n(x)$, and subtract the second from the first. We obtain

$$\big(\lambda_m - \lambda_n\big)\varrho(x)y_m(x)y_n(x)$$
$$= \Delta\big\{\sigma(x)\varrho(x)\big[y_m(x)\nabla y_n(x) - y_n(x)\nabla y_m(x)\big]\big\} \;.$$

If we now put $x = x_i$, $x_{i+1} = x_i + 1$ and sum over the values $x = x_i$ for which $a \le x_i \le b - 1$, we obtain

$$\big(\lambda_m - \lambda_n\big)\sum_{x_i=a}^{b-1} y_m\big(x_i\big)y_n\big(x_i\big)\varrho\big(x_i\big)$$
$$= \sigma(x)\varrho(x)\big[y_m(x)\nabla y_n(x) - y_n(x)\nabla y_m(x)\big]\Big|_a^b \;.$$

Since the expression $(y_m\nabla y_n - y_n\nabla y_m)$ is a polynomial in x, the polynomial solutions of (2.1.4) are orthogonal on $[a,\, b - 1]$ with weight $\varrho(x)$:

$$\sum_{x_i=a}^{b-1} y_m\big(x_i\big)y_n\big(x_i\big)\varrho\big(x_i\big) = \delta_{mn}d_n^2 \tag{2.3.1}$$

under the boundary conditions

$$\sigma(x)\varrho(x)x^l\big|_{x=a,b} = 0 \quad (l = 0,\, 1,\, \ldots) \;. \tag{2.3.2}$$

We call the polynomials $y_n(x)$ *classical orthogonal polynomials of a discrete variable*, provided that (2.3.1) is valid, the interval (a, b) is on the real axis and the function $\varrho(x)$ satisfies (2.1.17) and (2.3.2). They are usually considered under the additional condition $\varrho(x_i) > 0$ for $a \le x_i \le b - 1$.

Remark. Since the orthogonality relation (2.3.1) for the classical orthogonal polynomials of a discrete variable may be written in terms of the Stietjes integral it follows that all the general properties intrinsic to any orthogonal polynomials are preserved (see Sect. 1.4.2).

2.3.2. Let us consider the orthogonality property of difference derivatives of $y_n(x)$. The polynomials $\Delta y_n(x) = v_{1n}(x)$ satisfy the equation obtained from the equation for $y_n(x)$ by replacing $\varrho(x)$ by $\varrho_1(x) = \sigma(x+1)\varrho(x+1) = [\sigma(x)+\tau(x)]\varrho(x)$ and λ by $\mu_1 = \lambda + \tau'$. The function $\varrho_1(x)$ evidently satisfies a condition similar to (2.3.2):

$$\sigma(x)\varrho_1(x)x^l\big|_{x=a,b-1} = 0 \quad (l = 0,\, 1,\, \ldots) \;.$$

Hence the polynomials $\Delta y_n(x)$ have the orthogonality property

$$\sum_{x_i=a}^{b-2} v_{1m}(x_i)v_{1n}(x_i)\varrho_1(x_i) = \delta_{mn}d_{1n}^2 \ .$$

Proceeding similarly we can easily show that, provided (2.3.2) is satisfied, the condition

$$\sigma(x)\varrho_k(x)x^l\big|_{x=a,b-k} = 0 \quad (l = 0, 1, \ldots) \tag{2.3.3}$$

is also satisfied, and for the polynomials $\Delta^k y_n(x) = v_{kn}(x)$ we have the relations

$$\sum_{x_i=a}^{b-k-1} v_{km}(x_i)v_{kn}(x_i)\varrho_k(x_i) = \delta_{mn}d_{kn}^2 \ . \tag{2.3.4}$$

If we take $\varrho(a) > 0$, and

$$\begin{aligned}
\sigma(x_i) &> 0 && \text{for} \quad a+1 \le x_i \le b-1 \ , \\
\sigma(x_i) + \tau(x_i) &> 0 && \text{for} \quad a \le x_i \le b-2 \ ,
\end{aligned} \tag{2.3.5}$$

it follows from (2.1.17) written in the form

$$\frac{\varrho(x+1)}{\varrho(x)} = \frac{\sigma(x)+\tau(x)}{\sigma(x+1)}$$

and the explicit form of $\varrho_k(x)$ that

$$\varrho_k(x_i) > 0 \quad \text{for} \quad a \le x_i \le b-k-1 \quad (k = 0, 1, \ldots) \ .$$

2.3.3. We now discuss some considerations about the choice of a and b to satisfy the boundary conditions (2.3.2) and the positivity condition for the weight $\varrho(x_i)$ on the orthogonality interval $[a, b-1]$. If a is finite, then by hypothesis $\varrho(a) > 0$, i.e. a is a root of $\sigma(x)$. Since a linear change of variable $x \to x + a$ preserves the type of the equation, it is always possible, if $\sigma(x) \ne$ const, to take $\sigma(0) = 0$. That is, we may suppose that $a = 0$. If b is finite, we have by (2.1.17) that

$$\sigma(b)\varrho(b) = [\sigma(b-1)+\tau(b-1)]\varrho(b-1) \ .$$

Since $\varrho(b-1) > 0$, we have

$$\sigma(b-1) + \tau(b-1) = 0 \ . \tag{2.3.6}$$

When $b = +\infty$ the boundary conditions (2.3.2) will be satisfied if

$$\lim_{a \to +\infty} x^l \varrho(x) = 0 \quad (l = 0, 1, \ldots) \ .$$

A similar remark applies when $a = -\infty$.

2.3.4. To calculate the squared norms d_n^2 we first establish the connection between the squared norms d_{kn}^2 and $d_{k+1,n}^2$, where

$$d_{kn}^2 = \sum_{x_i=a}^{b-k-1} v_{kn}^2(x_i)\varrho_k(x_i)\,, \quad d_{0n}^2 = d_n^2\,, \quad v_{kn}(x) = \Delta^k y_n(x)\,.$$

To do this we write the difference equation for $v_{kn}(x)$

$$\Delta\big[\sigma(x)\varrho_k(x)\nabla v_{kn}(x)\big] + \mu_{kn}\varrho_k(x)v_{kn}(x) = 0\,,$$

where $\mu_{kn} = \mu_k(\lambda)\big|_{\lambda=\lambda_n} = \lambda_n - \lambda_k$, multiply by $v_{kn}(x)$ and sum over the values $x = x_i$ for which $a \le x_i \le b-k-1$:

$$\sum_i v_{kn}(x_i)\Delta\big[\sigma(x_i)\varrho_k(x_i)\nabla v_{kn}(x_i)\big] + \mu_{kn}d_{kn}^2 = 0\,.$$

By summation by parts and using the equations

$$\Delta v_{kn}(x) = v_{k+1,n}(x)\,, \quad \sigma(x+1)\varrho_k(x+1) = \varrho_{k+1}(x)$$

we find that

$$\sum_i v_{kn}(x_i)\Delta\big[\sigma(x_i)\varrho_k(x_i)\nabla v_{kn}(x_i)\big]$$
$$= \sigma(x)\varrho_k(x)\nabla v_{kn}(x)v_{kn}(x)\Big|_a^{b-k} - d_{k+1,n}^2\,.$$

Since the first part of the right-hand side is zero because of the boundary conditions (2.3.3), we have

$$d_{kn}^2 = \frac{1}{\mu_{kn}}d_{k+1,n}^2\,.$$

Hence, we obtain successively

$$d_n^2 = d_{0n}^2 = \frac{1}{\mu_{0n}}d_{1n}^2 = \frac{1}{\mu_{0n}}\frac{1}{\mu_{1n}}d_{2n}^2 = \ldots = \frac{d_{nn}^2}{\prod_{k=0}^{n-1}\mu_{kn}}\,.$$

Since

$$d_{nn}^2 = v_{nn}^2 S_n\,, \quad v_{nn} = A_{nn}B_n\,, \quad A_{nn} = (-1)^n\prod_{k=a}^{n-1}\mu_{kn}$$

[see (2.2.6) and (2.2.4)], we have

$$d_n^2 = (-1)^n A_{nn}B_n^2 S_n\,, \tag{2.3.7}$$

where

$$S_n = \sum_{x_i=a}^{b-n-1} \varrho_n(x_i)\,. \tag{2.3.8}$$

For $n = b - a - 1$ (in the case when $b - a = N$ is finite) the sum S_n contains only one term and hence is easily calculated:

$$S_{N-1} = \varrho_{N-1}(a)\,. \tag{2.3.9}$$

To calculate S_n for $n < N - 1$ it is enough to be able to calculate the ratio S_n/S_{n+1}. To do this, we transform the expression (2.3.8) for S_{n+1} by using the connection between $\varrho_n(x)$ and $\varrho_{n+1}(x)$:

$$S_{n+1} = \sum_{x_i=a}^{b-n-2} \varrho_{n+1}(x_i) = \sum_{x_i=a}^{b-n-2} \sigma(x_i+1)\varrho_n(x_i+1)$$
$$= \sum_{x_i=a}^{b-n-1} \sigma(x_i)\varrho_n(x_i) \ .$$

We expand $\sigma(x)$ in powers of the first-degree polynomial $\tau_n(x)$:

$$\sigma(x) = A\tau_n^2(x) + B\tau_n(x) + C \ .$$

Then, by using the equation for $\varrho_n(x)$ and summing by parts, we obtain

$$S_{n+1} = \sum_i [A\tau_n(x_i) + B]\tau_n(x_i)\varrho_n(x_i) + CS_n$$
$$= \sum_i [A\tau_n(x_i) + B]\Delta[\sigma(x_i)\varrho_n(x_i)] + CS_n$$
$$= -\sum_i \sigma(x_i+1)\varrho_n(x_i+1)\Delta[A\tau_n(x_i) + B] + CS_n$$
$$= -A\tau_n' S_{n+1} + CS_n \ .$$

Hence

$$\frac{S_n}{S_{n+1}} = \frac{1 + A\tau_n'}{C} = \frac{1 + \sigma''/(2\tau_n')}{\sigma(x_n^*)} \ , \tag{2.3.10}$$

where x_n^* is the root of the equation $\tau_n(x) = 0$ and we have used the equations $\sigma(x_n^*) = C$, $\sigma'' = 2A(\tau_n')^2$.

With the aid of formulas (2.3.7–10) we finally obtain

$$d_n^2 = (-1)^n A_{nn} B_n^2 \varrho_{N-1}(a) \prod_{k=n}^{N-2} \left[\frac{1 + \sigma''/(2\tau_k')}{\sigma(x_k^*)}\right] \ , \tag{2.3.11}$$

where $N = b - a$, $\tau_k' = \tau' + k\sigma''$, x_k^* is the root of the equation $\tau_k(x) = 0$, i.e.

$$\tau(x) + k\left[\sigma'(x) + \tau' + \tfrac{1}{2}k\sigma''\right] = 0 \ .$$

2.4 The Hahn, Chebyshev, Meixner, Kravchuk, and Charlier Polynomials

2.4.1. We have considered a general method of studying the classical orthogonal polynomials of a discrete variable as solutions of the difference equation of hypergeometric type on uniform lattices. In particular a representation of these solutions in the form of the Rodrigues formula was obtained and their orthogonality property under certain conditions was proved. Investigation of specific systems of polynomials is reduced to solving the difference first order equation (2.1.17) for the function $\varrho(x)$ which enters into the Rodrigues formula (2.2.7) and the orthogonality property (2.3.1). In order to find explicit expressions for $\varrho(x)$ we rewrite the difference equation (2.1.17) in the form

$$\frac{\varrho(x+1)}{\varrho(x)} = \frac{\sigma(x) + \tau(x)}{\sigma(x+1)} \ . \tag{2.4.1}$$

It is easily verified that the solution of the difference equation

$$\frac{\varrho(x+1)}{\varrho(x)} = f(x) \ ,$$

whose right-hand side can be expressed as a product or quotient of two functions, has the following simple property:

if the functions $\varrho_1(x)$ and $\varrho_2(x)$ are solutions of the equations

$$\frac{\varrho_1(x+1)}{\varrho_1(x)} = f_1(x) \ , \qquad \frac{\varrho_2(x+1)}{\varrho_2(x)} = f_2(x) \ ,$$

then the solution of the equation

$$\frac{\varrho(x+1)}{\varrho(x)} = f(x)$$

with $f(x) = f_1(x)f_2(x)$ *is* $\varrho(x) = \varrho_1(x)\varrho_2(x)$ *and with* $f(x) = f_1(x)/f_2(x)$ *it is* $\varrho(x) = \varrho_1(x)/\varrho_2(x)$.

Since the right-hand side of (2.4.1) is a rational function, it follows that its solution can be expressed in terms of the solutions of the difference equations

$$\frac{\varrho(x+1)}{\varrho(x)} = \gamma + x \ , \tag{2.4.2}$$

$$\frac{\varrho(x+1)}{\varrho(x)} = \gamma - x \ , \tag{2.4.3}$$

$$\frac{\varrho(x+1)}{\varrho(x)} = \gamma \ , \tag{2.4.4}$$

where γ is a constant. Since

$$\gamma + x = \frac{\Gamma(\gamma + x + 1)}{\Gamma(\gamma + x)} \ ,$$

a particular solution of (2.4.2) has the form $\varrho(x) = \Gamma(\gamma + x)$. Similarly, by using the equation

$$\gamma - x = \frac{\Gamma(\gamma - x + 1)}{\Gamma(\gamma - x)} = \frac{1}{\Gamma[(\gamma+1) - (x+1)]} : \frac{1}{\Gamma(\gamma + 1 - x)} \ ,$$

we obtain a particular solution of (2.4.3)

$$\varrho(x) = \frac{1}{\Gamma(\gamma + 1 - x)} \ .$$

It is easily verified that a particular solution of (2.4.4) is $\varrho(x) = \gamma^x$.

2.4.2. Let us now find solutions of (2.4.1) corresponding to the different degrees of the polynomial $\sigma(x)$.

2.4.2.1. Let $\sigma(x)$ be a polynomial of the second degree. We consider the following cases.

a) Let

$$\sigma(x) = x\left(\gamma_1 - x\right), \quad \sigma(x) + \tau(x) = \left(x + \gamma_2\right)\left(\gamma_3 - x\right) \ .$$

Here $\gamma_1, \gamma_2, \gamma_3$ are constants. With $a = 0$ and $b = N$, conditions (2.3.5) and (2.3.6), namely

$$\begin{aligned}
\sigma\left(x_i\right) &> 0, & 1 &\le x_i \le N - 1; \\
\sigma\left(x_i\right) + \tau\left(x_i\right) &> 0, & 0 &\le x_i \le N - 2; \\
\sigma(N - 1) + \tau(N - 1) &= 0,
\end{aligned}$$

(2.4.5)

will be satisfied if we take

$$\gamma_1 = N + \alpha, \quad \gamma_2 = \beta + 1 \quad (\alpha > -1, \ \beta > -1), \quad \gamma_3 = N - 1 \ .$$

In this case (2.4.1) assumes the form

$$\frac{\varrho(x + 1)}{\varrho(x)} = \frac{(x + \beta + 1)(N - 1 - x)}{(x + 1)(N + \alpha - 1 - x)} \ .$$

(2.4.6)

A solution of this equation is

$$\varrho(x) = \frac{\Gamma(N + \alpha - x)\Gamma(x + \beta + 1)}{\Gamma(x + 1)\Gamma(N - x)} \quad (\alpha > -1, \ \beta > -1) \ .$$

(2.4.7)

Let us discuss the reasons for choosing γ_1 and γ_2 in the forms $\gamma_1 = N + \alpha$, $\gamma_2 = \beta + 1$. It is natural to expect that a polynomial solution $y_n(x)$, after the linear change of variable $x = (N/2)(1 + s)$, which carries the interval $(0, N)$ to $(-1, 1)$, will tend to the Jacobi polynomial $P_n^{(\alpha,\beta)}(s)$ when $N \to \infty$ (that is, when $\Delta s = h = 2/N \to 0$), and that the weight function $\varrho(x)$ will tend, up to a constant multiplier, to the weight function $(1 - s)^\alpha(1 + s)^\beta$ for the Jacobi polynomials. A solution of (2.4.1) for

$$\sigma(x) = x\left(\gamma_1 - x\right), \quad \sigma(x) + \tau(x) = \left(x + \gamma_1\right)(N - 1 - x)$$

is given by

$$\begin{aligned}
\varrho(x) &= \frac{\Gamma(\gamma_1 - x_1)\Gamma(x + \gamma_2)}{\Gamma(x + 1)\Gamma(N - x)} \\
&= \frac{\Gamma[(N/2)(1 - s) + \gamma_1 - N]}{\Gamma[(N/2)(1 - s)]} \ \frac{\Gamma[(N/2)(1 + s) + \gamma_2]}{\Gamma[(N/2)(1 + s) + 1]} \ .
\end{aligned}$$

Since

$$\lim_{z \to \infty} \frac{\Gamma(z+a)}{\Gamma(z)z^a} = 1 \, , \qquad\qquad (2.4.8)$$

we have

$$\varrho(x) \approx \left[\frac{N}{2}(1-s) \right]^{\gamma_1 - N} \left[\frac{N}{2}(1+s) \right]^{\gamma_2 - 1}$$

as $N \to \infty$. Consequently it is natural to take $\gamma_1 - N = \alpha$, $\gamma_2 - 1 = \beta$.

The polynomials $y_n(x)$ obtained by the Rodrigues formula (2.2.7) when $B_n = (-1)^n/n!$, with the weight function $\varrho(x)$ defined by (2.4.7), are called the *Hahn polynomials* and are denoted by $h_n^{(\alpha,\beta)}(x, N)$. We shall also use the notation $h_n^{(\alpha,\beta)}(x)$ when N is fixed in the corresponding formulas. The Hahn polynomials are orthogonal on $[0, N-1]$ when $\alpha > -1$ and $\beta > -1$. These polynomials were introduced by P.L. Chebyshev in 1875 [T3].

An important special case of the Hahn polynomials are the *Chebyshev polynomials of a discrete variable* $t_n(x) = h_n^{(0,0)}(x, N)$, introduced in Ref. [T2], for which $\varrho(x) = 1$.

b) Let

$$\sigma(x) = x\big(x + \gamma_1\big) \, , \quad \sigma(x) + \tau(x) = \big(\gamma_2 - x\big)\big(\gamma_3 - x\big) \, .$$

Conditions (2.4.5) will be satisfied if

$$\gamma_1 > -1, \quad \gamma_2 > N - 2, \quad \gamma_3 = N - 1 \, .$$

In this case putting $\mu = \gamma_1$, $\nu = \gamma_2 - N + 1$ we obtain

$$\varrho(x) = \frac{C}{\Gamma(x+1)\Gamma(x+\mu+1)\Gamma(N+\nu-x)\Gamma(N-x)} \, , \qquad (2.4.9)$$
$$(\mu > -1, \nu > -1) \, .$$

The polynomials obtained by the Rodrigues formula with $B_n = 1/n!$, when $\varrho(x)$ is defined by (2.4.9) with $C = 1$, are also called the Hahn polynomials; they are denoted by $\bar{h}_n^{(\mu,\nu)}(x, N)$.

When

$$\mu = M(1-p) - m, \quad \nu = Mp - m, \quad m = N - 1 \, ,$$
$$C = \frac{1}{M!} m!(M-m)!(Mp)!(M-Mp)!$$

(M and Mp are integers, $0 < p < 1$), the weight function (2.4.9) coincides with the *hypergeometric distribution* known in probability theory [K26]. Let us consider a typical case when this distribution arises. Out of a series of M finished products, which contains Mp nondefective and $M(1-p)$ defective specimens, m specimens are randomly chosen. The probability that among m chosen specimens we have x (x is integer) nondefective specimens is described by the hypergeometric distribution $\varrho(x)$.

There is a simple connection between the polynomials $\tilde{h}_n^{(\mu,\nu)}(x)$ and $h_n^{(\alpha,\beta)}(x)$. If we formally set $\mu = -N - \alpha$ and $\nu = -N - \beta$, the expressions for $\sigma(x)$ and $\sigma(x) + \tau(x)$ corresponding to $\tilde{h}_n^{(\mu,\nu)}(x)$ and $h_n^{(\alpha,\beta)}(x)$ differ only in sign. Consequently the polynomials $\tilde{h}_n^{(-N-\alpha,-N-\beta)}(x)$ and $h_n^{(\alpha,\beta)}(x)$ satisfy the same difference equation.

It is easily verified that, with our normalization, these polynomials are the same. In fact by using Euler's reflection formula for the gamma-function

$$\Gamma(z)\Gamma(1-z) = \frac{\pi}{\sin \pi z}$$

we can show that when $\mu = -N - \alpha$, $\nu = -N - \beta$ the expressions (2.4.7) and (2.4.9) for $\varrho(x)$ differ only by the periodic factor

$$C(x) = \frac{\pi^2}{\sin[\pi(N + \alpha - x)] \sin[\pi(\beta + 1 - x)]} \, ,$$

which does not affect the explicit formulas for polynomials obtained by using the Rodrigues formula. Hence the formulas for $\tilde{h}_n^{(-N-\alpha,-N-\beta)}(x)$ and $h_n^{(\alpha,\beta)}(x)$ agree if the normalizing constants B_n differ by the factor $(-1)^n$, since the corresponding expressions for $\sigma(x)$ differ in sign:

$$\tilde{h}_n^{(-N-\alpha,-N-\beta)}(x) = h_n^{(\alpha,\beta)}(x) \, .$$

Consequently the polynomials $\tilde{h}_n^{(-N-\alpha,-N-\beta)}(x)$ can be obtained by using analytic continuation of $h_n^{(\alpha,\beta)}(x)$ with respect to the parameters α and β from the domain $\alpha > -1$, $\beta > -1$ into the domain $\alpha < 1 - N$, $\beta < 1 - N$.

Remark. The polynomials $\tilde{h}_n^{(\mu,\nu)}(x, N)$ and $h_n^{(\alpha,\beta)}(x, N)$ can be expressed in terms of the polynomials $p_n(x, \beta, \gamma, \delta)$ discussed in [E7] for which

$$\sigma(x) = x(\delta - 1 + x), \quad \varrho(x) = \frac{(\beta)_x(\gamma)_x}{\Gamma(x+1)(\delta)_x}, \quad a = 0, \quad b = \infty, \quad B_n = \frac{1}{n!} \, ,$$

where $(\alpha)_x = \Gamma(\alpha + x)/\Gamma(\alpha)$. Since the functions $\sigma(x)$ for the polynomials $\tilde{h}_n^{(\mu,\nu)}(x, N)$ and $p_n(x, \beta, \gamma, \delta)$ coincides for $\mu = \delta - 1$, $\nu = \gamma - \beta$ and $N = 1 - \gamma$, and the weight functions $\varrho(x)$ when $x = x_i$ differ only by a constant factor, we have

$$\tilde{h}_n^{(\mu,\nu)}(x, N) = p_n(x, 1 - N - \nu, 1 - N, 1 + \mu) \, .$$

By the relation

$$h_n^{(\alpha,\beta)}(x, N) = \tilde{h}_n^{(-N-\alpha,-N-\beta)}(x, N)$$

that we had above we also obtain

$$h_n^{(\alpha,\beta)}(x, N) = p_n(x, \beta + 1, 1 - N, 1 - N - \alpha) \, .$$

2.4.2.2. Let $\sigma(x)$ be a polynomial of the first degree: $\sigma(x) = x$. We consider three cases:

$$\sigma(x) + \tau(x) = \begin{cases} \mu(\gamma + x) \\ \mu(\gamma - x) \\ \mu\,. \end{cases}$$

Here μ and γ are constants. Then (2.4.1) has the following solutions:

$$\varrho(x) = \begin{cases} C\dfrac{\mu^x \Gamma(\gamma + x)}{\Gamma(x + 1)}\,, \\[2ex] C\dfrac{\mu^x}{\Gamma(x + 1)\Gamma(\gamma + 1 - x)}\,, \\[2ex] C\dfrac{\mu^x}{\Gamma(x + 1)}\,. \end{cases}$$

a) In the first case we can satisfy the boundary conditions (2.3.2) and the positivity of the weight function $\varrho(x_i)$ by taking

$$a = 0, \quad b = +\infty, \quad 0 < \mu < 1, \quad \gamma > 0\,.$$

It is convenient to take C to be $1/\Gamma(\gamma)$. We then obtain *the Pascal distribution* from probability theory

$$\varrho(x) = \frac{\mu^x(\gamma)_x}{\Gamma(x + 1)}\,. \tag{2.4.10}$$

With $B_n = \mu^{-n}$ the corresponding polynomials are the *Meixner polynomials* $m_n^{(\gamma,\mu)}(x)$, introduced in [M8].

b) Arguing similarly in the second case, we take

$$a = 0, \quad b = N + 1, \quad \gamma = N\,,$$

$$\mu = \frac{p}{q} \ (p > 0,\ q > 0,\ p + q = 1), \quad C = q^N N!\,.$$

The numbers $\varrho(x_i)$ become the familiar *binomial distribution* from probability theory,

$$\varrho(x_i) = C_N^i p^i q^{N-i}, \quad C_N^i = \frac{N!}{i!(N - i)!}\,. \tag{2.4.11}$$

With $B_n = (-1)^n q^n/n!$ the corresponding polynomials are the *Kravchuk polynomials* $k_n^{(p)}(x, N)$, introduced in [K29].

c) In the third case, with $a = 0$, $b = +\infty$, $C = e^{-\mu}$ we have the *Poisson distribution*

$$\varrho(x_i) = \frac{e^{-\mu}\mu^i}{i!}\,. \tag{2.4.12}$$

The corresponding orthogonal polynomials of a discrete variable, with $B_n = \mu^{-n}$, are the Charlier polynomials $c_n^{(\mu)}(x)$ introduced in [C6].

The case $\sigma(x) = 1$ is not of interest, since it does not lead to any new polynomials.

2.4.3. From (2.2.9) we obtain the following formulas for the Hahn, Meixner, Kravchuk, and Charlier polynomials:

$$\Delta h_n^{(\alpha,\beta)}(x, N) = (\alpha + \beta + n + 1) h_{n-1}^{(\alpha+1,\beta+1)}(x, N - 1) , \tag{2.4.13}$$

$$\Delta \tilde{h}_n^{(\mu,\nu)}(x, N) = -(\mu + \nu + 2N - n - 1)\tilde{h}_{n-1}^{(\mu,\nu)}(x, N - 1) , \tag{2.4.14}$$

$$\Delta k_n^{(p)}(x, N) = k_{n-1}^{(p)}(x, N - 1) , \tag{2.4.15}$$

$$\Delta m_n^{(\gamma,\mu)}(x) = -\frac{n(1 - \mu)}{\mu} m_{n-1}^{(\gamma+1,\mu)}(x) , \tag{2.4.16}$$

$$\Delta c_n^{(\mu)}(x) = -\frac{n}{\mu} c_{n-1}^{(\mu)}(x) . \tag{2.4.17}$$

2.4.4. Let us consider the *symmetry properties* of the orthogonal polynomials of a discrete variable that follow from the symmetry of $\varrho(x)$. For the Hahn polynomials $h_n^{(\alpha,\beta)}(x, N)$ the weight function $\varrho(x)$ has the following symmetries:

$$\varrho(x) \equiv \varrho(x, \alpha, \beta) = \varrho(N - 1 - x, \beta, \alpha) .$$

Hence by replacing i by $N - 1 - i$ we can rewrite the orthogonality relation

$$\sum_{i=0}^{N-1} h_n^{(\alpha,\beta)}(x_i) h_m^{(\alpha,\beta)}(x_i) \varrho(x_i, \alpha, \beta) = 0 \quad (m \neq n) ,$$

in the form

$$\sum_{i=0}^{N-1} h_n^{(\alpha,\beta)}(N - 1 - x_i) h_m^{(\alpha,\beta)}(N - 1 - x_i) \varrho(x_i, \alpha, \beta) = 0 \quad (m \neq n) .$$

Since the weight function $\varrho(x)$ and the interval of orthogonality (a, b) determine the polynomials uniquely, up to a constant multiple, we have

$$h_n^{(\alpha,\beta)}(N - 1 - x) = C_n h_n^{(\beta,\alpha)}(x) ,$$

where C_n is a constant. Equating the coefficients of x^n on both sides, by using (2.2.11) we obtain $C_n = (-1)^n$, i.e.

$$h_n^{(\alpha,\beta)}(N - 1 - x) = (-1)^n h_n^{(\beta,\alpha)}(x) . \tag{2.4.18}$$

Similarly the Kravchuk polynomials satisfy

$$k_n^{(p)}(x) = (-1)^n k_n^{(q)}(N - x), \quad p + q = 1 . \tag{2.4.19}$$

Relation (2.4.18) remains valid for any complex values of x, α, β, N. For a proof it is sufficient for the Hahn polynomials $y_n(x) = h_n^{(\alpha,\beta)}(x, N)$ to use the difference equation

$$x(N + \alpha - x)\Delta\nabla y_n(x) + [(\beta + 1)(N - 1) - (\alpha + \beta + 2)x]\Delta y_n(x)$$
$$+ n(n + \alpha + \beta + 1)y_n(x) = 0 \; .$$

It is easy to verify that on replacing x by $N - 1 - x$, α by β, and β by α this equation keeps its form. Since at such a replacement $y_n(x)$ remains the polynomial of the same degree, then owing to the uniqueness of polynomial solutions for difference equations of hypergeometric type we come to the relation

$$h_n^{(\alpha,\beta)}(x, N) = C_n h_n^{(\beta,\alpha)}(N - 1 - x, N) \; ,$$

where C_n is a constant which may be found by equating the coefficients of x^n. The obtained relation, obviously, is equivalent to (2.4.18). In a similar way by using the Rodrigues formula we may obtain (2.4.19) for any complex values of x, p, N as well as the following relations:

$$h_n^{(\alpha,\beta)}(x, N) = (-1)^n h_n^{(\alpha,\beta)}(-\beta - x - 1, -\alpha - \beta - N) \; , \tag{2.4.20}$$

$$m_n^{(\gamma,\mu)}(x) = \mu^{-n} m_n^{(\gamma,1/\mu)}(-\gamma - x) \; , \tag{2.4.21}$$

$$k_n^{(p)}(x, N) = \frac{p^n}{n!} m_n^{(-N,-p/q)}(x) \; . \tag{2.4.22}$$

2.4.5. By using the Rodrigues formula, it is easy to find *the values of the Hahn, Meixner, Kravchuk, and Charlier polynomials at the endpoints of the interval of orthogonality.* Let us use the formula

$$\nabla^n f(x) = \sum_{k=0}^{n} (-1)^k \frac{n!}{k!(n - k)!} f(x - k) \; ,$$

which can be proved by induction. Since under the condition $\sigma(0) = 0$ the function $\varrho_n(x) = \varrho(x + n)\sigma(x + 1) \ldots \sigma(x + n)$ is zero at $x = -1, -2, \ldots, -n$, we have $\nabla^n \varrho_n(0) = \varrho_n(0)$ and by the Rodrigues formula (2.2.7)

$$y_n(0) = \frac{B_n}{\varrho(0)} \varrho_n(0) \; . \tag{2.4.23}$$

Hence for the Hahn, Meixner, Kravchuk and Charlier polynomials we obtain

$$h_n^{(\alpha,\beta)}(0) = (-1)^n \frac{(N - 1)!}{n!(N - n - 1)!} \frac{\Gamma(n + \beta + 1)}{\Gamma(\beta + 1)} \; ,$$

$$m_n^{(\gamma,\mu)}(0) = \frac{\Gamma(n + \gamma)}{\Gamma(\gamma)} \; ,$$

$$k_n^{(p)}(0) = (-1)^n \frac{N!}{n!(N - n)!} p^n \; ,$$

$$c_n^{(\mu)}(0) = 1 \; .$$

By using their symmetry properties we can easily find expressions for

$h_n^{(\alpha,\beta)}(N-1)$ and $k_n^{(p)}(N)$:

$$h_n^{(\alpha,\beta)}(N-1) = \frac{(N-1)!}{n!(N-n-1)!}\,\frac{\Gamma(n+\alpha+1)}{\Gamma(\alpha+1)}\ ,$$

$$k_n^{(p)}(N) = \frac{n!}{n!(N-n)!}\,q^n\ .$$

$$(2.4.24)$$

2.4.6. Finally we observe that for any polynomials that satisfy an orthogonality relation of the form

$$\sum_{i=0}^{N-1} p_n(x_i)p_m(x_i)\varrho_i = d_n^2\delta_{mn}\ ,$$

$$(2.4.25)$$

where the points x_i do not necessarily belong to the uniform lattice, there is another orthogonality relation. In fact, if we consider the matrix C whose elements are

$$C_{ni} = \frac{1}{d_n}p_n(x_i)\sqrt{\varrho_i}\ ,$$

the orthogonality property for the polynomials $p_n(x)$ is equivalent to the unitary property of matrix C:

$$\sum_{i=0}^{N-1} C_{ni}C_{mi} = \delta_{mn}\ .$$

Hence the matrix C also satisfies

$$\sum_{n=0}^{N-1} C_{ni}C_{nj} = \delta_{ij}\ ,$$

which is equivalent to the "dual" orthogonality relation for the $p_n(x)$:

$$\sum_{n=0}^{N-1} p_n(x_i)p_n(x_j)\tilde{\varrho}_n = \frac{1}{\varrho_i}\delta_{ij}\ ,$$

$$(2.4.26)$$

where $\tilde{\varrho}_n = 1/d_n^2$.

Let us show that the dual orthogonality relation for the Hahn polynomials $h_n^{(\alpha,\beta)}(x)$ leads to another system of orthogonal polynomials. We need to know the dependence on n of the values of $h_n^{(\alpha,\beta)}(x)$ at $x = i$ ($i = 0, 1, \ldots$). For this we rewrite a difference equation for the Hahn polynomials in the form

$$y_{i+1} = \left(-A_i\lambda_n + B_i\right)y_i + C_i y_{i-1} \quad \left(C_0 = 0\right)\ ,$$

where

$$y_i = h_n^{(\alpha,\beta)}(i)\ , \quad y_0 = (-1)^n\frac{(N-1)!\,\Gamma(n+\beta+1)}{n!(N-n-1)!\,\Gamma(\beta+1)}\ ,$$

$$A_i = \frac{1}{\sigma(i) + \tau(i)} = \frac{1}{(N - i - 1)(i + \beta + 1)} \, ,$$

B_i and C_i are constants independent of n.

Hence by induction we can obtain that y_i/y_0 is a polynomial of degree i in λ_n with the leading coefficient equal to $(-i)^i \prod_{k=0}^{i-1} A_k$. Since

$$\lambda_n = t_n - \tfrac{1}{4}(\alpha + \beta)(\alpha + \beta + 2) \, ,$$

where $t_n = s_n(s_n + 1)$, $s_n = n + \tfrac{1}{2}(\alpha + \beta)$, as a result we obtain

$$h_n^{(\alpha,\beta)}(i) = (-1)^i h_n^{(\alpha,\beta)}(0) i! w_i^{(\alpha,\beta)}(t_n) \prod_{k=0}^{i-1} A_k$$

$$= (-1)^{n+i} \frac{i!(N - i - 1)! \Gamma(n + \beta + 1)}{n!(N - n - 1)! \Gamma(i + \beta + 1)} w_i^{(\alpha,\beta)}(t_n) \, ,$$

where $w_i^{(\alpha,\beta)}(t)$ is a polynomial of degree i in t with leading coefficient $1/i!$. Hence the dual orthogonality relation for the Hahn polynomials $h_n^{(\alpha,\beta)}(x)$ leads to the following orthogonality relation for the polynomials $w_i^{(\alpha,\beta)}(t)$, which we naturally call *the dual Hahn polynomials*:

$$\sum_{n=0}^{N-1} w_i^{(\alpha,\beta)}(t_n) w_j^{(\alpha,\beta)}(t_n) \tilde{\varrho}_n = \tilde{d}_i^2 \delta_{ij} \, , \tag{2.4.27}$$

where

$$\tilde{\varrho}_n = \frac{\alpha + \beta + 2n + 1}{n!(N - n - 1)} \frac{\Gamma(\alpha + \beta + n + 1)\Gamma(\beta + n + 1)}{\Gamma(\alpha + \beta + n + N + 1)\Gamma(\alpha + n + 1)} \, , \tag{2.4.28}$$

$$\tilde{d}_i^2 = \frac{\Gamma(\beta + i + 1)}{i!(N - i - 1)! \Gamma(N + \alpha - i)} \tag{2.4.29}$$

($\tilde{\varrho}_n$ and $\tilde{d}_i^2$ are obtained by taking d_n^2 for the Hahn polynomials – see Sect. 2.5.1, Table 2.1).

Consequently the dual orthogonality relation for the Hahn polynomials $h_n^{(\alpha,\beta)}(x_i)$ leads to the dual Hahn polynomials $w_i^{(\alpha,\beta)}(t_n)$ which are orthogonal on a quadratic in n lattice $t_n = s_n(s_n + 1)$, $s_n = n + (\alpha + \beta)/2$. The theory of these polynomials as well as the polynomials which are orthogonal on some classes of nonuniform lattices will be discussed in the next chapter.

We note also that the dual orthogonality relation for the Kravchuk polynomials does not lead to a new system of orthogonal polynomials, i.e. the Kravchuk polynomials are self-dual.

2.5 Calculation of Main Characteristics

2.5.1. Let us obtain the main data for the Hahn, Meixner, Kravchuk and Charlier polynomials. We first find the leading coefficients a_n and b_n in the expansions

$$y_n(x) = a_n x^n + b_n x^{n-1} + \dots .$$

Here it will be sufficient to use (2.2.11) and (2.2.12).

The squared norm d_n^2 can be found from (2.3.7). Its calculation leads to the evaluation of the sums

$$S_n = \sum_{x_i=a}^{b-n-1} \varrho_n(x_i) .$$

The calculation of S_n is especially simple for the Meixner, Kravchuk and Charlier polynomials. For these polynomials

$$\varrho_n(x) = \varrho(x+n) \prod_{k=1}^{n} \sigma(x+k) = \varrho(x+n)\frac{\Gamma(1+x+n)}{\Gamma(1+x)} .$$

1) For the Meixner polynomials

$$\sum_i \varrho_n(x_i) = \sum_{i=0}^{\infty} \frac{\mu^{i+n}\Gamma(\gamma+i+n)}{i!\Gamma(\gamma)} .$$

Since for $|\mu| < 1$ by the Taylor formula we have

$$(1-\mu)^{-(\gamma+n)} = \sum_{i=0}^{\infty} \frac{\Gamma(\gamma+i+n)}{\Gamma(\gamma+n)} \frac{\mu^i}{i!} ,$$

so for the polynomials $m_n^{(\gamma,\mu)}(x)$ we obtain

$$d_n^2 = \frac{n!(\gamma)_n}{\mu^n(1-\mu)^\gamma} . \tag{2.5.1}$$

2) For the Kravchuk polynomials

$$\sum_i \varrho_n(x_i) = \sum_{i=0}^{N-n} \frac{N!p^{i+n}q^{N-i-n}}{i!\Gamma(N+1-i-n)}$$

$$= \frac{N!p^n}{(N-n)!} \sum_{i=0}^{N-n} C_{N-n}^i p^i q^{N-n-i} = \frac{N!p^n}{(N-n)!} ,$$

whence

$$d_n^2 = C_N^n (pq)^n . \tag{2.5.2}$$

3) For the Charlier polynomials,

$$\sum_i \varrho_n(x_i) = \sum_{i=0}^{\infty} \frac{e^{-\mu}\mu^{i+n}}{i!} = \mu^n \,,$$

whence

$$d_n^2 = \frac{n!}{\mu^n} \,. \tag{2.5.3}$$

4) For the Hahn polynomials the squared norms can be found from (2.3.11):

$$d_n^2 = \begin{cases} \dfrac{\Gamma(\alpha+n+1)\Gamma(\beta+n+1)(\alpha+\beta+n+1)_N}{(\alpha+\beta+2n+1)n!(N-n-1)!} \\[2mm] \left(\text{for}\quad h_n^{(\alpha,\beta)}(x,N)\right) ; \\[4mm] \dfrac{(\mu+\nu+N-n)_N}{(\mu+\nu+2N-2n-1)n!\,\Gamma(\mu+N-n)\Gamma(\nu+N-n)(N-n-1)!} \\[2mm] \left(\text{for}\quad \tilde{h}_n^{(\mu,\nu)}(x)\right) . \end{cases} \tag{2.5.4}$$

2.5.2. For the Hahn, Meixner, Kravchuk and Charlier polynomials we have the recursion relation

$$xy_n(x) = \alpha_n y_{n+1}(x) + \beta_n y_n(x) + \gamma_n y_{n-1}(x) \,, \tag{2.5.5}$$

whose coefficients can be found from the known values of b_n, and d_n^2 by the formulas

$$\alpha_n = \frac{a_n}{a_{n+1}} \,, \quad \beta_n = \frac{b_n}{a_n} - \frac{b_{n+1}}{a_{n+1}} \,, \quad \gamma_n = \frac{a_{n-1}}{a_n}\frac{d_n^2}{d_{n-1}^2} \,. \tag{2.5.6}$$

The basic information about the Hahn, Chebyshev, Meixner, Kravchuk and Charlier polynomials is displayed in Tables 2.1–3.

Table 2.1. Data for the Hahn polynomials $h_n^{\alpha,\beta}(x, N)$ and the Chebyshev polynomials $t_n(x)$

$y_n(x)$	$h_n^{(\alpha,\beta)}(x, N)$	$t_n(x)$
(a, b)	$(0, N)$	$(0, N)$
$\varrho(x)$	$\dfrac{\Gamma(N + \alpha - x)\Gamma(\beta + 1 + x)}{\Gamma(x + 1)\Gamma(N - x)}$ $(\alpha > -1,\ \beta > -1)$	1
$\sigma(x)$ $\tau(x)$ λ_n	$x(N + \alpha - x)$ $(\beta + 1)(N - 1) - (\alpha + \beta + 2)x$ $n(\alpha + \beta + n + 1)$	$x(N - x)$ $N - 1 - 2x$ $n(n + 1)$
B_n	$(-1)^n/n!$	$(-1)^n/n!$
$\varrho_n(x)$	$\dfrac{\Gamma(N + \alpha - x)\Gamma(n + \beta + 1 + x)}{\Gamma(x + 1)\Gamma(N - n - x)}$	$\dfrac{\Gamma(N - x)\Gamma(n + 1 + x)}{\Gamma(N - n - x)\Gamma(x + 1)}$
a_n	$\dfrac{1}{n!}(\alpha + \beta + n + 1)_n$	$\dfrac{1}{n!}(n + 1)_n$
b_n	$-\dfrac{(\alpha + \beta + n + 1)_{n-1}}{(n - 1)!}\Big[(\beta + 1)(N - 1)$ $+ \dfrac{n - 1}{2}(\alpha - \beta + 2N - 2)\Big]$	$-\dfrac{N - 1}{(n - 1)!}(n)_n$
d_n^2	$\dfrac{\Gamma(\alpha + n + 1)\Gamma(\beta + n + 1)(\alpha + \beta + n + 1)_N}{(\alpha + \beta + 2n + 1)n!(N - n - 1)!}$	$\dfrac{(N + n)!}{(2n + 1)(N - n - 1)!}$
α_n	$\dfrac{(n + 1)(\alpha + \beta + n + 1)}{(\alpha + \beta + 2n + 1)(\alpha + \beta + 2n + 2)}$	$\dfrac{n + 1}{2(2n + 1)}$
β_n	$\dfrac{\alpha - \beta + 2N - 2}{4}$ $+ \dfrac{(\beta^2 - \alpha^2)(\alpha + \beta + 2N)}{4(\alpha + \beta + 2n)(\alpha + \beta + 2n + 2)}$	$\dfrac{N - 1}{2}$
γ_n	$\dfrac{(\alpha + n)(\beta + n)(\alpha + \beta + N + n)(N - n)}{(\alpha + \beta + 2n)(\alpha + \beta + 2n + 1)}$	$\dfrac{n(N^2 - n^2)}{2(2n + 1)}$

Table 2.2. Data for the Hahn polynomials $\tilde{h}_n^{(\mu,\nu)}(x, N)$

$y_n(x)$	$\tilde{h}_n^{(\mu,\nu)}(x, N)$
(a, b)	$(0, N)$
$\varrho(x)$	$[\Gamma(x+1)\Gamma(x+\mu+1)\Gamma(N+\nu-x)\Gamma(N-x)]^{-1}$ $(\mu > -1,\ \nu > -1)$
$\sigma(x)$ $\tau(x)$ λ_n	$x(x+\mu)$ $(N+\nu-1)(N-1) - (2N+\mu+\nu-2)x$ $n(2N+\mu+\nu-n-1)$
B_n	$1/n!$
$\varrho_n(x)$	$[\Gamma(x+1)\Gamma(x+\mu+1)\Gamma(N+\nu-n-x)\Gamma(N-n-x)]^{-1}$
a_n	$\dfrac{(-1)^n}{n!}(2N+\mu+\nu-2n)_n$
b_n	$\dfrac{(-1)^{n-1}}{(n-1)!}[(N+\nu-1)(N-1)$ $-\dfrac{n-1}{2}(2N+\nu-\mu-2)](2N+\mu+\nu-2n+1)_{n-1}$
d_n^2	$\dfrac{(N+\mu+\nu-n)_N}{(2N+\mu+\nu-2n-1)n!\,\Gamma(N+\mu-n)\Gamma(N+\nu-n)(N-n-1)!}$
α_n	$-\dfrac{(n+1)(2N+\mu+\nu-n-1)}{(2N+\mu+\nu-2n-1)(2N+\mu+\nu-2n-2)}$
β_n	$\dfrac{2(N-1)+\nu-\mu}{4}$ $+\dfrac{(\mu^2-\nu^2)(2N+\mu+\nu)}{4(2N+\mu+\nu-2n)(2N+\mu+\nu-2n-2)}$
γ_n	$-\dfrac{(N-n)(N-n+\mu)(N-n+\nu)(N-n+\mu+\nu)}{(2N+\mu+\nu-2n)(2N+\mu+\nu-2n-1)}$

Table 2.3. Data for the Meixner, Kravchuk and Charlier polynomials

$y_n(x)$	$m_n^{(\gamma,\mu)}(x)$	$k_n^{(p)}(x)$	$c_n^{(\mu)}(x)$
(a, b)	$(0, \infty)$	$(0, N+1)$	$(0, \infty)$
$\varrho(x)$	$\dfrac{\mu^x \Gamma(\gamma + x)}{\Gamma(x+1)\Gamma(\gamma)}$ $(\gamma > 0, 0 < \mu < 1)$	$\dfrac{N! p^x q^{N-x}}{\Gamma(x+1)\Gamma(N+1-x)}$ $(p > 0, q > 0, p + q = 1)$	$\dfrac{e^{-\mu}\mu^x}{\Gamma(x+1)}$ $(\mu > 0)$
$\sigma(x)$ $\tau(x)$ λ_n	x $\gamma\mu - x(1-\mu)$ $n(1-\mu)$	x $(Np - x)/q$ n/q	x $\mu - x$ n
B_n	$\dfrac{1}{\mu^n}$	$\dfrac{(-1)^n q^n}{n!}$	$\dfrac{1}{\mu^n}$
$\varrho_n(x)$	$\dfrac{\mu^{x+n}\Gamma(n+\gamma+x)}{\Gamma(\gamma)\Gamma(x+1)}$	$\dfrac{N! p^{x+n} q^{N-n-x}}{\Gamma(x+1)\Gamma(N+1-n-x)}$	$\dfrac{e^{-\mu}\mu^{x+n}}{\Gamma(x+1)}$
a_n	$\left(\dfrac{\mu-1}{\mu}\right)^n$	$\dfrac{1}{n!}$	$\dfrac{1}{(-\mu)^n}$
b_n	$n\left(\gamma + \dfrac{n-1}{2}\dfrac{\mu+1}{\mu}\right)$ $\times \left(\dfrac{\mu-1}{\mu}\right)^{n-1}$	$-\dfrac{Np + (n-1)(1/2 - p)}{(n-1)!}$	$\dfrac{n[1 + (n-1)/2\mu]}{(-\mu)^{n-1}}$
d_n^2	$\dfrac{n!(\gamma)_n}{\mu^n(1-\mu)^\gamma}$	$\dfrac{N!}{n!(N-n)!}(pq)^n$	$\dfrac{n!}{\mu^n}$
α_n	$\dfrac{\mu}{\mu-1}$	$n+1$	$-\mu$
β_n	$\dfrac{n + \mu(n+\gamma)}{1-\mu}$	$n + p(N - 2n)$	$n + \mu$
γ_n	$\dfrac{n(n-1+\gamma)}{\mu-1}$	$pq(N - n + 1)$	$-n$

2.6 Asymptotic Properties. Connection with the Jacobi, Laguerre, and Hermite Polynomials

The difference equation (2.1.2) approximates a differential equation for the classical orthogonal polynomials (2.1.1) to the second order of accuracy with respect to the step $\Delta x = h$. Therefore it is natural to expect that when $h \to 0$ the polynomial solutions of (2.1.2), properly normalized, will converge to the Jacobi, Laguerre, and Hermite polynomials. The validity of this proposition is easily established by induction if we use the recursion relations (2.5.5) for the respective polynomials.

2.6.1. As an example we carry out the limiting process *for the Hahn and Jacobi polynomials*. To begin with, the linear change of variable $x = N(1+s)/2$ carries the orthogonality interval $(0, N)$ for the Hahn polynomials to $(-1, 1)$. Then the difference equation (2.1.4) for the polynomials $h_n^{(\alpha,\beta)}(x) = u(s)$ takes the form

$$(1 + s)(1 - s + \alpha h)\frac{u(s + h) - 2u(s) + u(s - h)}{h^2}$$

$$- [(\alpha + \beta + 2)s + \alpha - \beta + (\beta + 1)h]\frac{u(s + h) - u(s)}{h}$$

$$+ n(n + \alpha + \beta + 1)u(s) = 0 \tag{2.6.1}$$

with $h = 2/N$.

As $h \to 0$, this equation goes over formally to the differential equation for the Jacobi polynomials $P_n^{(\alpha,\beta)}(s)$. Hence we expect the limit relation

$$\lim_{n \to \infty} C_n(N)h_n^{(\alpha,\beta)}\left[\frac{N}{2}(1 + s)\right] = P_n^{(\alpha,\beta)}(s) , \tag{2.6.2}$$

where $C_n(N)$ is a normalizing factor.

To establish the validity of (2.6.2) and find the factor $C_n(N)$ we compare the recursion relations for $p_n(s) = P_n^{(\alpha,\beta)}(s)$ and $v_n(s, N) = C_n(N)h_n^{(\alpha,\beta)}[N(1+s)/2]$ (see Tables 1.1 and 2.1):

$$sp_n = \alpha_n p_{n+1} + \beta_n p_n + \gamma_n p_{n-1} ,$$

$$sv_n = \alpha_n \frac{C_n}{NC_{n+1}}v_{n+1} + \left[\left(1 + \frac{\alpha + \beta}{2N}\right)\beta_n + \frac{\alpha - \beta - 2}{2N}\right]v_n$$

$$+ \gamma_n\left(1 - \frac{n}{N}\right)\left(1 + \frac{n + \alpha + \beta}{N}\right)\frac{NC_n}{C_{n-1}}v_{n-1} .$$

If we compare these recursion relations, it is clear that (2.6.2) will hold for all n if it is satisfied for $n = 0$ and if $C_n/C_{n+1} = N$. This yields $C_n = N^{-n}$.

Hence we obtain the following limit relation:

$$\frac{1}{N^n}h_n^{(\alpha,\beta)}\left[\frac{N}{2}(1 + s)\right] = P_n^{(\alpha,\beta)}(s) + O\left(\frac{1}{N}\right) . \tag{2.6.3}$$

In particular, for $t_n(x)$, the Chebyshev polynomials of a discrete variable, (2.6.3) takes the form [S38]

$$\frac{1}{N_n} t_n \left[\frac{N}{2}(1+s) \right] = P_n(s) + O\left(\frac{1}{N}\right), \tag{2.6.4}$$

where $P_n(s)$ are the Legendre polynomials.

By the same method we can obtain a more precise asymptotic formula for the Hahn polynomials [N10]:

$$\frac{1}{\tilde{N}^n} h_n^{(\alpha,\beta)} \left[\frac{\tilde{N}}{2}(1+s) - \frac{\beta+1}{2}, N \right] = P_n^{(\alpha,\beta)}(s) + O\left(\frac{1}{\tilde{N}^2}\right), \tag{2.6.5}$$

where $\tilde{N} = N + \frac{1}{2}(\alpha + \beta)$ $(N \to \infty)$.

In particular, for $t_n(x)$, the Chebyshev polynomials of a discrete variable, (2.6.5) takes the form

$$\frac{1}{N^n} t_n \left[\frac{N}{2}(1+s) - \frac{1}{2} \right] = P_n(s) + O\left(\frac{1}{N^2}\right). \tag{2.6.6}$$

Asymptotic formulas (2.6.5) and (2.6.6) may be derived in the following way. Since the coefficient β_n in the recursion relation for $h_n(x)$ contains the summand $(\alpha - \beta + 2N - 2)/4$, which does not depend on n (see Table 2.1), this relation may be rewritten in the form

$$\left(x - \frac{\alpha - \beta + 2N - 2)}{4} \right) h_n(x) = \frac{\alpha_n}{2} h_{n+1}(x) + \frac{\alpha + \beta + 2N}{4} \beta_n h_n(x)$$
$$+ \frac{\gamma_n}{2}(\alpha + \beta + N + n)(N - n) h_{n-1}(x),$$

where α_n, β_n and γ_n are the coefficients of the recursion relation for the Jacobi polynomials. Putting

$$\frac{x - (\alpha - \beta + 2N - 2)/4}{(\alpha + \beta + 2N)/4} = s, \quad N + \frac{\alpha + \beta}{2} = \tilde{N}, \quad v_n(s) = C_n h_n(x)$$

and taking into account that

$$(\alpha + \beta + N + n)(N - n)$$
$$= \left[\left(N + \frac{\alpha + \beta}{2} \right) + \left(n + \frac{\alpha + \beta}{2} \right) \right] \left[\left(N + \frac{\alpha + \beta}{2} \right) - \left(n + \frac{\alpha + \beta}{2} \right) \right]$$
$$= \tilde{N}^2 - \left(n + \frac{\alpha + \beta}{2} \right)^2,$$

we obtain

$$s v_n(s) = \alpha_n \frac{C_n}{\tilde{N} C_{n+1}} v_{n+1}(s) + \beta_n v_n(s)$$
$$+ \gamma_n \left[1 - \left(\frac{n + (\alpha + \beta)/2}{\tilde{N}} \right)^2 \right] \frac{\tilde{N} C_n}{C_{n-1}} v_{n-1}(s),$$

whence at $C_n = \tilde{N}^{-n}$ we have the relations (2.6.5) and (2.6.6).

46

The linear transformation

$$x = \frac{\tilde{N}}{2}(1+s) + \frac{\beta+1}{2}$$

converts the interval $[0,\ N-1]$ into $[-1+(\beta+1)/\tilde{N},\ 1-(\alpha+1)/\tilde{N}]$.

For the function $v_n(s) = v(s)$ the difference equation (2.1.3) takes the form

$$\left[1 - s^2 - \frac{(\alpha+1)(\beta+1)}{4}h^2\right] \frac{v(s+h) - 2v(s) + v(s-h)}{h^2}$$

$$+ [\beta - \alpha - (\alpha+\beta+2)s]\frac{v(s+h) - v(s-h)}{h} + n(\alpha+\beta+n+1)v(s) = 0\ ,$$

where $h = 2/\tilde{N}$. This difference equation approximates the differential equation for the Jacobi polynomials $u(s) = P_n^{(\alpha,\beta)}(s)$

$$(1 - s^2)u'' + [\beta - \alpha - (\alpha+\beta+2)s]u' + n(\alpha+\beta+n+1)u = 0$$

to the second order of accuracy on the lattice with the step $h = 2/\tilde{N}$.

To obtain an asymptotic expression corresponding to (2.6.5) for the weight function $\varrho(x)$ and the squared norm d_n^2 of the Hahn polynomials $h_n^{(\alpha,\beta)}(x)$ it is convenient to use the following asymptotic representation of the Γ-function [A1]:

$$\frac{\Gamma(x+a+1)}{\Gamma(x-a)} = x^{2a+1}\left[1 + O\left(\frac{1}{x^2}\right)\right]\ ,\qquad x \to \infty\ .$$

Then we obtain

$$\varrho(x) = \left(\frac{\tilde{N}}{2}\right)^{\alpha+\beta}(1-s)^\alpha(1+s)^\beta\left[1 + O\left(\frac{1}{\tilde{N}^2}\right)\right]\ ,$$

$$d_n^2 = \tilde{N}^{\alpha+\beta+2n+1}\frac{\Gamma(\alpha+n+1)\Gamma(\beta+n+1)}{(\alpha+\beta+2n+1)n!\Gamma(\alpha+\beta+n+1)}\left[1 + O\left(\frac{1}{\tilde{N}^2}\right)\right]\ ,$$

$$\tilde{N} = N + \tfrac{1}{2}(\alpha+\beta)\ ,\qquad N \to \infty\ .$$

The difference equation (2.1.17) for $\varrho(x)$ can be written in the form:

$$\frac{1}{h}[\tilde{\sigma}(s+h)\tilde{\varrho}(s+h) - \tilde{\sigma}(s)\tilde{\varrho}(s)] \tag{2.6.7}$$

$$= \tfrac{1}{2}[\tilde{\tau}(s+h)\tilde{\varrho}(s+h) + \tilde{\tau}(s)\tilde{\varrho}(s)]\ ,$$

where

$$\tilde{\sigma}(s) = 1 - s^2 - \tfrac{1}{4}(\alpha+1)(\beta+1)h^2\ ,$$

$$\tilde{\tau}(s) = \beta - \alpha - (\alpha+\beta+2)s\ ,$$

$$\tilde{\varrho}(s) = \left(\frac{2}{\tilde{N}}\right)^{\alpha+\beta}\varrho\left[\frac{\tilde{N}}{2}(1+s) - \frac{\beta+1}{2}\right]\ ,\qquad h = \frac{2}{\tilde{N}}\ .$$

It is easy to see that equation (2.6.7) approximates, to the second order of accuracy, the differential equation for the weight function of the Jacobi polynomials.

2.6.2. In just the same way if we put $y(x) = m_n^{(\gamma,\mu)}(x) = u(s)$, $x = s/h$, $h = 1-\mu$ in the equation for the Meixner polynomials we obtain the difference equation

$$s\frac{u(s+h) - 2u(s) + u(s-h)}{h^2} + [\gamma(1-h) - s]\frac{u(s+h) - u(s)}{h} + nu(s) = 0 \; ,$$

which for $h \to 0$ goes over to the differential equation

$$su'' + (\gamma - s)u' + nu = 0 \; .$$

Polynomial solutions of this equation have the form $L_n^{\gamma-1}(s)$. Hence we expect the limit relation

$$\lim_{h \to 0} C_n m_n^{(\gamma,1-h)}\left(\frac{s}{h}\right) = L_n^{\gamma-1}(s) \; .$$

Equating the coefficients at higher degrees s we obtain $C_n = 1/n!$. By using the recursion relations *for the Meixner and Laguerre polynomials* we obtain

$$\frac{1}{n!}m_n^{(\alpha+1,1-h)}\left(\frac{s}{h}\right) = L_n^{\alpha}(s) + \mathrm{O}(h) \; , \quad h \to 0 \; . \tag{2.6.8}$$

2.6.3. We now find the limit relation *for the Kravchuk polynomials* $k_n^{(p)}(x)$. Here it is convenient to appeal to a well known limit theorem from probability theory on the binomial distribution, namely that a $\quad l \to \infty$ we have

$$\varrho(x_i) = C_N^i p^i q^{N-i} \approx \frac{1}{\sqrt{2\pi N pq}} \exp\left[-\frac{(i - Np)^2}{2Npq}\right] \; ,$$

i.e. the weight function $\varrho(x)$ for the Kravchuk polynomials with $x = x_i = i$, tends, except for a normalizing factor, to the weight function of the Hermite polynomials

$$\varrho(s) = \exp\left(-s^2\right) \quad \text{with} \quad s = \frac{x - Np}{\sqrt{2Npq}} \; .$$

Corresponding to this, we put

$$x = Np + \sqrt{2Npq}\,s \; , \quad y(x) = u(s) \; , \quad h = \frac{1}{\sqrt{2Npq}}$$

in the equation for the Kravchuk polynomials. Then this equation takes the form

$$\left(1 + \sqrt{\frac{2q}{Np}}s\right)\frac{u(s+h) - 2u(s) + u(s-h)}{h^2}$$

$$- 2s\frac{u(s+h) - u(s)}{h} + 2nu(s) = 0 \; .$$

As $N \to \infty$ this equation goes over formally to the differential equation

$$u'' - 2su' + 2nu = 0 \; ,$$

whose polynomial solutions are the Hermite polynommials. Repeating the arguments used for (2.6.3), we obtain

$$\lim_{N \to \infty} \left(\frac{2}{Npq} \right)^{n/2} n! k_n^{(p)} \left(Np + \sqrt{2Npqs} \right) = H_n(s) \; . \tag{2.6.9}$$

Analogously *for the Charlier polynomials* we may obtain the limit relation

$$\lim_{\mu \to \infty} (2\mu)^{n/2} c_n^{(\mu)} \left(\mu - \sqrt{2\mu s} \right) = H_n(s) \; . \tag{2.6.10}$$

2.7 Representation in Terms of Generalized Hypergeometric Functions

2.7.1. Many special functions in mathematical physics can be expressed in terms of a generalized hypergeometric function $_pF_q(z)$ which is determined by the series

$$_pF_q \left(\begin{matrix} \alpha_1, \alpha_2, \ldots, \alpha_p \\ \beta_1, \beta_2, \ldots, \beta_q \end{matrix} \middle| z \right) = \sum_{k=0}^{\infty} \frac{(\alpha_1)_k(\alpha_2)_k \ldots (\alpha_p)_k z^k}{(\beta_1)_k(\beta_2)_k \ldots (\beta_q)_k k!} \; , \tag{2.7.1}$$

where $(a)_0 = 1$, $(a)_k = a(a+1) \ldots (a+k-1) = \Gamma(a+k)/\Gamma(a)$.

Important particular cases of t function $_pF_q(z)$ are the hypergeometric function $F(\alpha, \beta, \gamma, z)$ and the confluent hypergeometric function $F(\alpha, \gamma, z)$:

$$F(\alpha, \beta, \gamma, z) = \sum_{k=0}^{\infty} \frac{(\alpha)_k(\beta)_k z^k}{(\gamma)_k k!} \; , \quad F(\alpha, \gamma, z) = \sum_{k=0}^{\infty} \frac{(\alpha)_k z^k}{(\gamma)_k k!} \; . \tag{2.7.2}$$

2.7.2. For the Jacobi, Laguerre and Hermite polynomials we have [S38, E7]

$$P_n^{(\alpha,\beta)}(x) = \frac{(\alpha+1)_n}{n!} F \left(-n, \alpha+\beta+n+1, \alpha+1, \frac{1-x}{2} \right)$$
$$= (-1)^n \frac{(\beta+1)_n}{n!} F \left(-n, \alpha+\beta+n+1, \beta+1, \frac{1+x}{2} \right) \; , \tag{2.7.3}$$

$$L_n^{\alpha}(x) = \frac{(\alpha+1)_n}{n!} F(-n, \alpha+1, x) \; , \tag{2.7.4}$$

$$H_{2n}(x) = (-1)^n 2^{2n} \left(-\tfrac{1}{2} \right)_n F \left(-n, -\tfrac{1}{2}, x^2 \right) \; , \tag{2.7.5}$$

$$H_{2n+1}(x) = (-1)^n 2^{2n+1} \left(\tfrac{3}{2} \right)_n x F \left(-n, \tfrac{3}{2}, x^2 \right) \; . \tag{2.7.6}$$

Formulas (2.7.3–6) give expansions of the Jacobi, Laguerre and Hermite polynomials in powers x.

2.7.3. Let us obtain the relations similar to (2.7.3–6) for the Hahn, Meixner, Kravchuk and Charlier polynomials. For this purpose we transform the Rodrigues formula (2.2.7). Since

$$\nabla^n f(x) = \sum_{k=0}^{n} (-1)^k \frac{n!}{k!(n-k)!} f(x-k) \, ,$$

we have

$$y_n(x) = B_n \sum_{k=0}^{n} \frac{(-n)_k \varrho_n(x-k)}{k! \varrho(x)} \, . \tag{2.7.7}$$

Taking $k = n - s$ we can rewrite Eq. (2.7.7) in the form:

$$y_n(x) = (-1)^n B_n \sum_{s=0}^{n} \frac{(-n)_s \varrho_n(x-n+s)}{s! \varrho(x)} \, . \tag{2.7.8}$$

The relations (2.7.7) and (2.7.8) lead to representations of classical orthogonal polynomials of a discrete variable in terms of hypergeometric functions.

2.7.3.1. The Charlier polynomials. For the Charlier polynomials the following is valid:

$$\frac{\varrho_n(x-k)}{\varrho(x)} = \frac{\Gamma(x+1)}{\Gamma(x-k+1)} \mu^{n-k} = \mu^n (-x)_k \left(-\frac{1}{\mu} \right)^k$$

(the functions $\varrho(x)$ and $\varrho_n(x)$ are given in Table 2.4). Hence according to (2.7.7) and (2.7.8) we obtain

$$\begin{aligned}
c_n^{(\mu)}(x) &= {}_2F_0 \left(-n, -x; -\frac{1}{\mu} \right) \\
&= \frac{(-1)^n}{\mu^n} (x-n+1)_n F(-n, x-n+1, \mu) \, .
\end{aligned} \tag{2.7.9}$$

By comparing this relation with (2.7.4) we find the relationship between the Charlier and Laguerre polynomials in the form

$$c_n^{(\mu)}(x) = \frac{(-1)^n}{\mu^n} n! L_n^{x-n}(\mu) \quad . \tag{2.7.10}$$

In accordance with (2.7.9) we have the duality relation for the Charlier polynomials:

$$c_n^{(\mu)}(m) = c_m^{(\mu)}(n) \, ; \quad m, n = 0, 1, \ldots \, , \tag{2.7.10a}$$

whence

$$L_n^{m-n}(\mu) = \frac{m!}{n!} (-\mu)^{n-m} L_m^{n-m}(\mu) \, .$$

2.7.3.2. The Kravchuk and Meixner polynomials. From (2.7.7) for the Kravchuk polynomials we obtain

$$k_n^{(p)}(x, N) = \frac{(-1)^n p^n}{n!} \frac{\Gamma(N - x + 1)}{\Gamma(N - x - n + 1)}$$
$$\times F\left(-n, -x, N - x - n + 1, -\frac{q}{p}\right) . \qquad (2.7.11)$$

Hence by using the known relation

$$F(-n, \beta, \gamma, z) = \frac{\Gamma(\gamma)\Gamma(\gamma - \beta + n)}{\Gamma(\gamma + n)\Gamma(\gamma - \beta)} F(-n, \beta, \beta - \gamma - n + 1, 1 - z)$$

we obtain another representation of the Kravchuk polynomials in terms of the hypergeometric function:

$$k_n^{(p)}(x, N) = (-1)^n C_N^n p^n F\left(-n, -x, -N, \frac{1}{p}\right) , \qquad (2.7.11a)$$

where $C_N^n = N!/[n!(N - n)!]$. Similarly we can obtain for the Meixner polynomials

$$m_n^{(\gamma, \mu)}(x) = (\gamma)_n F\left(-n, -x, \gamma, 1 - \frac{1}{\mu}\right) . \qquad (2.7.12)$$

Comparison between (2.7.11) and (2.7.12) yields

$$k_n^{(p)}(x, N) = \frac{p^n}{n!} m_n^{(-N, -p/q)}(x) , \qquad (2.7.13)$$

which coincides with (2.4.22).

By using the transformation

$$F(-n, \beta, \gamma, z) = \frac{\Gamma(1 - \beta)\Gamma(1 - \gamma - n)}{\Gamma(1 - \beta - n)\Gamma(1 - \gamma)}(-z)^n$$
$$\times F\left(-n, 1 - \gamma - n, 1 - \beta - n, \frac{1}{z}\right)$$

we obtain from (2.7.3) and (2.7.11) the connection between the Jacobi and Kravchuk polynomials:

$$k_n^{(p)}(x, N) = P_n^{(x-n, N-n-x)}(1 - 2p) . \qquad (2.7.14)$$

Similarly in accordance with (2.7.3) and (2.7.12) one can derive

$$m_n^{(\gamma, \mu)}(x) = n! \left(1 - \frac{1}{\mu}\right)^n P_n^{(x-n, -\gamma-x-n)}\left(\frac{1 + \mu}{1 - \mu}\right) . \qquad (2.7.15)$$

By virtue of (2.7.12) and (2.7.11a) the following dual symmetry relations for the Meixner and Kravchuk polynomials are valid:

$$m_n^{(\gamma,\mu)}(l) = \frac{\Gamma(\gamma+n)}{\Gamma(\gamma+l)} m_l^{(\gamma,\mu)}(n) \quad (n,l=0,1,2,\ldots)\,, \tag{2.7.16}$$

$$k_n^{(p)}(m,N) = (-p)^{n-m} \frac{m!(N-m)!}{n!(N-n)!} k_m^{(p)}(n,N) \quad (m,n=0,1,\ldots,N)\,. \tag{2.7.17}$$

2.7.3.3. The Hahn polynomials. Equation (2.7.7) leads to a representation of the Hahn polynomials in terms of a generalized hypergeometric function $_3F_2(x)$ of an argument equal to unity:

$$\begin{aligned}
h_n^{(\alpha,\beta)}(x,N) =& \frac{(-1)^n \Gamma(\beta+n+x+1)\Gamma(N-x)}{n!\,\Gamma(\beta+x+1)\Gamma(N-x-n)} \\
& \times {}_3F_2\left(\begin{array}{c} -n,-x,\alpha+N-x \\ N-x-n,-\beta-x-n \end{array}\middle|1\right)\,.
\end{aligned}$$

Hence by using the transformation [W2]

$$\begin{aligned}
{}_3F_2\left(\begin{array}{c} -n,a,b \\ c,d \end{array}\middle|1\right) =& \frac{\Gamma(c)\Gamma(d)\Gamma(c-a+n)\Gamma(d-a+n)}{\Gamma(c+n)\Gamma(d+n)\Gamma(c-a)\Gamma(d-a)} \\
& \times {}_3F_2\left(\begin{array}{c} -n,a,1+a+b-c-d-n \\ 1+a-c-n,\,1+a-d-n \end{array}\middle|1\right)\,, \tag{2.7.18}
\end{aligned}$$

we obtain

$$\begin{aligned}
h_n^{(\alpha,\beta)}(x,N) =& \frac{(-1)^n}{n!}(N-n)_n(\beta+1)_n \\
& \times {}_3F_2\left(\begin{array}{c} -n,\alpha+\beta+n+1,-x \\ \beta+1,1-N \end{array}\middle|1\right)\,. \tag{2.7.19}
\end{aligned}$$

We can derive the transformation (2.7.18) for $\mathrm{Re}\,b>0$, $\mathrm{Re}(d-b)>0$ from the relation

$$_3F_2\left(\begin{array}{c} -n,a,b \\ c,d \end{array}\middle|1\right) = \frac{\Gamma(d)}{\Gamma(b)\Gamma(d-b)}\int_0^1 F(-n,a,c,t)t^{b-1}(1-t)^{d-b-1}\,dt\,,$$

which can be easily verified by using an expansion of the function $F(-n,a,c,t)$ in powers of t and integrating by terms. Replacing t by $1-t$ in this relation, using the known functional relation

$$F(-n,a,c,1-t) = \frac{\Gamma(c)\Gamma(c-a+n)}{\Gamma(c+n)} F(-n,a,a-c-n+1,t)$$

as well as the function $F(-n,a,a-c-n+1,t)$ expanded in powers t, we obtain after integration by terms:

$$_3F_2\left(\begin{array}{c} -n,a,b \\ c,d \end{array}\middle|1\right) = \frac{(c-a)_n}{(c)_n}\,{}_3F_2\left(\begin{array}{c} -n,a,d-b \\ a-c-n+1,d \end{array}\middle|1\right) \tag{2.7.20}$$

with Re $b > 0$, Re$(d - b) > 0$. By the principle of analytic continuation this functional relation remains valid for any values of parameters. By exchanging c and d and using again the transformation obtained we obtain relation (2.7.18).

Representation of the Hahn polynomials $\tilde{h}_n^{(\mu,\nu)}(x, N)$ in terms of the function $_3F_2(1)$ arises at the analytical continuation of (2.7.19) according to the equality

$$\tilde{h}_n^{(\mu,\nu)}(x, N) = h_n^{(-\mu-N,-\nu-N)}(x, N) \ .$$

By using the relations

$$h_n^{(\alpha,\beta)}(x, N) = (-1)^n h_n^{(-N,\alpha+\beta+N)}(-\beta - x - 1, -\beta) \ ,$$

$$P_n^{(\alpha,\beta)}(x) = (-1)^n \frac{(\beta + 1)_n}{n!} F\left(-n, \alpha + \beta + n + 1, \beta + 1, \frac{1+x}{2}\right) \ ,$$

$$\int_0^1 x^{p-1}(1 - x)^{q-1} F(a, b, c, x)dx = \frac{\Gamma(p)\Gamma(q)}{\Gamma(p+q)} \, _3F_2\left(\begin{matrix} a, b, p \\ c, p+q \end{matrix} \bigg| 1\right) \ ,$$

we can rewrite Eq. (2.7.19) for the Hahn polynomials in the form of an *integral representation*

$$h_n^{(\alpha,\beta)}(x, N) = \frac{2^{-\alpha-\beta-N} \Gamma(\alpha + \beta + N + n + 1)}{\Gamma(\alpha + N - x)\Gamma(\beta + x + 1)}$$

$$\times \int_{-1}^1 (1 - s)^{\alpha+N-x-1}(1 + s)^{\beta+x} P_n^{(\alpha,\beta)}(s)ds \ , \qquad (2.7.21)$$

which will be used below (see Sect. 5.4).

Equations (2.7.9, 11a, 12) and (2.7.19) give the classical orthogonal polynomials of a discrete variable expanded in terms of $(-x)_k = (-x)(-x+1) \ldots (-x+k-1)$. In order to obtain the Hahn, Meixner, Kravchuk and Charlier polynomials developed as a Taylor series it is sufficient to use the relation

$$(-x)_k = (-1)^k \sum_{m=0}^k S_k^{(m)} x^m \ ,$$

where $S_k^{(m)}$ are *the Stirling numbers of the first kind* [A1].

The above representations of classical orthogonal polynomials of a discrete variable in terms of hypergeometric functions are given in Table 2.4.

Table 2.4. Representations of the Hahn polynomials $h_n^{(\alpha,\beta)}(x, N)$, the Meixner polynomials $m_n^{(\gamma,\mu)}(x)$, the Kravchuk polynomials $k_n^{(p)}(x, N)$ and the Charlier polynomials $c_n^{(\mu)}(x)$ in terms of hypergeometric functions

$y_n(x)$	$_pF_q(z)$
$h_n^{(\alpha,\beta)}(x, N)$	$\dfrac{(-1)^n}{n!}\,(\beta+1)_n\,(N-n)_n\,{}_3F_2\left(\begin{array}{c} -n, \alpha+\beta+n+1, -x \\ \beta+1, 1-N \end{array}\middle\vert\, 1\right)$
$m_n^{(\gamma,\mu)}(x)$	$(\gamma)_n\, F\left(-n, -x, \gamma, 1-\dfrac{1}{\mu}\right)$
$k_n^{(p)}(x, N)$	$(-1)^n C_N^n p^n F\left(-n, -x, -N, \dfrac{1}{p}\right),\quad C_N^n = \dfrac{N!}{n!(N-n)!}$
$c_n^{(\mu)}(x)$	$_2F_0\left(-n, -x, -\dfrac{1}{\mu}\right)$

3. Classical Orthogonal Polynomials
of a Discrete Variable on Nonuniform Lattices

3.1 The Difference Equation of Hypergeometric Type
on a Nonuniform Lattice

In Chap. 2 we considered the generalization of the theory of polynomial solutions
of the hypergeometric type differential equation

$$\tilde{\sigma}(x)y'' + \tilde{\tau}(x)y' + \lambda y = 0 \tag{3.1.1}$$

to the difference equation

$$\tilde{\sigma}(x)\frac{1}{h}\left[\frac{y(x+h)-y(x)}{h} - \frac{y(x)-y(x-h)}{h}\right]$$
$$+ \frac{\tilde{\tau}(x)}{2}\left[\frac{y(x+h)-y(x)}{h} + \frac{y(x)-y(x-h)}{h}\right] + \lambda y(x) = 0 , \tag{3.1.2}$$

which approximates (3.1.1) on a lattice of constant mesh $\Delta x = h$. After a change
of independent variable, $x = x(s)$ we can obtain a further generalization to the
case when (3.1.1) is replaced by a difference equation on a class of lattices with
variable mesh $\Delta x = x(s+h) - x(s)$:

$$\tilde{\sigma}[x(s)]\frac{1}{x(s+h/2)-x(s-h/2)}\left[\frac{y(s+h)-y(s)}{x(s+h)-x(s)} - \frac{y(s)-y(s-h)}{x(s)-x(s-h)}\right]$$
$$+ \frac{\tilde{\tau}[x(s)]}{2}\left[\frac{y(s+h)-y(s)}{x(s+h)-x(s)} + \frac{y(s)-y(s-h)}{x(s)-x(s-h)}\right] + \lambda y(s) = 0 . \tag{3.1.3}$$

Equation (3.1.3) approximates (3.1.1) to second order in h, as is easily seen by
expanding $x(s \pm h)$, $x(s \pm h/2)$ and $y(s \pm h)$ by Taylor's formula.

3.1.1. Let us show that under certain requirements for the function $x(s)$ the
difference equation (3.1.3) has a property similar to the fundamental property of
the differential equation (3.1.1): *the difference derivative*

$$v_1(s) = \frac{y(s+h)-y(s)}{x(s+h)-x(s)} ,$$

*which is approximately equal to the derivative dy/dx at the point $x(s+h/2)$,
satisfies an equation of the form (3.1.3) with $x(s)$ replaced by $x_1(s) = x(s+h/2)$,
i.e. the equation*

$$\frac{\tilde{\sigma}_1[x_1(s)]}{x_1(s+h/2)-x_1(s-h/2)}\left[\frac{v_1(s+h)-v_1(s)}{x_1(s+h)-x_1(s)}-\frac{v_1(s)-v_1(s-h)}{x_1(s)-x_1(s-h)}\right]$$
$$+\frac{\tilde{\tau}_1[x_1(s)]}{2}\left[\frac{v_1(s+h)-v_1(s)}{x_1(s+h)-x_1(s)}+\frac{v_1(s)-v_1(s-h)}{x_1(s)-x_1(s-h)}\right]$$
$$+\mu_1 v_1(s)=0\,. \tag{3.1.4}$$

Here $\tilde{\tau}_1(x_1)$ and $\tilde{\sigma}(x_1)$ are polynomials of at most degree 1 and 2, respectively, in x_1; μ_1 is a constant.

For the proof of this statement it is convenient to replace s by hs in (3.1.3) and (3.1.4); as a result equations (3.1.3) and (3.1.4) become analogous equations with $h=1$:

$$\tilde{\sigma}[x(s)]\frac{\Delta}{\Delta x(s-1/2)}\left[\frac{\nabla y(s)}{\nabla x(s)}\right]+\frac{\tilde{\tau}[x(s)]}{2}\left[\frac{\Delta y(s)}{\Delta x(s)}+\frac{\nabla y(s)}{\nabla x(s)}\right]$$
$$+\lambda y(s)=0\,, \tag{3.1.5}$$

$$\tilde{\sigma}_1[x_1(s)]\frac{\Delta}{\Delta x_1(s-1/2)}\left[\frac{\nabla v_1(s)}{\nabla x_1(s)}\right]+\frac{\tilde{\tau}_1[x_1(s)]}{2}\left[\frac{\Delta v_1(s)}{\Delta x_1(s)}+\frac{\nabla v_1(s)}{\nabla x_1(s)}\right]$$
$$+\mu_1 v_1(s)=0\,, \tag{3.1.6}$$

where

$$x_1(s)=x\left(s+\tfrac{1}{2}\right)\,,\quad \Delta f(s)=f(s+1)-f(s)\,,\quad \nabla f(s)=f(s)-f(s-1)\,,$$
$$\frac{\Delta}{\Delta x(s-1/2)}f(s)=\frac{\Delta f(s)}{\Delta x(s-1/2)}\,,\quad v_1(s)=\frac{\Delta y(s)}{\Delta x(s)}\,.$$

Let us apply the operator $\Delta/\Delta x(s)$ to Eq. (3.1.5). It is convenient to use the relation

$$\Delta[f(s)g(s)]=\frac{g(s+1)+g(s)}{2}\Delta f(s)+\frac{f(s+1)+f(s)}{2}\Delta g(s)\,, \tag{3.1.7}$$

which follows from (2.1.8). Since

$$\tilde{\sigma}[x(s)]\frac{\Delta}{\Delta x(s-1/2)}\left[\frac{\nabla y(s)}{\nabla x(s)}\right]=\tilde{\sigma}[x(s)]\frac{\nabla v_1(s)}{\nabla x_1(s)}\,,$$

applying the operator $\Delta/\Delta x(s)$ to the first summand in (3.1.5) yields

$$\frac{\Delta}{\Delta x(s)}\left\{\tilde{\sigma}[x(s)]\frac{\nabla v_1(s)}{\nabla x_1(s)}\right\}=\frac{1}{2}\left[\frac{\Delta v_1(s)}{\Delta x_1(s)}+\frac{\nabla v_1(s)}{\nabla x_1(s)}\right]\frac{\Delta\tilde{\sigma}[x(s)]}{\Delta x(s)}$$
$$+\frac{1}{2}\{\tilde{\sigma}[x(s+1)]+\tilde{\sigma}[x(s)]\}\frac{\Delta}{\Delta x_1(s-1/2)}\left[\frac{\nabla v_1(s)}{\nabla x_1(s)}\right]\,.$$

The expression obtained will have a form analogous to the left-hand side of (3.1.6) if we require that the functions

$$\frac{\Delta\tilde{\sigma}[x(s)]}{\Delta x(s)}\,,\quad \frac{1}{2}\{\tilde{\sigma}[x(s+1)]+\tilde{\sigma}[x(s)]\}$$

are, respectively, polynomials of degrees at most 1 and 2 in $x_1(s) = x(s + \frac{1}{2})$. Since

$$\frac{\Delta}{\Delta x(s)} x^2(s) = x(s+1) + x(s) \, ,$$

these requirements are satisfied if the functions

1) $\quad x(s+1) + x(s)$, $\qquad$ 2) $\quad x^2(s+1) + x^2(s)$

are polynomials of degrees 1 and 2 in $x_1(s)$.

Let us show that if these two requirements for the function $x(s)$ are satisfied then applying the operator $\Delta/\Delta x(s)$ to Eq. (3.1.5) actually leads to an equation of the form (3.1.6). Applying the operator $\Delta/\Delta x(s)$ to the remaining terms of Eq. (3.1.5) we obtain

$$\frac{\Delta}{\Delta x(s)}[\lambda y(s)] = \lambda v_1(s) \, ,$$

$$\frac{\Delta}{\Delta x(s)} \left\{ \tilde{\tau}[x(s)] \left[\frac{\Delta y(s)}{\Delta x(s)} + \frac{\nabla y(s)}{\nabla x(s)} \right] \right\}$$

$$= \frac{\Delta}{\Delta x(s)} \left\{ \tilde{\tau}[x(s)] \left[v_1(s) + v_1(s-1) \right] \right\}$$

$$= \frac{1}{2} \left[v_1(s+1) + 2v_1(s) + v_1(s-1) \right] \frac{\Delta \tilde{\tau}[x(s)]}{\Delta x(s)}$$

$$+ \frac{\tilde{\tau}[x(s+1)] + \tilde{\tau}[x(s)]}{2} \frac{\Delta v_1(s) + \nabla v_1(s)}{\Delta x(s)} \, .$$

Now we express the functions

$$\frac{1}{2} \left[v_1(s+1) + 2v_1(s) + v_1(s-1) \right] \, , \qquad \frac{\Delta v_1(s) + \nabla v_1(s)}{\Delta x(s)}$$

in terms of the difference derivatives

$$\frac{\Delta}{\Delta x_1(s-1/2)} \left[\frac{\nabla v_1(s)}{\nabla x_1(s)} \right] \, , \qquad \frac{1}{2} \left[\frac{\Delta v_1(s)}{\Delta x_1(s)} + \frac{\nabla v_1(s)}{\nabla x_1(s)} \right]$$

which appear in Eq. (3.1.6). For this purpose we use the easily verified relations:

$$\frac{1}{2} \left[\frac{\Delta v_1(s)}{\Delta x_1(s)} + \frac{\nabla v_1(s)}{\nabla x_1(s)} \right] \pm \frac{1}{2} \Delta \left[\frac{\nabla v_1(s)}{\nabla x_1(s)} \right] = \left\{ \begin{array}{l} \left[\dfrac{\Delta v_1(s)}{\Delta x_1(s)} ; \right] \\ \dfrac{\nabla v_1(s)}{\nabla x_1(s)} \end{array} \right. \, .$$

As a result we have

$$\Delta v_1(s)$$

$$= \Delta x_1(s) \left\{ \frac{1}{2} \left[\frac{\Delta v_1(s)}{\Delta x_1(s)} + \frac{\nabla v_1(s)}{\nabla x_1(s)} \right] + \frac{\Delta x(s)}{2} \frac{\Delta}{\Delta x_1(s-1/2)} \left[\frac{\nabla v_1(s)}{\nabla x_1(s)} \right] \right\} \, ,$$

$$\nabla v_1(s)$$

$$= \nabla x_1(s) \left\{ \frac{1}{2} \left[\frac{\Delta v_1(s)}{\Delta x_1(s)} + \frac{\nabla v_1(s)}{\nabla x_1(s)} \right] - \frac{\Delta x(s)}{2} \frac{\Delta}{\Delta x_1(s - 1/2)} \left[\frac{\nabla v_1(s)}{\nabla x_1(s)} \right] \right\} ,$$

$$v_1(s+1) + 2v_1(s) + v_1(s-1) = \Delta v_1(s) - \nabla v_1(s) + 4v_1(s) .$$

Thus we obtain the equation

$$\bar{\sigma}_1(s) \frac{\Delta}{\Delta x_1(s - 1/2)} \left[\frac{\nabla v_1(s)}{\nabla x_1(s)} \right] + \frac{\bar{\tau}_1(s)}{2} \left[\frac{\Delta v_1(s)}{\Delta x_1(s)} + \frac{\nabla v_1(s)}{\nabla x_1(s)} \right]$$

$$+ \mu_1 v_1(s) = 0 , \tag{3.1.8}$$

where

$$\bar{\sigma}_1(s) = \frac{\tilde{\sigma}[x(s+1)] + \tilde{\sigma}[x(s)]}{2} + \frac{1}{4} \frac{\Delta \tilde{\tau}[x(s)]}{\Delta x(s)} \frac{\Delta x_1(s) + \nabla x_1(s)}{2\Delta x(s)} [\Delta x(s)]^2$$

$$+ \frac{\tilde{\tau}[x(s+1)] + \tilde{\tau}[x(s)]}{2} \frac{\Delta x_1(s) - \nabla x_1(s)}{4} , \tag{3.1.9}$$

$$\bar{\tau}_1(s) = \frac{\Delta \tilde{\sigma}[x(s)]}{\Delta x(s)} + \frac{\Delta \tilde{\tau}[x(s)]}{\Delta x(s)} \frac{\Delta x_1(s) - \nabla x_1(s)}{4}$$

$$+ \frac{\tilde{\tau}[x(s+1)] + \tilde{\tau}[x(s)]}{2} \frac{\Delta x_1(s) + \nabla x_1(s)}{2\Delta x(s)} , \tag{3.1.10}$$

$$\mu_1 = \lambda + \frac{\Delta \tilde{\tau}[x(s)]}{\Delta x(s)} . \tag{3.1.11}$$

By virtue of

$$\frac{\Delta \tilde{\tau}[x(s)]}{\Delta x(s)} = \text{const} ,$$

and since $[\Delta x(s)]^2 = 2[x^2(s+1) + x^2(s)] - [x(s+1) + x(s)]^2$ is a polynomial of at most degree 2 in $x_1(s)$ and $\{\tilde{\tau}[x(s+1)] + \tilde{\tau}[x(s)]\}/2$ is a polynomial of at most degree 1 in $x_1(s)$, Eq. (3.1.8) for $v_1(s)$ will have the form (3.1.6) if

$$\frac{\Delta x_1(s) + \nabla x_1(s)}{2\Delta x(s)} = \text{const} ,$$

and $[\Delta x_1(s) - \nabla x_1(s)]/4$ is a polynomial of at most degree 1 in $x_1(s)$.

According to the first requirement for $x(s)$ we have

$$\frac{x(s+1) + x(s)}{2} = \alpha x \left(s + \frac{1}{2} \right) + \beta , \tag{3.1.12}$$

where α and β are constants. Hence

$$\frac{\Delta x_1(s) + \nabla x_1(s)}{2\Delta x(s)} = \frac{\Delta [x(s + 1/2) + x(s - 1/2)]}{2\Delta x(s)}$$

$$= \frac{\Delta [\alpha x(s) + \beta]}{\Delta x(s)} = \alpha , \tag{3.1.13}$$

$$\frac{\Delta x_1(s) - \nabla x_1(s)}{4}$$

$$= \frac{1}{2}\left[\frac{x_1(s+1) + x_1(s)}{2} + \frac{x_1(s) + x_1(s-1)}{2}\right] - x_1(s)$$

$$= \tfrac{1}{2}\left[\alpha x_1\left(s+\tfrac{1}{2}\right) + \beta + \alpha x_1\left(s-\tfrac{1}{2}\right) + \beta\right] - x_1(s)$$

$$= \alpha\left[\alpha x_1(s) + \beta\right] + \beta - x_1(s) \,. \tag{3.1.14}$$

From this we see that $\bar{\tau}_1(s) = \tilde{\tau}_1[x_1(s)]$ and $\bar{\sigma}_1(s) = \tilde{\sigma}_1[x_1(s)]$, where $\tilde{\tau}_1(x_1)$ and $\tilde{\sigma}_1(x_1)$ are polynomials of at most degree 1 and 2, respectively, in x_1.

Therefore Eq. (3.1.5) may be called *the difference equation of hypergeometric type* because it has a property, analogous to the property of the differential equation of hypergeometric type (3.1.1) of retaining its form after differentiation. Let us recall that for the difference equation this property is satisfied only on lattices $x(s)$ that satisfy equation (3.1.12) under the additional condition: the function $x^2(s+1) + x^2(s)$ is a quadratic polynomial in $x_1(s) = x(s+\tfrac{1}{2})$. Let us determine the possible forms of the function $x(s)$.

When $\alpha \neq 1$ the general solution of (3.1.12) is

$$x(s) = c_1\kappa_1^{2s} + c_2\kappa_2^{2s} + c_3 \,,$$

where κ_1 and κ_2 are the roots of the equation

$$\kappa^2 - 2\alpha\kappa + 1 = 0$$

and c_1, c_2 are arbitrary functions of period $\tfrac{1}{2}$, $c_3 = \beta/(1-\alpha)$. Let us show that the additional condition on $x(s)$ will be satisfied if c_1 and c_2 are constants. Taking into account that $\kappa_1\kappa_2 = 1$ we have

$$x^2(s+1) + x^2(s) = \left(\kappa_1^2 + \kappa_2^2\right) x^2\left(s+\tfrac{1}{2}\right)$$
$$+ 2c_3\left[\kappa_1 + \kappa_2 - \left(\kappa_1^2 + \kappa_2^2\right)\right] x\left(s+\tfrac{1}{2}\right) + \text{const} \,,$$

i.e. $x^2(s+1) + x^2(s)$ is really a quadratic polynomial in $x\left(s+\tfrac{1}{2}\right)$.

When $\alpha = 1$, the general solution of (3.1.12) has the form

$$x(s) = c_1 s^2 + c_2 s + c_3 \,,$$

where $c_1 = 4\beta$; c_2 and c_3 are arbitrary functions of period $\tfrac{1}{2}$. It is easy to verify that the additional condition on $x(s)$ is satisfied if c_2 and c_3 are constants.

Thus, putting $\kappa_1^2 = q$, $\kappa_2^2 = 1/q$, we come to an important theorem by means of which we can now construct a theory of classical orthogonal polynomials of a discrete variable on nonuniform lattices.

Theorem 1. *Let the lattice function $x(s)$ have the form*

$$x(s) = c_1 q^s + c_2 q^{-s} + c_3 \quad or \quad x(s) = c_1 s^2 + c_2 s + c_3 \,,$$

where q, c_1, c_2, c_3 are constants. Then the difference derivative

$$v_1(s) = \frac{\Delta y(s)}{\Delta x(s)}$$

of the solution $y(s)$ of Eq. (3.1.5) satisfies a difference equation of the form (3.1.6), where $\tilde{\sigma}_1(x_1)$ and $\tilde{\tau}_1(x_1)$ are polynomials of at most second and first degrees in $x_1(s)$, respectively, and $\mu_1 = $ const.

The converse is also valid: every solution of equation (3.1.6) provided that $\lambda \neq 0$ can be represented in the form $v_1(s) = \Delta y(s)/\Delta x(s)$, where $y(s)$ is a solution of Eq. (3.1.5). For the proof, in accordance with (3.1.5), we assume

$$y(s) = -\frac{1}{\lambda}\left\{ \tilde{\sigma}[x(s)]\frac{\nabla v_1(s)}{\nabla x_1(s)} + \frac{\tilde{\tau}[x(s)]}{2}\left[v_1(s) + v_1(s-1)\right]\right\}. \qquad (3.1.15)$$

By applying the operator $\Delta/\Delta x(s)$ to both sides of this equality and repeating the transformations considered above we obtain with the aid of Eq. (3.1.6) for $v_1(s)$ with

$$\mu_1 = \lambda + \frac{\Delta\tilde{\tau}[x(s)]}{\Delta x(s)}\ , \quad \tilde{\sigma}_1[x(s)] = \tilde{\sigma}_1(s)\ , \quad \tilde{\tau}_1[x(s)] = \tilde{\tau}_1(s)$$

[$\tilde{\sigma}_1(s)$ and $\tilde{\tau}_1(s)$ are determined by Eq. (3.1.9) and (3.1.10)] that $v_1(s) = \Delta y(s)/\Delta x(s)$. Substitution of this relation into expression (3.1.15) yields Eq. (3.1.5) for the function $y(s)$, which was to be proved.

3.1.2. We may show by induction that the functions $v_k(s)$ connected with the solution $y = y(s)$ of Eq. (3.1.5) by the relations

$$v_k(s) = \frac{\Delta v_{k-1}(s)}{\Delta x_{k-1}(s)}\ , \quad v_0(s) = y(s)\ ,$$

$$x_k(s) = x\left(s + \frac{k}{2}\right) \quad k = 1, 2, \ldots \qquad (3.1.16)$$

satisfy the equations

$$\tilde{\sigma}_k\big[x_k(s)\big]\frac{\Delta}{\Delta x_k(s-1/2)}\left[\frac{\nabla v_k(s)}{\nabla x_k(s)}\right] + \frac{\tilde{\tau}_k\big[x_k(s)\big]}{2}\left[\frac{\Delta v_k(s)}{\Delta x_k(s)} + \frac{\nabla v_k(s)}{\nabla x_k(s)}\right]$$
$$+ \mu_k v_k(s) = 0\ , \qquad (3.1.17)$$

where $\tilde{\sigma}_k(x_k)$ and $\tilde{\tau}_k(x_k)$ are polynomials of at most second and first degrees in x_k, respectively; $\mu_k = $ const; and

$$\tilde{\sigma}_k\big[x_k(s)\big] = \frac{\tilde{\sigma}_{k-1}\big[x_{k-1}(s+1)\big] + \tilde{\sigma}_{k-1}\big[x_{k-1}(s)\big]}{2}$$
$$+ \frac{1}{4}\frac{\Delta\tilde{\tau}_{k-1}(s)}{\Delta x_{k-1}(s)}\frac{\Delta x_k(s) + \nabla x_k(s)}{2\Delta x_{k-1}(s)}\big[\Delta x_{k-1}(s)\big]^2$$
$$+ \frac{\tilde{\tau}_{k-1}\big[x_{k-1}(s+1)\big] + \tilde{\tau}_{k-1}\big[x_{k-1}(s)\big]}{2}\frac{\Delta x_k(s) - \nabla x_k(s)}{4}\ ,$$
$$\tilde{\sigma}_0(x_0) = \tilde{\sigma}(x)\ ; \qquad (3.1.18)$$

$$\tilde{\tau}_k\big[x_k(s)\big] = \frac{\Delta\tilde{\sigma}_{k-1}\big[x_{k-1}(s)\big]}{\Delta x_{k-1}(s)} + \frac{\Delta\tilde{\tau}_{k-1}\big[x_{k-1}(s)\big]}{\Delta x_{k-1}(s)}\,\frac{\Delta x_k(s) - \nabla x_k(s)}{4}$$

$$+ \frac{\tilde{\tau}_{k-1}\big[x_{k-1}(s+1)\big] + \tilde{\tau}_{k-1}\big[x_{k-1}(s)\big]}{2}\,\frac{\Delta x_k(s) + \nabla x_k(s)}{2\Delta x_{k-1}(s)}\,,$$

$$\tilde{\tau}_0\big(x_0\big) = \tilde{\tau}(x)\,; \tag{3.1.19}$$

$$\mu_k = \mu_{k-1} + \frac{\Delta\tilde{\tau}_{k-1}\big[x_{k-1}(s)\big]}{\Delta x_{k-1}(s)}\,, \qquad \mu_0 = \lambda\,. \tag{3.1.20}$$

The converse is also valid: every solution of Eq. (3.1.17) with $\mu_{k-1} \neq 0$ can be represented in the form

$$v_k(s) = \frac{\Delta v_{k-1}(s)}{\Delta x_{k-1}(s)}\,,$$

where $v_{k-1}(s)$ is a solution of the equation obtained from (3.1.17) by replacing k by $k-1$.

3.1.3. To study additional properties of solutions of (3.1.5) it is convenient to use the equation

$$\frac{1}{2}\left[\frac{\Delta y(s)}{\Delta x(s)} + \frac{\nabla y(s)}{\nabla x(s)}\right] = \frac{\Delta y(s)}{\Delta x(s)} - \frac{1}{2}\Delta\left[\frac{\nabla y(s)}{\nabla y(s)}\right]$$

and to rewrite (3.1.5) in the equivalent form

$$\sigma(s)\frac{\Delta}{\Delta x(s - 1/2)}\left[\frac{\nabla y(s)}{\nabla x(s)}\right] + \tau(s)\frac{\Delta y(s)}{\Delta x(s)} + \lambda y(s) = 0\,, \tag{3.1.21}$$

where

$$\sigma(s) = \tilde{\sigma}[x(s)] - \tfrac{1}{2}\tilde{\tau}[x(s)]\Delta x\left(s - \tfrac{1}{2}\right)\,, \tag{3.1.22}$$

$$\tau(s) = \tilde{\tau}[x(s)]\,. \tag{3.1.23}$$

An analogous form for Eq. (3.1.17) is

$$\sigma_k(s)\frac{\Delta}{\Delta x_k(s - 1/2)}\left[\frac{\nabla v_k(s)}{\nabla x_k(s)}\right] + \tau_k(s)\frac{\Delta v_k(s)}{\Delta x_k(s)} + \mu_k v_k(s) = 0\,, \tag{3.1.24}$$

$$\sigma_k(s) = \tilde{\sigma}_k\big[x_k(s)\big] - \tfrac{1}{2}\tilde{\tau}_k\big[x_k(s)\big]\Delta x_k\left(s - \tfrac{1}{2}\right)\,, \tag{3.1.25}$$

$$\tau_k(s) = \tilde{\tau}_k\big[x_k(s)\big]\,. \tag{3.1.26}$$

From (3.1.18) and (3.1.19) it follows that $\sigma_k(s) = \sigma_{k-1}(s)$, i.e.

$$\sigma_k(s) = \sigma(s)\,. \tag{3.1.27}$$

In addition, from (3.1.18) and (3.1.19) we can also obtain the relation

$$\sigma(s) + \tau_k(s)\Delta x_k \left(s - \tfrac{1}{2}\right) = \sigma(s+1) + \tau_{k-1}(s+1)\Delta x_{k-1}\left(s + \tfrac{1}{2}\right) , \qquad (3.1.28)$$

which enables us, by using the functions $\sigma(s)$, $\tau(s)$ and $x(s)$, to find $\tau_k(s)$ in the form

$$\tau_k(s) = \frac{\sigma(s+k) - \sigma(s) + \tau(s+k)\Delta x(s+k-1/2)}{\Delta x[s+(k-1)/2]} . \qquad (3.1.29)$$

An expression for μ_k may be obtained directly from (3.1.20) as

$$\mu_k = \lambda + \sum_{m=0}^{k-1} \frac{\Delta\tau_m(s)}{\Delta x_m(s)} . \qquad (3.1.30)$$

3.2 The Difference Analogs of Hypergeometric Type Polynomials. The Rodrigues Formula

Theorem 1 enables us to construct a family of polynomial solutions corresponding to certain values of λ in the same way as in the case of the differential equation. To do this we need the obvious equality

$$\frac{[\Delta x(s)]^2}{4} = \frac{x^2(s+1) + x^2(s)}{2} - \left[\frac{x(s+1) + x(s)}{2}\right]^2 , \qquad (3.2.1)$$

and the relations

$$\frac{\Delta x^n(s)}{\Delta x(s)} = \frac{x(s+1) + x(s)}{2} \frac{\Delta x^{n-1}(s)}{\Delta x(s)} + \frac{x^{n-1}(s+1) + x^{(n-1)}(s)}{2} , \qquad (3.2.2)$$

$$\frac{x^n(s+1) + x^n(s)}{2} = \frac{x(s+1) + x(s)}{2} \frac{x^{n-1}(s+1) + x^{n-1}(s)}{2}$$
$$+ \frac{[\Delta x(s)]^2}{4} \frac{\Delta x^{n-1}(s)}{\Delta x(s)} . \qquad (3.2.3)$$

Note that relations (3.2.2) and (3.2.3) may be obtained from (3.1.7) by putting $f(s) = x^{n-1}(s)$, $g(s) = x(s)$ and $f(s) = e^{i\pi s}x^{n-1}(s)$, $g(s) = x(s)$, respectively.

By using the properties of the lattice function $x(s)$ and (3.2.1–3), it is easy to prove by induction for an arbitrary polynomial $p_n(x)$ of degree n that

$$\frac{\Delta p_n[x(s)]}{\Delta x(s)} = q_{n-1}\big[x_1(s)\big] , \qquad (3.2.4)$$

$$\frac{p_n[x(s+1)] + p_n[x(s)]}{2} = r_n\big[x_1(s)\big] , \qquad (3.2.5)$$

where $q_{n-1}(x_1)$ and $r_n(x_1)$ are polynomials of the appropriate degrees in $x_1(s)$. It is obvious that equalities (3.2.4) and (3.2.5) remain valid under replacement of $x(s)$ and $x_1(s)$ by $x_k(s)$ and $x_{k+1}(s)$.

Further on, we shall need the relationship between leading coefficients of the polynomials $p_n(x)$ and $q_{n-1}(x)$ in (3.2.4). For this it is sufficient to find the coefficient A_n in the expansion

$$\frac{\Delta}{\Delta x(s)}\left[x^n(s)\right] = A_n x_1^{n-1}(s) + \dots .$$

We have

$$\Delta\left[x^n(s)\right] = \left[A_n x^{n-1}\left(s+\tfrac{1}{2}\right)+\dots\right]\Delta x(s).$$

If

$$x(s) = c_1 q^s + c_2 q^{-s} + c_3 ,$$

then

$$\Delta x(s) = c_1(q-1)q^s + c_2\left(\frac{1}{q}-1\right)q^{-s},$$

$$\Delta x^n(s) = \left(c_1 q^{s+1} + c_2 q^{-s-1} + c_3\right)^n - \left(c_1 q^s + c_2 q^{-s} + c_3\right)^n .$$

Comparing the coefficients of q^{ns} in the considered expansion yields

$$A_n = \frac{q^{n/2} - q^{-n/2}}{q^{1/2} - q^{-1/2}} .$$

The function

$$\psi_q(s) = \frac{q^{s/2} - q^{-s/2}}{q^{1/2} - q^{-1/2}}$$

will further be widely used for arbitrary values of s.

Thus in the case $x(s) = c_1 q^s + c_2 q^{-s} + c_3$ we obtain for the difference derivative of $x^n(s)$ the expansion

$$\frac{\Delta}{\Delta x(s)}\left[x^n(s)\right] = \psi_q(n)x_1^{n-1}(s) + \dots . \tag{3.2.4a}$$

For the lattice function $x(s) = c_1 s^2 + c_2 s + c_3$, which corresponds to $\alpha = 1$ in (3.1.12) we obtain a similar expansion with $\psi_q(n)$ replaced by n, which should be expected since $q \to 1$ when $\alpha \to 1$ and $\lim_{q\to 1}\psi_q(s) = s$.

3.2.1. To find a particular solution of Eq. (3.1.5) we observe that Eq. (3.1.17) with $k = n$, $\mu_n = 0$ has the particular solution $v_n(s) = \text{const}$. Let us show that when $k < n$ the functions $v_k(s)$ connected by the relations

$$v_{k+1}(s) = \frac{\Delta v_k(s)}{\Delta x_k(s)}$$

will be polynomials of degree $n - k$ in $x_k(s)$ if $v_n(s) = \text{const}$, provided that $\mu_k \neq 0$ for $k = 0, 1, \dots, n - 1$. We shall prove this by induction, assuming that the function $v_{k+1}(s)$ is a polynomial of degree $n - k - 1$ in $x_{k+1}(s)$. From Eq. (3.1.17) we have

$$v_k(s) = -\frac{1}{\mu_k}\left\{\tilde{\sigma}_k\big[x_k(s)\big]\frac{\nabla v_{k+1}(s)}{\nabla x_{k+1}(s)} + \frac{\tilde{\tau}_k\big[x_k(s)\big]}{2}\big[v_{k+1}(s) + v_{k+1}(s-1)\big]\right\}.$$

By virtue of (3.2.4) and (3.2.5) the functions

$$\frac{\nabla v_{k+1}(s)}{\nabla x_{k+1}(s)} = \frac{\Delta v_{k+1}(t)}{\Delta x_{k+1}(t)}\bigg|_{t=s-1}$$

and

$$\tfrac{1}{2}\big[v_{k+1}(s) + v_{k+1}(s-1)\big] = \tfrac{1}{2}\big[v_{k+1}(t+1) + v_{k+1}(t)\big]\bigg|_{t=s-1}$$

are, respectively, polynomials of degrees $n - k - 2$ and $n - k - 1$ in $x_{k+2}(s - 1)$. Since $x_{k+2}(s-1) = x_k(s)$ the function $v_k(s)$ is, obviously, a polynomial of degree $n - k$ in $x_k(s)$.

By applying this argument successively for $k = n-1$, $n-2$, etc. we find that there exists a solution $y(s) = v_0(s)$ of (3.1.5) which is a polynomial of degree n in $x(s)$ for those $\lambda = \lambda_n$ for which $\mu_n = 0$. We see from (3.1.30) that

$$\lambda_n = -\sum_{m=0}^{n-1}\frac{\Delta \tau_m(s)}{\Delta x_m(s)} = -\sum_{m=0}^{n-1}\tilde{\tau}'_m. \tag{3.2.6}$$

A simpler expression for λ_n will be obtained in Sect. 3.7.1.

3.2.2. To obtain an explicit expression for the polynomial $y(s) = \tilde{y}_n[x(s)]$ we rewrite Eqs. (3.1.5) and (3.1.17) in the form of (3.1.21) and (3.1.24) by taking account of (3.1.27). Multiplying Eqs. (3.1.21) and (3.1.24) by appropriate functions $\varrho(s)$ and $\varrho_k(s)$, and using (2.1.8), we reduce these equations to self-adjoint form:

$$\frac{\Delta}{\Delta x(s - 1/2)}\left[\sigma(s)\varrho(s)\frac{\nabla y(s)}{\nabla x(s)}\right] + \lambda\varrho(s)y(s) = 0, \tag{3.2.7}$$

$$\frac{\Delta}{\Delta x_k(s - 1/2)}\left[\sigma(s)\varrho_k(s)\frac{\nabla v_k(s)}{\nabla x_k(s)}\right] + \mu_k\varrho_k(s)v_k(s) = 0. \tag{3.2.8}$$

Here $\varrho(s)$ and $\varrho_k(s)$ satisfy the difference equations

$$\frac{\Delta}{\Delta x(s - 1/2)}[\sigma(s)\varrho(s)] = \tau(s)\varrho(s), \tag{3.2.9}$$

$$\frac{\Delta}{\Delta x_k(s - 1/2)}[\sigma(s)\varrho_k(s)] = \tau_k(s)\varrho_k(s). \tag{3.2.10}$$

By using (3.2.10) and (3.1.28) we can determine the connection between $\varrho_k(s)$ and $\varrho(s)$:

$$\frac{\sigma(s+1)\varrho_k(s+1)}{\varrho_k(s)} = \sigma(s) + \tau_k(s)\Delta x_k\left(s - \tfrac{1}{2}\right)$$

$$= \sigma(s+1) + \tau_{k-1}(s+1)\Delta x_{k-1}\left(s+\tfrac{1}{2}\right) = \frac{\sigma(s+2)\varrho_{k-1}(s+2)}{\varrho_{k-1}(s+1)} \ .$$

From this we obtain

$$\frac{\sigma(s+1)\varrho_{k-1}(s+1)}{\varrho_k(s)} = \frac{\sigma(s+2)\varrho_{k-1}(s+2)}{\varrho_k(s+1)} = C(s) \ ,$$

where $C(s)$ is an arbitrary function of period 1. By assuming $C(s) = 1$, we obtain

$$\varrho_k(s) = \sigma(s+1)\varrho_{k-1}(s+1) \ . \tag{3.2.11}$$

Hence

$$\varrho_k(s) = \varrho(s+k)\prod_{i=1}^{k}\sigma(s+i) \ . \tag{3.2.12}$$

By means of (3.2.11), Eq. (3.2.8) can be rewritten as a recurrence relation between the functions $v_k(s)$ and $v_{k+1}(s)$. In fact,

$$\varrho_k(s)v_k(s) = -\frac{1}{\mu_k}\frac{\nabla}{\nabla x_{k+1}(s)}\left[\varrho_{k+1}(s)v_{k+1}(s)\right] \ . \tag{3.2.13}$$

From this we obtain

$$\varrho_k(s)v_k(s) = \frac{A_k}{A_n}\nabla_n^{(n-k)}\left[\varrho_n(s)v_n(s)\right] \ , \tag{3.2.14}$$

where

$$A_k = (-1)^k\prod_{i=0}^{k-1}\mu_i \ , \quad A_0 = 1 \ , \tag{3.2.15}$$

$$\nabla_n^{(m)}[f(s)] = \nabla_{n-m+1}\ \ldots\ \nabla_{n-1}\nabla_n[f(s)] \ , \quad \nabla_k = \frac{\nabla}{\nabla x_k(s)} \ . \tag{3.2.16}$$

Let us recall that, by definition (3.1.16),

$$v_k(s) = \frac{\Delta v_{k-1}(s)}{\Delta x_{k-1}(s)} \ ,$$

i.e.

$$v_k(s) = \Delta^{(k)}[y(s)] \ , \tag{3.2.17}$$

where

$$\Delta^{(k)}[f(s)] = \Delta_{k-1}\Delta_{k-2}\ \ldots\ \Delta_0[f(s)] \ , \quad \Delta_k = \frac{\Delta}{\Delta x_k(s)} \ .$$

If $v_n(s) = C_n$, where C_n is a constant, then, as was shown above, $y(s)$ is a polynomial of degree n in $x(s)$, i.e. $y = y_n(s) \equiv \tilde{y}_n[x(s)]$. In this case, for the function

$$v_{kn}(s) = \Delta^{(k)}\left[y_n(s)\right]$$

in (3.2.14) we obtain

$$v_{kn}(s) = \frac{A_{kn}B_n}{\varrho_k(s)} \nabla_n^{(n-k)}\big[\varrho_n(s)\big] \,, \tag{3.2.18}$$

where

$$A_{kn} = A_k(\lambda)\big|_{\lambda=\lambda_n} = (-1)^k \prod_{m=0}^{k-1} \mu_{mn} \,;$$

$$\mu_{mn} = \mu_m(\lambda)\big|_{\lambda=\lambda_n} = \lambda_n - \lambda_m \,;$$

$$A_{0n} = 1 \,, \quad B_n = \frac{C_n}{A_{nn}} \,.$$

Here, in particular, from (3.2.18) with $k = 0$ we obtain the *difference analog of the Rodrigues formula* for the polynomial $\tilde{y}_n[x(s)] \equiv y_n(s)$:

$$\begin{aligned}
y_n(s) &= \frac{B_n}{\varrho(s)} \nabla_n^{(n)}\big[\varrho_n(s)\big] \\
&= \frac{B_n}{\varrho(s)} \frac{\nabla}{\nabla x_1(s)} \cdots \frac{\nabla}{\nabla x_{n-1}(s)} \frac{\nabla}{\nabla x_n(s)} \big[\varrho_n(s)\big] \,.
\end{aligned} \tag{3.2.19}$$

It can easily be seen that the arbitrariness in the choice of a periodic multiplier for the function $\varrho(s)$ has no influence on the explicit form of the polynomial $\tilde{y}_n(x)$ obtained by the Rodrigues formula (3.2.19).

Thus the polynomial solutions $y = \tilde{y}_n[x(s)]$ of Eq. (3.1.5) are uniquely determined by Eq. (3.2.19) up to the normalizing factor B_n. These solutions correspond to the values $\lambda = \lambda_n$ from (3.2.6) for which $\mu_n = 0$.

3.2.3. By using the Rodrigues formula (3.2.19) we obtain for the polynomials $y_n(s) = \tilde{y}_n[x(s)]$ an explicit expression similar to (2.7.8). To derive this formula let us note that the expression $\nabla_n^{(n)}[f(s)]$ for the arbitrary function $f(s)$ is a linear combination of functions $f(s - n)$, $f(s - n + 1)$, ..., $f(s)$:

$$\nabla_n^{(n)}[f(s)] = \sum_{k=0}^{n} a_{kn}(s) f(s - n + k) \,. \tag{3.2.20}$$

In order to evaluate the coefficients $a_{kn}(s)$ at fixed values of $k = m$ and $s = \xi$ we use Eq. (3.2.20) for the function $f(s) = f_m(s)$ such that it is zero at the points $s = s_k = \xi - n + k$ when $k \neq m$, $0 \leq k \leq n$. These conditions are satisfied, for example, by the function

$$f_m(s) = \prod_{\substack{l=0 \\ (l \neq m)}}^{n} \big[x_n(s) - x_n(s_l)\big] \,, \tag{3.2.21}$$

which is a polynomial of degree n in $x_n(s)$ with the leading coefficient equal to unity. By virtue of (3.2.4a) we have

$$\nabla_n^{(n)}[f(s)] = \nabla_n^{(n)}\big[x_n^n(s) + \ldots\big] = \psi_q(n)\psi_q(n-1)\ldots\psi_q(1) \,.$$

Therefore at $s = \xi$ we obtain from (3.2.20)

$$\prod_{k=1}^{n} \psi_q(k) = a_{mn}(\xi)f_m(\xi - n + m) \,,$$

whence

$$a_{mn}(\xi) = \frac{\displaystyle\prod_{k=1}^{n} \psi_q(k)}{\displaystyle\prod_{\substack{l=0 \\ (l \neq m)}}^{n} [x_n(\xi - n + m) - x_n(\xi - n + l)]} \,.$$

The formula obtained is obviously valid for any ξ and $m = 0, 1, \ldots, n$. Therefore the expansion (3.2.20) may be written in the form

$$\nabla_n^{(n)}[f(s)] = \left[\prod_{k=1}^{n} \psi_q(k)\right] \sum_{k=0}^{n} \frac{f(s - n + k)}{\displaystyle\prod_{\substack{l=0 \\ (l \neq k)}}^{n}[x_n(s - n + k) - x_n(s - n + l)]} \,.$$

$$(3.2.22)$$

Since $\lim_{q \to 1} \psi_q(k) = k$, the product $\prod_{k=1}^{n} \psi_q(k)$ for the linear and quadratic lattices must be replaced by

$$\lim_{q \to 1} \prod_{k=1}^{n} \psi_q(k) = n! = \Gamma(n+1) \,.$$

For $q \neq 1$ the product $\prod_{k=1}^{n} \psi_q(k)$ may be expressed through the function $\Gamma_q(s)$, which is called the q-gamma function; it is generalization of Euler's gamma-function $\Gamma(s)$ [J1]. The function $\Gamma_q(s)$ satisfies the equation

$$\frac{\Gamma_q(s+1)}{\Gamma_q(s)} = \frac{q^s - 1}{q - 1} \quad (\Gamma_q(1) = 1) \,. \tag{3.2.23}$$

Since

$$\frac{q^s - 1}{q - 1} = q^{(s-1)/2}\psi_q(s) \,,$$

we have

$$\prod_{k=1}^{n} \psi_q(k) = q^{-n(n-1)/4}\Gamma_q(n+1) \,.$$

Further instead of function $\Gamma_q(s)$ we shall use the function

$$\tilde{\Gamma}_q(s) = q^{-(s-1)(s-2)/4}\Gamma_q(s) \,, \tag{3.2.24}$$

for which

$$\prod_{k=1}^{n} \psi_q(k) = \tilde{\Gamma}_q(n+1) \, . \tag{3.2.25}$$

Obviously the function $\tilde{\Gamma}_q(s)$ satisfies the equation

$$\frac{\tilde{\Gamma}_q(s+1)}{\tilde{\Gamma}_q(s)} = \psi_q(s) \quad \left(\tilde{\Gamma}_q(1) = 1\right) \, . \tag{3.2.26}$$

For $q \neq 1$ Eq. (3.2.22) may be simplified if for the function $x(s) = c_1 q^s + c_2 q^{-s} + c_3$ we use the easily verified relation

$$x(s) - x(s - \xi) = \psi_q(\xi)\nabla x \left(s - \frac{\xi - 1}{2}\right) \, , \tag{3.2.27}$$

whence

$$x_n(s - n + k) - x_n(s - n + l) = x \left(s - \frac{n}{2} + k\right) - x \left(s - \frac{n}{2} + l\right)$$
$$= \psi_q(k - l)\nabla x \left(s - \frac{n}{2} + \frac{k + l + 1}{2}\right) \, .$$

Since $\psi_q(-s) = -\psi_q(s)$, we obtain as a result:

$$\prod_{\substack{l=0 \\ (l \neq k)}}^{n} \left[x_n(s - n + k) - x_n(s - n + l)\right]$$

$$= \prod_{\substack{l=0 \\ (l \neq k)}}^{n} \psi_q(k - l)\nabla x \left(s - \frac{n}{2} + \frac{k + l + 1}{2}\right) \tag{3.2.28}$$

$$= \prod_{l=0}^{k-1} \psi_q(k - l) \prod_{l=k+1}^{n} \psi_q(k - l) \frac{\prod_{l=0}^{n} \nabla x[s - n/2 + (k + l + 1)/2]}{\nabla x[s - n/2 + (2k + 1)/2]}$$

$$= \tilde{\Gamma}_q(k + 1)(-1)^{n-k}\tilde{\Gamma}_q(n - k + 1)\frac{\prod_{l=0}^{n} \nabla x[s + (k - l + 1)/2]}{\nabla x[s + k - (n - 1)/2]} \, ,$$

whence

$$\nabla_n^{(n)}[f(s)] = \sum_{k=0}^{n} \frac{(-1)^{n-k}\tilde{\Gamma}_q(n + 1)}{\tilde{\Gamma}_q(k + 1)\tilde{\Gamma}_q(n - k + 1)}$$

$$\times \prod_{l=0}^{n} \frac{\nabla x[s + k - (n - 1)/2]}{\nabla x[s + (k - l + 1)/2]} f(s - n + k) \, . \tag{3.2.29}$$

Therefore the Rodrigues formula (3.2.19) may be written in a form similar to (2.7.8):

$$y_n(s) = B_n \sum_{k=0}^{n} \frac{(-1)^{n-k}\tilde{\Gamma}_q(n+1)}{\tilde{\Gamma}_q(k+1)\tilde{\Gamma}_q(n-k+1)}$$
$$\times \frac{\nabla x[s+k-(n-1)/2]}{\prod_{l=0}^{n}\nabla x[s+(k-l+1)/2]} \frac{\varrho_n(s-n+k)}{\varrho(s)} . \qquad (3.2.30)$$

Since by virtue of (3.2.12) and (3.2.9),

$$\varrho_n(s-n+k) = \varrho(s+k) \prod_{l=1}^{n} \sigma(s-n+k+l) ,$$

$$\frac{\varrho(s+1)}{\varrho(s)} = \frac{\sigma(s)+\tau(s)\Delta x(s-1/2)}{\sigma(s+1)} ,$$

we have

$$\frac{\varrho(s+k)}{\varrho(s)} = \frac{\prod_{l=0}^{k-1}[\sigma(s+l)+\tau(s+l)\Delta x(s+l-1/2)]}{\prod_{l=0}^{k-1}\sigma(s+l+1)}$$

and hence

$$\frac{\varrho_n(s-n+k)}{\varrho(s)}$$
$$= \frac{\prod_{l=1}^{n}\sigma(s-n+k+l)\prod_{l=0}^{k-1}[\sigma(s+l)+\tau(s+l)\Delta x(s+l-1/2)]}{\prod_{l=0}^{k-1}\sigma(s+l+1)}$$
$$= \prod_{l=0}^{n-k-1}\sigma(s-l)\prod_{l=0}^{k-1}\left[\sigma(s+l)+\tau(s+l)\Delta x\left(s+l-\tfrac{1}{2}\right)\right] ,$$

whence

$$y_n(s) = B_n \sum_{k=0}^{n} \frac{(-1)^{n-k}\tilde{\Gamma}_q(n+1)}{\tilde{\Gamma}_q(k+1)\tilde{\Gamma}_q(n-k+1)} \frac{\nabla x[s+k-(n-1)/2]}{\prod_{l=0}^{n}\nabla x[s+(k-l+1)/2]}$$
$$\times \prod_{l=0}^{n-k-1}\sigma(s-l)\prod_{l=0}^{k-1}\left[\sigma(s+l)+\tau(s+l)\Delta x\left(s+l-\tfrac{1}{2}\right)\right] \qquad (3.2.31)$$

(the products of the form $\prod_{l=0}^{-1}$ should be assumed equal to unity).

3.2.4. One can rewrite the Rodrigues-type formula (3.2.19) and the difference equation (3.2.7) for polynomials $y(s) = \tilde{y}_n[x(s)]$ in another equivalent form by using instead of Δ and ∇ a more symmetric operator

$$\delta f(s) = f\left(s+\tfrac{1}{2}\right) - f\left(s-\tfrac{1}{2}\right) = \Delta f\left(s-\tfrac{1}{2}\right) = \nabla f\left(s+\tfrac{1}{2}\right) .$$

Since

$$\frac{\nabla}{\nabla x_n}\left[\varrho_n(s)\right] = \frac{\delta \varrho_n(s-1/2)}{\delta x[s+(n-1)/2]} ,$$

$$\left(\frac{\nabla}{\nabla x_{n-1}}\right)\left(\frac{\nabla}{\nabla x_n}\right)[\varrho_n(s)] = \frac{\delta}{\delta x[s+(n-2)/2]}\left[\frac{\delta\varrho_n(s-1)}{\delta x[s+(n-2)/2]}\right],$$

etc., we have

$$\nabla_n^{(n)}[\varrho_n(s)] = \left(\frac{\delta}{\delta x(s)}\right)^n \left[\varrho_n\left(s - \frac{n}{2}\right)\right] .$$

Therefore

$$y_n(s) = \frac{B_n}{\varrho(s)}\left(\frac{\delta}{\delta x(s)}\right)^n \left[\varrho_n\left(s - \frac{n}{2}\right)\right] , \tag{3.2.32}$$

where $\varrho_n(s)$ is defined in (3.2.12).

Equation (3.2.7) may be rewritten in the form [G7a]

$$\frac{\delta}{\delta x(s)}\left[\sigma\left(s + \frac{1}{2}\right)\varrho\left(s + \frac{1}{2}\right)\frac{\delta y(s)}{\delta x(s)}\right] + \lambda\varrho(s)y(s) = 0 . \tag{3.2.33}$$

3.3 The Orthogonality Property

3.3.1. We now prove that the polynomial solutions of Eqs. (3.1.5) have the orthogonality property. For this purpose we write the equations for $y_m(s)$ and $y_n(s)$ in the self-adjoint form (3.2.7):

$$\frac{\Delta}{\Delta x(s-1/2)}\left[\sigma(s)\varrho(s)\frac{\nabla y_m(s)}{\nabla x(s)}\right] + \lambda_m\varrho(s)y_m(s) = 0 ,$$

$$\frac{\Delta}{\Delta x(s-1/2)}\left[\sigma(s)\varrho(s)\frac{\nabla y_n(s)}{\nabla x(s)}\right] + \lambda_n\varrho(s)y_n(s) = 0 .$$

Multiplying the first equation by $y_n(s)$ and the second by $y_m(s)$, and subtracting the second from the first we obtain:

$$(\lambda_m - \lambda_n)y_m(s)y_n(s)\varrho(s)\Delta x\left(s - \tfrac{1}{2}\right)$$
$$= \Delta\left\{\sigma(s)\varrho(s)\left[y_m(s)\frac{\nabla y_n(s)}{\nabla x(s)} - y_n(s)\frac{\nabla y_m(s)}{\nabla x(s)}\right]\right\} . \tag{3.3.1}$$

If we now put $s = s_i$ and $s_{i+1} = s_i + 1$, where $i = 0, 1, \ldots$, and sum over the indices i for which $a \leq s_i \leq b - 1$ we obtain

$$(\lambda_m - \lambda_n)\sum_{s_i=a}^{b-1} y_m(s_i)y_n(s_i)\varrho(s_i)\Delta x\left(s_i - \tfrac{1}{2}\right)$$

$$= \sigma(s)\varrho(s)\left[y_m(s)\frac{\nabla y_n(s)}{\nabla x(s)} - y_n(s)\frac{\nabla y_m(s)}{\nabla x(s)}\right]\Bigg|_a^b . \tag{3.3.2}$$

Since

$$y_m(s) = \frac{y_m(s) + y_m(s-1)}{2} + \frac{1}{2}\nabla y_m(s) \,,$$

$$y_n(s) = \frac{y_n(s) + y_n(s-1)}{2} + \frac{1}{2}\nabla y_n(s)$$

and the functions $y_m(s)$ and $y_n(s)$ are polynomials in $x(s)$, by virtue of (3.2.4) and (3.2.5) the expression

$$y_m(s)\frac{\nabla y_n(s)}{\nabla x(s)} - y_n(s)\frac{\nabla y_m(s)}{\nabla x(s)}$$
$$= \frac{y_m(s) + y_m(s-1)}{2}\frac{\nabla y_n(s)}{\nabla x(s)} - \frac{y_n(s) + y_n(s-1)}{2}\frac{\nabla y_m(s)}{\nabla x(s)}$$

is a polynomial in $x(s - \frac{1}{2})$. Hence under the boundary conditions

$$\sigma(s)\varrho(s)x^l\left(s - \tfrac{1}{2}\right)\Big|_{s=a,b} = 0 \quad (l = 0, 1, \ldots) \tag{3.3.3}$$

the right-hand side of (3.3.2) is zero. As a result we obtain

$$\sum_{s_i=a}^{b-1} y_m(s_i)y_n(s_i)\varrho(s_i)\Delta x\left(s_i - \tfrac{1}{2}\right) = \delta_{mn}d_n^2 \,. \tag{3.3.4}$$

The polynomial solutions of Eq. (3.1.5) satisfying the orthogonality relations (3.3.4) under the additional conditions

$$\varrho(s_i)\Delta x\left(s_i - \tfrac{1}{2}\right) > 0 \quad (a \le s_i \le b-1) \,, \tag{3.3.5}$$

will be called *classical orthogonal polynomials of a discrete variable on nonuniform lattices*.

Since, by the Rodrigues formula (3.2.19),

$$\tilde{y}_1(x) = B_1\tilde{\tau}[x(s)] \equiv B_1\tau(s)$$

for classical orthogonal polynomials, the function $\tau(s)$ is a polynomial of the first degree in $x = x(s)$ with a nonzero coefficient for $x(s)$.

If a and b are finite, the boundary conditions (3.3.3) can be presented in a simpler form,

$$\sigma(a)\varrho(a) = 0 \,, \quad \sigma(b)\varrho(b) = 0 \,, \tag{3.3.6}$$

because $x(s-\frac{1}{2})$ is bounded. If we take $\varrho(s_i) \ne 0$ for $a \le s_i \le b-1$, the boundary condition at $s = a$ is satisfied, provided that

$$\sigma(a) = 0 \,. \tag{3.3.7}$$

On the other hand, by virtue of the equality

$$\sigma(s+1)\varrho(s+1) = \varrho(s)\left[\sigma(s) + \tau(s)\Delta x\left(s - \tfrac{1}{2}\right)\right] \,,$$

the boundary condition $\sigma(b)\varrho(b) = 0$ is satisfied if

$$\left. \sigma(s) + \tau(s)\Delta x \left(s - \tfrac{1}{2}\right) \right|_{s=b-1} = 0 \ . \tag{3.3.8}$$

3.3.2. Proceeding similarly for Eq. (3.2.8) with $k = 1$ we can show that for the functions $v_{1n}(s)$ the orthogonality relation

$$\sum_{s_i=a_1}^{b_1-1} v_{1m}(s_i) v_{1n}(s_i) \varrho_1(s_i) \Delta x_1 \left(s_i - \tfrac{1}{2}\right) = \delta_{mn} d_{1n}^2 \tag{3.3.9}$$

is valid if the boundary conditions

$$\left. \sigma(s)\varrho_1(s) x_1^l \left(s - \tfrac{1}{2}\right) \right|_{s=a_1,b_1} = 0 \quad (l = 0, 1, \ldots)$$

are satisfied. For finite a and b, because of the relations

$$\sigma(a) = 0 \ , \quad \sigma(b)\varrho(b) = 0 \ ,$$
$$\sigma(s)\varrho_1(s) = \sigma(s)\sigma(s + 1)\varrho(s + 1)$$

the boundary conditions (3.3.10) are satisfied for $a_1 = a$ and $b_1 = b - 1$. In a similar way, by induction, we find that, provided the boundary conditions

$$\left. \sigma(s)\varrho_k(s) x_k^l \left(s - \tfrac{1}{2}\right) \right|_{s=a,b-k} = 0 \quad (l = 0, 1, \ldots) \tag{3.3.10}$$

are valid, the polynomials $v_{kn}(s)$ satisfy the orthogonality relations

$$\sum_{s_i=a}^{b-k-1} v_{km}(s_i) v_{kn}(s_i) \varrho_k(s_i) \Delta x_k \left(s_i - \tfrac{1}{2}\right) = \delta_{mn} d_{kn}^2 \ , \tag{3.3.11}$$

where d_{kn}^2 is the squared norm of the polynomial $v_{kn}(s)$.

We shall also assume that the polynomials $v_{kn}(s)$, which satisfy the orthogonality relation (3.3.11), also satisfy the conditions

$$\varrho_k(s_i) \Delta x_k \left(s_i - \tfrac{1}{2}\right) > 0 \quad (a \le s_i \le b - k - 1) \ , \tag{3.3.12}$$

coinciding with (3.3.5) for $k = 0$.

Remark. It can be shown that from (3.2.31) with the aid of (3.3.7) and (3.3.8) it follows that

$$y_N(s_i) = 0 \quad (s_i = a + i, \ b = a + N, \ i = 0, 1, \ldots, N - 1) \ . \tag{3.3.13}$$

3.4 Classification of Lattices

The form of Eq. (3.1.5) is preserved under the linear transformations

$$x(s) \rightarrow A x(s) + B , \quad s \rightarrow \pm s + s_0 . \tag{3.4.1}$$

By using the transformation (3.4.1) we reduce the expressions for the functions $x(s)$ to simpler forms. We shall assume that in Eq. (3.1.12) the constants α and β are real, while the function $x(s)$ takes real values when s is real.

1) Let $\alpha = 1$. Then

$$x(s) = c_1 s^2 + c_2 s + c_3 , \tag{3.4.2}$$

where c_1, c_2 and c_3 are constants.

If $c_1 = 0$, the transformation $x(s) \rightarrow A x(s) + B$ with $A = c_2$, $B = c_3$ carries $x(s)$ to the form $x(s) = s$. On the other hand, if $c_1 \neq 0$, the transformation $x(s) \rightarrow A x(s) + B$, $s \rightarrow s + s_0$, with $A = c_2 + 2 c_1 s_0 = c_1$, $B = c_1 s_0^2 + c_2 s_0 + c_3$, will carry $x(s)$ to the form $x(s) = s(s+1)$.

In the last case, the function $x(s)$ was chosen in the form $x(s) = s(s+1)$, rather than $x(s) = s^2$, since the polynomials of a discrete variable on the lattice $x(s) = s(s+1)$ are connected in a simple way with the Racah coefficients, which are extensively used in atomic physics.

2) Let $\alpha \neq 1$, and κ_1, κ_2 be the roots of the equation $\kappa^2 - 2\alpha\kappa + 1 = 0$. Then

$$x(s) = c_1 q^s + c_2 q^{-s} + c_3 , \tag{3.4.3}$$

where

$$q = \kappa_1^2 = \left(\alpha + \sqrt{\alpha^2 - 1} \right)^2 . \tag{3.4.4}$$

If $\alpha > 1$ we have $\kappa_1 > 1$, $\kappa_1 \kappa_2 = 1$ and we may put $\kappa_1 = e^\omega$, $\kappa_2 = e^{-\omega}(\omega > 0)$. If $c_1 c_2 > 0$, the function $x(s)$ can be reduced to the form $x(s) = \cosh(2\omega s)$ by using the transformations $s \rightarrow s + s_0$, $x(s) \rightarrow A x(s) + B$, provided that the constants s_0, A and B are chosen to satisfy the conditions:

$$c_1 e^{2\omega s_0} = c_2 e^{-2\omega s_0} = A/2 , \quad B = c_3 .$$

If, however, $c_1 c_2 < 0$, the function $x(s)$ can be represented, in a similar way, in the form $x(s) = \sinh(2\omega s)$. Now suppose that $c_2 = 0$. Then if $s_0 = 0$, and $A = c_1$, $B = c_3$, the function $x(s)$ can be represented in the form $x(s) = e^{2\omega s}$. If $c_1 = 0$, by using the transformation $x(s) \rightarrow c_2 x(s) + c_3$, $s \rightarrow -s$ we can represent $x(s)$ in the form $x(s) = e^{2\omega s}$.

If $\alpha < 1$, then $\kappa_1 = e^{i\omega}$, $\kappa_2 = e^{-i\omega}$, $c_2 = \bar{c}_1 = |c_1| e^{i\delta}$ (the bar denotes the complex conjugate), and $x(s)$ has the form

$$x(s) = 2|c_1| \cos(2\omega s - \delta) + c_3 .$$

The transformation $x(s) \rightarrow 2|c_1| x(s) + c_3$, $s \rightarrow s + s_0$, $s_0 = \delta/2\omega$ carries $x(s)$ to the form $x(s) = \cos(2\omega s)$.

Thus we come to the following cannonical forms of the functions $x(s)$:

$$\text{I}. \quad x(s) = s \qquad\qquad (\alpha = 1,\ \beta = 0)\,; \tag{3.4.5}$$

$$\text{II}. \quad x(s) = s(s+1) \qquad \left(\alpha = 1,\ \beta = \tfrac{1}{4}\right)\,; \tag{3.4.6}$$

$$\text{III}. \quad x(s) = e^{2\omega s} \qquad\qquad (\alpha > 1,\ \alpha = \cosh\omega,\ \beta = 0)\,; \tag{3.4.7}$$

$$\text{IV}. \quad x(s) = \sinh(2\omega s) \quad (\alpha > 1,\ \alpha = \cosh\omega,\ \beta = 0)\,; \tag{3.4.8}$$

$$\text{V}. \quad x(s) = \cosh(2\omega s) \quad (\alpha > 1,\ \alpha = \cosh\omega\ \beta = 0) \tag{3.4.9}$$

$$\text{VI}. \quad x(s) = \cos(2\omega s) \qquad (0 < \alpha < 1,\ \alpha = \cos\omega,\ \beta = 0)\,. \tag{3.4.10}$$

The case $\alpha \le 0$ is usually not of interest. The form of the lattices for $x(s)$ is chosen so that the function $x(s)$ will be real at real s.

3.5 Classification of Polynomial Systems on Linear and Quadratic Lattices. The Racah and the Dual Hahn Polynomials

Let us consider the basic systems of classical orthogonal polynomials on lattices (3.4.5–6). In order to find explicit expressions for the weight functions $\varrho(s)$ for which the polynomials (3.2.19) are orthogonal, we rewrite (3.2.9) in the form

$$\frac{\varrho(s+1)}{\varrho(s)} = \frac{\sigma(s) + \tau(s)\Delta x(s - 1/2)}{\sigma(s+1)}\,. \tag{3.5.1}$$

So that a one-to-one correspondence will exist between $x = x(s)$ and s we shall assume that $x(s)$ is monotonic on the interval $a \le s \le b$.

3.5.1 The Lattice $x(s) = s$

The case of a linear lattice $x(s) = s$ was discussed in detail in Chap. 2. Depending on the degrees of the polynomials $\sigma(s)$ and $\sigma(s)+\tau(s)\Delta x(s-\tfrac{1}{2}) = \sigma(s)+\tau(s)$, by solving (3.5.1) we obtain the Hahn polynomials $h_n^{(\alpha,\beta)}(x, N)$ and $\bar{h}_n^{(\mu,\nu)}(x, N)$, the Meixner polynomials $m_n^{(\gamma,\mu)}(x)$, the Kravchuk polynomials $k_n^{(p)}(x, N)$, and the Charlier polynomials $c_n^{(\mu)}(x)$, the basic data for which are given in Tables 2.1–4.

3.5.2 The Lattice $x(s) = s(s + 1)$

For $x(s) = s(s+1)(s > -\tfrac{1}{2})$ Eq. (3.5.1) can be transformed into a more convenient form. Under the transformation $s \rightarrow -s - 1$ we have

$$x(s) = x(-s-1), \quad \Delta x\left(s - \tfrac{1}{2}\right) = -\Delta x\left(t - \tfrac{1}{2}\right)\Big|_{t=-s-1};$$

then according to (3.1.22) and (3.1.23) we obtain

$$\sigma(s) + \tau(s)\Delta x\left(s - \tfrac{1}{2}\right) = \sigma(-s-1), \tag{3.5.2}$$

$$\bar{\sigma}[x(s)] = \frac{1}{2}[\sigma(s) + \sigma(-s-1)], \quad \bar{\tau}[x(s)] = \frac{\sigma(-s-1) - \sigma(s)}{\Delta x(s - 1/2)}. \tag{3.5.3}$$

The equation for $\varrho(s)$ has the form

$$\frac{\varrho(s+1)}{\varrho(s)} = \frac{\sigma(-s-1)}{\sigma(s+1)}. \tag{3.5.4}$$

By virtue (3.1.22) and (3.1.23) $\sigma(s)$ is a polynomial of the fourth or third degree in s in this case.

3.5.2.1. Let $\sigma(s)$ be a polynomial of the fourth degree:

$$\sigma(s) = A \prod_{j=1}^{4} (s - s_j). \tag{3.5.5}$$

Then (3.5.4) has the form

$$\frac{\varrho(s+1)}{\varrho(s)} = \frac{\prod_{j=1}^{4}(s+1+s_j)}{\prod_{j=1}^{4}(s+1-s_j)}. \tag{3.5.6}$$

Since $\sigma(a) = 0$, $\sigma(-b) = 0$ according to (3.3.7, 8) and (3.5.2), we may take $s_1 = a$ and $s_2 = -b$.

a) Let

$$\sigma(s) = (s - a)(s + b)(s - c)(d - s) \tag{3.5.7}$$

[in (3.5.5) we put $A = -1$, $s_3 = c$, $s_4 = d$]. Then

$$\varrho(s) = \frac{\Gamma(s+a+1)\Gamma(s+c+1)\Gamma(s+d+1)\Gamma(d-s)}{\Gamma(s-a+1)\Gamma(s-c+1)\Gamma(s+b+1)\Gamma(b-s)}. \tag{3.5.8}$$

Since $\Delta x(s - \tfrac{1}{2}) = 2s + 1 > 0$ for $s > -\tfrac{1}{2}$, the condition (3.3.5) will be satisfied when

$$-\tfrac{1}{2} < a < b < 1 + d, \quad |c| < 1 + a.$$

b) Let

$$\sigma(s) = (s - a)(s + b)(s - c)(s + d) \quad (A = 1, \; s_3 = c, \; s_4 = -d). \tag{3.5.9}$$

Then

$$\varrho(s) = \frac{\Gamma(s+a+1)\Gamma(s+c+1)}{\Gamma(s-a+1)\Gamma(s-c+1)\Gamma(s+b+1)\Gamma(b-s)\Gamma(s+d+1)\Gamma(d-s)}$$
$$(-\tfrac{1}{2} < a < b < 1 + d, |c| < 1 + a). \tag{3.5.10}$$

We denote $u_n^{(c,d)}(x)$ and $\tilde{u}_n^{(c,d)}(x)$, respectively, the polynomials $\tilde{y}_n(x)$, with $B_n = (-1)^n/n!$ and $B_n = 1/n!$ corresponding to the weight functions in (3.5.8) and (3.5.10). We call these *the Racah polynomials*, because they are connected by a simple relation with the Racah coefficients which are widely used in atomic physics.

3.5.2.2. Let $\sigma(s)$ be a cubic polynomial, i.e.

$$\sigma(s) = (s - a)(s + b)(s - c) . \tag{3.5.11}$$

Then

$$\frac{\varrho(s + 1)}{\varrho(s)} = \frac{(s + 1 + a)(b - s - 1)(s + 1 + c)}{(s + 1 - a)(s + 1 + b)(s + 1 - c)} , \tag{3.5.12}$$

whence

$$\varrho(s) = \frac{\Gamma(s + a + 1)\Gamma(s + c + 1)}{\Gamma(s - a + 1)\Gamma(s + b + 1)\Gamma(b - s)\Gamma(s - c + 1)}$$
$$\left(-\tfrac{1}{2} < a < b, \ |c| < 1 + a\right) . \tag{3.5.13}$$

We denote by $w_n^{(c)}(x)$ the orthogonal polynomials with $B_n = (-1)^n/n!$. Comparing the corresponding orthogonality relations and the coefficients of the leading terms of the polynomials $w_n^{(c)}(x) \equiv w_n^{(c)}(x, a, b)$ with those of the dual Hahn polynomials $w_n^{(\alpha,\beta)}(x)$ (see Sect. 2.4.6), we see that they coincide if

$$a = \tfrac{1}{2}(\alpha + \beta), \ b = a + N, \ c = \tfrac{1}{2}(\beta - \alpha) ,$$

i.e. the Hahn polynomials and the $w_n^{(c)}(x, a, b)$ are connecto by

$$h_n^{(\alpha,\beta)}(i) = (-1)^{n+i} \frac{i!(N - i - 1)!\Gamma(\beta + n + 1)}{n!(N - n - 1)!\Gamma(\beta + i + 1)}$$
$$\times w_i^{\left(\frac{\beta - \alpha}{2}\right)} \left(t_n, \frac{\alpha + \beta}{2}, \frac{\alpha + \beta}{2} + N\right) \tag{3.5.14}$$
$$\left(t_n = s_n(s_n + 1), \ s_n = \frac{\alpha + \beta}{2} + n; \ i, n = 0, 1, \ldots, N - 1\right) .$$

3.5.3. We obtained the difference equation (3.1.3) from the differential equation (3.1.1) for the classical orthogonal polynomials. Consequently it is natural to expect that the polynomial solutions of (3.1.3) and the weight functions will, in the limit $h \to 0$, become (with appropriate normalization) the polynomial solutions of (3.1.1) and the corresponding weight functions.

Let us consider this limiting process for the Racah polynomials. Setting $h \to 0$ in (3.1.3) corresponds to $N = b - a \to \infty$ for the Racah polynomials. It is easy to show that the weight function $\varrho(s)$ for the Racah polynomials $u_n^{(c,d)}[x(s)]$

becomes, in the limit $N \to \infty$, the weight function $(1-t)^{\alpha}(1+t)^{\beta}$ for the Jacobi polynomials $P_n^{(\alpha,\beta)}(t)$, where

$$t = 2\frac{x(s) - x(a)}{x(b) - x(a)} - 1, \quad \alpha = d - b, \quad \beta = a + c.$$

For the proof it is sufficient to use the relation

$$\frac{\Gamma(z + \gamma)}{\Gamma(z + \delta)} \to z^{\gamma - \delta} \quad \text{as} \quad z \to \infty.$$

In fact, for a fixed $t \in (-1, 1)$ and $N \to \infty$ we have

$$b = N + a \approx N, \quad 1 + t = 2\frac{(s + 1/2)^2 - (a + 1/2)^2}{(b + 1/2)^2 - (a + 1/2)^2} \approx \frac{2}{N^2}s^2,$$

$$1 - t \approx \frac{2}{N^2}(b^2 - s^2),$$

$$\frac{\Gamma(s + a + 1)\Gamma(s + c + 1)}{\Gamma(s - a + 1)\Gamma(s - c + 1)} \approx s^{2(a+c)} = \left(s^2\right)^{\beta} \approx \left[\frac{N^2}{2}(1 + t)\right]^{\beta},$$

$$\frac{\Gamma(s + d + 1)\Gamma(d - s)}{\Gamma(s + b + 1)\Gamma(b - s)} \approx (s + b)^{d-b}(b - s)^{d-b}$$

$$= (b^2 - s^2)^{\alpha} \approx \left[\frac{N^2}{2}(1 - t)\right]^{\alpha}.$$

Consequently

$$\lim_{N \to \infty} \left(\frac{2}{N^2}\right)^{\alpha + \beta} \varrho(s) = (1 - t)^{\alpha}(1 + t)^{\beta}. \tag{3.5.15}$$

A similar limit relation must connect the Racah polynomials $u_n^{(c,d)}[x(s)]$ and the Jacobi polynomials $P_n^{(\alpha,\beta)}(t)$:

$$\lim_{N \to \infty} c_n(N)u_n^{(\beta - a, b + \alpha)}[x(s)] = P_n^{(\alpha,\beta)}(t). \tag{3.5.16}$$

The constants $c_n(N)$ are easily determined by equating the coefficients of the leading terms on the two sides of (3.5.16):

$$c_n(N) = N^{-2n}.$$

Because of the limit relation (3.5.16) we shall now refer to the Racah polynomials $u_n^{(c,d)}(x)$ as $u_n^{(\alpha,\beta)}(x)$, taking $\alpha = d - b$, $\beta = a + c$.

Remark. By using the limiting process $a \to \infty$, $b - a = N = \text{const}$ we may derive also the relations that connect the polynomials on quadratic and linear lattices. For example, by using the asymptotic formula

$$\frac{\Gamma(z + \gamma)}{\Gamma(z + \delta)} \to z^{\gamma - \delta}, \quad z \to \infty$$

for the weight function of the Racah polynomials we have

$$\varrho(s) \approx (2a)^{\alpha+\beta} \frac{\Gamma(\alpha + N - t)\Gamma(\beta + t + 1)}{\Gamma(N - t)\Gamma(t + 1)} \, ,$$

$$s = a + t \, , \quad a \to \infty \, .$$

Hence we obtain from this the limit relation between the Racah polynomials $u_n^{(\alpha,\beta)}(x)$ and the Hahn polynomials $h_n^{(\alpha,\beta)}(t)$ in the form

$$\lim_{a \to \infty} (2a)^{-n} u_n^{(\alpha,\beta)}(x) = h_n^{(\alpha,\beta)}(t) \, , \tag{3.5.17}$$

where $x = s(s + 1)$, $s = a + t$.

3.6 q-Analogs of Polynomials Orthogonal on Linear and Quadratic Lattices

We have considered the systems of polynomials orthogonal on the lattices $x(s) = s$ (the Hahn, Meixner, Kravchuk and Charlier polynomials) and $x(s) = s(s + 1)$ (the Racah and dual Hahn polynomials). In constructing a theory of classical orthogonal polynomials of a discrete variable we used the difference equations of hypergeometric type, which retain this form after difference differentiation. It is possible to introduce difference equations of this kind only for certain types of lattices $x(s)$. As shown in Sect. 3.4, besides the linear and quadratic lattices, the lattice functions (3.4.7–10) also satisfy this requirement:

$$x(s) = \begin{cases} q^s = e^{2\omega s} & \left(q = e^{2\omega}\right) ; \\ \frac{1}{2}\left(q^s - q^{-s}\right) = \sinh(2\omega s) & \left(q = e^{2\omega}\right) ; \\ \frac{1}{2}\left(q^s + q^{-s}\right) = \cosh(2\omega s) & \left(q = e^{2\omega}\right) ; \\ \frac{1}{2}\left(q^s + q^{-s}\right) = \cos(2\omega s) & \left(q = e^{2i\omega}\right) . \end{cases}$$

When $q \to 1 \, (\omega \to 0)$ we have

$$e^{2\omega s} \approx 1 + 2\omega s \, , \quad \sinh(2\omega s) \approx 2\omega s \, ;$$

$$\cosh(2\omega s) = 1 + 2\omega^2 s^2 \, , \quad \cos(2\omega s) = 1 - 2\omega^2 s^2 \, ,$$

i.e. the lattices (3.4.7–10) become either linear or quadratic in s. The polynomials whose limits as $q \to 1$ are polynomials which are orthogonal on linear or quadratic lattices $x(s) = s$ or $x(s) = s(s + 1)$ are called q-analogs of the corresponding polynomials.

Let us consider the methods of constructing weight functions for the q-analogs of the Hahn, Meixner, Kravchuk and Charlier polynomials on the lattices $x(s) = \exp(2\omega s)$ and $x(s) = \sinh(2\omega s)$ as well as for the q-analogs of the Racah and dual Hahn polynomials on the lattices $x(s) = \cosh(2\omega s)$ and $x(s) = \cos(2\omega s)$. In constructing the weight functions $\varrho(s)$ we shall proceed from Eq. (3.5.1).

3.6.1. The q-Analogs of the Hahn, Meixner, Kravchuk, and Charlier Polynomials on the Lattices $x(s) = \exp(2\omega s)$ and $x(s) = \sinh(2\omega s)$

3.6.1.1. *The lattice* $x(s) = q^s$ $(q = e^{2\omega})$. Replacing s by $s - a$ does not change the form of (3.1.5) with $x(s) = q^s$ and consequently, when considering the orthogonality property, we may take $a = 0$, as in the case $x(s) = s$. Since

$$\sigma(s) = \tilde{\sigma}[x(s)] - \frac{1}{2}\tilde{\tau}[x(s)]\Delta x\left(s - \frac{1}{2}\right), \quad \Delta x\left(s - \frac{1}{2}\right) = \frac{q-1}{\sqrt{q}}x(s),$$

where $\tilde{\sigma}(x)$ and $\tilde{\tau}(x)$ are polynomials of at most the second and first degrees, respectively, the functions $\sigma(s)$ and $\sigma(s) + \tau(s)\Delta x(s - \frac{1}{2})$ are polynomials of at most degree 2 in $x = x(s)$, and the right-hand side of (3.5.1) is a rational function in $x(s)$. We are now going to decompose the functions $\sigma(s + 1)$ and $\sigma(s) + \tau(s)\Delta x(s - \frac{1}{2})$ into simple factors, linear in $x = q^s$, and use the property of the solution of the equation

$$\frac{\varrho(s+1)}{\varrho(s)} = F(s) \quad \text{with} \quad F(s) = \frac{\prod_i f_i(s)}{\prod_i g_i(s)},$$

which was considered in Sect. 2.4. Then the solutions of (3.5.1) for the lattice $x = q^s$ can always be determined if we know particular solutions of the following equations

$$\frac{\varrho(s+1)}{\varrho(s)} = \begin{cases} \dfrac{q^{s+\gamma} - 1}{q - 1}, & \dfrac{q^{\gamma-s} - 1}{q - 1}; \\[2mm] \dfrac{q^{s+\gamma} + 1}{q + 1}, & \dfrac{q^{\gamma-s} + 1}{q + 1}; \\[2mm] \alpha^s & \beta, \end{cases} \tag{3.6.1}$$

where α, β and γ are constants. These solutions are

$$\varrho(s+1) = \begin{cases} \Gamma_q(s + \gamma), & \dfrac{1}{\Gamma_q(\gamma - s + 1)}; \\[2mm] \Pi_q(s + \gamma), & \dfrac{1}{\Pi_q(\gamma - s + 1)}; \\[2mm] \alpha^{s(s-1)/2} & \beta^s. \end{cases} \tag{3.6.2}$$

The q-gamma function $\Gamma_q(s)$ is a generalization of Euler's gamma-function $\Gamma(s)$ [J1]. It is defined by

$$\Gamma_q(s) = \begin{cases} (1 - q)^{1-s}\dfrac{\prod_{k=0}^{\infty}(1 - q^{k+1})}{\prod_{k=0}^{\infty}(1 - q^{s+k})}, & |q| < 1; \\[3mm] q^{(s-1)(s-2)/2}\Gamma_{1/q}(s), & |q| > 1. \end{cases} \tag{3.6.3}$$

For the function $\Gamma_q(s)$ we have the relations

$$\Gamma_q(s + 1) = \frac{q^s - 1}{q - 1}\Gamma_q(s), \tag{3.6.4}$$

$$\lim_{q \to 1} \Gamma_q(s) = \Gamma(s) \ . \tag{3.6.5}$$

The functions $\Pi_q(s)$ and $\Gamma_q(s)$ are connected by the relation

$$\Pi_q(s) = \frac{\Gamma_{q^2}(s)}{\Gamma_{q(s)}} \ . \tag{3.6.6}$$

From (3.6.4) and (3.6.5) we have

$$\Pi_q(s+1) = \frac{q^s + 1}{q + 1} \Pi_q(s) \ , \tag{3.6.7}$$

$$\lim_{q \to 1} \Pi_q(s) = 1 \ . \tag{3.6.8}$$

Instead of $\Gamma_q(s)$ we shall use the function $\tilde{\Gamma}_q(s)$ introduced in Sect. 3.2 [see (3.2.24) and (3.2.26)]:

$$\tilde{\Gamma}_q(s) = q^{-(s-1)(s-2)/4} \Gamma_q(s) \ .$$

The function $\tilde{\Gamma}_q(s)$ satisfies the equation

$$\frac{\tilde{\Gamma}_q(s+1)}{\tilde{\Gamma}_q(s)} = \psi_q(s) , \quad \psi_q(s) = \frac{q^{s/2} - q^{-s/2}}{q^{1/2} - q^{-1/2}} = \frac{\sinh \omega s}{\sinh \omega} \ . \tag{3.6.9}$$

In a similar way instead of $\Pi_q(s)$ we shall use the function $\tilde{\Pi}_q(s)$ satisfying the equation

$$\frac{\tilde{\Pi}_q(s+1)}{\tilde{\Pi}_q(s)} = \varphi_q(s) , \quad \varphi_q(s) = \frac{q^{s/2} + q^{-s/2}}{q^{1/2} + q^{-1/2}} = \frac{\cosh \omega s}{\cosh \omega} \ , \tag{3.6.10}$$

where

$$\tilde{\Pi}_q(s) = \frac{\tilde{\Gamma}_{q^2}(s)}{\tilde{\Gamma}_q(s)} \ . \tag{3.6.11}$$

These functions have more symmetry than $\Gamma_q(s)$ and $\Pi_q(s)$. For example, from the relation $\psi_q(-s) = -\psi_q(s)$ it follows that

$$\left. \frac{\tilde{\Gamma}_q(t+1)}{\tilde{\Gamma}_q(t)} \right|_{t=-s} = - \frac{\tilde{\Gamma}_q(s+1)}{\tilde{\Gamma}_q(s)} \ ,$$

which corresponds to a similar relation for Euler's gamma-function:

$$\left. \frac{\Gamma(t+1)}{\Gamma(t)} \right|_{t=-s} = - \frac{\Gamma(s+1)}{\Gamma(s)} \ .$$

For the function $\Gamma_q(s)$ the analogous equality has a more complicated form:

$$\left.\frac{\Gamma_q(t+1)}{\Gamma_q(t)}\right|_{t=-s} = -q^{-s}\frac{\Gamma_q(s+1)}{\Gamma_q(s)} \; .$$

Furthermore, the relation

$$\Gamma_q(s) = q^{(s-1)(s-2)/2}\Gamma_{1/q}(s)$$

becomes the symmetric relation:

$$\tilde{\Gamma}_q(s) = \tilde{\Gamma}_{1/q}(s) \; .$$

This relation allows a natural generalization of the definition of $\tilde{\Gamma}_q(s)$ and $\tilde{\Pi}_q(s)$ to the case when $q = \exp(2i\omega)$ $(\omega > 0)$, i.e. to the case $|q| = 1$:

$$\begin{aligned}
\tilde{\Gamma}_q(s) &= \lim_{q' \to q} \tilde{\Gamma}_{q'}(s) \; , \quad |q'| \neq 1 \; ; \\
\tilde{\Pi}_q(s) &= \lim_{q' \to q} \tilde{\Pi}_{q'}(s) \; , \quad |q'| \neq 1 \; .
\end{aligned} \tag{3.6.12}$$

Note that for the function $\tilde{\Gamma}_q(s)$ we have

$$\lim_{q' \to q, |q'|<1} \tilde{\Gamma}_{q'}(s) = \lim_{q' \to q, |q'|>1} \tilde{\Gamma}_{q'}(s) \; .$$

We shall also use the following asymptotic properties of $\tilde{\Gamma}_q(s)$ when $s \to +\infty$, $0 < q < 1$:

$$\begin{aligned}
\tilde{\Gamma}_q(s+1) &\approx q^{-s(s-1)/4}(1-q)^{-s}e^{-c_q} \; , \\
\frac{\tilde{\Gamma}_q(s+a)}{\tilde{\Gamma}_q(s)} &\approx q^{-a(a+2s-3)/4}(1-q)^{-a} \; .
\end{aligned}$$

Here $c_q = -\sum_{k=0}^{\infty} \ln\left(1 - q^{k+1}\right)$.

In fact, according to (3.6.3), when $0 < q < 1$ we have

$$\ln\Gamma_q(s+1) + s\ln(1-q) + c_q = \sum_{k=0}^{\infty} \ln\frac{1}{1-q^{k+s+1}} \; .$$

Since $\ln(1+x) < x$ for $x > 0$, the following inequalities

$$\begin{aligned}
\ln\frac{1}{1-q^{k+s+1}} &= \ln\left(1 + \frac{q^{k+s+1}}{1-q^{k+s+1}}\right) \\
&< \frac{q^{k+s+1}}{1-q^{k+s+1}} < \frac{q^{s+1}}{1-q^{s+1}}q^k
\end{aligned}$$

are valid. Therefore

$$\sum_{k=0}^{\infty} \ln\frac{1}{1-q^{k+s+1}} < \sum_{k=0}^{\infty} \frac{q^{s+1}}{1-q^{s+1}}q^k = \frac{1}{1-q}\frac{q^{s+1}}{1-q^{s+1}} \; .$$

As a result we obtain

$$\ln \Gamma_q(s+1) + s\ln(1-q) + c_q < \frac{1}{1-q}\frac{q^{s+1}}{1-q^{s+1}} \ .$$

When $s \to \infty$ the right-hand side of this inequality tends to zero if $0 < q < 1$, which leads to the considered asymptotic expressions.

For the function $\tilde{\Pi}_q(s)$ when $s \to +\infty$, $0 < q < 1$ by using the preceding properties and (3.6.11) we obtain

$$\tilde{\Pi}_q(s+1) \approx q^{-s(s-1)/4}(1+q)^{-s}\exp\left[-\left(c_{q^2}-c_q\right)\right] ,$$

$$\frac{\tilde{\Pi}_q(s+a)}{\tilde{\Pi}_q(s)} \approx q^{-a(a+2s-3)/4}(1+q)^{-a} \ .$$

The analogous relations for $q > 1$ can easily be obtained by using the connections between the functions $\tilde{\Gamma}_q(s)$ and $\tilde{\Gamma}_{1/q}(s)$, and the functions $\tilde{\Pi}_q(s)$ and $\tilde{\Pi}_{1/q}(s)$:

$$\tilde{\Gamma}_{1/q}(s) = \tilde{\Gamma}_q(s) , \qquad \tilde{\Pi}_{1/q}(s) = \tilde{\Pi}_q(s) \ .$$

By means of the simple relation

$$\lim_{q\to 1}\psi_q(s) = s \tag{3.6.13}$$

we can construct systems of orthogonal polynomials which when $q \to 1$ become the polynomials on the lattice $x(s) = s$ considered above. In accordance with (3.6.13) we shall choose the form of the functions $\sigma(s)$ and $\sigma(s)+\tau(s)\Delta x(s-\frac{1}{2})$ so that the weight functions $\varrho(s)$ and the polynomials $\tilde{y}_n[x(s)]$ (with a specific choice of the constant B_n in the Rodrigues formula) become, when $q \to 1$, the weight functions and polynomials on the lattice $x(s) = s$. The polynomials on the lattice $x = q^s$ corresponding to the Hahn polynomials $h_n^{(\alpha,\beta)}(s)$ and $\tilde{h}_n^{(\mu,\nu)}(s)$, the Meixner polynomials $m_n^{(\gamma,\mu)}(s)$, the Kravchuk polynomials $k_n^{(p)}(s)$, and the Charlier polynomials $c_n^{(\mu)}(s)$ will be denoted by $h_n^{(\alpha,\beta)}(x,q)$ and $\tilde{h}_n^{(\mu,\nu)}(x,q)$, $m_n^{(\gamma,\mu)}(x,q)$, $k_n^{(p)}(x,q)$ and $c_n^{(\mu)}(x,q)$ respectively. In order to guess the form of the functions $\sigma(s)$ and $\sigma(s)+\tau(s)\Delta x(s-\frac{1}{2})$ in (3.5.1) for these polynomials, we replace the factors of the form $\pm(s+\gamma)$ in the corresponding expressions on the lattice $x(s) = s$ by the function $q^{(s+\gamma)/2}\psi_q[\pm(s+\gamma)]$, which is a polynomial of first degree in $x(s) = q^s$. Under this substitution the functions $\sigma(s)$ and $\sigma(s)+\tau(s)\Delta x(s-\frac{1}{2})$ will be the polynomials of at most second degree in q^s.

In addition, in the process of choosing the form of the polynomials $\sigma(s)$ and $\sigma(s)+\tau(s)\Delta x(s-\frac{1}{2})$ it should be kept in mind that their constant terms must obviously coincide, since the product

$$\tau(s)\Delta x\left(s-\tfrac{1}{2}\right) = \tilde{\tau}[x(s)]\left[\left(q^{1/2}-q^{-1/2}\right)x(s)\right]$$

is a polynomial of second degree in $x = x(s)$ with constant term zero. In order to satisfy this condition in the case when the functions $\sigma(s)$ and $\sigma(s) + \tau(s)$ on the lattice $x(s) = s$ are polynomials of at most first degree, we multiply the assumed expressions for $\sigma(s+1)$ and $\sigma(s) + \tau(s)\Delta x(s - 1/2)$ in (3.5.1) by q^s [the introduction of this multiplier does not change Eq. (3.5.1) for $\varrho(s)$]. But if $\sigma(s)$ is a polynomial of second degree on the lattice $x(s) = s$, under the preceding transformations the constant terms in the polynomials $\sigma(s)$ and $\sigma(s) + \tau(s)\Delta x(s - 1/2)$ coincide, as can easily be verified.

In order to express the function $\varrho(s)$ in terms of functions of the form $\tilde{\Gamma}_q[\pm(s - \gamma)]$ and $\tilde{\Pi}_q[\pm(s - \gamma)]$ we shall use, instead of (3.6.1) and (3.6.2), particular solutions of the equations

$$\frac{\varrho(s+1)}{\varrho(s)} = \begin{cases} \psi_q(s + \gamma), & \psi_q(\gamma - s); \\ \varphi_q(s + \gamma), & \varphi_q(\gamma - s); \\ \alpha^s, & \beta, \end{cases} \tag{3.6.14}$$

which have the form

$$\varrho(s) = \begin{cases} \tilde{\Gamma}_q(s + \gamma), & \dfrac{1}{\tilde{\Gamma}_q(\gamma - s + 1)}; \\[2ex] \tilde{\Pi}_q(s + \gamma), & \dfrac{1}{\tilde{\Pi}_q(\gamma - s + 1)}; \\[2ex] \alpha^{s(s-1)/2}, & \beta^s \ . \end{cases} \tag{3.6.15}$$

Let us consider some specific cases of applying these methods.

1) *The Hahn polynomials $h_n^{(\alpha,\beta)}(x, q)$ and $\tilde{h}_n^{(\mu,\nu)}(x, q)$.*

For the Hahn polynomials $h_n^{(\alpha,\beta)}(s)$ we have

$$\sigma(s) = s(N + \alpha - s), \quad \sigma(s) + \tau(s) = (s + \beta + 1)(N - 1 - s) \ .$$

Therefore on the lattice $x(s) = q^s$ we take

$$\sigma(s) = q^{s/2}\psi_q(s)q^{(s-N-\alpha)/2}\psi_q(N + \alpha - s) \ ,$$
$$\sigma(s) + \tau(s)\Delta x(s - \tfrac{1}{2}) = q^{s+\beta+1)/2}\psi_q(s + \beta + 1)q^{(s-N+1)/2}\psi_q(N - 1 - s) \ .$$

Equation (3.5.1) for $\varrho(s)$ takes the form

$$\frac{\varrho(s+1)}{\varrho(s)} = q^{(\alpha+\beta)/2}\frac{\psi_q(s + \beta + 1)\psi_q(N - 1 - s)}{\psi_q(s + 1)\psi_q(N + \alpha - 1 - s)} \ .$$

By using (3.6.14) and (3.6.15) we obtain

$$\varrho(s) = q^{(\alpha+\beta)s/2}\frac{\tilde{\Gamma}_q(s + \beta + 1)\tilde{\Gamma}_q(N + \alpha - s)}{\tilde{\Gamma}_q(s + 1)\tilde{\Gamma}_q(N - s)} \ . \tag{3.6.16}$$

Since, when $q \to 1$, the function $\varrho(s)$ takes in the limit the form of the weight function for the Hahn polynomials $h_n^{(\alpha,\beta)}(s)$, it can be seen from the Rodrigues formula that the limiting relation

$$\lim_{q\to 1} h_n^{(\alpha,\beta)}(x(s),q) = h_n^{(\alpha,\beta)}(s)$$

will be satisfied if in the Rodrigues formula for $h_n^{(\alpha,\beta)}[x(s),q]$ we take $B_n = (1-q)^n/\tilde{\Gamma}_q(n+1)$.

In a similar way for the Hahn polynomials $\tilde{h}_n^{(\mu,\nu)}(x(s),q)$ we obtain:

$$\sigma(s) = q^{s/2}\psi_q(s)q^{(s+\mu)/2}\psi_q(s+\mu)\,,$$

$$\sigma(s) + \tau(s)\Delta x\left(s - \tfrac{1}{2}\right)$$
$$= q^{(s-N-\nu+1)/2}\psi_q(N+\nu-1-s)q^{(s-N+1)/2}\psi_q(N-1-s)\,,$$

$$\varrho(s) = \frac{q^{-[N+(\mu+\nu)/2]s}}{\tilde{\Gamma}_q(N+\nu-s)\tilde{\Gamma}_q(N-s)\tilde{\Gamma}_q(s+1)\tilde{\Gamma}_q(s+\mu+1)}\,;$$

$$\lim_{q\to 1} \tilde{h}_n^{(\mu,\nu)}(x(s),q) = \tilde{h}_n^{(\mu,nu)}(s)\,, \quad B_n = (q-1)^n/\tilde{\Gamma}_q(n+1)\,. \tag{3.6.17}$$

2) *The Meixner, Kravchuk and Charlier q-polynomials.* For the Meixner polynomials we have

$$\sigma(s) = s\,, \quad \sigma(s) + \tau(s) = \mu(s+\gamma) \quad (0 < \mu < 1\,, \gamma > 0)\quad.$$

Hence for the polynomials $m_n^{(\gamma,\mu)}(x,q)$, by assuming

$$\sigma(s) = q^{s/2}\psi_q(s)q^{s-1}\quad,$$

$$\sigma(s) + \tau(s)\Delta x\left(s - \tfrac{1}{2}\right) = \mu q^{(s+\gamma)/2}\psi_q(s+\gamma)q^s\quad,$$

we obtain

$$\frac{\varrho(s+1)}{\varrho(s)} = \mu q^{(\gamma-1)/2}\frac{\psi_q(s+\gamma)}{\psi_q(s+1)}\quad.$$

A solution of this equation has the form

$$\varrho(s) = C\mu^s q^{s(\gamma-1)/2}\frac{\tilde{\Gamma}_q(s+\gamma)}{\tilde{\Gamma}_q(s+1)}\quad, \tag{3.6.18}$$

where C is a constant. In order for the function $\varrho(s)$, when $q \to 1$, to take the form of the weight function for the polynomials $m_n^{(\gamma,\mu)}(s)$, we suppose that $C = 1/\tilde{\Gamma}_q(\gamma)$. The limit relation

$$\lim_{q\to 1} m_n^{(\gamma,\mu)}\left(x(s),q\right) = m_n^{(\gamma,\mu)}(s)$$

is satisfied for $B_n = \left((q-1)/\mu\right)^n$.

For the Kravchuk polynomials

$$\sigma(s) = s\,, \quad \sigma(s) + \tau(s) = \frac{p}{1-p}(N-s)\quad.$$

Therefore for the polynomials $k_n^{(p)}(x,q)$ we take

$$\sigma(s) = q^{s/2}\psi_q(s)q^{s-1} \quad ,$$

$$\sigma(s) + \tau(s)\Delta x\left(s + \tfrac{1}{2}\right) = \frac{p}{1-p}q^{(s-N)/2}\psi_q(N-s)q^s \quad ,$$

from which

$$\frac{\varrho(s+1)}{\varrho(s)} = \frac{p}{1-p}q^{-(N+1)/2}\frac{\psi_q(N-s)}{\psi_q(s+1)} \quad ,$$

and hence

$$\varrho(s) = C\left(\frac{p}{1-p}\right)^s \frac{q^{-s(N+1)/2}}{\tilde{\Gamma}_q(s+1)\tilde{\Gamma}_q(N-s+1)} \quad , \tag{3.6.19}$$

where C is a constant. The limit relations will be satisfied if

$$C = (1-p)^N\tilde{\Gamma}_q(N+1)\,, \quad B_n = \frac{(1-p)^n(1-q)^n}{\tilde{\Gamma}_q(n+1)} \quad .$$

Similarly for the polynomials $c_n^{(\mu)}(x,q)$ we obtain

$$\sigma(s) = q^{s/2}\psi_q(s)q^{s-1}\,, \quad \sigma(s) + \tau(s)\Delta x\left(s - \tfrac{1}{2}\right) = \mu q^s \quad ,$$

$$\varrho(s) = e^{-\mu}\frac{\mu^s}{\tilde{\Gamma}_q(s+1)}q^{-s(s+1)/4} \quad , \tag{3.6.20}$$

$$\lim_{q\to 1} c_n^{(\mu)}(x(s),q) = c_n^{(\mu)}(s)\,, \quad B_n = \left(\frac{q-1}{\mu}\right)^n \quad .$$

All the q-polynomials considered on the lattice $x(s) = q^s$ are orthogonal (in s) for the same values of a and b as for the corresponding polynomials on the lattice $x(s) = s$ (under the additional restriction $q < 1$ for $m_n^{(\gamma,\mu)}(x,q)$ which follows from the asymptotic behavior of the function $\varrho(s)$ when $s \to +\infty$).

In addition to the polynomials that we have considered on the lattice $x(s) = q^s$ we can also construct polynomial systems for which there are no analogs on the lattice $x(s) = s$. For example, for

$$\sigma(s) = q^{s/2}\psi_q(s)q^{(s-N-\alpha)/2}\psi_q(N+\alpha-s) \quad ,$$

$$\sigma(s) + \tau(s)\Delta x\left(s - \tfrac{1}{2}\right) = \frac{q^{1/2}}{1-q}q^{(s-N+1)/2}\psi_q(N-1-s)$$

$$(\alpha > 0, \quad 0 < q < 1)$$

we obtain

$$\varrho(s) = \frac{q^{s(2\alpha+1-s)/4}}{(1-q)^s}\frac{\tilde{\Gamma}_q(N+\alpha-s)}{\tilde{\Gamma}_q(s+1)\tilde{\Gamma}_q(N-s)} \quad . \tag{3.6.21}$$

The corresponding polynomials are orthogonal for $a = 0$, $b = N$. In this case these polynomials have no analogs on the lattice $x(s) = s$, since, when $q \to 1$, the function $\sigma(s)$ has a limit; however,

$$\sigma(s) + \tau(s)\Delta x \left(s - \tfrac{1}{2}\right) \to \infty \, , \quad \varrho(s) \to \infty \quad .$$

3.6.1.2. *The lattice* $x(s) = \sinh(2\omega s)$. Let $q = \mathrm{e}^{2\omega}$, then

$$x(s) = \tfrac{1}{2}\left(q^s - q^{-s}\right) , \quad \Delta x \left(s - \tfrac{1}{2}\right) = \tfrac{1}{2}\left(q^{1/2} - q^{-1/2}\right)\left(q^s + q^{-s}\right) .$$

It is also convenient to use the symmetry property of the lattice $x(s)$:

$$x(s) = x(t), \quad \Delta x \left(s - \tfrac{1}{2}\right) = -\Delta x \left(t - \tfrac{1}{2}\right) \, ,$$

where $t = -s - \mathrm{i}\pi/\ln q$.

In accordance with (3.1.22) and (3.1.23)

$$\sigma(s) + \tau(s)\Delta x \left(s - \tfrac{1}{2}\right) = \sigma \left(-s - \frac{\mathrm{i}\pi}{\ln q}\right) \, , \tag{3.6.22}$$

from which

$$\tilde{\sigma}[x(s)] = \frac{1}{2}\left[\sigma(s) + \sigma \left(-s - \frac{\mathrm{i}\pi}{\ln q}\right)\right] \, , \tag{3.6.23}$$

$$\tilde{\tau}[x(s)] = \frac{\sigma \left(-s - \mathrm{i}\pi/\ln q\right) - \sigma(s)}{\Delta x(s - 1/2)} \, . \tag{3.6.24}$$

We note that by virtue of (3.1.22)

$$\sigma(s) = q^{-2s}p_4\left(q^s\right) \, , \tag{3.6.25}$$

where $p_4(q^s)$ is a polynomial of fourth degree in q^s. By using (3.6.23) and (3.6.24) it can be easily verified by means of (3.6.25) that the functions $\tilde{\tau}[x(s)]$ and $\tilde{\sigma}[x(s)]$ are polynomials of at most first and second degrees in $x(s) = (q^s - q^{-s})/2$, rexpectively, for an arbitrary form of $p_4(q^s)$. Let us construct the systems of orthogonal polynomials which, when $q \to 1$, i.e. $\omega \to 0$, take the form of the Hahn, Meixner, Kravchuk and Charlier polynomials on the lattice $x(s) = s$. To do this we use the limit relations

$$\lim_{q \to 1} \psi_q(s) = s \, , \quad \lim_{q \to 1} \varphi_q(s) = 1 \, , \tag{3.6.26}$$

where the functions $\psi_q(s)$ and $\varphi_q(s)$ are determined by (3.6.9) and (3.6.10). In order to guess the form of the functions $\sigma(s)$ and $\sigma(s) + \tau(s)\Delta x(s - 1/2)$ for these polynomials we observe the following rule: if on the lattice $x(s) = s$ the expression for $\sigma(s)$ has the factor $s - \gamma$ and the expression for $\sigma(s) + \tau(s)$ has the factor $s - \gamma_1$, then on the lattice $x(s) = \sinh(2\omega s)$ we shall take the multiplier $\psi_q(s - \gamma)\varphi_q(s + \gamma_1)$ in $\sigma(s)$. Then, in accordance with (3.6.22), the factor $\varphi_q(s + \gamma)\psi_q(s - \gamma_1)$ will appear in $\sigma(s) + \tau(s)\Delta x(s - 1/2)$. This is because under the transformation $s \to -s - \mathrm{i}\pi/\ln q$ we have

$$\psi_q(s-\gamma) \to -i\frac{q^{1/2}+q^{-1/2}}{q^{1/2}-q^{-1/2}}\varphi_q(s+\gamma)\,,$$

$$\varphi_q(s+\gamma_1) \to i\frac{q^{1/2}-q^{-1/2}}{q^{1/2}+q^{-1/2}}\psi_q(s-\gamma_1)\,,$$

$$\psi_q(s-\gamma)\varphi_q(s+\gamma_1) \to \varphi_q(s+\gamma)\psi_q(s-\gamma_1)\,.$$

In addition, by virtue of (3.6.26),

$$\lim_{q\to 1}\psi_q(s-\gamma)\varphi_q(s+\gamma_1) = s-\gamma\,,$$

$$\lim_{q\to 1}\varphi_q(s+\gamma)\psi_q(s-\gamma_1) = s-\gamma_1\,.$$

In our further discussion, when the cases cannot be reduced to the one considered, we shall choose the form of the factors in $\sigma(s)$ specifically in each case.

1) *Analogs of the Hahn polynomials $h_n^{(\alpha,\beta)}(s-a)$ and $\tilde{h}_n^{(\mu,\nu)}(s-a)$ on the lattice $x(s) = \sinh(2\omega s)$.* For the Hahn polynomials $h_n^{(\alpha,\beta)}(s-a)$ $(a \le s \le b-1)$ we have

$$\sigma(s) = (s-a)(b+\alpha-s)\,,$$

$$\sigma(s)+\tau(s) = (s-a+\beta+1)(b-1-s)\,.$$

Consequently on the lattice $x(s) = \sinh(2\omega s)$ we suppose that

$$\sigma(s) = \psi_q(s-a)\varphi_q(s+a-\beta-1)\psi_q(b+\alpha-s)\varphi_q(b-1+s)\,,$$

$$\sigma(s)+\tau(s)\Delta x\left(s-\tfrac{1}{2}\right)$$
$$= \varphi_q(s+a)\psi_q(s-a+\beta+1)\varphi_q(b+\alpha+s)\psi_q(b-1-s)\,.$$

It is evident that the expression for $\sigma(s)$ has the form (3.6.25), and that the functions $\tilde{\sigma}(x)$ and $\tilde{\tau}(x)$ will be polynomials of at most second and first degrees in x, respectively.

Equation (3.5.1) takes the form

$$\frac{\varphi(s+1)}{\varphi(s)} = \frac{\varphi_q(s+a)\psi_q(s-a+\beta+1)\varphi_q(b+\alpha+s)\psi_q(b-1-s)}{\psi_q(s+1-a)\varphi_q(s+a-\beta)\psi_q(b+\alpha-1-s)\varphi_q(s+b)}\,.$$

By using (3.6.14) and (3.6.15), we obtain

$$\varrho(s) = \frac{\tilde{\Pi}_q(s+a)\tilde{\Gamma}_q(s-a+\beta+1)\tilde{\Pi}_q(b+\alpha+s)\tilde{\Gamma}_q(b+\alpha-s)}{\tilde{\Gamma}_q(s+1-a)\tilde{\Gamma}_q(b-s)\tilde{\Pi}_q(s+a-\beta)\tilde{\Pi}_q(s+b)}\,. \tag{3.6.27}$$

By the same argument for analogs of the polynomials $\tilde{h}_n^{(\mu,\nu)}(s-a)$ we obtain

$$\sigma(s) = \psi_q(s-a)\varphi_q(s+b+\nu-1)\psi_q(s-a+\mu)\varphi_q(s+b-1)\,,$$

$$\sigma(s)+\tau(s)\Delta x\left(s-\tfrac{1}{2}\right)$$
$$= \varphi_q(s+a)\psi_q(b+\nu-1-s)\varphi_q(s+a-\mu)\psi_q(b-1-s)\,,$$

$$\varrho(s) = \frac{\tilde{\Pi}_q(s+a)\tilde{\Pi}_q(s+a-\mu)}{\tilde{\Gamma}_q(s+1-a)\tilde{\Gamma}_q(b-s)\tilde{\Gamma}_q(b+\nu-s)\tilde{\Gamma}_q(s-a+\mu+1)\tilde{\Pi}_q(s+b)\tilde{\Pi}_q(s+b+\nu)} \ .$$

$$(3.6.28)$$

2) *Analogs of the Meixner polynomials on the lattice* $x(s) = \sinh(2\omega s)$.
For the Meixner polynomials $m_n^{(\gamma,\mu)}(s-a)$ $(s \geq a)$ we have

$$\sigma(s) = s-a \, , \quad \sigma(s)+\tau(s) = \mu(s+\gamma-a) \quad (0 < \mu < 1) \ .$$

The factors $s-a$ and $s+\gamma-a$ in the expression for $\sigma(s)$ on the lattice $x(s) = \sinh(2\omega s)$ correspond to the factor $\psi_q(s-a)\varphi_q(s-\gamma+a)$. Furthermore, in the expression for $\sigma(s)$ we have to choose a factor which is a polynomial of first degree in q^s and such that, in the limit $q \to 1$, it will be equal to unity and, under the transformation $s \to -s - i\pi/\ln q$, to μ. The factor can be taken in the form

$$\frac{(1+\mu)+(1-\mu)q^{s+a}}{q^{1/2}+q^{-1/2}} \ .$$

It corresponds to the factor

$$\frac{(1+\mu)-(1-\mu)q^{-(s-a)}}{q^{1/2}+q^{-1/2}}$$

in the expression for $\sigma(s)+\tau(s)\Delta x(s-1/2)$. As a result we obtain

$$\sigma(s) = \psi_q(s-a)\varphi_q(s-\gamma+a)\frac{(1+\mu)+(1-\mu)q^{s+a}}{q^{1/2}+q^{-1/2}} \ ,$$

$$\sigma(s)+\tau(s)\Delta x\left(s-\tfrac{1}{2}\right) = \varphi_q(s+a)\psi_q(s+\gamma-a)\frac{(1+\mu)-(1-\mu)q^{-(s-a)}}{q^{1/2}+q^{-1/2}} \ .$$

By assuming

$$\frac{1-\mu}{1+\mu} = q^{-\delta} \quad \left(\delta = \delta(q)\right) \, ,$$

we obtain an equation for $\varrho(s)$ in the form

$$\frac{\varrho(s+1)}{\varrho(s)} = \frac{q-1}{q+1}q^{-(s+1/2)}\frac{\varphi_q(s+a)\psi_q(s+\gamma-a)\psi_q(s+\delta-a)}{\psi_q(s+1-a)\varphi_q(s-\gamma+a+1)\varphi_q(s+a+1-\delta)} \ .$$

Using (3.6.14) and (3.6.15) yields

$$\varrho(s) = \left(\frac{q-1}{q+1}\right)^s q^{-s^2/2}\frac{\tilde{\Pi}_q(s+a)\tilde{\Gamma}_q(s+\gamma-a)\tilde{\Gamma}_q(s+\delta-a)}{\tilde{\Gamma}_q(s+1-a)\tilde{\Pi}_q(s-\gamma+a+1)\tilde{\Pi}_q(s+a+1-\delta)} \ .$$

$$(3.6.29)$$

3) *Analogs of the Kravchuk polynomials on the lattice* $x(s) = \sinh(2\omega s)$. For the Kravchuk polynomials $k_n^{(p)}(s-a)$ $(a \leq s \leq b-1, \ b-a = N+1)$ we have

$$\sigma(s) = s - a, \quad \sigma(s) + \tau(s) = \mu(b - 1 - s), \quad \mu = \frac{p}{1 - p}.$$

By using the same argument as for the analogs of the Meixner polynomials we can choose the functions $\sigma(s)$ and $\sigma(s) + \tau(s)\Delta x \left(s - \frac{1}{2}\right)$ in the form

$$\sigma(s) = \psi_q(s - a)\varphi_q(s + b - 1)F_1(s, \mu, q),$$

$$\sigma(s) + \tau(s)\Delta x \left(s - \frac{1}{2}\right) = \varphi_q(s + a)\psi_q(b - 1 - s)F_2(s, \mu, q),$$

$$F_1(s, \mu, q) = \begin{cases} \dfrac{(1 - \mu) + (1 + \mu)q^{s+b-1}}{q^{1/2} + q^{-1/2}} & \left(0 < \mu < 1, \text{ i.e. } 0 < p < \frac{1}{2}\right), \\ q^s & \left(\mu = 1, \text{ i.e. } p = \frac{1}{2}\right), \\ \dfrac{(\mu + 1)q^{s-a} - (\mu - 1)}{q^{1/2} + q^{-1/2}} & \left(\mu > 1, \text{ i.e. } \frac{1}{2} < p < 1\right); \end{cases}$$

$$F_2(s, \mu, q) = \begin{cases} \dfrac{(1 + \mu)q^{b-1-s} - (1 - \mu)}{q^{1/2} + q^{-1/2}} & (0 < \mu < 1), \\ q^{-s} & (\mu = 1), \\ \dfrac{(\mu + 1)q^{-(s+a)} + (\mu - 1)}{q^{1/2} + q^{-1/2}} & (\mu > 1). \end{cases}$$

By solving Eq. (3.5.1) we obtain

a) $\quad \varrho(s) = \left(\dfrac{q - 1}{q + 1}\right)^s$

$$\times \frac{q^{-s^2/2}\tilde{\Pi}_q(s + a)}{\tilde{\Gamma}_q(b - s)\tilde{\Gamma}_q(b - s + \delta)\tilde{\Gamma}_q(s - a + 1)\tilde{\Pi}_q(s + b)\tilde{\Pi}_q(s + b + \delta)}$$

$$\tag{3.6.30}$$

$$\left(0 < \mu < 1, \; q^\delta = \frac{1 + \mu}{1 - \mu} = \frac{1}{1 - 2p}\right);$$

b) $\quad \varrho(s) = \dfrac{q^{-s^2}\tilde{\Pi}_q(s + a)}{\tilde{\Gamma}_q(b - s)\tilde{\Gamma}_q(s - a + 1)\tilde{\Pi}_q(s + b)} \quad (\mu = 1); \tag{3.6.31}$

c) $\quad \varrho(s) = \left(\dfrac{q + 1}{q - 1}\right)^s$

$$\frac{q^{-s^2/2}\tilde{\Pi}_q(s + \tilde{a})\tilde{\Pi}_q(s + a - \delta)}{\tilde{\Gamma}_q(b - s)\tilde{\Gamma}_q(s - a + 1)\tilde{\Pi}_q(s + b)\tilde{\Gamma}_q(s - a + \delta + 1)}$$

$$\tag{3.6.32}$$

$$\left(\mu > 1, \; q^\delta = \frac{\mu + 1}{\mu - 1} = \frac{1}{2p - 1}\right).$$

4) *Analogs of the Charlier polynomials on the lattice* $x(s) = \sinh(2\omega s)$. For The Charlier polynomials $c_n^{(\mu)}(s - a)$ $(s \geq a)$ we have

$$\sigma(s) = s - a, \quad \sigma(s) + \tau(s) = \mu.$$

We shall choose the functions $\sigma(s)$ and $\sigma(s) + \tau(s)\Delta x \left(s - \frac{1}{2}\right)$ in the form

$$\sigma(s) = q^{(s-a)/2}\psi_q(s-a)F_1(s,\mu,q)\,,$$

$$\sigma(s) + \tau(s)\Delta x\left(s-\tfrac{1}{2}\right) = q^{-(s+a)/2}\varphi_q(s+a)F_2(s,\mu,q)\,,$$

$$F_1(s,\mu,q) = \frac{q^s + q^{-s} + 2 + 2\mu(q-1)q^{-s}}{(q^{1/2} + q^{-1/2})^2}\,,$$

$$\begin{aligned}
F_2(s,\mu,q) &= -\frac{q^{1/2} + q^{-1/2}}{q^{1/2} - q^{-1/2}}F_1\left(-s - \frac{i\pi}{\ln q},\mu,q\right)\\
&= \frac{q+1}{q-1}\frac{q^s + q^{-s} - 2 + 2\mu(q-1)q^s}{(q^{1/2} + q^{-1/2})^2}\,.
\end{aligned}$$

In order to rewrite (3.5.1) in a simpler form we transform the expression for $F_1(s,\mu,q)$ into

$$\begin{aligned}
F_1(s,\mu,q) &= \frac{q^{s/2} + \left(1 + i\sqrt{2\mu(q-1)}\right)q^{-s/2}}{q^{1/2} + q^{-1/2}}\frac{q^{s/2} + \left(1 - i\sqrt{2\mu(q-1)}\right)q^{-s/2}}{q^{1/2} + q^{-1/2}}\\
&= q^{(\alpha+i\beta)/2}\varphi_q(s - \alpha - i\beta)q^{(\alpha-i\beta)/2}\varphi_q(s - \alpha + i\beta)\\
&= q^{\alpha}|\varphi_q(s - \alpha + i\beta)|^2\,,
\end{aligned}$$

where

$$1 \pm i\sqrt{2\mu(q-1)} = q^{\alpha \pm i\beta}\,.$$

By using the connection between the functions $F_2(s,\mu,q)$ and $F_1(s,\mu,q)$ we obtain

$$F_2(s,\mu,q) = \frac{q-1}{q+1}q^{\alpha}|\psi_q(s + \alpha + i\beta)|^2\,.$$

Solving Eq. (3.5.1) yields

$$\varrho(s) = \left(\frac{q-1}{q+1}\right)^s q^{-s^2/2}\frac{\tilde{\Pi}_q(s+a)|\tilde{\Gamma}_q(s+\alpha+i\beta)|^2}{\tilde{\Gamma}_q(s-a+1)|\tilde{\Pi}_q(s+1-\alpha+i\beta)|^2}\,. \tag{3.6.33}$$

In order for the polynomials corresponding to these forms of $\varrho(s)$ to become the analogous polynomials on the lattice $x(s) = s$ when $q \to 1$ it is enough to use the same values of the constants B_n as on the lattice $x(s) = q^s$ in the Rodrigues formula.

3.6.2. The q-Analogs of the Racah and Dual Hahn Polynomials on the Lattices $x(s) = \cosh(2\omega s)$ and $x(s) = \cos 2\omega s$.

3.6.2.1. *The lattice $x(s) = \cosh(2\omega s)$.* We suppose that $q = e^{2\omega}$; then

$$x(s) = \tfrac{1}{2}\left(q^s + q^{-s}\right)\,,\quad \Delta x\left(s-\tfrac{1}{2}\right) = \tfrac{1}{2}\left(q^{1/2} - q^{-1/2}\right)\left(q^s - q^{-s}\right)\,.$$

Since

$$x(-s) = x(s)\,,\quad \Delta x\left(t-\tfrac{1}{2}\right)\Big|_{t=-s} = -\Delta x\left(s-\tfrac{1}{2}\right)\,,$$

then according to (3.1.22) and (3.1.23) we have

$$\sigma(s) + \tau(s)\Delta x\left(s - \tfrac{1}{2}\right) = \sigma(-s) .$$

From this,

$$\tilde{\sigma}[x(s)] = \frac{\sigma(s) + \sigma(-s)}{2} , \tag{3.6.34}$$

$$\tilde{\tau}[x(s)] = \frac{\sigma(-s) - \sigma(s)}{\Delta x(s - 1/2)} . \tag{3.6.35}$$

By virtue of (3.1.22)

$$\sigma(s) = q^{-2s} p_4\left(q^s\right) , \tag{3.6.36}$$

where $p_4(q^s)$ is a polynomial of fourth degree in q^s. By using (3.6.34) and (3.6.35), we can easily verify by means of (3.6.36) that the functions $\tilde{\tau}[x(s)]$ and $\tilde{\sigma}[x(s)]$ are polynomials of at most first and second degrees, respectively, in $x(s) = (q^s + q^{-s})/2$, for an arbitrary form of the polynomial $p_4(q^s)$.

Since

$$2\frac{x(s + 1/2) - 1}{(q - 1)^2} - \frac{1}{4} \to s(s + 1) \quad \text{as} \quad q \to 1 ,$$

then on the lattice $x(s) = \cosh(2\omega s)$ we may obtain analogs of the polynomials on the square lattice $x(s) = s(s + 1)$ if, starting from the expressions for $\sigma(s)$ and $\sigma(s) + \tau(s)\Delta x(s - 1/2)$ on the lattice $s(s + 1)$, we replace s by $s - 1/2$, a by $a - 1/2$ and b by $b - 1/2$.

1) *Analogs of the Racah polynomials* $u_n^{(\alpha,\beta)}(x)$ *on the lattice* $x(s) = \cosh(2\omega s)$. For the Racah polynomials under the transformation $s \to s - 1/2$ we have

$$\sigma(s) = (s - a)(s + b - 1)(s + a - \beta - 1)(b + \alpha - s) .$$

For their analogs we choose the functions $\sigma(s)$ and $\sigma(s) + \tau(s)\Delta x(s - 1/2)$ in the form

$$\sigma(s) = \psi_q(s - a)\psi_q(s + b - 1)\psi_q(s + a - \beta - 1)\psi_q(b + \alpha - s) ,$$

$$\sigma(s) + \tau(s)\Delta x\left(s - \tfrac{1}{2}\right) = \sigma(-s)$$

$$= \psi_q(s + a)\psi_q(b - s - 1)\psi_q(s - a + \beta + 1)\psi_q(b + \alpha + s) .$$

Solving Eq. (3.5.1) yields

$$\varrho(s) = \frac{\tilde{\Gamma}_q(s + a)\tilde{\Gamma}_q(s - a + \beta + 1)\tilde{\Gamma}_q(s + b + \alpha)\tilde{\Gamma}_q(b + \alpha - s)}{\tilde{\Gamma}_q(s - a + 1)\tilde{\Gamma}_q(b - s)\tilde{\Gamma}_q(s + b)\tilde{\Gamma}_q(s + a - \beta)} . \tag{3.6.37}$$

2) *Analogs of the dual Hahn polynomials on the lattice* $x(s) = \cosh(2\omega s)$. For the dual Hahn polynomials we obtain, after the transformation $s \to s - 1/2$,

$$\sigma(s) = (s - a)(s + b - 1)\left(s - c - \tfrac{1}{2}\right) .$$

For their analogs we choose the functions $\sigma(s)$ and $\sigma(s) + \tau(s)\Delta x(s - 1/2)$ in the form

$$\sigma(s) = \psi_q(s - a)\psi_q(s + b - 1)\psi_q\left(s - c - \tfrac{1}{2}\right)\varphi_q(s - a) ,$$

$$\sigma(s) + \tau(s)\Delta x\left(s - \tfrac{1}{2}\right) = \psi_q(s + a)\psi_q(b - s - 1)\psi_q(s + c + 1)\varphi_q(s + a) .$$

Solving Eq. (3.5.1) yields

$$\varrho(s) = \frac{\tilde{\Gamma}_q(s + a)\tilde{\Pi}_q(s + a)\tilde{\Gamma}_q(s + c + 1/2)}{\tilde{\Gamma}_q(s - a + 1)\tilde{\Pi}_q(s - a + 1)\tilde{\Gamma}_q(s + b)\tilde{\Gamma}_q(b - s)\tilde{\Gamma}_q(s - c + 1/2)} .$$

$$(3.6.38)$$

Since

$$\frac{\Delta x(s - 1/2)}{(q - 1)^2} \to s \quad \text{as} \quad q \to 1 ,$$

which coincides with $\tfrac{1}{2}\Delta x(s - 1/2)$ on the lattice

$$x(s) = [t(t + 1) - \tfrac{1}{4}]|_{t=s-1/2} ,$$

in both cases we should take

$$B_n = \frac{(-1)^n}{\tilde{\Gamma}_q(n + 1)}\left(\frac{q - 1}{q + 1}\right)^{2n} .$$

3.6.2.2. *The lattice $x(s) = \cos(2\omega s)$.* In order to obtain expressions for the weight function with which the analogs of the Racah polynomials and the dual Hahn polynomials are orthogonal on the lattice $x(s) = \cos(2\omega s)$, it is natural to replace, in the formulas for the lattice $x(s) = \cosh(2\omega s)$, the parameter ω by $i\omega$,

$$\psi_q(s) = \frac{\sinh \omega s}{\sinh \omega}\ \left(q = e^{2\omega}\right) \quad \text{by} \quad \psi_q(s) = \frac{\sin \omega s}{\sin \omega}\ \left(q = e^{2i\omega}\right) ,$$

$$\varphi_q(s) = \frac{\cosh \omega s}{\cosh \omega} \quad \text{by} \quad \varphi_q(s) = \frac{\cos \omega s}{\cos \omega} ,$$

and then to use the definitions of the functions $\tilde{\Gamma}_q(s)$, $\tilde{\Pi}_q(s)$ for $q = e^{2i\omega}$, when $|q| = 1$:

$$\tilde{\Gamma}_q(s) = \lim_{q' \to q} \tilde{\Gamma}_{q'}(s) , \quad \tilde{\Pi}_q(s) = \lim_{q' \to q} \tilde{\Pi}_{q'}(s) \quad (|q'| \neq 1) .$$

1) *Analogs of the Racah polynomials $u_n^{(\alpha,\beta)}(x)$ on the lattice $x(s) = \cos(2\omega s)$.*

$$\sigma(s) = \psi_q(s - a)\psi_q(s + b - 1)\psi_q(s + a - \beta - 1)\psi_q(b + \alpha - s) ,$$

$$\sigma(s) + \tau(s)\Delta x\left(s - \tfrac{1}{2}\right) = \sigma(-s)$$

$$= \psi_q(s + a)\psi_q(b - s - 1)\psi_q(s - a + \beta + 1)\psi_q(b + \alpha + s) ,$$

$$\varrho(s) = \frac{\tilde{\Gamma}_q(s + a)\tilde{\Gamma}_q(s - a + \beta + 1)\tilde{\Gamma}_q(s + \beta + \alpha)\tilde{\Gamma}_q(b + \alpha - s)}{\tilde{\Gamma}_q(s - a + 1)\tilde{\Gamma}_q(b - s)\tilde{\Gamma}_q(s + b)\tilde{\Gamma}_q(s + a - \beta)} .$$

$$(3.6.39)$$

Table 3.1. Lattice I, $x(s) = s$. Basic Data for the Hahn, Meixner, Kravchuk and Charlier polynomials

No.	$\tilde{y}_n(x)$	(a, b)	$\varrho(s)$	$\sigma(s)$	$\sigma(s) + \tau(s)\varDelta x(s - 1/2)$	B_n
I–H_1	$h_n^{(\alpha,\beta)}(s)$	$(0, N)$	$\dfrac{\Gamma(n + \alpha - s)\Gamma(\beta + 1 + s)}{\Gamma(s + 1)\Gamma(N - s)}$ $(\alpha > -1, \beta > -1)$	$s(N + \alpha - s)$	$(s + \beta + 1)(N - 1 - s)$	$\dfrac{(-1)^n}{n!}$
I–H_2	$\tilde{h}_n^{(\mu,v)}(s)$	$(0, N)$	$\dfrac{1}{\Gamma(s + 1)\Gamma(s + \mu + 1)\Gamma(N + v - s)\Gamma(N - s)}$ $(\mu > -1, v > -1)$	$s(s + \mu)$	$(N + v - 1 - s)(N - 1 - s)$	$\dfrac{1}{n!}$
I–M	$m_n^{(\gamma,\mu)}(s)$	$(0, +\infty)$	$\dfrac{\mu^s \Gamma(s + \gamma)}{\Gamma(s + 1)\Gamma(\gamma)}$ $(\gamma > 0, 0 < \mu < 1)$	s	$\mu(\gamma + s)$	$\dfrac{1}{\mu^n}$
I–K	$k_n^{(p)}(s)$	$(0, N + 1)$	$\dfrac{N!p^s(1 - p)^{N-s}}{\Gamma(s + 1)\Gamma(N + 1 - s)}$ $(0 < p < 1)$	s	$\dfrac{p}{1 - p}(N - s)$	$\dfrac{(-1)^n(1 - p)^n}{n!}$
I–C	$c_n^{(\mu)}(s)$	$(0, +\infty)$	$\dfrac{e^{-\mu}\mu^s}{\Gamma(s + 1)}$ $(\mu > 0)$	s	μ	$\dfrac{1}{\mu^n}$

Table 3.2. Lattice II, $x(s) = s(s + 1)$. Basic Data for the Racah and dual Hahn polynomials

No.	$\tilde{y}_n(x)$	(a, b)	$\varrho(s)$	$\sigma(s)$	$\sigma(s) + \tau(s)\varDelta x(s - 1/2)$	B_n
II–R_1	$u_n^{(\alpha,\beta)}(x)$	(a, b)	$\dfrac{\Gamma(s+a+1)\Gamma(s-a+\beta+1)\Gamma(b+\alpha-s)\Gamma(b+\alpha+s+1)}{\Gamma(s+a-\beta+1)\Gamma(s-a+1)\Gamma(b-s)\Gamma(b+s+1)}$ $(-1/2 < a < b, \alpha > -1, -1 < \beta < 2a+1, b = a+N)$	$(s-a)(s+b)(s+a-\beta)(b+\alpha-s)$	$(s+a+1)(s+1-a+\beta)$ $\times (b-s-1)(b+\alpha+s+1)$	$\dfrac{(-1)^n}{n!}$
II–R_2	$\tilde{u}_n^{(c,d)}(x)$	(a, b)	$\dfrac{\Gamma(s+a+1)\Gamma(s+c+1)}{\Gamma(s-a+1)\Gamma(s-c+1)\Gamma(s+d+1)\Gamma(s+b+1)\Gamma(b-s)\Gamma(d-s)}$ $(-1/2 < a < b < 1+d, |c| < 1+a, b = a+N)$	$(s-a)(s+b)(s-c)(s+d)$	$(s+a+1)(s+c+1)$ $\times (b-s-1)(d-s-1)$	$\dfrac{1}{n!}$
II–H_d	$w_n^{(c)}(x)$	(a, b)	$\dfrac{\Gamma(s+a+1)\Gamma(s+c+1)}{\Gamma(s-a+1)\Gamma(s-c+1)\Gamma(b-s)\Gamma(b+s+1)}$ $(-1/2 < a < b, |c| < a+1, b = a+N)$	$(s-a)(s+b)(s-c)$	$(s+a+1)(s+c+1)(b-s-1)$	$\dfrac{(-1)^n}{n!}$

Table 3.3. Lattice III, $x(s) = q^s$, $q = e^{2\omega}$. Basic Data for the q-analogs of the Hahn, Meixner, Kravchuk and Charlier polynomials

No.	$\tilde{y}_n(x)$	(a, b)	$\varrho(s)$	$\sigma(s)$	$\sigma(s) + \tau(s)\Delta x(s - 1/2)$	B_n
III–H_1	$h_n^{(\alpha,\beta)}(x, q)$	$(0, N)$	$q^{s(\alpha+\beta)/2} \dfrac{\tilde{\Gamma}_q(s + \beta + 1)\tilde{\Gamma}_q(N + \alpha - s)}{\tilde{\Gamma}_q(s + 1)\tilde{\Gamma}_q(N - s)}$ $(\alpha > -1, \beta > -1)$	$q^{s/2}\Psi_q(s)q^{(s-N-\alpha)/2}$ $\times \Psi_q(N + \alpha - s)$	$q^{(s+\beta+1)/2}\Psi_q(s + \beta + 1)$ $\times q^{(s-N+1)/2}\Psi_q(N - 1 - s)$	$\dfrac{(1 - q)^n}{n!}$
III–H_2	$\tilde{h}_n^{(\mu,\nu)}(x, q)$	$(0, N)$	$\dfrac{q^{-(N+(\mu+\nu)/2)s}}{\tilde{\Gamma}_q(N + \nu - s)\tilde{\Gamma}_q(N - s)\tilde{\Gamma}_q(s + 1)\tilde{\Gamma}_q(s + \mu + 1)}$ $(\mu > -1, \nu > -1)$	$q^{s/2}\Psi_q(s)q^{(s+\mu)/2}\Psi_q(s + \mu)$	$q^{(s-N-\nu+1)/2}$ $\times \Psi_q(N + \nu - 1 - s)$ $\times q^{(s-N+1)/2}\Psi_q(N-1-s)$	$\dfrac{(q - 1)^n}{n!}$
III–M	$m_n^{(\gamma,\mu)}(x, q)$	$(0, +\infty)$	$q^{s(\gamma-1)/2} \dfrac{\mu^s\tilde{\Gamma}_q(s + \gamma)}{\tilde{\Gamma}_q(s + 1)\tilde{\Gamma}_q(\gamma)}$ $(0 < q < 1, 0 < \mu < 1, \gamma > 0)$	$q^{s/2}\Psi_q(s)q^{s-1}$	$\mu q^{(s+\gamma)/2}\Psi_q(s + \gamma)q^s$	$\left(\dfrac{q - 1}{\mu}\right)^n$
III–K	$k_n^{(p)}(x, q)$	$(0, N + 1)$	$\dfrac{\tilde{\Gamma}_q(N + 1)p^s(1 - p)^{N-s}q^{-(N+1)s/2}}{\tilde{\Gamma}_q(s + 1)\tilde{\Gamma}_q(N - s + 1)}$ $(0 < p < 1)$	$q^{s/2}\Psi_q(s)q^{s-1}$	$\dfrac{p}{1 - p}q^{(s-N)/2}\Psi_q(N - s)q^s$	$\dfrac{(1 - p)^n(1 - q)^n}{n!}$
III–C	$c_n^{(\mu)}(x, q)$	$(0, +\infty)$	$q^{-s(s+1)/4} \dfrac{e^{-\mu}\mu^s}{\tilde{\Gamma}_q(s + 1)}$ $\begin{pmatrix} 0 < q < 1, & \mu < \dfrac{1}{(1 - q)q^{1/2}}; \\ q > 1, & \mu > 0 \end{pmatrix}$	$q^{s/2}\Psi_q(s)q^{s-1}$	μq^s	$\left(\dfrac{q - 1}{\mu}\right)^n$

Table 3.4. Lattice IV, $x(s) = \sinh(2\omega s) = \frac{1}{2}(q^s - q^{-s})$, $q = e^{2\omega}$. Basic Data for the q-analogs of the Hahn, Meixner, Kravchuk and Charlier polynomials

No.	(a, b)	$\varrho(s)$
IV–H_1	(a, b)	$\dfrac{\tilde{\Pi}_q(s + a)\tilde{\Gamma}_q(s - a + \beta + 1)\tilde{\Pi}_q(b + \alpha + s)\tilde{\Gamma}_q(b + \alpha - s)}{\tilde{\Gamma}_q(s - a + 1)\tilde{\Gamma}_q(b - s)\tilde{\Pi}_q(s + a - \beta)\tilde{\Pi}_q(s + b)}$ $(\alpha > -1, \beta > -1, b = a + N)$
IV–H_2	(a, b)	$\dfrac{\tilde{\Pi}_q(s + a)\tilde{\Pi}_q(s + a - \mu)}{\tilde{\Gamma}_q(s + 1 - a)\tilde{\Gamma}_q(b - s)\tilde{\Gamma}_q(b + v - s)\tilde{\Gamma}_q(s - a + \mu + 1)\tilde{\Pi}_q(s + b)\tilde{\Pi}_q(s + b + v)}$ $(\mu > -1, v > -1, b = a + N)$
IV–M	$(a, +\infty)$	$\left(\dfrac{q - 1}{q + 1}\right)^s q^{-s^2/2} \dfrac{\tilde{\Pi}_q(s + a)\tilde{\Gamma}_q(s + \gamma - a)\tilde{\Gamma}_q(s + \delta - a)}{\tilde{\Gamma}_q(s + 1 - a)\tilde{\Pi}_q(s + a + 1 - \gamma)\tilde{\Pi}_q(s + a + 1 - \delta)}$ $\left(0 < \mu < 1, \gamma > 0, q^{-\delta} = \dfrac{1 - \mu}{1 + \mu}\right)$
IV–K	(a, b)	$\dfrac{\tilde{\Pi}_q(s + a)F_1(s, q)}{\tilde{\Gamma}_q(s - a + 1)\tilde{\Gamma}_q(b - s + 1)\tilde{\Pi}_q(s + b + 1)}$ $(b = a + N + 1),$ $$F_1(s, q) = \begin{cases} \left(\dfrac{q - 1}{q + 1}\right)^s \dfrac{q^{-s^2/2}}{\tilde{\Gamma}_q(b - s + \delta + 1)\tilde{\Pi}_q(s + b + \delta + 1)} \\ \quad (0 < p < 1/2, q^\delta = 1/(1 - 2p)); \\[2mm] q^{-s^2}(p = 1/2); \\[2mm] \left(\dfrac{q + 1}{q - 1}\right)^s \dfrac{q^{-s^2/2}\tilde{\Pi}_q(s + a - \delta)}{\tilde{\Gamma}_q(s - a + \delta + 1)} \\ \quad \left(1/2 < p < 1, q^\delta = \dfrac{1}{2p - 1}\right) \end{cases}$$
IV–C	$(a, +\infty)$	$\left(\dfrac{q - 1}{q + 1}\right)^s q^{-s^2/2} \dfrac{\tilde{\Pi}_q(s + a)|\tilde{\Gamma}_q(s + \alpha + i\beta)|^2}{\tilde{\Gamma}_q(s - a + 1)|\tilde{\Pi}_q(s + 1 - \alpha + i\beta)|^2}$ $(q^{\alpha + i\beta} = 1 + i\sqrt{2\mu(q - 1)})$

Table 3.4 (cont).

$\sigma(s)$	$\sigma(s) + \tau(s)\Delta x(s - 1/2)$	B_n
$\Psi_q(s - a)\phi_q(s + a - \beta - 1)$ $\times \Psi_q(b + \alpha - s)\phi_q(b - 1 + s)$	$\phi_q(s + a)\Psi_q(s - a + \beta + 1)$ $\times \phi_q(b + \alpha + s)\Psi_q(b - 1 - s)$	$\dfrac{(1 - q)^n}{n!}$
$\Psi_q(s - a)\phi_q(s + b + v - 1)$ $\times \Psi_q(s - a + \mu)\phi_q(b - 1 + s)$	$\phi_q(s + a)\Psi_q(b + v - 1 - s)$ $\times \phi_q(s + a - \mu)\Psi_q(b - 1 - s)$	$\dfrac{(q - 1)^n}{n!}$
$\Psi_q(s - a)\phi_q(s - \gamma + a)$ $\times \dfrac{(1 + \mu) + (1 - \mu)q^{s+a}}{q^{1/2} + q^{-1/2}}$	$\phi_q(s + a)\Psi_q(s + \gamma - a)$ $\times \dfrac{(1 + \mu) - (1 - \mu)q^{-(s-a)}}{q^{1/2} + q^{-1/2}}$	$\dfrac{(q - 1)^n}{\mu^n}$
$\Psi_q(s - a)\phi_q(s + b)F_2(s, q)$	$\phi_q(s + a)\Psi_q(b - s)F_3(s, q)$	$\dfrac{(1 - p)^n(1 - q)^n}{n!}$
$F_2(s, q) = \begin{cases} \dfrac{(1-\mu)+(1+\mu)q^{s+b}}{q^{1/2}+q^{-1/2}} \\ \left(0 < p < 1/2,\ \mu = \dfrac{p}{1-p}\right); \\ q^s\ (p = 1/2); \\ \dfrac{(\mu+1)q^{(s-a)}-(\mu-1)}{q^{1/2}+q^{-1/2}} \\ \left(1/2 < p < 1,\ \mu = \dfrac{p}{1-p}\right) \end{cases}$	$F_3(s, q) = \begin{cases} \dfrac{(1+\mu)q^{b-s}-(1-\mu)}{q^{1/2}+q^{-1/2}} \\ \left(0 < p < 1/2,\ \mu = \dfrac{p}{1-p}\right); \\ q^{-s}\ (p = 1/2); \\ \dfrac{(\mu+1)q^{-(s+a)}+(\mu-1)}{q^{1/2}+q^{-1/2}} \\ \left(1/2 < p < 1,\ \mu = \dfrac{p}{1-p}\right) \end{cases}$	
$q^{(s-a)/2}\Psi_q(s - a)$ $\times \dfrac{q^s + q^{-s} + 2 + 2\mu(q-1)q^{-s}}{(q^{1/2} + q^{-1/2})^2}$	$q^{-(s+a)/2}\phi_q(s + a)\left(\dfrac{q+1}{q-1}\right)$ $\times \dfrac{q^s + q^{-s} - 2 + 2\mu(q-1)q^s}{(q^{1/2} + q^{-1/2})^2}$	$\left(\dfrac{q-1}{\mu}\right)^n$

Table 3.5. Lattice V, $x(s) = \cosh(2\omega s)$, $q = e^{2\omega}$ and lattice VI, $x(s) = \cos(2\omega s)$, $q = e^{2i\omega}$. Basic Data for the q-analogs of the Racah and dual Hahn polynomials

No.	(a, b)	$\varrho(s)$
V–R_1	(a, b)	$\dfrac{\tilde{\Gamma}_q(s + a)\tilde{\Gamma}_q(s - a + \beta + 1)\tilde{\Gamma}_q(s + b + \alpha)\tilde{\Gamma}_q(b + \alpha - s)}{\tilde{\Gamma}_q(s - a + 1)\tilde{\Gamma}_q(b - s)\tilde{\Gamma}_q(s + b)\tilde{\Gamma}_q(s + a - \beta)}$ $(\alpha > -1, \beta > -1, a > 0, b = a + N)$
V–H_d	(a, b)	$\dfrac{\tilde{\Gamma}_q(s + a)\tilde{\Pi}_q(s + a)\tilde{\Gamma}_q(s + c + 1/2)}{\tilde{\Gamma}_q(s - a + 1)\tilde{\Pi}_q(s - a + 1)\tilde{\Gamma}_q(s + b)\tilde{\Gamma}_q(s - c + 1/2)\tilde{\Gamma}_q(b - s)}$ $(a > 0, |c| < a + 1/2, b = a + N)$
VI–R_1	(a, b)	$\dfrac{\tilde{\Gamma}_q(s + a)\tilde{\Gamma}_q(s - a + \beta + 1)\tilde{\Gamma}_q(s + b + \alpha)\tilde{\Gamma}_q(b + \alpha - s)}{\tilde{\Gamma}_q(s - a + 1)\tilde{\Gamma}_q(b - s)\tilde{\Gamma}_q(s + b)\tilde{\Gamma}_q(s + a - \beta)}$ $(\alpha > -1, -1 < \beta < 2a, b = a + N, a > 0)$
VI–H_d	(a, b)	$\dfrac{\tilde{\Gamma}_q(s + a)\tilde{\Pi}_q(s + a)\tilde{\Gamma}_q(s + c + 1/2)}{\tilde{\Gamma}_q(s - a + 1)\tilde{\Pi}_q(s - a + 1)\tilde{\Gamma}_q(s + b)\tilde{\Gamma}_q(b - s)\tilde{\Gamma}_q(s - c + 1/2)}$ $(a > 0, |c| < a + 1/2, b = a + N)$

2) *Analogs of the dual Hahn polynomials on the lattice* $x(s) = \cos(2\omega s)$.

$$\sigma(s) = \psi_q(s - a)\psi_q(s + b - 1)\psi_q\left(s - c - \tfrac{1}{2}\right)\varphi_q(s - a),$$

$$\sigma(s) + \tau(s)\Delta x\left(s - \tfrac{1}{2}\right)$$
$$= \psi_q(s + a)\psi_q(b - s - 1)\psi_q\left(s + c + \tfrac{1}{2}\right)\varphi_q(s + a),$$

$$\varrho(s) = \frac{\tilde{\Gamma}_q(s + a)\tilde{\Gamma}_q(s + c + 1/2)\tilde{\Pi}_q(s + a)}{\tilde{\Gamma}_q(s - a + 1)\tilde{\Gamma}_q(b - s)\tilde{\Gamma}_q(s + b)\tilde{\Pi}_q(s - a + 1)\tilde{\Gamma}_q(s - c + 1/2)}.$$

$$(3.6.40)$$

In both cases we should set $B_n = 2^n \omega^{2n}/\tilde{\Gamma}_q(n + 1)$.

For the lattice $x(s) = \cos(2\omega s)$ the conditions (3.3.5) may fail to be satisfied with certain values of a, b, ω. If, for example, $\varrho(s)\Delta x(s - 1/2) < 0$ with $a \leq s \leq b - 1$, then we must replace $\varrho(s_i)\Delta x(s_i - 1/2)$ by the absolute value $|\varrho(s_i)\Delta x(s_i - 1/2)|$ in the orthogonality relation (3.3.4), so that the squared norms d_n^2 of the polynomials will be positive.

Remark. In Sect. 3.11.3 we shall consider the general method of constructing q-analogs with the aid of q-hypergeometric series.

Table 3.5 (cont).

$\sigma(s)$	$\sigma(s) + \tau(s)\Delta x(s - 1/2)$	B_n
$\Psi_q(s - a)\psi_q(s + b - 1)$ $\times \Psi_q(s + a - \beta - 1)\Psi_q(b + \alpha - s)$	$\psi_q(s + a)\Psi_q(b - s - 1)$ $\times \Psi_q(s - a + \beta + 1)\Psi_q(b + \alpha + s)$	$\dfrac{(-1)^n}{2^n n!}$ $\times (q - 1)^{2n}$
$\Psi_q(s - a)\Psi_q(s + b - 1)$ $\times \Psi_q(s - c - 1/2)\phi_q(s - a)$	$\Psi_q(s + a)\Psi_q(b - s - 1)$ $\times \psi_q(s + c + 1/2)\phi_q(s + a)$	$\dfrac{(-1)^n}{2^n n!}$ $\times (q - 1)^{2n}$
$\Psi_q(s - a)\Psi_q(s + b - 1)$ $\times \Psi_q(s + a - \beta - 1)\Psi_q(b + \alpha - s)$	$\Psi_q(s + a)\Psi_q(b - s - 1)$ $\times \psi_q(s - a + \beta + 1)\Psi_q(b + \alpha + s)$	$\dfrac{2^n \omega^{2n}}{n!}$
$\Psi_q(s - a)\Psi_q(s + b - 1)$ $\times \Psi_q(s - c - 1/2)\phi_q(s - a)$	$\Psi_q(s + a)\Psi_q(b - s - 1)$ $\times \psi_q(s + c + 1/2)\phi_q(s + a)$	$\dfrac{2^n \omega^{2n}}{n!}$

3.6.3 Tables of Basic Data for q-Analogs

In conclusion we give summary tables which contain the basic data about the Hahn, Meixner, Kravchuk, Charlier, Racah, dual Hahn polynomials and their q-analogs. We use the following notations: first we give the lattice number (I–VI) in (3.4.5–10) and then the conditional notation of the polynomial q-analog. The notations H_1, H_2, H_d correspond to the Hahn polynomials $h_n^{(\alpha,\beta)}(x)$, $\tilde{h}_n^{(\mu,\nu)}(x)$ and the dual Hahn polynomials $w_n^{(c)}(x)$. The notations M, K, C, R_1, R_2 correspond to the Meixner, Kravchuk, Charlier and Racah polynomials: $m_n^{(\gamma,\mu)}(x)$, $k_n^{(p)}(x)$, $c_n^{(\mu)}(x)$, $u_n^{(\alpha,\beta)}(x)$ and $\tilde{u}_n^{(c,d)}(x)$. For example, the systems of polynomials defined by the weight-functions (3.6.16–20) will be denoted by $III - H_1$, $III - H_2$, $III - M$, $III - K$, $III - C$.

3.7 Calculation of the Leading Coefficients and Squared Norms. Tables of Data

We obtain the basic data for the classical orthogonal polynomials of a discrete variable on nonuniform lattices, supposing that the functions $\sigma(s)$, $\tau(s)$ and $\varrho(s)$ are given for each form of the lattices (3.4.5–10).

3.7.1. For calculating the coefficients a_n, b_n in the expansion

$$\tilde{y}_n(x) = a_n x^n + b_n x^{n-1} + \dots$$

we use the Rodrigues formula (3.2.18) with $k = n - 1$. In accordance with (3.2.11) and (3.2.10) we have

$$\begin{aligned}
v_{n-1,n}(s) &= \frac{A_{n-1,n} B_n}{\varrho_{n-1}(s)} \frac{\nabla}{\nabla x_n(s)} \left[\varrho_n(s)\right] \\
&= \frac{A_{n-1,n} B_n}{\varrho_{n-1}(s)} \frac{\nabla}{\nabla x_n(s)} \left[\sigma(s+1)\varrho_{n-1}(s+1)\right] \\
&= \frac{A_{n-1,n} B_n}{\varrho_{n-1}(s)} \frac{\Delta}{\Delta x_{n-1}(s - 1/2)} \left[\sigma(s)\varrho_{n-1}(s)\right] \\
&= A_{n-1,n} B_n \tilde{\tau}_{n-1} \left[x_{n-1}(s)\right] .
\end{aligned}$$

By Eq. (3.1.29) the first-degree polynomial $\tilde{\tau}_{n-1}[x_{n-1}(s)]$ can be expressed in terms of $\tau(s)$ and $\sigma(s)$. On the other hand, by virtue of (3.2.17) we have

$$v_{n-1,n}(s) = \Delta^{(n-1)}\tilde{y}_n[x(s)] = a_n \Delta^{(n-1)}\left[x^n(s)\right] + b_n \Delta^{(n-1)}\left[x^{n-1}(s)\right] .$$

The operator $\Delta^{(k)}$ carries every polynomial of degree n in $x(s)$ to a polynomial of degree $n - k$ in $x_k(s)$. Hence

$$\Delta^{(n-1)}\left[x^n(s)\right] = \alpha_n \left[x_{n-1}(s) + \beta_n\right] \quad (n = 2, 3, \dots) . \tag{3.7.1}$$

Equating coefficients for different powers of $x_{n-1}(s)$ in the equality

$$a_n \Delta^{(n-1)}\left[x^n(s)\right] + b_n \Delta^{(n-1)}\left[x^{n-1}(s)\right] = A_{n-1,n} B_n \tilde{\tau}_{n-1}\left[x_{n-1}(s)\right] ,$$

which yields

$$a_n \alpha_n \left[x_{n-1}(s) + \beta_n\right] + b_n \alpha_{n-1} = A_{n-1,n} B_n \tilde{\tau}_{n-1}\left[x_{n-1}(s)\right] ,$$

we can find a_n and b_n in the form

$$a_n = \frac{A_{n-1,n} B_n}{\alpha_n} \tilde{\tau}'_{n-1} , \qquad \frac{b_n}{a_n} = \frac{\alpha_n}{\alpha_{n-1}} \left(\frac{\tilde{\tau}_{n-1}(0)}{\tilde{\tau}'_{n-1}} - \beta_n\right) . \tag{3.7.2}$$

Let us determine the coefficients α_n and β_n. We have

$$\begin{aligned}
\alpha_{n+1}\left[x_n(s) + \beta_{n+1}\right] &= \Delta^{(n)}\left[x^{n+1}(s)\right] \\
&= \frac{\Delta}{\Delta x_{n-1}(s)} \frac{\Delta}{\Delta x_{n-2}(s)} \cdots \frac{\Delta}{\Delta x_1(s)} \left\{\frac{\Delta}{\Delta x(s)}\left[x^{n+1}(s)\right]\right\} \\
&= \frac{\Delta}{\Delta x_{n-2}(t)} \cdots \frac{\Delta}{\Delta x_1(t)} \frac{\Delta}{\Delta x(t)}[f(t)] = \Delta^{(n-1)} f(t) ,
\end{aligned}$$

where

$$t = s + \tfrac{1}{2} , \quad f(t) = \frac{\Delta}{\Delta x(s)}\left[x^{n+1}(s)\right] = C_{n+1} x^n(t) + D_{n+1} x^{n-1}(t) + \dots \tag{3.7.3}$$

$(C_n$ and D_n evidently depend on a form of the lattice). From this it follows that

$$\alpha_{n+1}\left[x_{n-1}(t)+\beta_{n+1}\right]=C_{n+1}\alpha_n\left[x_{n-1}(t)+\beta_n\right]$$
$$+\,D_{n+1}\frac{\Delta}{\Delta x_{n-2}(t)}\left\{\alpha_{n-1}\left[x_{n-2}(t)+\beta_{n-1}\right]\right\},$$

which with the given C_n and D_n yields a system of equations for α_n and β_n:

$$\alpha_{n+1}=C_{n+1}\alpha_n\,,$$
$$\alpha_{n+1}\beta_{n+1}=C_{n+1}\alpha_n\beta_n+D_{n+1}\alpha_{n-1}\,. \tag{3.7.4}$$

System (3.7.4) may be represented in the form of separate equations for each value of interest:

$$\alpha_{n+1}=C_{n+1}\alpha_n\,, \tag{3.7.5}$$

$$\beta_{n+1}=\beta_n+\frac{D_{n+1}}{C_nC_{n+1}}\,. \tag{3.7.6}$$

It is not difficult to solve these equations.

In order to find the coefficients C_n and D_n it is convenient to use the relations (3.2.2) and (3.2.3).

a) In the case of the quadratic lattice $x(s)=s(s+1)$, i.e. lattice II [see (3.4.6)] relations (3.2.2) and (3.2.3) have the form

$$\frac{\Delta x^n(s)}{\Delta x(s)}=\left[x\left(s+\tfrac{1}{2}\right)+\tfrac{1}{4}\right]\frac{\Delta x^{n-1}(s)}{\Delta x(s)}+\frac{x^{n-1}(s+1)+x^{n-1}(s)}{2}\,, \tag{3.7.7}$$

$$\frac{x^n(s+1)+x^n(s)}{2}$$
$$=\left[x\left(s+\tfrac{1}{2}\right)+\tfrac{1}{4}\right]\left\{\frac{x^{n-1}(s+1)+x^{n-1}(s)}{2}+\frac{\Delta x^{n-1}(s)}{\Delta x(s)}\right\}\,. \tag{3.7.8}$$

By supposing in accordance with (3.7.3) and (3.2.5) that

$$\frac{\Delta x^n(s)}{\Delta x(s)}=C_nx^{n-1}\left(s+\tfrac{1}{2}\right)+D_nx^{n-2}\left(s+\tfrac{1}{2}\right)+\dots\,,$$
$$\frac{x^n(s+1)+x^n(s)}{2}=A_nx^n\left(s+\tfrac{1}{2}\right)+B_nx^{n-1}\left(s+\tfrac{1}{2}\right)+\dots\,,$$

and equating the coefficients of the same powers of x in relations (3.7.7) and (3.7.8) we obtain

$$C_n=C_{n-1}+A_{n-1}\,,$$
$$D_n=D_{n-1}+C_{n-1}/4+B_{n-1}\,,$$
$$A_n=A_{n-1}\,,$$
$$B_n=A_{n-1}/4+B_{n-1}+C_{n-1}$$
$$\left(A_1=1\,,\ B_1=\tfrac{1}{4}\,,\ C_1=1\,,\ D_1=0\right)\,.$$

From this it is easy to find that

$$A_n = 1, \ C_n = n, \ B_n = \frac{n(2n-1)}{4}, \ D_n = \frac{n(n-1)(2n-1)}{12}.$$

By means of (3.7.5) and (3.7.6) we obtain

$$\frac{\alpha_{n+1}}{\alpha_n} = n+1, \ \beta_{n+1} - \beta_n = \frac{2n+1}{12} \quad (\alpha_1 = 1, \ \beta_1 = 0),$$

whence

$$\alpha_n = n!, \quad \beta_n = \frac{n^2 - 1}{12}.$$

b) In the remaining cases, i.e. for lattices III–VI, we have $x(s) = Aq^s + Bq^{-s}$ (A and B are constants). Therefore

$$x(s+1) + x(s) = 2\alpha x \left(s + \tfrac{1}{2}\right) \quad (\alpha \neq 0);$$

$$x(s)x(s+1) = \left(Aq^s + Bq^{-s}\right)\left(Aq^{s+1} + Bq^{-(s+1)}\right)$$

$$= \left(Aq^{s+1/2} + Bq^{-(s+1/2)}\right)^2 + \text{const} = x^2 \left(s + \tfrac{1}{2}\right) + \text{const},$$

$$[\Delta x(s)]^2 = [x(s+1) + x(s)]^2 - 4x(s+1)x(s)$$

$$= \left[2\alpha x \left(s + \tfrac{1}{2}\right)\right]^2 - 4x^2 \left(s + \tfrac{1}{2}\right) + \text{const}.$$

As a result the relations (3.2.2) and (3.2.3) take the form

$$\frac{\Delta x^n(s)}{\Delta x(s)} = \alpha x \left(s + \tfrac{1}{2}\right) \frac{\Delta x^{n-1}(s)}{\Delta x(s)} + \frac{x^{n-1}(s+1) + x^{n-1}(s)}{2}, \tag{3.7.9}$$

$$\frac{x^n(s+1) + x^n(s)}{2} = \alpha x \left(s + \tfrac{1}{2}\right) \frac{x^{n-1}(s+1) + x^{n-1}(s)}{2}$$

$$+ \left[(\alpha^2 - 1)x^2 \left(s + \tfrac{1}{2}\right) + \text{const}\right] \frac{\Delta x^{n-1}(s)}{\Delta x(s)}. \tag{3.7.10}$$

From these relations we can obtain, by induction, that

$$\frac{\Delta x^n(s)}{\Delta x(s)} = C_n x^{n-1} \left(s + \tfrac{1}{2}\right) + E_n x^{n-3} \left(s + \tfrac{1}{2}\right) + \ldots \tag{3.7.11}$$

$$\frac{x^n(s+1) + x^n(s)}{2} = A_n x^n \left(s + \tfrac{1}{2}\right) + F_n x^{n-2} \left(s + \tfrac{1}{2}\right) + \ldots, \tag{3.7.12}$$

where A_n, C_n, E_n and F_n are constants. Substituting the expansions (3.7.11) and (3.7.12) in (3.7.9) and (3.7.10) yields a linear homogeneous system of first-order difference equations with the constant coefficients for A_n and C_n

$$C_n = \alpha C_{n-1} + A_{n-1}, \quad A_n = \alpha A_{n-1} + (\alpha^2 - 1)C_{n-1},$$

or

$$\begin{pmatrix} C_n \\ A_n \end{pmatrix} = \begin{pmatrix} \alpha & 1 \\ \alpha^2 - 1 & \alpha \end{pmatrix} \begin{pmatrix} C_{n-1} \\ A_{n-1} \end{pmatrix}. \tag{3.7.13}$$

As usual in solving a system of the form $X_n = AX_{n-1}$, where X_n is a column vector and A is a matrix, we seek particular solutions in the form $X_n = \lambda^n X_0$. After substituting $X_n = \lambda^n X_0$ into the equation we obtain $AX_0 = \lambda X_0$. From this it is seen that X_0 is an eigenvector and λ is an eigenvalue of matrix A. In order to find λ we must solve the secular equation $\det(A - \lambda E) = 0$, where E is a unit matrix. In our case

$$\det(A - \lambda E) = \begin{vmatrix} \alpha - \lambda & 1 \\ \alpha^2 - 1 & \alpha - \lambda \end{vmatrix} = \lambda^2 - 2\alpha\lambda + 1 = 0 , \tag{3.7.14}$$

whence $\lambda_{1,2} = \alpha \pm \sqrt{\alpha^2 - 1}$, while $\lambda_1 \lambda_2 = 1$, and $\lambda_1 + \lambda_2 = 2\alpha$. We note that Eq. (3.7.14) coincides with the equation for κ in Sect. 3.1.1, i.e. $\lambda_1 = \kappa_1$, $\lambda_2 = \kappa_2$.

The general solution of system (3.7.13) is a linear combination of partial solutions. Since $C_1 = 1$ and $A_1 = \alpha$ the solution of the difference equations for C_n and A_n has the form

$$C_n = \frac{q^{n/2} - q^{-n/2}}{q^{1/2} - q^{-1/2}} = \psi_q(n) = \frac{\sinh n\omega}{\sinh \omega} ,$$
$$A_n = \tfrac{1}{2}\left(q^{n/2} + q^{-n/2}\right) = \alpha\varphi_q(n) = \cosh n\omega ,$$

where

$$q^{1/2} = e^{\omega} = \kappa_1 , \quad q^{-1/2} = e^{-\omega} = \kappa_2 , \quad \alpha = \cosh \omega .$$

In this case we have $D_n = 0$ in (3.7.3), whence by means of (3.7.5) and (3.7.6) we obtain

$$\frac{\alpha_{n+1}}{\alpha_n} = \psi_q(n + 1) , \quad \beta_{n+1} = \beta_n \ \left(\alpha_1 = 1 , \ \beta_1 = 0\right) .$$

From this, $\alpha_n = C(q)\tilde{\Gamma}_q(n + 1)$, $\beta_n = 0$. Since $\tilde{\Gamma}_q(2) = 1$, $\alpha_1 = 1$, we have $C(q) = 1$, i.e. $\alpha_n = \tilde{\Gamma}_q(n + 1)$. With the aid of α_n and β_n it is easy to find a_n and b_n from (3.7.2).

However, we may avoid solving system (3.7.13) if use the fact that $C_n = \psi_q(n)$ by virtue of (3.2.4a) and $D_n = 0$ by virtue of (3.7.3), (3.7.11). By means of (3.7.6) and (3.7.5) we obtain

$$\beta_{n+1} = \beta_n = \beta_1 = 0 , \quad \alpha_n = \tilde{\Gamma}_q(n + 1) ,$$

whence

$$A_n = C_{n+1} - \alpha C_n = \frac{\kappa_1^{n+1} - \kappa_2^{n+1}}{\kappa_1 - \kappa_2} - \frac{\kappa_1 + \kappa_2}{2} \frac{\kappa_1^n - \kappa_2^n}{\kappa_1 - \kappa_2}$$
$$= \tfrac{1}{2}\left(\kappa_1^n + \kappa_2^n\right) = \alpha\varphi_q(n) = \cosh n\omega .$$

By using formulas (see (3.2.4a), (3.7.12))

$$\frac{\Delta}{\Delta x(s)} \left[x^n(s)\right] = \psi_q(n)x^{n-1} \left(s + \tfrac{1}{2}\right) + \dots ,$$

$$\frac{\nabla}{\nabla x(s)}\left[x^n(s)\right] = \psi_q(n)x^{n-1}\left(s - \tfrac{1}{2}\right) + \ldots ,$$

$$\tfrac{1}{2}\left[x^n(s+1) + x^n(s)\right] = \tfrac{1}{2}\left(q^{n/2} + q^{-n/2}\right)x^n\left(s + \tfrac{1}{2}\right) + \ldots$$

and expansions of polynomials $\tilde{\sigma}[x(s)]$, $\tilde{\tau}[x(s)]$ and $y(s) = \tilde{y}_n[x(s)]$ in powers of $x(s)$, after equating the coefficients of $x^n(s)$ on the left-hand and right-hand sides in (3.1.5) we obtain the following expression for $\lambda = \lambda_n$:

$$\lambda_n = -\tfrac{1}{2}\psi_q(n)\left[\left(q^{\frac{n-1}{2}} + q^{-\frac{n-1}{2}}\right)\tilde{\tau}' + \psi_q(n-1)\tilde{\sigma}''\right] .$$

3.7.2. To determine the squared norm $d_n^2 = d_{0n}^2$ in (3.3.11) we first need to know the connection between d_{kn}^2 and $d_{k+1,n}^2$ (cf. Sect. 2.3.4). By multiplying both sides of Eq. (3.2.8), where we put $v_k(s) = v_{kn}(s)$, $s = s_i$, $\lambda = \lambda_n$, by the product $v_{kn}(s_i)\Delta x(s_i - 1/2)$ and using the equation for summation by parts

$$\sum_{s_i=a}^{b-k-1} f(s_i)\Delta g(s_i) = f(s_i)g(s_i)\Big|_a^{b-k} - \sum_{s_i=a}^{b-k-1} g(s_{i+1})\Delta f(s_i)$$

we obtain

$$\mu_{kn}d_{kn}^2 = -\sum_i v_{kn}(s_i)\Delta\left[\sigma(s_i)\varrho_k(s_i)\frac{\nabla v_{kn}(s_i)}{\nabla x_k(s_i)}\right]$$

$$= -\nabla_{kn}(s_i)\sigma(s_i)\varrho_k(s_i)\frac{\nabla v_{kn}(s_i)}{\nabla x_k(s_i)}\Bigg|_a^{b-k}$$

$$+ \sum_i \sigma(s_i+1)\varrho_k(s_i+1)\left[\frac{\Delta v_{kn}(s_i)}{\Delta x_k(s_i)}\right]^2\Delta x_k(s_i)$$

$$= \sum_i v_{k+1,n}^2(s_i)\varrho_{k+1}(s_i)\Delta x_{k+1}\left(s_i - \tfrac{1}{2}\right) = d_{k+1,n}^2 .$$

The terms evaluated at the limits are zero by virtue of the boundary conditions (3.3.10). From this we successively obtain [cf. (2.3.7)]

$$d_n^2 = d_{0n}^2 = \frac{1}{\mu_{0n}}d_{1n}^2 = \frac{1}{\mu_{0n}}\frac{1}{\mu_{1n}}d_{2n}^2 = \ldots = \frac{d_{nn}^2}{\prod_{k=0}^{n-1}\mu_{kn}} = \frac{v_{nn}^2}{\prod_{k=0}^{n-1}\mu_{kn}}S_n ,$$

where

$$S_n = \sum_{s_i=a}^{b-n-1} \varrho_n(s_i)\Delta x_n\left(s_i - \tfrac{1}{2}\right) .$$

By using (3.2.18) we finally obtain

$$d_n^2 = (-1)^n A_{nn}B_n^2 S_n . \tag{3.7.15}$$

If a and b are finite, the squared norm is calculated very simply. In this case we have, in fact, $b - a = N$, where N is a positive integer. For $n = N - 1$ the sum S_n contains only one summand:

$$S_{N-1} = \varrho_{N-1}(a)\Delta x_{N-1}\left(a - \tfrac{1}{2}\right) . \tag{3.7.16}$$

To determine S_n, when $n < N - 1$, it is sufficient to know how to calculate the ratio S_n/S_{n+1}. For this purpose we transform the expression for S_{n+1}, using the connection between $\varrho_n(s)$ and $\varrho_{n+1}(s)$ and (3.2.10):

$$\varrho_{n+1}(s) = \sigma(s+1)\varrho_n(s+1) = \varrho_n(s)\left[\sigma(s) + \tau_n(s)\Delta x_n\left(s - \tfrac{1}{2}\right)\right] .$$

From this, on the one hand

$$S_{n+1} = \sum_i \sigma(s_i + 1)\varrho_n(s_i + 1)\Delta x_{n+1}\left(s_i - \tfrac{1}{2}\right)$$
$$= \sum_i \sigma(s_i)\varrho_n(s_i)\Delta x_{n+1}\left(s_i - \tfrac{3}{2}\right) .$$

On the other hand,

$$S_{n+1} = \sum_i \varrho_n(s_i)\left[\sigma(s_i) + \tau_n(s_i)\Delta x_n\left(s_i - \tfrac{1}{2}\right)\right]\Delta x_{n+1}\left(s_i - \tfrac{1}{2}\right) .$$

We take half the sum of these expressions and by appealing to (3.1.27, 25) and (3.1.26) we obtain

$$S_{n+1} = \frac{1}{2}\sum_i \varrho_n(s_i)\Delta x_n\left(s_i - \tfrac{1}{2}\right)$$
$$\times \left\{ \tilde{\sigma}_n\left[x_n(s_i)\right]\frac{\Delta x_{n+1}(s_i - 1/2) + \Delta x_{n+1}(s_i - 3/2)}{\Delta x_n(s_i - 1/2)}\right.$$
$$\left. + \tilde{\tau}_n\left[x_n(s_i)\right]\frac{\Delta x_{n+1}(s_i - 1/2) - \Delta x_{n+1}(s_i - 3/2)}{2}\right\} .$$

Using the relations (3.1.13) and (3.1.14) we obtain an expression for S_{n+1} in the form

$$S_{n+1} = \sum_i \varrho_n(s_i)\Delta x_n\left(s_i - \tfrac{1}{2}\right) Q_n\left[x_n(s_i)\right] ,$$

where

$$Q_n\left[x_n(s)\right] = \alpha\tilde{\sigma}_n\left[x_n(s)\right] + \tilde{\tau}_n\left[x_n(s)\right]\left[(\alpha^2 - 1)x_n(s) + (\alpha + 1)\beta\right] \tag{3.7.17}$$

is a polynomial of at most second degree in $x_n(s)$. We decompose the polynomial $Q_n(x_n)$ into powers of the first-degree polynomial $\tilde{\tau}_n(x_n)$:

$$Q_n(x_n) = A\tilde{\tau}_n^2(x_n) + B\tilde{\tau}_n(x_n) + C . \tag{3.7.18}$$

Then $S_{n+1} = S_n^{(1)} + CS_n$, where

$$S_n^{(1)} = \sum_i \left\{ A\tilde{\tau}_n[x_n(s_i)] + B \right\} \tilde{\tau}_n[x_n(s_i)] \varrho_n(s_i) \Delta x_n\left(s_i - \tfrac{1}{2}\right)$$

$$= \sum_i \left\{ A\tilde{\tau}_n[x_n(s_i)] + B \right\} \Delta[\sigma(s_i)\varrho_n(s_i)] \ .$$

By using summation by parts, since the terms evaluated at the limits are zero and

$$\frac{\Delta\tilde{\tau}_n[x_n(s)]}{\Delta x_n(s)} = \tilde{\tau}_n' = \text{const} \ ,$$

we obtain

$$S_n^{(1)} = -A\tilde{\tau}_n' \sum_i \sigma(s_i+1)\varrho_n(s_i+1)\Delta x_n(s_i) = -A\tilde{\tau}_n' S_{n+1} \ .$$

Thus $S_{n+1} = -A\tilde{\tau}_n' S_{n+1} + CS_n$, whence

$$\frac{S_n}{S_{n+1}} = \frac{1 + A\tilde{\tau}_n'}{C} \ .$$

To calculate the constant A it is sufficient to compare the coefficients of $x_n^2(s)$ in Eqs. (3.7.17) and (3.7.18):

$$A = \frac{1}{\left(\tilde{\tau}_n'\right)^2} \left[\alpha\frac{\tilde{\sigma}_n''}{2} + \left(\alpha^2 - 1\right)\tilde{\tau}_n' \right] \ .$$

For calculating C we set $x_n = x_n^*$ in (3.7.17) and (3.7.18), where x_n^* is a root of the equation

$$\tilde{\tau}_n\left(x_n\right) = 0 \ ;$$

this yields $C = \alpha\tilde{\sigma}_n\left(x_n^*\right)$. As a result we obtain

$$\frac{S_n}{S_{n+1}} = \frac{\alpha + \left(\tilde{\sigma}_n''/(2\tilde{\tau}_n')\right)}{\tilde{\sigma}_n\left(x_n^*\right)} \ . \tag{3.7.19}$$

By using the values of a_n, b_n, d_n^2 we can construct *the recursion relation*

$$xy_n(x) = \alpha_n y_{n+1}(x) + \beta_n y_n(x) + \gamma_n y_{n-1}(x) \ , \tag{3.7.20}$$

where

$$\alpha_n = \frac{a_n}{a_{n+1}} \ , \quad \beta_n = \frac{b_n}{a_n} - \frac{b_{n+1}}{a_{n+1}} \ , \quad \gamma_n = \frac{a_{n-1}}{a_n}\frac{d_n^2}{d_{n-1}^2} \ .$$

3.7.3. We obtain *the basic data* for the classical orthogonal polynomials of a discrete variable on nonuniform lattices for the examples *of the Racah polynomials* $u_n^{(\alpha,\beta)}(x)$ and *the dual Hahn polynomials* $w_n^{(c)}(x)$.

106

For the Racah polynomials $\sigma(s) = (s - a)(s + b)(s + a - \beta)(b + \alpha - s)$. In accordance with (3.5.3)

$$\tau(s) = \tilde{\tau}[x(s)] = \frac{\sigma(-s - 1) - \sigma(s)}{2s + 1}$$
$$= (\alpha + 1)(a - \beta)a + (\beta + 1)(b - 1)(b + \alpha + 1)s(s + 1) \,.$$

By using Eqs. (3.1.29) and (3.2.18) for $\tilde{\tau}_{n-1}(x_{n-1})$ and $A_{n-1,n}$ we can calculate the constants a_n and b_n from (3.7.2). To determine the squared norm d_n^2 we first need to find the value $x_n^* = -\tilde{\tau}_n(0)/\tilde{\tau}_n'$ for which $\tilde{\tau}_n(x_n) = 0$ and then use (3.7.15, 16) and (3.7.19).

Similarly we can calculate the leading coefficients and squared norms of the dual Hahn polynomials. The results for the Racah and dual Hahn polynomials are presented in Tables 3.6 and 3.7, where we also give the coefficients of the recursion relations (3.7.20).

3.7.4. Finally, we discuss the difference derivatives of the Racah and dual Hahn polynomials. Using (3.2.18), we obtain the following formulas for the Racah polynomials $u_n^{(\alpha,\beta)}(x) \equiv u_n^{(\alpha,\beta)}(x, a, b)$ and the dual Hahn polynomials $w_n^{(c)}(x) \equiv w_n^{(c)}(x, a, b)$:

$$\frac{\Delta u_n^{(\alpha,\beta)}[x(s), a, b]}{\Delta x(s)}$$
$$= (\alpha + \beta + n + 1)u_{n-1}^{(\alpha+1,\beta+1)}\left[x\left(s + \tfrac{1}{2}\right), a + \tfrac{1}{2}, b - \tfrac{1}{2}\right] , \tag{3.7.21}$$

$$\frac{\Delta w_n^{(c)}[x(s), a, b]}{\Delta x(s)} = w_{n-1}^{(c-1/2)}\left[x\left(s + \tfrac{1}{2}\right), a + \tfrac{1}{2}, b - \tfrac{1}{2}\right] . \tag{3.7.22}$$

Let us note the relations which connect the Racah polynomials $u_n^{(\alpha,\pm 1/2)}(x, \pm\tfrac{1}{2}, b)$ and the Hahn polynomials $h_n^{(\alpha,\alpha)}(x, N)$ as well as the dual Hahn polynomials $w_n^{(0)}(x, \pm\tfrac{1}{2}, b)$ and the Kravchuk polynomials $k_n^{(1/2)}(x, N)$:

$$u_n^{(\alpha,-1/2)}\left[s(s + 1), -\tfrac{1}{2}, b\right] = \frac{(2n)!\,\Gamma(\alpha + n + 1)}{2^{2n}n!\,\Gamma(\alpha + 2n + 1)}h_{2n}^{(\alpha,\alpha)}(b + s, 2b) , \tag{3.7.23}$$

$$(2s + 1)u_n^{(\alpha,1/2)}\left[s(s + 1), \tfrac{1}{2}, b\right]$$
$$= \frac{(2n + 1)!\,\Gamma(\alpha + n + 1)}{2^{2n}n!\,(\alpha + 2n + 2)}h_{2n+1}^{(\alpha,\alpha)}(b + s, 2b) , \tag{3.7.24}$$

$$w_n^{(0)}\left[s(s + 1), -\tfrac{1}{2}, b\right] = \frac{(2n)!}{n!}k_{2n}^{(1/2)}(b + s, 2b - 1) , \tag{3.7.25}$$

$$(2s + 1)w_n^{(0)}\left[s(s + 1), \tfrac{1}{2}, b\right] = \frac{2(2n + 1)!}{n!}k_{2n+1}^{(1/2)}(b + s, 2b - 1) . \tag{3.7.26}$$

For the proof of (3.7.23–26) it is sufficient to verify that the polynomials in the left- and right-hand sides of these equalities are orthogonal with the same weights, and to compare the leading coefficients of these polynomials.

Table 3.6. Data for the Racah polynomials $u_n^{(\alpha,\beta)}(x)$

$\tilde{y}_n(x)$	$u_n^{(\alpha,\beta)}[x(s)]$, $x(s) = s(s+1)$
(a, b)	(a, b)
$\varrho(s)$	$\dfrac{\Gamma(a+s+1)\Gamma(s-a+\beta+1)\Gamma(b+\alpha-s)\Gamma(b+\alpha+s+1)}{\Gamma(a-\beta+s+1)\Gamma(s-a+1)\Gamma(b-s)\Gamma(b+s+1)}$ $(-1/2 < a \le b-1, \alpha > -1, -1 < \beta < 2a+1)$
$\sigma(s)$	$(s-a)(s+b)(s+a-\beta)(b+\alpha-s)$
$\tau(s)$	$(\alpha+1)a(a-\beta) + (\beta+1)b(b+\alpha) - (\alpha+1)(\beta+1) - (\alpha+\beta+2)x(s)$
λ_n	$n(\alpha+\beta+n+1)$
B_n	$\dfrac{(-1)^n}{n!}$
$\varrho_n(s)$	$\dfrac{\Gamma(a+s+n+1)\Gamma(s-a+\beta+n+1)\Gamma(b+\alpha-s)\Gamma(b+\alpha+s+n+1)}{\Gamma(a-\beta+s+1)\Gamma(s-a+1)\Gamma(b-s-n)\Gamma(b+s+1)}$
a_n	$\dfrac{1}{n!}(\alpha+\beta+n+1)_n$
b_n	$\dfrac{(\alpha+\beta+n+1)_{n-1}}{(n-1)!}\big[-ab(b-a+\alpha+\beta+n) + (a-\beta)(b+\alpha)(b-a-n)$ $- (b-a)(b-a+\alpha+\beta)n + \tfrac{1}{3}(2n^2+1)(\alpha+\beta) + \tfrac{1}{3}n(n^2+2)\big]$
d_n^2	$\dfrac{\Gamma(\alpha+n+1)\Gamma(\beta+n+1)\Gamma(b-a+\alpha+\beta+n+1)\Gamma(a+b+\alpha+n+1)}{(\alpha+\beta+2n+1)n!\,\Gamma(\alpha+\beta+n+1)(b-a-n-1)!\,\Gamma(a+b-\beta-n)}$
α_n	$\dfrac{(n+1)(\alpha+\beta+n+1)}{(\alpha+\beta+2n+1)(\alpha+\beta+2n+2)}$
β_n	$\tfrac{1}{4}[a^2+b^2+(a-\beta)^2+(b+\alpha)^2-2] - \tfrac{1}{8}(\alpha+\beta+2n)$ $\times (\alpha+\beta+2n+2) + \dfrac{(\beta^2-\alpha^2)[(b+\alpha/2)^2 - (a-\beta/2)^2]}{2(\alpha+\beta+2n)(\alpha+\beta+2n+2)}$
γ_n	$\dfrac{(\alpha+n)(\beta+n)}{(\alpha+\beta+2n)(\alpha+\beta+2n+1)}\left[\left(a+b+\dfrac{\alpha-\beta}{2}\right)^2 - \left(n+\dfrac{\alpha+\beta}{2}\right)^2\right]$ $\times \left[\left(b-a+\dfrac{\alpha+\beta}{2}\right)^2 - \left(n+\dfrac{\alpha+\beta}{2}\right)^2\right]$

Table 3.7. Data for the dual Hahn polynomials $w_n^{(c)}(x)$

$\tilde{y}_n(x)$	$w_n^{(c)}[x(s)]$, $x(s) = s(s+1)$
(a, b)	(a, b)
$\varrho(s)$	$\dfrac{\Gamma(a+s+1)\Gamma(c+s+1)}{\Gamma(s-a+1)\Gamma(b-s)\Gamma(b+s+1)\Gamma(s-c+1)}$ $(-1/2 < a \le b-1, \, |c| < a+1)$
$\sigma(s)$	$(s-a)(s+b)(s-c)$
$\tau(s)$	$ab - ac + bc - a + b - c - 1 - x(s)$
λ_n	n
B_n	$\dfrac{(-1)^n}{n!}$
$\varrho_n(s)$	$\dfrac{\Gamma(a+s+n+1)\Gamma(c+s+n+1)}{\Gamma(s-a+1)\Gamma(b-s-n)\Gamma(b+s+1)\Gamma(s-c+1)}$
a_n	$\dfrac{1}{n!}$
b_n	$-\dfrac{1}{(n-1)!}\left[ab - ac + bc - \dfrac{1}{3} + (b-a-c)n - \dfrac{2n^2}{3} \right]$
d_n^2	$\dfrac{\Gamma(a+c+n+1)}{n!(b-a-n-1)!\,\Gamma(b-c-n)}$
α_n	$n+1$
β_n	$ab - ac + bc + (b-a-c-1)(2n+1) - 2n^2$
γ_n	$(a+c+n)(b-a-n)(b-c-n)$

3.8 Asymptotic Properties of the Racah and Dual Hahn Polynomials

The classical orthogonal polynomials of a discrete variable, which are solutions of the difference equation (3.1.3), tend, as we have seen, to the polynomial solutions of (3.1.1) as $h \to 0$. For the Racah polynomials this property was established in Sect. 3.5.

It turns out that the Racah and Jacobic polynomials, $u_n^{(\alpha,\beta)}(x)$ and $P_n^{(\alpha,\beta)}(t)$, are connected by an asymptotic equation of higher precision than (3.5.16). To derive this formula we rewrite the recursion relation for $u_n \equiv u_n^{(\alpha,\beta)}(x)$ in the following way (see Table 2.1):

$$x u_n = \frac{\alpha_n}{2} u_{n+1} + \left\{ \frac{1}{2} \left[\left(b + \frac{\alpha}{2}\right)^2 - \left(a - \frac{\beta}{2}\right)^2 \right] \beta_n - \frac{1}{4} \right.$$

$$+ \frac{1}{2} \left[\left(b + \frac{\alpha}{2}\right)^2 - \left(a - \frac{\beta}{2}\right)^2 \right] - \frac{1}{4}(\alpha + 1)(\beta + 1) - \frac{n}{2}(\alpha + \beta + n + 1) \right\} u_n$$

$$+ \frac{\gamma_n}{2}(b - a + \alpha + \beta + n)(b - a - n)(b + a + \alpha + n)(b + a - \beta - n) u_{n-1} ,$$

$$(3.8.1)$$

where α_n, β_n, γ_n are the coefficients of the recursion relation for the Jacobi polynomials (see Table 1.1). For deriving (3.8.1) we used the identity

$$\frac{1}{4}\left[a^2 + b^2 + (a - \beta)^2 + (b + \alpha)^2 - 2 \right] - \frac{1}{8}(\alpha + \beta)(\alpha + \beta + 2)$$

$$= -\frac{1}{4} + \frac{1}{2} \left[\left(b + \frac{\alpha}{2}\right)^2 + \left(a - \frac{\beta}{2}\right)^2 \right] - \frac{1}{4}(\alpha + 1)(\beta + 1) .$$

Putting

$$\frac{x + 1/4 - [(b + \alpha/2)^2 + (a - \beta/2)^2]/2}{[(b + \alpha/2)^2 - (a - \beta/2)^2]/2} = t ,$$

$$\left(b + \frac{\alpha}{2}\right)^2 - \left(a - \frac{\beta}{2}\right)^2 = \tilde{N}^2 , \quad u_n^{(\alpha,\beta)}(x) = \frac{1}{C_n} p_n(t) ,$$

where C_n is a constant, and taking into account that

$$(b - a + \alpha + \beta + n)(b - a - n)(b + a + \alpha + n)(b + a - \beta - n)$$

$$= \tilde{N}^4 \left(1 - \frac{\left(n + \frac{\alpha+\beta}{2}\right)\left[\left(n + \frac{\alpha+\beta}{2}\right) - 2\left(a - \frac{\beta}{2}\right)\right]}{\tilde{N}^2} \right)$$

$$\times \left(1 - \frac{\left(n + \frac{\alpha+\beta}{2}\right)\left[\left(n + \frac{\alpha+\beta}{2}\right) + 2\left(a - \frac{\beta}{2}\right)\right]}{\tilde{N}^2} \right)$$

from (3.8.1) we obtain

$$t p_n(t) = \frac{C_n}{\tilde{N}^2 C_{n+1}} \alpha_n p_{n+1}(t) + \left[\beta_n - \frac{(\alpha + 1)(\beta + 1) + 2n(\alpha + \beta + n + 1)}{2\tilde{N}^2} \right] p_n(t)$$

$$+ \left(1 - \frac{\left(n + \frac{\alpha+\beta}{2}\right)\left[\left(n + \frac{\alpha+\beta}{2}\right) - 2\left(a - \frac{\beta}{2}\right)\right]}{\tilde{N}^2} \right)$$

$$\times \left(1 - \frac{\left(n + \frac{\alpha+\beta}{2}\right)\left[\left(n + \frac{\alpha+\beta}{2}\right) + 2\left(a - \frac{\beta}{2}\right)\right]}{\tilde{N}^2} \right) \frac{\tilde{N}^2 C_n}{C_{n-1}} \gamma_n p_{n-1}(t) .$$

Hence at $C_n = \tilde{N}^{-2n}$ by using the equality $p_0(t) = P_0^{(\alpha,\beta)}(t) = 1$ we obtain by induction the desired asymptotic formula

$$u_n^{(\alpha,\beta)}(x) = \tilde{N}^{2n}\left[P_n^{(\alpha,\beta)}(t) + \mathrm{O}\left(\frac{1}{\tilde{N}^2}\right)\right] \quad (b - a = N \to \infty), \qquad (3.8.2)$$

where

$$x = -\frac{1}{4} + \left(a - \frac{\beta}{2}\right)^2 \frac{1-t}{2} + \left(b + \frac{\alpha}{2}\right)^2 \frac{1+t}{2},$$

$$\tilde{N}^2 = \left(b + \frac{\alpha}{2}\right)^2 - \left(a - \frac{\beta}{2}\right)^2.$$

Under the same conditions, by means of the asymptotic representation

$$\frac{\Gamma(s+a+1)}{\Gamma(s-a)} = s^{2a+1}\left[1 + \mathrm{O}\left(s^{-2}\right)\right], \quad s \to \infty,$$

we obtain equations for $\varrho(s)$ and the squared norm d_n^2 of $u_n^{(\alpha,\beta)}(x)$ (cf. Table 1.1):

$$\varrho(s) = \left(\tfrac{1}{2}\tilde{N}^2\right)^{\alpha+\beta}(1-t)^\alpha(1+t)^\beta\left[1 + \mathrm{O}\left(\tilde{N}^{-2}\right)\right], \qquad (3.8.3)$$

$$d_n^2 = \left(\tilde{N}^2\right)^{\alpha+\beta+2n+1}\frac{\Gamma(\alpha+n+1)\Gamma(\beta+n+1)}{(\alpha+\beta+2n+1)n!\,\Gamma(\alpha+\beta+n+1)}\left[1 + \mathrm{O}\left(\tilde{N}^{-2}\right)\right]. \quad (3.8.4)$$

In a similar way we can obtain an asymptotic equation for the dual Hahn polynomials as $b \to \infty$:

$$(-1)^n b^{-n} w_n^{(\alpha-a)}(x) = L_n^\alpha(t) + \mathrm{O}\left(\frac{1}{b}\right), \qquad (3.8.5)$$

where

$$x = a(\alpha - a) + (b - 1)t.$$

3.9 Construction of Some Orthogonal Polynomials on Nonuniform Lattices by Means of the Darboux-Christoffel Formula

We have considered a class of lattices for which it is possible to construct a rather simple theory of orthogonal polynomials of a discrete variable by using the difference equation of hypergeometric type.

3.9.1 We now consider another method of constructing lattices for orthogonal polynomials of a discrete variable by using the Darboux-Christoffel formula. Let

$\{p_n(x)\}$ be any system of polynomials for which the recurrence relation

$$xp_n(x) = \alpha_n p_{n+1}(x) + \beta_n p_n(x) + \gamma_n p_{n-1}(x) \tag{3.9.1}$$

takes place with $n = 0, 1, \ldots, N - 1$ and the additional conditions

$$p_{-1}(x) = 0, \quad p_0(x) = a_0 . \tag{3.9.2}$$

Here the coefficients α_n, β_n and γ_n are real numbers, and $\alpha_n \neq 0$, $\alpha_{n-1}\gamma_n > 0$ for $1 \leq n \leq N - 1$. We show that in this case for the polynomials $p_n(x)$ the orthogonality relation of the form

$$\sum_{i=0}^{N-1} p_n(x_i) p_m(x_i) \varrho_i = \delta_{mn} d_n^2 \tag{3.9.3}$$

holds if $d_n^2 > 0$, $\varrho_i > 0$, $m < N$, $n < N$ and x_i are the real different roots of the equation

$$p_N(x_i) = 0 . \tag{3.9.4}$$

For the proof we preliminarily obtain the Darboux-Christoffel formula for the polynomials $\{p_n(x)\}$. By comparing the coefficients for powers x in the relation (3.9.1) we obtain

$$\alpha_n = \frac{a_n}{a_{n+1}}, \quad \beta_n = \frac{b_n}{a_n} - \frac{b_{n+1}}{a_{n+1}},$$

where a_n and b_n are coefficients in the expansion

$$p_n(x) = a_n x^n + b_n x^{n-1} + \ldots .$$

Let d_0^2 be an arbitrary positive number, and the constants $d_n^2 > 0$ are determined from the relation

$$\frac{d_n^2}{d_{n-1}^2} = \frac{\gamma_n}{\alpha_{n-1}} . \tag{3.9.5}$$

Then the recurrence relation for $p_n(x)$ can be written in the form

$$xp_n(x) = \frac{a_n}{a_{n+1}} p_{n+1}(x) + \left(\frac{b_n}{a_n} - \frac{b_{n+1}}{a_{n+1}} \right) p_n(x) + \frac{a_{n-1}}{a_n} \frac{d_n^2}{d_{n-1}^2} p_{n-1}(x) . \tag{3.9.6}$$

From the recurrence relation (3.9.6) we obtain the Darboux-Christoffel formula [N18]:

$$\sum_{n=0}^{N-1} \frac{p_n(x)p_n(y)}{d_n^2} = \frac{a_{N-1}}{a_N} \frac{1}{d_{N-1}^2} \frac{p_N(x)p_{N-1}(y) - p_{N-1}(x)p_N(y)}{x - y} . \tag{3.9.7}$$

Let x_i be the roots of Eq. (3.9.4). Then for $x = x_i$ and $y = x_j$ it follows from

(3.9.7) that

$$\sum_{n=0}^{N-1} \frac{p_n(x_i)p_n(x_j)}{d_n^2} = \delta_{ij}D_i^2 \,, \tag{3.9.8}$$

$$D_i^2 = \sum_{n=0}^{N-1} \frac{p_n^2(x_i)}{d_n^2} = \frac{a_{N-1}}{a_N d_{N-1}^2} p_N'(x_i)p_{N-1}(x_i) \,. \tag{3.9.9}$$

Since the polynomial $p_N(x)$ has real coefficients for powers x, each complex root x_i in (3.9.4) corresponds to the complex conjugate root x_i^* in this equation. By putting $x_j = x_i^*$ in (3.9.8), we come to a contradiction when $x_i^* \neq x_i$ (in this case the left-hand part of (3.9.8) is positive while the right-hand part is zero). For $x_j = x_i$ it follows from (3.9.8) that $p_N'(x_i) \neq 0$, i.e. Eq. (3.9.4) does not have multiple roots. Let us show that by using (3.9.8) we can find the constants $\varrho_i > 0$ such that the polynomials $p_n(x)$ satisfy the orthogonality relation of the form (3.9.3). The relation (3.9.8) in fact can be rewritten in the form

$$\sum_{n=0}^{N-1} c_{ni}c_{nj} = \delta_{ij} \tag{3.9.10}$$

with

$$c_{ni} = \frac{p_n(x_i)}{d_n D_i} \,.$$

It follows from (3.9.10) that the matrix C with elements $c_{ni}(n, i = 0, 1, \ldots,$ $N-1)$ is unitary, and hence there is another orthogonality relation for C:

$$\sum_{i=0}^{N-1} c_{mi}c_{ni} = \delta_{mn} \quad (m, n = 0, 1, \ldots, N-1) \,. \tag{3.9.11}$$

It is evident that (3.9.11) is equivalent to the orthogonality relation (3.9.3) for the polynomials $p_n(x)$ if $\varrho_i = 1/D_i^2$.

Remark. Since for the polynomials $p_n(x)$ that satisfy the orthogonality relations of the form

$$\int_a^b p_n(x)p_m(x)\varrho(x)dx = \delta_{mn}d_n^2$$

with $\varrho(x) > 0$, $x \in (a, b)$ the Darboux-Christoffel formula holds, the orthogonality relations of the form (3.9.3) are valid for these polynomials too.

We have discussed the method of constructing an orthogonality relation of the form (3.9.3) for the polynomials $p_n(x)$ in the case when the lattice $\{x_i\}$ is determined by using the equations $p_N(x_i) = 0$. The entire discussion can be carried over if $\{x_i\}$ is determined by using the more general equation $\alpha p_N(x_i) + \beta p_{N-1}(x_i) = 0$, where α and β are real coefficients, not both zero.

3.9.2. As an example we consider an orthogonality relation of the form (3.9.3) for the Chebyshev polynomials of the first kind, $T_n(x) = \cos(n \arccos x)$. In this case

$$a_n = \frac{1}{2^{n-1}}, \quad d_n^2 = \begin{cases} \pi & \text{for} \quad n = 0, \\ \frac{\pi}{2} & \text{for} \quad n \neq 0; \end{cases}$$

$$T_n(x_i) = 0 \quad \text{for}$$

$$x_i = \cos\left[\frac{\pi}{N}\left(i + \frac{1}{2}\right)\right] \quad (i = 0, 1, \ldots, N - 1),$$

whence

$$D_i^2 = \frac{4}{\pi} T_N'(x_i) T_{N-1}(x_i) = \frac{4N}{\pi}.$$

Hence we can write (3.9.3) in the form

$$\sum_{i=0}^{N-1} T_n[x(s_i)] T_n[x(s_i)] \frac{\pi}{4N} = \delta_{mn} d_n^2, \tag{3.9.12}$$

where $x(s) = \cos(\pi s/N)$, $s_i = i + 1/2$ $(0, 1, \ldots, N - 1)$. The lattice $x_i = \cos(\pi s_i/N)$ is a special case of the lattice (3.4.10) with $\omega = \pi/2N$. Consequently it is natural to expect that the Chebyshev polynomials $T_n(x)$ coincide, up to a normalizing factor, with the q-analogs of the Racah polynomials $u_n^{(\alpha, \beta)}[x(s), q]$ which are orthogonal on the lattice with $x(s) = \cos(2\omega s)$, $\omega = \pi/2N$, $a = 1/2$, $b = N + 1/2$ and some values of α and β.

By comparing (3.9.12) and (3.3.4) we see that our expection is fulfilled if

$$\varrho(s_i)\Delta x\left(s_i - \tfrac{1}{2}\right) = \text{const}, \tag{3.9.13}$$

where $\varrho(s)$ is defined by (3.6.39).

Let us verify that (3.9.13) is satisfied with $\alpha = \beta = -1/2$. In fact (3.9.13) will be satisfied if

$$\frac{\varrho(s_i + 1)}{\varrho(s_i)} = \frac{\Delta x(s_i - 1/2)}{\Delta x(s_i + 1/2)}. \tag{3.9.14}$$

Since (see Table 3.5)

$$\frac{\Delta x(s_i - 1/2)}{\Delta x(s_i + 1/2)} = \frac{\sin(\pi s_i/N)}{\sin[\pi(s_i + 1)/N]},$$

$$\frac{\varrho(s_i + 1)}{\varrho(s_i)} = \frac{\sigma(s_i) + \tau(s_i)\Delta x(s_i - 1/2)}{\sigma(s_i + 1)}$$

$$= \frac{\psi_q(N - 1/2 - s_i)\psi_q(s_i)\psi_q(N + s_i)}{\psi_q(N + 1/2 + s_i)\psi_q(s_i + 1)\psi_q(N - 1 - s_i)}$$

$$= \frac{\sin[\pi(N - 1/2 - s_i)/2N] \sin[\pi s_i/2N] \sin[\pi(N + s_i)/2N]}{\sin[\pi(N + 1/2 + s_i)/2N] \sin[\pi(s_i + 1)/2N] \sin[\pi(N - 1 - s_i)/2N]}$$

$$= \frac{\sin\left[\pi s_i/2N\right]\cos\left[\pi s_i/2N\right]}{\sin\left[\pi(s_i+1)/2N\right]\cos\left[\pi(s_i+1)/2N\right]} = \frac{\sin\left[\pi s_i/N\right]}{\sin\left[\pi(s_i+1)/N\right]} ,$$

the validity of (3.9.13) is established.

Consequently

$$T_n(x) = A_n u_n^{(-1/2,\,-1/2)}[x(s), q] \tag{3.9.15}$$

for $x(s) = \cos(2\omega s)$, $\omega = \pi/2N$, $a = 1/2$, $b = N + 1/2$. The constant A_n can be determined by comparing coefficients of the leading terms in (3.9.15).

By using (3.9.15) one can show that the Chebyshev polynomials of the second kind

$$U_n(x) = \frac{\sin[(n+1)\varphi]}{\sin\varphi} \quad (\varphi = \arccos x)$$

coincide, up to a normalizing factor, with the q-analogs of the Racah polynomials $u_n^{(1/2,\,1/2)}[x(s), q]$ on the lattice $x(s) = \cos(2\omega s)$, $\omega = \pi/2N$ for $a = 1$, $b = N$. This follows from the easily verified relation

$$\frac{\Delta T_{n+1}[x(s)]}{\Delta x(s)} = U_n\left[x_1(s)\right]U_n\left[x_1(0)\right] \quad \left(x_1(s) = x\left(s + \tfrac{1}{2}\right)\right)$$

and the orthogonality property of the polynomials

$$\frac{\Delta u_{n+1}^{(-1/2,\,-1/2)}[x(s), q]}{\Delta x(s)}$$

on the lattice under consideration.

3.10 Continuous Orthogonality

In Chap. 1 it was shown that the polynomial solutions $y_n(x)$ of differential equations of hypergeometric type (1.1.1) have the orthogonality property determined by means of the integral

$$\int_a^b y_n(x)y_m(x)\varrho(x)dx = 0 \quad (m \neq n) . \tag{3.10.1}$$

At the same time by applying the Darboux-Christoffel formula to these polynomials (see Sect. 3.9) the orthogonality property may be obtained in the form of a sum over some discrete set of points:

$$\sum_i y_n(x_i)y_m(x_i)\tilde{\varrho}(x_i) = 0 \quad (m \neq n) . \tag{3.10.2}$$

In Sect. 3.3 for the polynomial solutions of difference equations of hypergeometric type (3.1.5) the orthogonality relation was naturally obtained in the form of the sum (3.10.2). In this section it will be shown that under certain conditions the polynomial solutions of the difference equation (3.1.5) also have the orthogonality property in the form analogous to (3.10.1), i.e.

$$\int_C y_n(s)y_m(s)\varrho(s)ds = 0 \quad (m \neq n)\,,\tag{3.10.3}$$

where integration is performed in the complex plane of variable s over a contour C $\left(y_n(s) = \tilde{y}_n[x(s)]\right)$.

Further on, the orthogonality properties in the form of (3.10.2) or (3.10.1, 3) will be called, for brevity, discrete or continuous orthogonality, respectively.

3.10.1. For obtaining the orthogonality property of polynomial solutions of the difference equation (3.1.5) in the form of (3.10.3) we use the above derived relation (3.3.1) and rewrite it in the following way:

$$\begin{aligned}
\left(\lambda_m - \lambda_n\right)&y_m(s)y_n(s)\varrho(s)\Delta x \left(s - \tfrac{1}{2}\right)\\
&= \Delta\{\sigma(s)\varrho(s)W\left[y_m(s),\, y_n(s)\right]\}\,,
\end{aligned}\tag{3.10.4}$$

where

$$W\left[y_m(s),\, y_n(s)\right] = y_m(s)\frac{\nabla y_n(s)}{\nabla x(s)} - y_n(s)\frac{\nabla y_m(s)}{\nabla x(s)}$$

is the difference analog of Wronskian. As shown in Sect. 3.3.1 $W\left[y_m(s),\, y_n(s)\right]$ is a polynomial in $x(s - 1/2)$:

$$W\left[y_m(s),\, y_n(s)\right] = \sum_{k=0}^{m+n-1} C_k x^k \left(s - \tfrac{1}{2}\right)\,,$$

where C_k are constants. Therefore the relation (3.10.4) may be rewritten in the form

$$\begin{aligned}
\left(\lambda_m - \lambda_n\right)&y_m(s)y_n(s)\varrho(s)\Delta x \left(s - \tfrac{1}{2}\right)\\
&= \sum_k C_k \Delta \left[\sigma(s)\varrho(s)x^k \left(s - \tfrac{1}{2}\right)\right]\,.
\end{aligned}\tag{3.10.5}$$

Since

$$\begin{aligned}
\Delta \left[\sigma(s)\varrho(s)x^k \left(s - \tfrac{1}{2}\right)\right] &= \nabla \left[\sigma(s+1)\varrho(s+1)x^k \left(s + \tfrac{1}{2}\right)\right]\\
&= \nabla \left[\varrho_1(s)x_1^k(s)\right]\,,
\end{aligned}$$

by integrating the both sides of (3.10.5) over a contour C in the complex plane of variable s we come to the orthogonality relation

$$\int_C \tilde{y}_m[x(s)]\tilde{y}_n[x(s)]\varrho(s)\Delta x \left(s - \tfrac{1}{2}\right) ds = 0 \quad (m \neq n)\,,\tag{3.10.6}$$

if the conditions

$$\int_C \nabla \left[\varrho_1(s) x_1^l(s) \right] ds = 0 \quad (l = 0, 1, \ldots) \tag{3.10.7}$$

are satisfied.

In a similar way, by using Eq. (3.2.8) for the polynomials

$$v_{kn}(s) = \frac{\Delta v_{k-1,n}(s)}{\Delta x_{k-1}(s)} \, ,$$

where $v_{0n}(s) = y_n(s) \big(v_{kn}(s) = \tilde{v}_{kn} \left[x_k(s) \right], \ k = 1, 2, \ldots \big)$ we may obtain the orthogonality relation

$$\int_C v_{kn}(s) v_{km}(s) \varrho_k(s) \Delta x_k \left(s - \tfrac{1}{2} \right) ds = 0 \quad (m \neq 0) \, , \tag{3.10.8}$$

if the conditions

$$\int_C \nabla \left[\varrho_{k+1}(s) x_{k+1}^l(s) \right] ds = 0 \quad (l = 0, 1, \ldots) \tag{3.10.9}$$

are satisfied. These conditions are usually satisfied simultaneously with the conditions (3.10.7).

The conditions (3.10.7) may be rewritten in the form

$$\int_C f(s) ds = \int_{C'} f(s') ds' \, , \tag{3.10.10}$$

where $f(s) = \varrho_1(s) x_1^l(s)$, while the contour C' may be obtained from the contour C by the shift $s' = s - 1$.

In many cases considered below we may take the straight line $\operatorname{Re} s = s_1$ as the contour C. Let us show that in this case the condition (3.10.10) is satisfied if the function $f(s) = \varrho_1(s) x_1^l(s)$ does not have singularities in the strip region $s_1 - 1 \leq \operatorname{Re} s \leq s_1$ and

$$\lim_{\operatorname{Im} s \to \pm \infty} f(s) = 0 \, . \tag{3.10.11}$$

For the proof it is sufficient to note that according to the Cauchy theorem the integral in s over the closed contour (Fig. 1) consisted of segments of straight lines $\left[s_1 - 1 - i\omega, \ s_1 - i\omega \right]$, $\left[s_1 - i\omega, \ s_1 + i\omega \right]$, $\left[s_1 + i\omega, \ s_1 - 1 + i\omega \right]$, $\left[s_1 - 1 + i\omega, \ s_1 - 1 - i\omega \right]$, where $\omega > 0$, is zero while the integrals over the straight lines $\left[s_1 - 1 - i\omega, \ s_1 - i\omega \right]$ and $\left[s_1 + i\omega, \ s_1 - 1 + i\omega \right]$ tend to zero when $\omega \to +\infty$ by virtue of (3.10.10). Hence

$$\int_{s_1 - i\infty}^{s_1 + i\infty} f(s) ds = \int_{s_1 - 1 - i\infty}^{s_1 - 1 + i\infty} f(s) ds \, ,$$

i.e. the condition (3.10.10) is really satisfied.

Sometimes the contour C may be chosen as a segment $\left[s_1, \ s_1 + i\omega \right]$ if instead of condition (3.10.11) we require that the function $f(s)$ be periodic with period

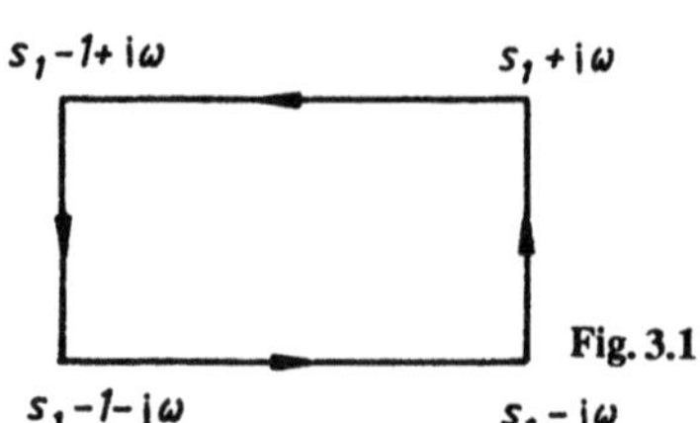

Fig. 3.1

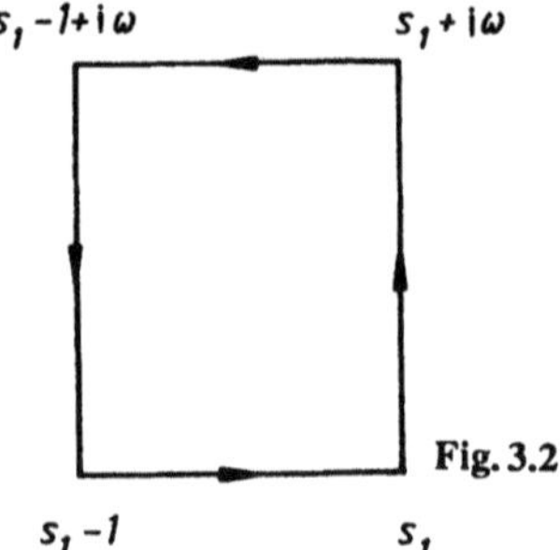

Fig. 3.2

$i\omega$. The proof is carried out in a similar manner: if the Cauchy theorem is used for the closed contour consisting of segments of the straight lines $[s_1 - 1, \, s_1]$, $[s_1, \, s_1 + i\omega]$, $[s_1 + i\omega, \, s_1 - 1 + i\omega]$, $[s_1 - 1 + i\omega, \, s_1 - 1]$, then a sum of integrals over the contours $[s_1 - 1, \, s_1]$ and $[s_1 + i\omega, \, s_1 - 1 + i\omega]$ will be zero because $f(s + i\omega) = f(s)$. Hence

$$\int_{s_1}^{s_1 + i\omega} f(s)ds = \int_{s_1 - 1}^{s_1 - 1 + i\omega} f(s)ds \, ,$$

i.e. the condition (3.10.11) is satisfied if the contour C is a segment of the straight line $\mathrm{Re}\, s = s_1$, $0 \le \mathrm{Im}\, s \le \omega$ (Fig. 3.2).

In addition to the cases considered above a case is also possible when the contour C is the straight line $-\infty < s < +\infty$ provided that the integral $\int_{-\infty}^{+\infty} \varrho_1(s)x_1^l(s)ds$ converges at $l = 0, 1, 2, \ldots$. In this case the contours C and C' coincide.

3.10.2. To calculate the squared norms

$$d_{kn}^2 = \int_C v_{kn}^2(s)\varrho_k(s)\Delta x_k \left(s - \tfrac{1}{2}\right) ds$$

in the considered cases we multiply Eq. (3.2.8) by $v_{kn}(s)\Delta x_k \left(s - \tfrac{1}{2}\right)$ and integrate the result over the contour C (cf. Sect. 1.3.2), which yields

$$\mu_{kn}^2 d_{kn}^2 = - \int_C v_{kn}(s)\nabla \left[\varrho_{k+1}(s)v_{k+1,n}(s)\right] ds \, .$$

Hence by using the relation

$$\nabla[f(s + 1)g(s)] = f(s)\nabla g(s) + g(s)\Delta f(s)$$

with $f(s) = v_{kn}(s)$, $g(s) = \varrho_{k+1}(s)v_{k+1,n}(s)$ we obtain

$$\mu_{kn} d_{kn}^2 = d_{k+1,n}^2 \tag{3.10.12}$$

under the conditions

$$\int_C \nabla\big[\varrho_{k+1}(s)v_{kn}(s+1)v_{k+1,n}(s)\big]\,ds = 0 \ , \tag{3.10.13}$$

which are usually satisfied if the conditions (3.10.9) are satisfied.

From (3.10.12) in accordance with (3.7.15) we find

$$\begin{aligned}
d_n^2 &= \int_C \tilde{y}_n^2[x(s)]\varrho(s)\Delta x \left(s - \tfrac{1}{2}\right) ds \\
&= (-1)^n A_{nn} B_n^2 \int_C \varrho_n(s)\Delta x_n \left(s - \tfrac{1}{2}\right) ds \ ,
\end{aligned} \tag{3.10.14}$$

where $A_{nn} = (-1)^n \prod_{k=0}^{n-1} \mu_{kn}$.

Let us consider some specific cases of continuous orthogonality for polynomial solutions of the difference equation of hypergeometric type (3.1.21)[1].

3.10.3. *The lattice* $x(s) = s$. We consider the property of continuous orthogonality for the Hahn and Meixner polynomials.

3.10.3.1. For the Meixner polynomials it is convenient to choose the function $\varrho(s)$ as the one differing from that chosen above by a periodic factor with period 1 (as it has been shown above such a factor takes no effect on the form of the polynomial obtained by the Rodrigues formula):

$$\varrho(s) = \mu^s e^{i\pi s} \Gamma(\gamma + s)\Gamma(-s) \ , \tag{3.10.15}$$

while

$$\varrho_1(s) = \mu^{s+1} e^{i\pi s} \Gamma(\gamma + 1 + s)\Gamma(-s) \ . \tag{3.10.16}$$

Since the function $\varrho_1(s)$ has its singularities (poles) at $s = -\gamma - 1 - k$ and $s = k$ $(k = 0, 1, 2, \dots)$ the function $f(s) = \varrho_1(s)x_1^k(s)$ has no singularities for $s_1 - 1 \le \operatorname{Re} s \le s_1$, where $s_1 = -\gamma/2$ if $\gamma > 0$. Since

$$|\Gamma(\lambda + it)| \approx (2\pi)^{1/2}|t|^{\lambda - 1/2} e^{-\pi|t|/2} \quad \left(\lambda = \frac{\gamma}{2}\right) \ ,$$

when $t \to \pm\infty$ [A1], the condition (3.10.11) is also satisfied if in the orthogonality relation (3.10.3) the contour C is given in the form $s = -it - \gamma/2$ $(-\infty < t < +\infty)$. In this case in (3.10.15) we have

$$\Gamma(\gamma + s)\Gamma(-s) = \left|\Gamma\left(it + \frac{\gamma}{2}\right)\right|^2 \ ,$$

$$\mu^s e^{i\pi s} = |\mu|^{it - \gamma/2} e^{-(\arg \mu + \pi)t} e^{-(\arg \mu + \pi)\gamma/2} \ ,$$

---

[1] Sections 3.10.3–5 are rather complex and may be omitted on a first reading. It may be better to become acquainted with this material reading after Sect. 3.11.5.

and at $|\mu| = 1$ for the polynomials $p_n(t) \equiv y_n(\mathrm{i}t - \gamma/2)$ that are orthogonal on the interval $(-\infty, \infty)$ the weight function $\varrho(s)$ differs from the function

$$\bar{\varrho}(t) = \frac{1}{2\pi} \mathrm{e}^{-(\arg \mu + \pi)t} \left| \Gamma\left(\mathrm{i}t + \frac{\gamma}{2}\right) \right|^2 ,$$

which is positive when $-\infty < t < +\infty$, only by a constant factor. At $\arg \mu = -2\varphi\,(0 < \varphi < \pi)$, $\gamma = 2\lambda$ the weight $\bar{\varrho}(t)$ coincides with the weight to which the Pollaczek polynomials $\mathcal{P}_n^\lambda(t, \varphi)$ are orthogonal (see [P10]):

$$\int_{-\infty}^{\infty} \mathcal{P}_n^\lambda(t, \varphi) \mathcal{P}_m^\lambda(t, \varphi) \bar{\varrho}(t)dt = 0 \quad (m \neq n) .$$

Since the Pollaczek polynomials $\mathcal{P}_n^\lambda(t, \varphi)$ and the Meixner polynomials $m_n^{(\gamma, \mu)}(\mathrm{i}t - \gamma/2)$ for $\mu = \mathrm{e}^{-2\mathrm{i}\varphi}$, $\lambda = \gamma/2$ satisfy the same orthogonality relation with the positive weight $\bar{\varrho}(t)$ they coincide to within a factor. The leading coefficient in the polynomial $\mathcal{P}_n^\lambda(t, \varphi)$ is equal to $(2 \sin \varphi)^n / n!$, so the resulting equality is

$$\mathcal{P}_n^\lambda(t, \varphi) = \frac{\mathrm{e}^{-\mathrm{i}n\varphi}}{n!} m_n^{(2\lambda, \mu)}(\mathrm{i}t - \lambda), \quad \mu = \mathrm{e}^{-2\mathrm{i}\varphi} . \tag{3.10.17}$$

With real t, λ, φ the Pollaczek polynomials take real values. Note that by using the asymptotic representations for the function $\Gamma(z)$ when $|z| \to \infty$, we may obtain that as $t \to \pm\infty$

$$\bar{\varrho}(t) \sim |t|^{2\lambda - 1} \mathrm{e}^{-[\pi \pm (\pi - 2\varphi)]|t|} .$$

The polynomials $\mathcal{P}_n^\lambda(t, \varphi)$ were introduced in [M8], and then they were rediscovered and generalized in [P10]. Sometimes they are called the Meixner-Pollaczek polynomials. The main data of the Pollaczek polynomials are given in Table 3.8.

3.10.3.2. In the case of the Hahn polynomials $h_n^{(\alpha, \beta)}(s, N)$ for which $\sigma(s) = s(N + \alpha - s)$, $\sigma(s) + \tau(s) = (\beta + 1 + s)(N - 1 - s)$ we choose the solution of the equation $\Delta[\sigma(s)\varrho(s)] = \tau(s)\varrho(s)$ in the form

$$\varrho(s) = \Gamma(\beta + s + 1)\Gamma(s - N + 1)\Gamma(N + \alpha - s)\Gamma(-s) . \tag{3.10.18}$$

This solution differs from the one chosen in Sect. 2.4.2 by the periodic factor with period equal to unity that does not affect an explicit form of the polynomial obtained with the use of the Rodrigues formula. In this case

$$\varrho_1(s) = -\Gamma(\beta + s + 2)\Gamma(s - N + 2)\Gamma(N + \alpha - s)\Gamma(-s) , \tag{3.10.18}$$

and it may be shown that the above considered conditions for the contour will be satisfied with complex values α, β, N if the contour C is the straight line $\mathrm{Re}\, s = s_1$ and

$$\operatorname{Re}\left(\beta + s_1 + 1\right) > 0, \qquad \operatorname{Re}\left(N + \alpha - s_1\right) > 0, \atop \operatorname{Re}\left(s_1 - N + 1\right) > 0, \qquad \operatorname{Re}\left(-s_1\right) > 0. \tag{3.10.19}$$

In this case the Hahn polynomials have the orthogonality property

$$\int_C h_n^{(\alpha,\beta)}(s,N) h_m^{(\alpha,\beta)}(s,N) \varrho(s)\,ds = 0 \quad (m \neq n), \tag{3.10.20}$$

where the contour C is the straight line $\operatorname{Re} s = s_1$, while $\varrho(s)$ is determined by Eq. (3.10.18).

Under certain parameter values we may obtain from (3.10.20) the orthogonality property with the positive weight $\bar\varrho(t)$ on the straight line $(-\infty, \infty)$ when on the contour C the arguments of two pairs of Γ-functions in (3.10.18) are complex conjugate, i.e.

$$\beta + s + 1 = -s^*, \quad s - N + 1 = (N + \alpha - s)^* \tag{3.10.21}$$

or

$$\beta + s + 1 = (N + \alpha - s)^*, \quad s - N + 1 = -s^*. \tag{3.10.22}$$

Let us consider these cases.

1) From the relations (3.10.21) it follows that α and β are real numbers, $s_1 = -(\beta + 1)/2$, $\operatorname{Re} N = -(\alpha + \beta)/2$. The inequalities (3.10.19) will be satisfied if $\alpha > -1$, $\beta > -1$. By assuming in (3.10.20) that $N = i\gamma - (\alpha + \beta)/2$, $s = i(t + \gamma)/2 - (\beta + 1)/2$, where γ is a real number, $-\infty < t < +\infty$, we obtain the orthogonality relation with positive weight

$$\bar\varrho(t) = \frac{1}{4\pi} \left| \Gamma\left(i\frac{t - \gamma}{2} + \frac{\alpha + 1}{2}\right) \Gamma\left(i\frac{t + \gamma}{2} + \frac{\beta + 1}{2}\right) \right|^2. \tag{3.10.23}$$

The polynomials

$$p_n^{(\alpha,\beta)}(t,\gamma) = i^{-n} h_n^{(\alpha,\beta)}(s,N), \tag{3.10.24}$$

where

$$s = i\frac{t + \gamma}{2} - \frac{\beta + 1}{2}, \quad N = i\gamma - \frac{\alpha + \beta}{2},$$

which are orthogonal on the real straight line $(-\infty, \infty)$, will be called *the Hahn polynomials of an imaginary argument*. They have a real leading coefficient. For the weight $\bar\varrho(t)$ we may obtain the estimate

$$\bar\varrho(t) \sim \frac{\pi}{2^{\alpha+\beta}} |t|^{\alpha+\beta} e^{-\pi|t|}$$

when $|t| \to \infty$. If $\gamma \to \infty$, we have the limiting relation

$$p_n^{(\alpha,\beta)}(\gamma\xi,\gamma) = \gamma^n \left[P_n^{(\alpha,\beta)}(\xi) + O\left(\frac{1}{\gamma^2}\right) \right],$$

where $P_n^{(\alpha,\beta)}(\xi)$ is the Jacobi polynomial. This relation follows from (2.6.5) and (3.10.24).

2) The conditions (3.10.22) are satisfied at $\alpha = \beta^*$, $N = N^*$. It may be shown that the case $\operatorname{Im}\alpha \neq 0$ is reduced to that mentioned above. At real values of $\alpha = \beta$ and by using the relation (2.4.18) we may obtain the orthogonality relation in a similar way with $t \in (-\infty, \infty)$ and positive weight $\bar{\varrho}(t)$ for the polynomials

$$q_n^\alpha(t, \delta) = \mathrm{i}^{-n} h_n^{(\alpha,\alpha)}(s, N) , \tag{3.10.25}$$

where $s = \mathrm{i}t/2 - (\alpha - \delta + 1)/2$, $N = \delta - \alpha\ (\delta > 0)$ (see Table 3.8). The polynomials $q_n^\alpha(t, \delta)$ and $p_n^{(\alpha,\beta)}(t, \gamma)$ are connected by the relationships

$$q_n^\alpha(t, \delta) = p_n^{(\alpha,\alpha)}(t, -\mathrm{i}\delta) = p_n^{(\alpha-\delta,\alpha+\delta)}(t, 0) , \tag{3.10.26}$$

since all of them have the same weight and leading coefficient (Table 3.8). In this case for the weight $\bar{\varrho}(t)$ we may obtain the estimate

$$\varrho(t) \sim \frac{\pi}{2^{2\alpha}} |t|^{2\alpha} \mathrm{e}^{-\pi|t|}$$

when $|t| \to \infty$. The polynomials $p_n^{(\alpha,\beta)}(t, \gamma)$ and $q_n^{(\alpha)}(t, \delta)$ will be called the Hahn polynomials of an imaginary argument.

For the considered polynomials we note the symmetry properties which may be obtained from analogous properties of the Hahn polynomials:

$$p_n^{(\alpha,\beta)}(t, \gamma) = p_n^{(\beta,\alpha)}(t, -\gamma) = (-1)^n p_n^{(\beta,\alpha)}(-t, \gamma) ,$$
$$q_n^{(\alpha)}(t, \delta) = q_n^{(\alpha)}(t, -\delta) = (-1)^n q_n^{(\alpha)}(-t, \delta) .$$

To evaluate the square norm d_n^2 in the case of the Pollaczek and Hahn polynomials of an imaginary argument we may use the relation (3.10.14) (the values of d_n^2 are given in Table 3.9; the values of integrals in (3.10.14) are given from [W5]).

3.10.3.3. For the polynomials $p_n^{(\alpha,\beta)}(t, \gamma)$ at $\alpha = \beta = 0$ by using the equation $\Gamma(s)\Gamma(1 - s) = \pi/\sin\pi s$ we obtain

$$\bar{\varrho}(t) = \frac{\pi/2}{\cosh\pi t + \cosh\pi\gamma} , \qquad d_n^2 = \frac{\pi\gamma}{(2n + 1)\sinh\pi\gamma} \prod_{k=0}^{n-1} [(n - k)^2 + \gamma^2] .$$

We note that the weight $\bar{\varrho}(t)$ satisfies not only the difference equation but the nonlinear differential equation

$$\frac{d}{dt}\bar{\varrho}(t) + (2\pi \sinh\pi t)^2 \bar{\varrho}^2(t) = 0 .$$

The polynomials $q_n^{(0)}(t, \delta)$ were considered in [B10–13] at $\delta = 0$ and then in [P3, T7, W10, B30, C3, C4, C18] for $-1 < \delta < 1$. For these polynomials

$$\bar{\varrho}(t) = \frac{\frac{\pi}{2}}{\cosh\pi t + \cos\pi\delta} ,$$

$$d_n^2 = \frac{\pi\delta}{(2n+1)\sin\pi\delta} \prod_{k=0}^{n-1} [(n-k)^2 - \delta^2]$$

when $-1 < \delta < 1$.

We note that for the polynomials $q_n^{(\alpha)}(t, 1/2)$ by using the duplication formula for the Γ-function we may obtain the following relation for the weight $\bar\varrho(t)$:

$$\bar\varrho(t) = \frac{1}{2^{2\alpha+1}} \left| \Gamma\left(\alpha + \tfrac{1}{2} + it\right) \right|^2 .$$

Since the polynomials $q_n^{(0)}(t, 1/2)$ and the Pollaczek polynomials $\mathcal{P}_n^{1/2}(t, \frac{\pi}{2})$ are orthogonal on the interval $(-\infty, \infty)$ with the weight $\bar\varrho(t) = \mathrm{const}/\cosh\pi t$, we may deduce that

$$q_n^{(0)}\left(t, \tfrac{1}{2}\right) = \left(\tfrac{1}{2}\right)_n \mathcal{P}_n^{1/2}\left(t, \tfrac{\pi}{2}\right) .$$

We may obtain a more generalized relation as

$$q_n^{(\alpha)}\left(t, \tfrac{1}{2}\right) = \frac{(2\alpha + n + 1)_n}{2^{2n}} \mathcal{P}_n^{\alpha+1/2}\left(t, \tfrac{\pi}{2}\right) .$$

The main characteristics for the considered polynomials are given in Table 3.8.

3.10.4. *The lattice* $x(s) = C_1 s^2 + C_2 s + C_3$. In order to obtain the property of continuous orthogonality for polynomial solutions of a difference equation of hypergeometric type on a quadratic lattice it is convenient by using the linear replacement of variables to use, instead of function $x(s) = s(s+1)$, as it was done in Sect. 3.6, the function $x(s) = s^2$. Since $x(-s) = x(s)$, $\Delta x(t-1/2)|_{t=-s} = -\Delta x(s-1/2)$, we have in this case $\sigma(s) + \tau(s)\Delta x(s-1/2) = \sigma(-s)$, and the equation for the function $\varrho(s)$ takes the form:

$$\frac{\varrho(s+1)}{\varrho(s)} = \frac{\sigma(-s)}{\sigma(s+1)} .$$

Two cases are possible:

(a) $\sigma(s)$ is the fourth degree polynomial:

$$\sigma(s) = (s-a)(s-b)(s-c)(s-d) ;$$

(b) $\sigma(s)$ is the third degree polynomial:

$$\sigma(s) = (s-a)(s-b)(s-c) .$$

Let us consider the first case. The solution of the equation for $\varrho(s)$ has the form:

$$\begin{aligned}
\varrho(s) =\, & \Gamma(s+a)\Gamma(s+b)\Gamma(s+c)\Gamma(s+d)\Gamma(a-s)\Gamma(b-s) \\
& \times \Gamma(c-s)\Gamma(d-s)f(s) ,
\end{aligned}$$

Table 3.8. Continuous orthogonality property for the Hahn and Meixner polynomials

$p_n(t)$	$p_n^{(\alpha,\beta)}(t,\gamma)$	$q_n^{(\alpha)}(t,\delta)$	$P_n^{\lambda}(t,\varphi)$
(a, b)	$(-\infty, \infty)$	$(-\infty, \infty)$	$(-\infty, \infty)$
$\varrho(t)$	$\dfrac{1}{4\pi}\left\|\Gamma\!\left(\mathrm{i}\dfrac{t-\gamma}{2}+\dfrac{\alpha+1}{2}\right)\Gamma\!\left(\mathrm{i}\dfrac{t+\gamma}{2}+\dfrac{\beta+1}{2}\right)\right\|^2$ $(\alpha > -1, \beta > -1)$	$\dfrac{1}{4\pi}\left\|\Gamma\!\left(\mathrm{i}\dfrac{t}{2}+\dfrac{\alpha+\delta+1}{2}\right)\Gamma\!\left(\mathrm{i}\dfrac{t}{2}+\dfrac{\alpha-\delta+1}{2}\right)\right\|^2$ $(|\delta| < \alpha+1)$	$\dfrac{1}{2\pi}|\Gamma(\lambda+\mathrm{i}t)|^2 e^{(2\varphi-\pi)t}$ $(\lambda > 0, 0 < \varphi < \pi)$
a_n	$\dfrac{1}{2^n n!}(\alpha+\beta+n+1)_n$	$\dfrac{1}{2^n n!}(2\alpha+n+1)_n$	$\dfrac{(2\sin\varphi)^n}{n!}$
d_n^2	$\dfrac{\Gamma(\alpha+n+1)\Gamma(\beta+n+1)\left\|\Gamma\!\left(\dfrac{\alpha+\beta}{2}+\mathrm{i}\gamma+n+1\right)\right\|^2}{n!(\alpha+\beta+2n+1)\Gamma(\alpha+\beta+n+1)}$	$\dfrac{\Gamma^2(\alpha+n+1)\Gamma(\alpha+\delta+n+1)\Gamma(\alpha-\delta+n+1)}{n!(2\alpha+2n+1)\Gamma(2\alpha+n+1)}$	$\dfrac{\Gamma(2\lambda+n)}{n!(2\sin\varphi)^{2\lambda}}$
α_n	$\dfrac{2(n+1)(\alpha+\beta+n+1)}{(\alpha+\beta+2n+1)(\alpha+\beta+2n+2)}$	$\dfrac{(n+1)(2\alpha+n+1)}{(2\alpha+2n+1)(\alpha+n+1)}$	$\dfrac{n+1}{2\sin\varphi}$
β_n	$\dfrac{\gamma(\beta^2-\alpha^2)}{(\alpha+\beta+n)(\alpha+\beta+2n+2)}$	0	$-(\lambda+n)\dfrac{\cos\varphi}{\sin\varphi}$
γ_n	$\dfrac{2(\alpha+n)(\beta+n)[\gamma^2+((\alpha+\beta)/2+n)^2]}{(\alpha+\beta+2n)(\alpha+\beta+2n+1)}$	$(\alpha+n)\dfrac{[(\alpha+n)^2-\delta^2]}{2\alpha+2n+1}$	$\dfrac{2\lambda+n-1}{2\sin\varphi}$
Definition	$\mathrm{i}^{-n}h_n^{(\alpha,\beta)}(z,N),$ $z=\mathrm{i}\dfrac{t+\gamma}{2}-\dfrac{\beta+1}{2},$ $N=\mathrm{i}\gamma-\dfrac{\alpha+\beta}{2}$	$p_n^{(\alpha,\alpha)}(t,-\mathrm{i}\delta)=p_n^{(\alpha-\delta,\alpha+\delta)}(t,0)$	$\dfrac{e^{-\mathrm{i}n\varphi}}{n!}m_n^{(2\lambda,\mu)}(\mathrm{i}t-\lambda),$ $\mu=e^{-2\mathrm{i}\varphi}$

where $f(s)$ is a periodical factor of period 1. Since

$$\varrho_1(s) = \Gamma(s+a+1)\Gamma(s+b+1)\Gamma(s+c+1)\Gamma(s+d+1)$$
$$\times \Gamma(a-s)\Gamma(b-s)\Gamma(c-s)\Gamma(d-s)f(s)$$

and $\Delta x(s-1/2) = 2s$, the conditions for contour C will be satisfied, for example, if $a > 0$, $b > 0$, $c > 0$, $d > 0$, while the contour C is the straight lines $s = is'$, $-\infty < s' < +\infty$.

In this case

$$\varrho(s)\Delta x\left(s - \tfrac{1}{2}\right) = \left|\Gamma(is' + a)\Gamma(is' + \beta)\Gamma(is' + c)\Gamma(is' + d)\right|^2 f(is')2is' \ .$$

In order that the function $\varrho(s)\Delta x(s - 1/2)$ take positive values when $-\infty < s' < \infty$ it is convenient to choose the periodic function $f(s)$ in the form $f(s) = \sin(2\pi s)$. Then for the polynomials $p_n(t; a, b, c, d) = y_n(-x)\,(t = s'^2)$ we come to the orthogonality relation

$$\int_0^\infty p_n(t; a, b, c, d)p_m(t; a, b, c, d)\bar{\varrho}(t)dt = 0 \quad (n \ne m) \ ,$$

where

$$\bar{\varrho}(t) = \left|\Gamma\left(it^{1/2} + a\right)\Gamma\left(it^{1/2} + b\right)\Gamma\left(it^{1/2} + c\right)\Gamma\left(it^{1/2} + d\right)\right|^2$$
$$\times \sinh\left(2\pi t^{1/2}\right) \ . \tag{3.10.27}$$

Comparing the functions $x(s)$ and $\sigma(s)$ for the polynomials $y_n(s)$ in this case with the same functions for the Racah polynomials $u_n^{(\alpha,\beta)}(x, a, b)$, we obtain that the polynomials $p_n(t; a, b, c, d)$ coincide with the Racah polynomials $u_n^{(\alpha,\beta)}(x, \bar{a}, \bar{b})$ to within a constant factor for $x = -t - 1/4$, $\bar{a} = a - 1/2$, $\bar{b} = 1/2 - b$, $\alpha = b + d - 1$, $\beta = a + c - 1$.

We may obtain the orthogonality property for the polynomials $p_n(t; a, b, c, d)$ with the weight $\bar{\varrho}(t)$ not only when $a > 0$, $b > 0$, $c > 0$, $d > 0$ but in other cases too, for example, when $b = a^*$, $c > 0$, $d > 0$. The orthogonal polynomials $p_n(t; a, b, c, d)$ with the weight $\bar{\varrho}(t)$ were considered by Willson in [W8]. A different method was used in [M14] to obtain the orthogonality property for the polynomials $p_n(t; a, b, c, d)$.

In a similar way one can introduce polynomials $p_n^{(\alpha,\beta)}(t, \gamma, \delta)$ for which parameters α and β are chosen by analogy with the Jacobi polynomials $P_n^{(\alpha,\beta)}(x)$ (see Table 3.9). The following limit relation is valid:

$$(\gamma^2 - \delta^2)^{-n} p_n^{(\alpha,\beta)}\left[\frac{\gamma^2 + \delta^2}{2} + \frac{\gamma^2 - \delta^2}{2}x, y, \delta\right] = P_n^{(\alpha,\beta)}(x) + O\left(\frac{1}{\gamma^2 - \delta^2}\right)$$

as $\gamma \to \infty$.

When $\sigma(s) = (s - a)(s - b)(s - c)$ a continuous orthogonality property may be obtained by analogy with the above (see Table 3.9). This property was established in [W8] as well as in [A32] (in the last publication it was considered in connection with applications).

Table 3.9. Continuous orthogonality property for the Racah and dual Hahn polynomials

$p_n(t)$	$p_n^{(\alpha,\beta)}(t, \gamma, \delta)$	$p_n(t, a, b, c)$
(a, b)	$(0, \infty)$	$(0, \infty)$
$\varrho(t)$	$\dfrac{1}{2\pi^2}\left\|\Gamma\left(it^{1/2} - i\gamma + \dfrac{\alpha+1}{2}\right)\Gamma\left(it^{1/2} + i\gamma + \dfrac{\alpha+1}{2}\right)\right\|^2$ $\times\left\|\Gamma\left(it^{1/2} - i\delta + \dfrac{\beta+1}{2}\right)\Gamma\left(it^{1/2} + i\delta + \dfrac{\beta+1}{2}\right)\right\|^2 \sinh(2\pi t^{1/2})$ $(\alpha > -1, \beta > -1)$	$\dfrac{1}{2\pi^2}\|\Gamma(it^{1/2} + a)\Gamma(it^{1/2} + b)\Gamma(it^{1/2} + c)\|^2$ $\times \sinh(2\pi t^{1/2})$ $(a > 0,\ b > 0,\ c > 0;$ $a = b^*,\ \operatorname{Re} a > 0,\ c > 0)$
a_n	$\dfrac{1}{n!}(\alpha + \beta + n + 1)_n$	$\dfrac{1}{n!}$
d_n^2	$\dfrac{\Gamma(\alpha + n + 1)\Gamma(\beta + n + 1)\|\Gamma[(\alpha+\beta)/2 + i\gamma + i\delta + n + 1]\|^2}{n!(\alpha + \beta + n + 1)\Gamma(\alpha + \beta + 2n + 1)}$ $\times\left\|\Gamma\left(\dfrac{\alpha+\beta}{2} - i\gamma + i\delta + n + 1\right)\right\|^2$	$\dfrac{1}{n!}\Gamma(a + b + n)\Gamma(a + c + n)\Gamma(b + c + n)$
α_n	$\dfrac{(n + 1)(\alpha + \beta + n + 1)}{(\alpha + \beta + 2n + 1)(\alpha + \beta + 2n + 2)}$	$n + 1$
β_n	$\dfrac{1}{4}(\alpha + 1)(\beta + 1) + \dfrac{1}{2}(\gamma^2 + \delta^2) + \dfrac{n}{2}(\alpha + \beta + n + 1)$ $+ \dfrac{(\beta^2 - \alpha^2)(\gamma^2 - \delta^2)}{2(\alpha + \beta + 2n)(\alpha + \beta + 2n + 2)}$	$ab + ac + bc + 2n^2$ $+ n(2a + 2b + 2c - 1)$
γ_n	$\dfrac{(\alpha + n)(\beta + n)\{[(\alpha + \beta)/2 + n]^2 + (\gamma - \delta)^2\}\{[(\alpha + \beta)/2 + n]^2 + (\gamma + \delta)^2\}}{(\alpha + \beta + 2n)(\alpha + \beta + 2n + 2)}$	$(a + b + n - 1)(a + c + n - 1)(b + c + n - 1)$
Definition	$(-1)^n u_n^{(\alpha,\beta)}(x, a, b),\ x = -t - \tfrac{1}{4},$ $a = \dfrac{\beta}{2} + i\delta,\ b = -\dfrac{\alpha}{2} + i\gamma$	$(-1)^n w_n^{(c-1/2)}(x, a - \tfrac{1}{2}, \tfrac{1}{2} - b),$ $x = -t - \tfrac{1}{4}$

3.10.5. *The lattice $x(s) = C_1 q^s + C_2 q^{-s}$.* Now we discuss the continuous orthogonality property for polynomial solutions of the difference equation of hypergeometric type on the lattice $x(s) = C_1 q^s + C_2 q^{-s}$.

3.10.5.1. Let $C_1 = 1$, $C_2 = 0$, i.e. $x(s) = q^s$. We consider the simplest case when

$$\sigma(s) = q^{2s}, \quad \sigma(s) + \tau(s)\Delta x\left(s - \tfrac{1}{2}\right) = q^{3/2} q^s .$$

In this case $\varrho(s) = q^{-s^2/2}$, and if $q > 1$ the conditions (3.10.7) are satisfied provided the contour C is the straight line $(-\infty < s < +\infty)$; for the polynomials $y_n(x)$ we obtain the following orthogonality property

$$\int_{-\infty}^{\infty} y_n[x(s)]y_m[x(s)]\varrho(s)\Delta x\left(s - \tfrac{1}{2}\right) ds = 0 \quad (n \neq m) .$$

At $q = \exp(1/2k^2)$ for the polynomials $y_n(x)$, which will be denoted by $p_n(x, k)$ (they are called the Stieltjes-Wigert polynomials) the orthogonality relation may be rewritten in the form

$$\int_0^{\infty} p_n(x, k)p_m(x, k)\bar\varrho(x)dx = 0 \quad (n \neq m) ,$$

where $\bar\varrho(x) = \exp(-k^2 \ln^2 x)$ (see [S24, W6, S38, C18]).

The polynomials $p_n(x, k)$ have also the discrete orthogonality property given by

$$\sum_i p_n\left(x_i, k\right)p_m\left(x_i, k\right)\bar\varrho\left(x_i\right)x_i = 0 \quad (n \neq m)$$

$$(i = 0, \pm 1, \pm 2, \dots)$$

with $x_i = \exp(i/2k^2)$.

3.10.5.2. For $x(s) = \left(q^s + q^{-s}\right)/2$ we have in the most general form[2]

$$\sigma(s) = q^{-2s}\left(q^s - a\right)\left(q^s - b\right)\left(q^s - c\right)\left(q^s - d\right) = q^{2s} \prod_{v=a,b,c,d} \left(1 - vq^{-s}\right) ,$$

$$\frac{\varrho(s + 1)}{\varrho(s)} = \frac{\sigma(-s)}{\sigma(s + 1)} = q^{-4s-2} \prod_{v=a,b,c,d} \frac{1 - vq^s}{(1 - vq^{-s-1})} . \tag{3.10.28}$$

In order to find the solutions of this equation that would satisfy the condition (3.10.1) with $0 < q < 1$, we use the periodicity property of $x(s)$ and the right-hand side of Eq. (3.10.28) with period $2\pi i/\ln q$. We shall search for the solutions of (3.10.28) such that they will be periodic with period $2\pi i/\ln q$, and moreover we require that the parameters a, b, c, d satisfy the condition when the function

---

[2] In Sects. 3.10.5.2 and 3.10.5.3 it would be better and more natural to use notations that were accepted in Sect. 3.6.2.1. However, in this case we decided to keep the notations of the original work [A29].

$\varrho_1(s)$ has no singularities at the band $s_1 - 1 \leq \mathrm{Re}\, s \leq s_1$ at a certain value of constant s_1. Then by considering the zero integral of function $\varrho_1(s)x_1^k(s)$ (periodic with period $2\pi \mathrm{i}/\ln q$) over a closed contour consisting of segments of the straight lines $\left[s_1 - 1,\, s_1\right]$, $\left[s_1,\, s_1 + 2\pi \mathrm{i}/\ln q\right]$, $\left[s_1 + 2\pi \mathrm{i}/\ln q,\, s_1 - 1 + 2\pi \mathrm{i}/\ln q\right]$, $\left[s_1 - 1 + 2\pi \mathrm{i}/\ln q,\, s_1\right]$ we obtain that a sum of integrals over the segments $\left[s_1 - 1,\, s_1\right]$ and $\left[s_1 + 2\pi \mathrm{i}/\ln q,\, s_1 - 1 + 2\pi \mathrm{i}/\ln q\right]$ is zero owing to the periodicity of function $\varrho_1(s)x_1^k(s)$. Therefore the integral over the segment $\left[s_1 - 1,\, s_1 - 1 + 2\pi \mathrm{i}/\ln q\right]$ is equal to the integral over $\left[s_1,\, s_1 + 2\pi \mathrm{i}/\ln q\right]$, i.e. the condition (3.10.10) holds if the contour C is the segment $\left[s_1,\, s_1 + 2\pi \mathrm{i}/\ln q\right]$. Now we shall construct the solution of Eq. (3.10.28) that is a periodic function with period $2\pi \mathrm{i}/\ln q$. The solution of this equation has the form:

$$\varrho(s) = f(s) \prod_{v=a,b,c,d} g(s,v)\,, \tag{3.10.29}$$

where $f(s)$ and $g(s,v)$ are periodic functions that satisfy the relations

$$\frac{f(s+1)}{f(s)} = q^{-4s-2}\,, \tag{3.10.30}$$

$$\frac{g(s+1,v)}{g(s,v)} = \frac{1 - vq^s}{1 - vq^{-s-1}}\,. \tag{3.10.31}$$

A particular solution of Eq. (3.10.31) may be expressed through the functions $\Gamma_q(s)$ if we put $v = q^u$:

$$g(s,v) = \mathrm{const}\, \Gamma_q(s+u)\Gamma_q(u-s)\,.$$

If we represent the functions $\Gamma_q(s+u)$ and $\Gamma_q(u-s)$ as infinite products and choose an appropriate constant factor, the expression for $g(s,v)$ may be rewritten in the form

$$g(s,v) = \frac{1}{\prod_{k=0}^{\infty}(1 - q^{k+u+s})(1 - q^{k+u-s})}\,,$$

i.e.

$$g(s,v) = \frac{1}{\prod_{k=0}^{\infty}(1 - vq^{k+s})(1 - vq^{k-s})}\,. \tag{3.10.32}$$

It is easy to verify that $g(s \pm 2\pi \mathrm{i}/\ln q,\, v) = g(s,v)$.

The simplest solution of the equation for $f(s)$ has the form $f(s) = q^{-2s^2}$. However, it does not satisfy the periodicity condition with period $2\pi \mathrm{i}/\ln q$. By using the representation of $\Gamma_q(s)$ in the form of an infinite product it is easy to verify that the solution of Eq. (3.10.30) in the form

$$f(s) = f_q(s) = \frac{1}{\Gamma_q(2s)\Gamma_q(-2s)(q^s - q^{-s})} \tag{3.10.33}$$

satisfies the above condition. So let the function $\varrho(s) = \varrho(s,a,b,c,d)$ be determined by Eq. (3.10.29), where $f(s)$ and $g(s,v)$ are given by (3.10.33) and (3.10.32). By means of the easily verified relations

$$q^{2s+2}f(s+1) = -q^{1/2}f\left(s+\tfrac{1}{2}\right) ,$$
$$\left(1 - vq^{-s-1}\right)g(s+1,v) = g\left(s+\tfrac{1}{2}, vq^{1/2}\right) \tag{3.10.34}$$

we may obtain that

$$\varrho_1(s) = \sigma(s+1)\varrho(s+1)$$
$$= -q^{1/2}\varrho\left(s+\tfrac{1}{2}, aq^{1/2}, bq^{1/2}, cq^{1/2}, dq^{1/2}\right) . \tag{3.10.35}$$

By representing the function $\Gamma_q(s)$ in the form of an infinite product we may deduce that the function $f(s)$ has no singularities. Meanwhile the function $g(s, vq^{1/2})$ has singularities only for the values of s such that $q^{k+s+1/2} = 1/v$ or $q^{k-s+1/2} = 1/v$ with certain values of k $(k = 0, 1, \ldots)$. Hence when $|v| < 1$, $0 < q < 1$ the function $g(s, vq^{1/2})$ has no singularities in the band $-1 \le \operatorname{Re} s \le 0$. Thus we obtain that for $-1 \le \operatorname{Re} s \le 0$, $|a| < 1$, $|b| < 1$, $|c| < 1$, $|d| < 1$ the function $\varrho_1(s)x_1^k(s)$ has no singularities. Moreover, this function is periodic with period $2\pi i/\ln q$. As a result we obtain the continuous orthogonality property on the contour C which is the straight line $s = is'$, $2\pi/\ln q \le s' < 0$. By putting $t = x(is')$, $y_n(is') = p_n(t, a, b, c, d|q)$, where $p_n(t, a, b, c, d|q)$ is a polynomial of degree n, and by presenting the functions $\Gamma_q(2s)$, $\Gamma_q(-2s)$ in the form of an infinite product we may rewrite the orthogonality relation as

$$\int_{-1}^{1} p_n(t, a, b, c, d|q)p_m(t, a, b, c, d|q)\bar{\varrho}(t)dt = 0 \quad (n \ne m) , \tag{3.10.36}$$

where

$$\bar{\varrho}(t) = \left(1 - t^2\right)^{-1/2} \frac{\prod_{k=0}^{\infty}\left[1 - 2\left(2t^2 - 1\right)q^k + q^{2k}\right]}{\prod_{v=a,b,c,d} \prod_{k=0}^{\infty}\left(1 - 2vtq^k + v^2q^{2k}\right)} . \tag{3.10.37}$$

The polynomials $p_n(t, a, b, c, d|q)$ that satisfy the orthogonality relation (3.10.37) were introduced in [A29]. They are called the Askey-Wilson or q-Wilson polynomials. The orthogonality property of these polynomials was first proved in [A29].

Remark. In an analogous manner we may establish the orthogonality property of the Askey-Wilson polynomials for $-1 < q < 0$ if $\max_{v=a,b,c,d} |v| < 1$. In the limit $q \to 0$, $x = \text{const}$ the weight function (3.10.37) takes the form:

$$\varrho(x) = 4\frac{\sqrt{1 - t^2}}{\prod_{v=a,b,c,d}\left(1 - 2vt + v^2\right)} .$$

In this case according to [S38, A29] we have

$$p_n(x, a, b, c, d|0) \equiv p_n(x) = v_n(x) - \sigma_1 U_{n-1}(x)$$
$$+ \sigma_2 U_{n-2}(x) - \sigma_3 U_{n-3}(x) + \sigma_4 U_{n-4}(x), \quad n \ge 3 ,$$

$$p_2(x) = U_2(x) - \sigma_1 U_1(x) + (\sigma_2 - \sigma_4)U_0(x) ,$$

$$p_1(x) = (1 - \sigma_4)U_1(x) + (\sigma_3 - \sigma_1)U_0(x) ,$$

$$p_0(x) = U_0(x) = 1 ,$$

where

$$U_n(\cos\vartheta) = \frac{\sin(n+1)\vartheta}{\sin\vartheta} , \quad U_{-1}(x) = 0 ,$$

$$\sigma_1 = a + b + c + d , \quad \sigma_2 = ab + ac + ad + bc + bd + cd ,$$

$$\sigma_3 = abc + abd + acd + bcd , \quad \sigma_4 = abcd .$$

According to (3.10.35) Eq. (3.2.33) for the Askey-Wilson polynomials $y_n(s) = p_n(t, a, b, c, d|q)$ may be rewritten in the form

$$\frac{\delta}{\delta x(s)}\left[\varrho\left(s, aq^{1/2}, bq^{1/2}, cq^{1/2}, dq^{1/2}\right)\frac{\delta y_n(s)}{\delta x(s)}\right]$$

$$= \lambda_n q^{-1/2}\varrho(s, a, b, c, d)y_n(s) , \tag{3.10.38}$$

where

$$\lambda_n = 4q^{3/2}\frac{(1 - q^{-n})(1 - abcdq^{n-1})}{(1 - q)^2}$$

(cf. [A29], in which the factor q is absent).

By generalizing Eq. (3.10.35) one can obtain that

$$\varrho_n(s) = (-1)^n q^{-n(n-2)/2}\varrho\left(s + \frac{n}{2}, aq^{n/2}, bq^{n/2}, cq^{n/2}, dq^{n/2}\right) .$$

Therefore the Rodrigues-type formula (3.2.32) for Askey-Wilson polynomials has the form

$$y_n(s) = \frac{(-1)^n B_n q^{-n(n-2)/2}}{\varrho(s, a, b, c, d)}\left(\frac{\delta}{\delta x(s)}\right)^n\left[\varrho\left(s, aq^{n/2}, bq^{n/2}, cq^{n/2}, dq^{n/2}\right)\right] ,$$

where $(3.10.39)$

$$B_n = \left(\frac{1-q}{2}\right)^n q^{n(3n-5)/4} \quad \text{(cf. [A29])} .$$

3.10.5.3. The different special cases of the Askey-Wilson polynomials $p_n(t, a, b, c, d|q)$ are the so called q-Jacobi polynomials $P_n^{(\alpha,\beta)}(t|q)$ [A29] and $P_n^{(\alpha,\beta)}(t; q)$ [R6], the continuous q-ultraspherical polynomials $C_n(t; \beta|q)$ [R22, A29], the q-Laguerre polynomials $L_n(t|q)$ [A7, A29] and the q-Hermite polynomials $H_n(t, q)$ [R22, S37, A26, A29] for which we have, respectively,

$$\sigma(s) = \begin{cases} q^{-2s}\left(q^s - q^{\alpha/2+1/4}\right)\left(q^s + q^{\beta/2+1/4}\right)\left(q^s - q^{\alpha/2+3/4}\right)\left(q^s + q^{\beta/2+3/4}\right) , \\ \left(1 - q^{1-2s}\right)\left(q^s - q^{\alpha+1/2}\right)\left(q^s + q^{\beta+1/2}\right) , \\ q^{-2s}\left(q^{2s} - \beta\right)\left(q^{2s} - \beta q\right) , \\ \left(q^s - q^{1-s}\right)\left(q^s - q^{\alpha+1/2}\right) , \\ q^{2s} ; \end{cases}$$

Table 3.10. Lattice $x(s) = \cosh(2\omega s) = (q^s + q^{-s})/2$, $q = e^{2\omega}$. The continuous orthogonality property for the Askey-Wilson polynomials $p_n(t, a, b, c, d|q)$, the q-Jacobi polynomials $P_n^{(\alpha,\beta)}(t|q)$ and $P_n^{(\alpha,\beta)}(t; q)$, the continuous q-ultraspherical polynomials $C_n(t; \beta|q)$, the q-Laguerre polynomials $L_n(t|q)$ and the q-Hermite polynomials $H_n(t|q)$.

$p_n(t)$	(a, b)	$\varrho(t)$	$\sigma(s)$
$p_n(t, a, b, c, d|q)$	$(-1, 1)$	$(1 - t^2)^{-1/2} \dfrac{\prod\limits_{k=0}^{\infty} (1 - 2(2t^2 - 1)q^k + q^{2k})}{\prod\limits_{v=a,b,c,d} \prod\limits_{k=0}^{\infty} (1 - 2vtq^k + v^2q^{2k})}$ $\max(|a|, |b|, |c|, |d|) < 1,\ -1 < q < 1$	$q^{-2s} \prod\limits_{v=a,b,c,d} (q^s - v)$
$P_n^{(\alpha,\beta)}(t|q)$	$(-1, 1)$	$(1 - t^2)^{-1/2} \left| \dfrac{\Gamma_{q^{1/2}}(\alpha + 2is' + \tfrac{1}{2}) \Pi_{q^{1/2}}(\beta + 2is' + \tfrac{1}{2})}{\Gamma_{q^{1/2}}(2is') \Pi_{q^{1/2}}(2is')} \right|^2$ $\alpha \geq -\tfrac{1}{2},\ \beta \geq \tfrac{1}{2},\ 0 < q < 1$	$q^{-2s}(q^s - q^{\alpha/2 + 1/4})(q^s + q^{\beta/2 + 1/4})$ $\times (q^s - q^{\alpha/2 + 3/4})(q^s + q^{\beta/2 + 3/4})$
$P_n^{(\alpha,\beta)}(t; q)$	$(-1, 1)$	$(1 - t^2)^{-1/2} \left| \dfrac{\Gamma_q(\alpha + is' + \tfrac{1}{2}) \Pi_q(\beta + is' + \tfrac{1}{2})}{\Gamma_q(is') \Pi_q(is')} \right|^2$ $\alpha \geq -\tfrac{1}{2},\ \beta \geq -\tfrac{1}{2},\ 0 < q < 1$ $s' = \dfrac{1}{2\omega} \arccos t$	$(1 - q^{1 - 2s})(q^s - q^{\alpha + 1/2})$ $\times (q^s + q^{\beta + 1/2})$
$C_n(t; \beta|q)$	$(-1, 1)$	$(1 - t^2)^{-1/2} \prod\limits_{k=0}^{\infty} \dfrac{1 - 2(2t^2 - 1)q^k + q^{2k}}{1 - 2(2t^2 - 1)\beta q^k + \beta^2 q^{2k}}$ $-1 < \beta < 1,\ -1 < q < 1$	$q^{-2s}(q^{2s} - \beta)(q^{2s} - \beta q)$
$L_n(t|q)$	$(-1; 1)$	$(1 - t^2)^{-1/2} \left| \dfrac{\Gamma_q(\alpha + is' + \tfrac{1}{2})}{\Gamma_q(is') \Pi_q(is')} \right|^2,\ \alpha \geq -\tfrac{1}{2},\ 0 < q < 1$ $s' = \dfrac{1}{2\omega} \arccos t$	$(q^s - q^{1 - s})(q^s - q^{\alpha + 1/2})$
$H_n(t|q)$	$(-1, 1)$	$(1 - t^2)^{-1/2} \prod\limits_{k=0}^{\infty} (1 - 2(2t^2 - 1)q^k + q^{2k}),$ $-1 < q < 1$	q^{2s}

$$\varrho(s) = \begin{cases} f_q(s)\Gamma_{q^{1/2}}\left(\alpha + 2s + \tfrac{1}{2}\right) \Gamma_{q^{1/2}}\left(\alpha - 2s + \tfrac{1}{2}\right) \Pi_{q^{1/2}}\left(\beta + 2s + \tfrac{1}{2}\right) \\ \quad \times \Pi_{q^{1/2}}\left(\beta - 2s + \tfrac{1}{2}\right), \\ f_{q^2}(s/2)\Gamma_q\left(\alpha + s + \tfrac{1}{2}\right) \Gamma_q\left(\alpha - s + \tfrac{1}{2}\right) \Pi_q\left(\beta + s + \tfrac{1}{2}\right) \\ \quad \times \Pi_q\left(\beta - s + \tfrac{1}{2}\right), \\ f_q(s)\Gamma_q(2\gamma + 2s)\Gamma_q(2\gamma - 2s),\ \beta = q^{2\gamma}, \\ f_{q^2}\left(\tfrac{s}{2}\right) \Gamma_q\left(\alpha + s + \tfrac{1}{2}\right) \Gamma_q\left(\alpha - s + \tfrac{1}{2}\right), \\ f_q(s). \end{cases}$$

Here

$$f_q(s) = \left[\Gamma_q(2s)\Gamma_q(-2s)\left(q^s - q^{-s}\right) \right]^{-1}, \quad \Pi_q(s) = \Gamma_{q^2}(s)/\Gamma_q(s).$$

Also the identities

$$\Gamma_q(s)\Gamma_q\left(s+\tfrac{1}{2}\right) = \Gamma_q\left(\tfrac{1}{2}\right)\left(1+q^{1/2}\right)^{1-2s}\Gamma_{q^{1/2}}(2s),$$

$$\Pi_q(s)\Pi_q\left(s+\tfrac{1}{2}\right) = \Pi_q\left(\tfrac{1}{2}\right)\left(\frac{1+q}{1+q^{1/2}}\right)^{1-2s}\Pi_{q^{1/2}}(2s)$$

were used. For details about these polynomials see [A29].

3.11 Representation in Terms of Generalized Hypergeometric and q-Hypergeometric Functions

This section concludes the discussion of the theory of polynomial solutions of difference equations of hypergeometric type on nonuniform lattices. It will be shown that any of the considered polynomials may be expressed through hypergeometric functions and their generalization, in which the functions $\tilde{\Gamma}_q(z)$ are used instead of gamma-functions $\Gamma(z)$. Generalization is constructed by replacing the Pochhammer symbol

$$(a)_k = \frac{\Gamma(a+k)}{\Gamma(a)} = \prod_{l=0}^{k-1}(a+l),$$

which enters into the definition of the hypergeometric function (2.7.1), by the expression

$$(a|q)_k = \frac{\tilde{\Gamma}_q(a+k)}{\tilde{\Gamma}_q(a)} = \prod_{l=0}^{k-1}\psi_q(a+l),\tag{3.11.1}$$

which transforms into $(a)_k$ when $q \to 1$, since $\lim_{q\to 1}\psi_q(a+l) = a+l$.

3.11.1. We consider first a simplest case of linear and quadratic lattices when

$$x(s) = C_1 s^2 + C_2 s + C_3,\tag{3.11.2}$$

where C_1, C_2 and C_3 are arbitrary constants. To obtain representations of the polynomials $y_n(s)$ in terms of hypergeometric functions we shall start from the Rodrigues formula by analogy with Sect. 2.7. In this case the Rodrigues formula (3.2.31) may essentially be simplified by using the symmetry property of the lattice function, i.e.

$$x(s) = x(-s-\mu), \quad \mu = \frac{C_2}{C_1}.\tag{3.11.3}$$

Since

$$\Delta x\left(s-\tfrac{1}{2}\right) = -\Delta x\left(t-\tfrac{1}{2}\right)\Big|_{t=-s-\mu},$$

from (3.1.22) and (3.1.23) we have

$$\sigma(s) + \tau(s)\Delta x\left(s-\tfrac{1}{2}\right) = \sigma(-s-\mu).\tag{3.11.4}$$

By assuming in accordance with (3.5.5) that

$$\sigma(s) = A \prod_{i=1}^{4} (s - s_i) \tag{3.11.5}$$

and by taking into account that for the lattice (3.11.2) $q = 1$ and the function $\tilde{\Gamma}_q(n+1)$ should be replaced by $n!$ we shall write (3.2.31) in the form

$$y_n(s) = (-1)^n B_n \sum_{k=0}^{n} \frac{(-1)^k n!}{k!(n-k)!} \frac{\nabla x[s + k - (n-1)/2]}{\prod_{l=0}^{n} \nabla x[s + (k-l+1)/2]}$$

$$\times A^n \prod_{i=1}^{4} \left\{ \prod_{l=0}^{n-k-1} (s - s_i - l) \prod_{l=0}^{k-1} (s + s_i + \mu + l) \right\} .$$

Since for the function (3.11.2)

$$\Delta x \left(s - \tfrac{1}{2} \right) = 2C_1 s + C_2 = (2s + \mu)C_1 ,$$

we have

$$\nabla x \left(s + k - \frac{n-1}{2} \right) = \left[2 \left(s + k - \frac{n}{2} \right) + \mu \right] C_1$$

$$= 2C_1 \left(s + \frac{\mu - n}{2} + k \right)$$

$$= C_1 \frac{[s + (\mu - n)/2 + 1]_k}{[s + (\mu - n)/2]_k} (2s + \mu - n) ,$$

where the formula

$$a + k = \frac{(a+1)_k}{(a)_k} a$$

was used. Analogously we may obtain the following relations:

$$\nabla x \left(s + \frac{k-l+1}{2} \right) = (2s + \mu + k - l)C_1 ;$$

$$\prod_{l=0}^{n} (2s + \mu + k - l) = (2s + \mu + k - n)_{n+1} = \frac{\Gamma(2s + \mu + k + 1)}{\Gamma(2s + \mu + k - n)}$$

$$= \frac{(2s + \mu + 1)_k}{(2s + \mu - n)_k} \frac{\Gamma(2s + \mu + 1)}{\Gamma(2s + \mu - n)} ;$$

$$\prod_{l=0}^{n-k-1} (s - s_i - l) = (s - s_i - n + k + 1)_{n-k}$$

$$= \frac{\Gamma(s - s_i + 1)}{\Gamma(s - s_i - n + k + 1)}$$

$$= \frac{\Gamma(s - s_i + 1)}{(s - s_i + 1 - n)_k \Gamma(s - s_i + 1 - n)} = \frac{(s - s_i + 1 - n)_n}{(s - s_i + 1 - n)_k} ;$$

$$\prod_{l=0}^{k-1}\left(s + s_i + \mu + l\right) = \left(s + s_i + \mu\right)_k .$$

Hence we have

$$y_n(s) = \frac{(-1)^n A^n B_n}{C_1^n} \frac{\Gamma(2s + \mu - n + 1)}{\Gamma(2s + \mu + 1)} \prod_{i=1}^{4}(s - s_i + 1 - n)_n$$

$$\times \sum_{k=0}^{n} \frac{(-n)_k}{k!} \frac{[s + (\mu - n)/2 + 1]_k}{[s + (\mu - n)/2]_k} \frac{(2s + \mu - n)_k}{(2s + \mu + 1)_k}$$

$$\times \prod_{i=1}^{4} \frac{(s + s_i + \mu)}{(s - s_i + 1 - n)_k} = \frac{(-1)^n A^n B_n}{C_1^n} \frac{\prod_{i=1}^{4}(s - s_i + 1 - n)_n}{(2s + \mu + 1 - n)_n}$$

$$\times {}_7F_6\left(\begin{array}{c} -n,\, v,\, \frac{v}{2}+1,\, a_1,\, a_2,\, a_3,\, a_4 \\ \frac{v}{2},\, 1+v+n,\, 1+v-a_1,\, 1+v-a_2,\, 1+v-a_3,\, 1+v-a_4 \end{array}\middle|\, 1\right),$$

$$(3.11.6)$$

where $v = 2s + \mu - n$, $a_i = s + s_i + \mu$. The expression (3.11.6) may be considerably simplified if we use first Whipple's transformation [B1]

$${}_7F_6\left(\begin{array}{c} -n,\, v,\, \frac{v}{2}+1,\, a_1,\, a_2,\, a_3,\, a_4 \\ \frac{v}{2},\, 1+v+n,\, 1+v-a_1,\, 1+v-a_2,\, 1+v-a_3,\, 1+v-a_4 \end{array}\middle|\, 1\right)$$

$$= \frac{(1+v)_n(1+v-a_1-a_2)_n}{(1+v-a_1)_n(1+v-a_2)_n}$$

$$\times {}_4F_3\left(\begin{array}{c} -n,\, 1+v-a_3-a_4,\, a_2,\, a_1 \\ a_1 + a_2 - v - n,\, 1+v-a_3,\, 1+v-a_4 \end{array}\middle|\, 1\right) \qquad (3.11.7)$$

and then Bailey's transformation [B1]

$${}_4F_3\left(\begin{array}{c} -n,\, b_1,\, b_2,\, b_3 \\ c_1,\, c_2,\, c_3 \end{array}\middle|\, 1\right) = \frac{(c_2 - b_3)_n(c_3 - b_3)_n}{(c_2)_n(c_3)_n}$$

$$\times {}_4F_3\left(\begin{array}{c} -n,\, c_1 - b_1,\, c_1 - b_2,\, b_3 \\ c_1,\, 1 - c_2 + b_3 - n,\, 1 - c_3 + b_3 - n \end{array}\middle|\, 1\right)$$

$$\left(\sum_{i=1}^{3} c_i = \sum_{i=1}^{3} b_i - n + 1\right). \qquad (3.11.8)$$

In our case

$$\begin{array}{ll} b_1 = 1 - \mu - n - s_3 - s_4 , & c_1 = s_1 + s_2 + \mu , \\ b_2 = \mu + s_2 + s , & c_2 = 1 - n - s_3 + s , \\ b_3 = \mu + s_1 + s , & c_3 = 1 - n - s_4 + s , \end{array}$$

and the condition $\sum_i c_i = \sum_i b_i - n + 1$ is satisfied. As a result of the above transformations we obtain

$$y_n(s) = \left(\frac{A}{C_1}\right)^n B_n (s_1 + s_2 + \mu)_n (s_1 + s_3 + \mu)_n (s_1 + s_4 + \mu)_n$$

$$\times {}_4F_3 \left(\begin{array}{c} -n,\ s_1 + s_2 + s_3 + s_4 + 2\mu + n - 1,\ s_1 - s,\ s_1 + s + \mu \\ s_1 + s_2 + \mu,\ s_1 + s_3 + \mu,\ s_1 + s_4 + \mu \end{array} \middle| 1 \right) .$$

$$(3.11.9)$$

This expression determines a general polynomial solution of Eq. (3.1.21) on the lattices (3.11.2) when $\mu = C_2/C_1$, and $\sigma(s)$ is given in the form of (3.11.5).

3.11.2. Now we consider a general case of the nonuniform lattice

$$x(s) = \tilde{c}_1 q^s + \tilde{c}_2 q^{-s} + \tilde{c}_3 . \tag{3.11.10}$$

We note that arbitrary constants $\tilde{c}_1$, $\tilde{c}_2$ and $\tilde{c}_3$ may, in general, depend on q so that in final expressions we should put $\tilde{c}_1 = \tilde{c}_1(q)$, $\tilde{c}_2 = \tilde{c}_2(q)$, $\tilde{c}_3 = \tilde{c}_3(q)$ in contrast to (3.11.2). By using the symmetry property of the function (3.11.10), i.e.

$$x(s) = x(-s - \mu) , \quad q^\mu = \frac{\tilde{c}_1}{\tilde{c}_2} , \tag{3.11.11}$$

after applying (3.1.22) and (3.1.23) we obtain the relation

$$\sigma(s) + \tau(s)\Delta x \left(s - \tfrac{1}{2} \right) = \sigma(-s - \mu) , \tag{3.11.11a}$$

which is similar to (3.11.4). Here

$$\Delta x \left(s - \tfrac{1}{2} \right) = \left(\tilde{c}_1 q^s - \tilde{c}_2 q^{-s} \right) \left(q^{1/2} - q^{-1/2} \right) = \tilde{c}_1 q^{-\mu/2} \kappa^2 \psi_q (2s + \mu) ,$$

$$\kappa = \kappa(q) = q^{1/2} - q^{-1/2} .$$

From (3.11.11a) and (3.1.22) we have

$$\tilde{\sigma}[x(s)] = \tfrac{1}{2}[\sigma(-s - \mu) + \sigma(s)] ,$$

$$\tilde{\tau}[x(s)] = \frac{\sigma(-s - \mu) - \sigma(s)}{\Delta x(s - 1/2)} .$$

For the classical orthogonal polynomials of a discrete variable the function $\tau(s) \equiv \tilde{\tau}[x(s)]$ is a polynomial of the first degree in $x(s)$ (see Sect. 3.3.1). Hence

$$\tau(s)\Delta x \left(s - \tfrac{1}{2} \right) = q^{-2s} \sum_{k=0}^{4} c_k q^{ks} ,$$

where c_k are constants with $c_4 \neq 0$. Since the expression for $\tilde{\sigma}[x(s)]$ has an analogous form, the function $\sigma(s)$ due to (3.1.22) may be presented as

$$\sigma(s) = q^{-2s} p_4(q^s) ,$$

where $p_4(q^s)$ is the polynomial of at most the fourth degree in q^s.

On the other hand, we may also prove the opposite statement: if

$$\sigma(s) = q^{-2s} p_4(q^s) ,$$

where $p_4(q^s)$ is an arbitrary polynomial of at most the fourth degree in q^s, then the functions

$$\frac{\sigma(-s-\mu)-\sigma(s)}{\Delta x(s-1/2)} \quad \text{and} \quad \tfrac{1}{2}[\sigma(-s-\mu)+\sigma(s)]$$

are polynomials of at most the first and the second degree, respectively, in

$$x(s) = \tilde{c}_1 q^s + \tilde{c}_2 q^{-s} + \tilde{c}_3 = \tilde{c}_1\left(q^s + q^{-s-\mu}\right) + \tilde{c}_3 \quad (\tilde{c}_1/\tilde{c}_2 = q^\mu).$$

Note that the leading coefficient of the polynomial $p_4(q^s)$ and a free term cannot simultaneously be zero because otherwise

$$\tau(s) = \frac{\sigma(-s-\mu)-\sigma(s)}{\Delta x(s-1/2)} = \text{const} ,$$

which may immediately be verified.

We consider a most general case when

$$p_4(q^s) = C \prod_{i=1}^{4} (q^s - q^{s_i}) .$$

Here s_i are roots of the equation $\sigma(s_i) = 0$ $(i = 1,2,3,4)$ and C is a constant. Since the leading coefficient of the polynomial $p_4(q^s)$ and the free term cannot simultaneously be zero, the remaining cases may be obtained from the general case under consideration by taking the limits when for all chosen values of i we have $q^{s_i} \to 0$ (or $q^{s_i} \to \infty$).

Using the equality

$$q^s - q^{s_i} = \kappa q^{(s+s_i)/2}\psi_q\left(s - s_i\right)$$

we may write the expression for $\sigma(s)$ in the general case in the form

$$\sigma(s) = A \prod_{i=1}^{4} \psi_q\left(s - s_i\right) \quad (A = \text{const}) . \tag{3.11.12}$$

The other possible rest cases may be obtained from (3.11.12) by using the limiting process when $q^{s_i} \to 0$ (or $q^{-s_i} \to 0$) for certain values of i with an appropriate choice of the constant $A = A(q)$ (see Sect. 3.11.5). In a general case (3.11.12) the Rodrigues formula (3.2.31) takes the form

$$y_n(s) = B_n \sum_{k=0}^{n} \frac{(-1)^{n-k}\tilde{\Gamma}_q(n+1)}{\tilde{\Gamma}_q(k+1)\tilde{\Gamma}_q(n-k-1)} \left(\frac{A}{\tilde{c}_1 q^{-\mu/2}\kappa^2}\right)^n$$

$$\times \frac{\psi_q(2s+2k-n+\mu)}{\prod_{l=0}^{n} \psi_q(2s+\mu+k-l)}$$

$$\times \prod_{i=1}^{4} \left\{ \prod_{l=0}^{n-k-1} \psi_q(s-l-s_i) \prod_{l=0}^{k-1} \psi_q(-s-l-\mu-s_i) \right\} . \tag{3.11.13}$$

For further transformations we shall use the relations which follow from (3.11.1) and the appropriate definitions. These relations are:

$$\frac{(-1)^k \tilde{\Gamma}_q(n+1)}{\tilde{\Gamma}_q(n-k+1)} = (-n|q)_k = (-1)^k (n-k+1|q)_k , \tag{3.11.14}$$

$$(1|q)_k = \tilde{\Gamma}_q(k+1) ;$$

$$\psi_q(a+k) = \frac{(a+1|q)_k}{(a|q)_k} \psi_q(a) , \quad \tilde{\Gamma}_q(a+1) = \psi_q(a)\tilde{\Gamma}_q(a) , \tag{3.11.15}$$

For example, we shall prove Eq. (3.11.14) like this:

$$(-n|q)_k = \frac{\tilde{\Gamma}_q(-n+k)}{\tilde{\Gamma}_q(-n)} = \prod_{l=0}^{k-1} \psi_q(-n+l) = (-1)^k \prod_{l=0}^{k-1} \psi_q(n-l)$$

$$= (-1)^k \prod_{l=0}^{k-1} \psi_q(n-k+1+l) = (-1)^k (n-k+1|q)_k$$

$$= \frac{(-1)^k \tilde{\Gamma}_q(n+1)}{\tilde{\Gamma}_q(n-k+1)} .$$

By using (3.11.1, 14, 15) we obtain

$$\prod_{l=0}^{n} \psi_q(2s+k+\mu-l) = \prod_{l=0}^{n} \psi_q(2s+k+\mu-n+l)$$

$$= \frac{\tilde{\Gamma}_q(2s+k+\mu+1)}{\tilde{\Gamma}_q(2s+k+\mu-n)} = \frac{(2s+\mu+1|q)_k \tilde{\Gamma}_q(2s+\mu+1)}{(2s+\mu-n|q)_k \tilde{\Gamma}_q(2s+\mu-n)} ;$$

$$\prod_{l=0}^{n-k-1} \psi_q(s-l-s_i) = \prod_{l=0}^{n-k-1} \psi_q(s-s_i-n+k+1+l)$$

$$= \frac{\tilde{\Gamma}_q(s-s_i+1)}{\tilde{\Gamma}_q(s-s_i-n+k+1)} = \frac{\tilde{\Gamma}_q(s-s_i+1)}{\tilde{\Gamma}_q(s-s_i+1-n)} \frac{1}{(s-s_i+1-n|q)_k}$$

$$= \frac{(s-s_i+1-n|q)_n}{(s-s_i+1-n|q)_k} ;$$

$$\prod_{l=0}^{k-1} \psi_q(-s-l-\mu-s_i)$$

$$= (-1)^k \prod_{l=0}^{k-1} \psi_q(s+s_i+\mu+l) = (-1)^k (s+s_i+\mu|q)_k .$$

Further on, since

$$\psi_q(2a) = \frac{1}{\kappa}(q^a - q^{-a}) = \frac{1}{\kappa}(q^{a/2} - q^{-a/2})(q^{a/2} + q^{-a/2})$$

$$= \psi_q(a)q^{-i\pi/2\ln q}(q^{(a+i\pi/\ln q)/2} - q^{-(a+i\pi/\ln q)/2})$$

$$= \kappa q^{-i\pi/2\ln q} \psi_q(a)\psi_q\left(a+\frac{i\pi}{\ln q}\right) ,$$

we have

$$\psi_q(2s + 2k - n + \mu)$$

$$= \kappa q^{-i\pi/2\ln q}\psi_q\left(s + \frac{\mu - n}{2} + k\right)\psi_q\left(s + \frac{\mu - n}{2} + k + \frac{i\pi}{\ln q}\right)$$

$$= \frac{[s + (\mu - n)/2 + 1|q]_k}{[s + (\mu - n)/2|q]_k}\frac{[s + (\mu - n)/2 + i\pi/\ln q + 1|q]_k}{[s + (\mu - n)/2 + i\pi/\ln q|q]_k}$$

$$\times \kappa q^{-i\pi/2\ln q}\psi_q\left(s + \frac{\mu - n}{2}\right)\psi_q\left(s + \frac{\mu - n}{2} + \frac{i\pi}{\ln q}\right)$$

$$= \frac{[s + (\mu - n)/2 + 1|q]_k}{[s + (\mu - n)/2|q]_k}\frac{[s + (\mu - n)/2 + i\pi/\ln q + 1|q]_k}{[s + (\mu - n)/2 + i\pi/\ln q|q]_k}\psi_q(2s + \mu - n)\ .$$

Therefore it follows that (3.11.13) may be written in the form

$$y_n(s) = (-1)^n\left(\frac{A}{\tilde{c}_1 q^{-\mu/2}\kappa^2}\right)^n B_n\frac{\prod_{i=1}^4(s - s_i + 1 - n|q)_n}{(2s + \mu + 1 - n|q)_n}$$

$$\times \sum_{k=0}^n\frac{(-n|q)_k}{(1|q)_k}\frac{[s + (\mu - n)/2 + 1|q]_k}{[s + (\mu - n)/2|q]_k}\frac{[s + (\mu - n)/2 + 1 + i\pi/\ln q|q]_k}{[s + (\mu - n)/2 + i\pi/\ln q|q]_k}$$

$$\times \frac{(2s + \mu - n|q)_k}{(2s + \mu + 1|q)_k}\prod_{i=1}^4\frac{(s + s_i + \mu|q)_k}{(s - s_i + 1 - n|q)_k}\ .$$

Thus

$$y_n(s) = (-1)^n\left(\frac{A}{\tilde{c}_1 q^{-\mu/2}\kappa^2}\right)^n B_n\frac{\prod_{i=1}^4(s - s_i + 1 - n|q)_n}{(2s + \mu + 1 - n|q)_n}$$

$$\times {}_8F_7\left(\begin{array}{c} -n,\ v,\ \frac{v}{2} + 1,\ \frac{v}{2} + 1 + \frac{i\pi}{\ln q},\ a_1,\ a_2,\ a_3,\ a_4 \\ \frac{v}{2},\ \frac{v}{2} + \frac{i\pi}{\ln q},\ 1 + v + n,\ 1 + v - a_1,\ 1 + v - a_2,\ 1 + v - a_3,\ 1 + v - a_4 \end{array}\middle|\ q, 1\right)$$

$$(v = 2s + \mu - n\ ,\quad a_i = s + s_i + \mu)\ . \tag{3.11.16}$$

Here by definition

$${}_{p+1}F_p\left(\begin{array}{c} a_1,\ a_2,\ \ldots,\ a_{p+1} \\ b_1,\ b_2,\ \ldots,\ b_p \end{array}\middle|\ q, z\right)$$

$$= \sum_{k=0}^\infty\frac{(a_1|q)_k(a_2|q)_k\ \ldots\ (a_{p+1}|q)_k}{(b_1|q)(b_2|q)_k\ \ldots\ (b_p|q)_k(1|q)_k}z^k\ , \tag{3.11.17}$$

where $(a|q)_k$ is introduced in (3.11.1). When $q \to 1$ the expression $(a|q)_k$ transforms into $(a)_k$. Therefore it is obvious that when $q \to 1$ the q-hypergeometric function (3.11.17) transforms into the function ${}_{p+1}F_p(z)$ [see (2.7.1)]. For the polynomial $y_n(s)$ the series in (3.11.16) contains only $(n + 1)$ first terms for which $k = 0, 1, \ldots, n$, since $a_1 = -n$. Indeed when $a_1 = -n$ and $k = n + m$ $(m = 1, 2, \ldots)$ the terms in the series are zero because

$$(-n|q)_{n+m} = (-n|q)_n \prod_{l=0}^{m-1} \psi_q(l) , \quad \psi_q(0) = 0 .$$

Remark. In mathematical literature another generalization of the hypergeometric function $_{p+1}F_p(z)$ is commonly used; it is called *the basic hypergeometric series* and has the form

$$_{p+1}\varphi_p \left(\begin{array}{c} a_1, a_2, \ldots, a_{p+1} \\ b_1, b_2, \ldots, b_p \end{array} \middle| q, z \right)$$

$$= \sum_{k=0}^{\infty} \frac{(a_1, q)_k (a_2, q)_k \ldots (a_{p+1}, q)_k}{(b_1, q)_k (b_2, q)_k \ldots (b_p, q)_k (q, q)_k} z^k , \tag{3.11.18}$$

where

$$(a, q)_k = \prod_{l=0}^{k-1} (1 - aq^l) , \quad (a, q)_0 = 1 .$$

The basic hypergeometric series (3.11.18) obviously differs from the *generalized q-hypergeometric series* (3.11.17) used here.

The expression (3.11.16) for $y_n(s)$ is analogous to (3.11.6). Therefore we shall try to simplify (3.11.16) by using transformations similar to (3.11.7) and (3.11.8) which were employed to reduce (3.11.6) to (3.11.9). Let us show that they may be obtained by starting from the Watson and Sears transformations ([B1] and [S4]) for the basic hypergeometric series. To do this we first establish the relationship between the functions (3.11.17) and (3.11.18). We have

$$(q^a, q)_k = \prod_{l=0}^{k-1} (1 - q^{a+l}) = (-1)^k \prod_{l=0}^{k-1} q^{(a+l)/2} \left(q^{(a+l)/2} - q^{-(a+l)/2} \right)$$

$$= (-1)^k \kappa^k q^{[\sum_{l=0}^{k-1}(a+l)]/2} \prod_{l=0}^{k-1} \psi_q(a+l) ,$$

where $\kappa = q^{1/2} - q^{-1/2}$. Hence, owing to (3.11.1), we obtain

$$(q^a, q)_k = (-1)^k \kappa^k q^{k(k-1)/4 + ak/2} (a|q)_k , \tag{3.11.19}$$

and hence

$$_{p+1}\varphi_p \left(\begin{array}{c} q^{a_1}, q^{a_2}, \ldots, q^{a_{p+1}} \\ q^{b_1}, q^{b_2}, \ldots, q^{b_p} \end{array} \middle| q, z \right)$$

$$= \sum_{k=0}^{\infty} \frac{(a_1|q)_k (a_2|q)_k \ldots (a_{p+1}|q)_k}{(b_1|q)_k (b_2|q)_k \ldots (b_p|q)_k (1|q)_k} \left[zq^{\left(\sum_{i=1}^{p+1} a_i - \sum_{i=1}^{p} b_i - 1 \right)/2} \right]^k$$

$$= {}_{p+1}F_p \left(\begin{array}{c} a_1, a_2, \ldots, a_{p+1} \\ b_1, b_2, \ldots, b_p \end{array} \middle| q, t \right) \Bigg|_{t = zq^{\left(\sum_{i=1}^{p+1} a_i - \sum_{i=1}^{p} b_i - 1 \right)/2}} . \tag{3.11.20}$$

The Watson and Sears transformations for the basic hypergeometric series have the forms

$$
{}_8\varphi_7\left(\begin{array}{c} q^{-n},\, qv^{1/2},\, -qv^{1/2},\, v,\, a_1,\, a_2,\, a_3,\, a_4 \\ v^{1/2},\, -v^{1/2},\, vq^{n+1},\, \dfrac{vq}{a_1},\, \dfrac{vq}{a_2},\, \dfrac{vq}{a_3},\, \dfrac{vq}{a_4} \end{array}\ \middle|\ q,\, \dfrac{v^2 q^{n+2}}{a_1 a_2 a_3 a_4}\right)
$$

$$
= \frac{(vq, q)_n (vq/a_1 a_2, q)_n}{(vq/a_1, q)_n (vq/a_2, q)_n}\ {}_4\varphi_3\left(\begin{array}{c} q^{-n},\, a_1,\, a_2,\, \dfrac{vq}{a_3 a_4} \\ \dfrac{a_1 a_2}{vq^n},\, \dfrac{vq}{a_3},\, \dfrac{vq}{a_4} \end{array}\ \middle|\ q,\, q\right),\qquad (3.11.21)
$$

$$
{}_4\varphi_3\left(\begin{array}{c} q^{-n},\, b_1,\, b_2,\, b_3 \\ c_1,\, c_2,\, c_3 \end{array}\ \middle|\ q,\, q\right)
$$

$$
= \left(\frac{b_1 b_2}{c_1}\right)^n \frac{\left(b_3 q^{1-n}/c_2, q\right)_n \left(b_3 q^{1-n}/c_3, q\right)_n}{(c_2, q)_n (c_3, q)_n}
$$

$$
\times\ {}_4\varphi_3\left(\begin{array}{c} q^{-n},\, \dfrac{c_1}{b_1},\, \dfrac{c_1}{b_2},\, b_3 \\ c_1,\, \dfrac{b_3 q^{1-n}}{c_2},\, \dfrac{b_3 q^{1-n}}{c_3} \end{array}\ \middle|\ q,\, q\right)
$$

$$
\left(c_1 c_2 c_3 = b_1 b_2 b_3 q^{1-n}\right). \qquad (3.11.22)
$$

By using (3.11.20) we may show that (3.11.21) and (3.11.22) are equivalent to the transformations for the q-hypergeometric functions that have the forms:

$$
{}_8F_7\left(\begin{array}{c} -n,\, v,\, \dfrac{v}{2}+1,\, \dfrac{v}{2}+1+\dfrac{i\pi}{\ln q},\, a_1,\, a_2,\, a_3,\, a_4 \\ \dfrac{v}{2},\, \dfrac{v}{2}+\dfrac{i\pi}{\ln q},\, 1+v+n,\, 1+v-a_1,\, 1+v-a_2,\, 1+v-a_3,\, 1+v-a_4 \end{array}\ \middle|\ q,1\right)
$$

$$
= \frac{(1+v|q)_n \left(1+v-a_1-a_2|q\right)_n}{\left(1+v-a_1|q\right)_n \left(1+v-a_2|q\right)_n}
$$

$$
\times\ {}_4F_3\left(\begin{array}{c} -n,\, 1+v-a_3-a_4,\, a_2,\, a_1 \\ a_1+a_2-v-n,\, 1+v-a_3,\, 1+v-a_4 \end{array}\ \middle|\ q,1\right),\qquad (3.11.23)
$$

$$
{}_4F_3\left(\begin{array}{c} -n,\, b_1,\, b_2,\, b_3 \\ c_1,\, c_2,\, c_3 \end{array}\ \middle|\ q,1\right)
$$

$$
= \frac{\left(b_3-c_2+1-n|q\right)_n \left(b_3-c_3+1-n|q\right)_n}{\left(c_2|q\right)_n \left(c_3|q\right)_n}
$$

$$
\times\ {}_4F_3\left(\begin{array}{c} -n,\, c_1-b_1,\, c_1-b_2,\, b_3 \\ c_1,\, b_3-c_2-n+1,\, b_3-c_3-n+1 \end{array}\ \middle|\ q,1\right)
$$

$$
\left(\sum_{i=1}^{3} c_i = \sum_{i=1}^{3} b_i - n + 1\right). \qquad (3.11.24)
$$

Using the transformations (3.11.23) and (3.11.24) successively lets us reduce (3.11.16) to the form analogous to (3.11.9), i.e.

$$y_n(s) = \left(\frac{A}{\tilde{c}_1 q^{-\mu/2}\kappa^2}\right)^n B_n \left(s_1 + s_2 + \mu | q\right)_n \left(s_1 + s_3 + \mu | q\right)_n \left(s_1 + s_4 + \mu | q\right)_n$$

$$\times\ {}_4F_3 \left(\begin{matrix} -n,\ \sum_i s_i + 2\mu + n - 1,\ s_1 - s,\ s_1 + s + \mu \\ s_1 + s_2 + \mu,\ s_1 + s_3 + \mu,\ s_1 + s_4 + \mu \end{matrix}\ \middle|\ q, 1\right) \quad (3.11.25)$$

This formula may have a bulky appearance but all the systems of polynomials which were obtained and discussed in Chaps. 2 and 3 as well as those introduced in [A9, A17, A27, A29, H1–6, N18, R22, S24, W6, W8] can be obtained from (3.11.25) or from (3.11.16), (3.11.23), which are equivalent to (3.11.25), by an appropriate choice of parameters and by taking some limits. For example, the Askey-Wilson polynomials [A29] can be obtained from (3.11.25) with the aid of (3.11.20) by setting $x(s) = \cos\theta$, i.e. $q^s = e^{i\theta}$, $\tilde{c}_1 = \tilde{c}_2 = 1/2(\mu = 0)$. Different limiting cases for polynomials $y_n(s)$ are considered in Sects. 3.11.4 and 3.11.5 (see Conclusion for their classification).

By involving the symmetry property (3.11.11a) we may reduce Eq. (3.2.9) for the function $\varrho(s)$ to the form

$$\frac{\varrho(s + 1)}{\varrho(s)} = \frac{\sigma(-s - \mu)}{\sigma(s + 1)} . \qquad (3.11.26)$$

Solution of this equation when $\sigma(s)$ is determined by Eq. (3.11.12) may be derived by applying the method considered in Sect. 3.6.1.

The symmetry property of (3.11.11a) enables us also to write down (3.1.21) for $y_n(s)$ in the form

$$\sigma(-s - \mu)\frac{\Delta y_n(s)}{\Delta x(s)} - \sigma(s)\frac{\nabla y_n(s)}{\nabla x(s)} = -\lambda_n \Delta x \left(s - \tfrac{1}{2}\right) y_n(s) . \qquad (3.11.26a)$$

As is seen from (3.11.12) and (3.11.16) the functions $\sigma(s)$ and $y_n(s)$ do not vary for any permutation of s_i $(i = 1, 2, 3, 4)$. This property will be used further in Sect. 3.11.5 for various limiting processes.

Remark. It is easy to verify that when $q \to 1$ the Watson transformation (3.11.21) takes the form of the Whipple transformation (3.11.7). This is due to the fact that when $q \to 1$ we have

$$\frac{(v/2 + i\pi/\ln q + 1 | q)_k}{(v/2 + i\pi/\ln q | q)_k} = \frac{\psi_q(v/2 + i\pi/\ln q + k)}{\psi_q(v/2 + i\pi/\ln q)} = \frac{q^{(v+k)/2} + q^{-k/2}}{q^{v/2} + 1} \to 1 .$$

Also the transformation (3.11.24) takes the form of the Bailey transformation (3.11.8) when $q \to 1$.

3.11.3. Now we show that the expression (3.11.9) for the polynomial $y_n(s)$ on the lattice

$$x(s) = C_1 s^2 + C_2 s + C_3 \quad \left(\frac{C_2}{C_1} = \mu\right) , \qquad (3.11.27)$$

corresponding to $q = 1$, may be obtained by taking the limit when $q \to 1$ from (3.11.25), if a normalizing constant $B_n = B_n(q)$ in the Rodrigues formula for $y_n(s)$ on the lattice (3.11.10) goes into the appropriate constant on the lattice (3.11.27). When $q \to 1$ at the fixed values of μ and s_i ($i = 1, 2, 3, 4$) the q-hypergeometric function in (3.11.25) goes into the hypergeometric function in (3.11.9) and the product

$$\prod_{i=2}^{4} (s_1 + s_i + \mu | q)_n$$

in (3.11.25) goes into

$$\prod_{i=2}^{4} (s_1 + s_i + \mu)_n \ .$$

Further, when $q \to 1$, the expression (3.11.12) for $\sigma(s)$ with fixed A and s_i goes into (3.11.5). After comparison of (3.11.25) and (3.11.9) it is easy to see that expression (3.11.25) really transforms into (3.11.9) if we put $\tilde{c}_1 q^{-\mu/2} \kappa^2 = C_1$.

Now let us verify that for the lattice function (3.11.10) with

$$\tilde{c}_1 q^{-\mu/2} \kappa^2 = C_1 \ , \quad \frac{\tilde{c}_1}{\tilde{c}_2} = q^\mu \quad \left(\kappa = q^{1/2} - q^{-1/2} \right)$$

we may choose the constant $\tilde{c}_3 = \tilde{c}_3(q)$ such that the expression (3.11.10) transforms into (3.11.27) when $q \to 1$. We check that the above condition is fulfilled at $\tilde{c}_3 = C_3 - \tilde{c}_1 - \tilde{c}_2$. In this case the expression (3.11.10) may be rewritten in the form

$$x(s) = \frac{C_1}{\kappa^2} q^{\mu/2} \left(q^s - 1 \right) + \frac{C_1}{\kappa^2} q^{-\mu/2} \left(q^{-s} - 1 \right) + C_3$$

$$= C_1 \frac{q^{s+\mu/2} + q^{-(s+\mu/2)} - 2}{\left(q^{1/2} - q^{-1/2} \right)^2} - C_1 \frac{q^{\mu/2} + q^{-\mu/2} - 2}{\left(q^{1/2} - q^{-1/2} \right)^2} + C_3 \ . \qquad (3.11.28)$$

Since at any t

$$\lim_{q \to 1} \frac{q^t + q^{-t} - 2}{\left(q^{1/2} - q^{-1/2} \right)^2} = \lim_{q \to 1} \psi_q^2(t) = t^2 \ ,$$

the expression (3.11.28) really transforms into (3.11.27) when $q \to 1$.

For $q \to 1$ the polynomials (3.11.25) transform into (3.11.9) with fixed s_i and μ, so the q-analogs for the polynomials of hypergeometric type may be obtained by the method similar to the one considered in 3.6, i.e. with the given lattice (3.11.27) for $x(s)$ we may use the expression (3.11.28) with the same value of μ, while for the function $\sigma(s)$ the values of s_i chosen for the lattice (3.11.27) may be used.

3.11.4. By using Eq. (3.11.9) we can obtain explicit expressions in terms of hypergeometric functions for the above considered systems of polynomials $y_n(s)$ on lattices $x(s) = s(s+1)$ and $x(s) = s$.

1) Let $x(s) = s(s+1)$. In this case the function $x(s)$ may be obtained from (3.11.27) if we put $C_1 = 1$, $\mu = 1$, $C_3 = 0$. For the Racah polynomials $u_n^{(\alpha,\beta)}(x, a, b)$ we have (see Table 3.6)

$$B_n = \frac{(-1)^n}{n!} \,, \quad \sigma(s) = A \prod_{i=1}^{4} (s - s_i) \,,$$

where

$$s_1 = a \,, \quad s_2 = -b \,, \quad s_3 = \beta - a \,, \quad s_4 = b + \alpha \,, \quad A = -1 \,.$$

Hence from (3.11.9) it follows that

$$u_n^{(\alpha,\beta)}\big(x(s), a, b\big) = \frac{1}{n!}(a + 1 - b)_n(\beta + 1)_n(a + b + \alpha + 1)_n$$
$$\times\ {}_4F_3\left(\begin{array}{c} -n,\ \alpha + \beta + n + 1,\ a - s,\ a + s + 1 \\ a + 1 - b,\ \beta + 1,\ a + b + \alpha + 1 \end{array} \middle|\ 1 \right).$$

$$(3.11.29)$$

Since the Racah polynomials (3.11.29) are obtained by means of the limiting process with $q \to 1$ from (3.11.25), as shown in Sect. 3.11.3, the polynomials (3.11.25) are sometimes called the *q-Racah polynomials*.

For the dual Hahn polynomials $w_n^{(c)}(x)$ (see Table 3.7) the function $\sigma(s) = (s - a)(s + b)(s - c)$ may be obtained from $\sigma(s) = A \prod_{i=1}^{4}(s - s_i)$ by taking the limit $s_4 = d \to \infty$, if we put $s_1 = a$, $s_2 = -b$, $s_3 = c$, $A = -1/d$. By letting $B_n = (-1)^n/n!$ we obtain from (3.11.9) that

$$w_n^{(c)}\big(x(s)\big) = \lim_{d \to \infty} \frac{1}{d^n n!}(a + 1 - b)_n(a + 1 + c)_n(a + 1 + d)_n$$
$$\times\ {}_4F_3\left(\begin{array}{c} -n,\ a - b + c + d + n + 1,\ a - s,\ a + 1 + s \\ a + 1 - b,\ a + 1 + c,\ a + 1 + d \end{array} \middle|\ 1 \right)$$
$$= \frac{(a + 1 - b)_n(a + 1 + c)_n}{n!}\ {}_3F_2\left(\begin{array}{c} -n,\ a - s,\ a + 1 + s \\ a + 1 - b,\ a + 1 + c \end{array} \middle|\ 1 \right) \,.$$

$$(3.11.30)$$

2) Let $x(s) = s$. The function $x(s) = s$ may be deduced from (3.11.27) for $C_3 = 0$, $C_2 = 1$, $C_1 = 1/\mu$ if $\mu \to \infty$. Then the expressions for polynolmials $y_n(s)$ in terms of hypergeometric functions with $x(s) = s$ may be obtained from (3.11.9) if in the formulas for

$$\sigma(s) = A \prod_{i=1}^{4} (s - s_i)$$

and

$$\sigma(s) + \tau(s)\Delta x\left(s - \tfrac{1}{2}\right) = \sigma(-s-\mu) = A \prod_{i=1}^{4}(s + s_i + \mu)$$

we choose the constants $A = A(\mu)$, $s_i = s_i(\mu)$ $(i = 1, 2, 3, 4)$ such that the functions $\sigma(s) \equiv \sigma(s,\mu)$ and $\sigma(s)+\tau(s)\Delta x(s-1/2) \equiv \sigma(-s-\mu,\mu)$ for $\mu \to \infty$ go into the corresponding ones on the lattice $x(s) = s$ when for the most general case $\sigma(s)$ and $\sigma(s)+\tau(s)$ are polynomials of the second degree with equal leading coefficients:

$$\sigma(s) = \tilde{A}(s - s_1)(s - s_2) \, ,$$

$$\sigma(s) + \tau(s) = \tilde{A}(s - \bar{s}_1)(s - \bar{s}_2)$$

($\tilde{A}$, $\bar{s}_1$, $\bar{s}_2$ are constants).

These conditions will be satisfied if we put $A(\mu) = \tilde{A}/\mu^2$, $s_1(\mu) = s_1$, $s_2(\mu) = s_2$, $s_3 = -\mu - \bar{s}_1$, $s_4(\mu) = -\mu - \bar{s}_2$. Substituting this into (3.11.9) we obtain

$$y_n(s) = \lim_{\mu \to \infty} \left(\frac{\tilde{A}}{\mu}\right)^n B_n\,(s_1 + s_2 + \mu)_n\,(s_1 - \bar{s}_1)_n\,(s_1 - \bar{s}_2)_n$$

$$\times {}_4F_3\left(\begin{array}{c} -n,\ s_1 + s_2 - \bar{s}_1 - \bar{s}_2 + n - 1,\ s_1 - s,\ s_1 + s + \mu \\ s_1 + s_2 + \mu,\ s_1 - \bar{s}_1,\ s_1 - \bar{s}_2 \end{array}\middle|\ 1\right)$$

$$= (\tilde{A})^n B_n\,(s_1 - \bar{s}_1)_n\,(s_1 - \bar{s}_2)_n$$

$$\times {}_3F_2\left(\begin{array}{c} -n,\ s_1 + s_2 - \bar{s}_1 - \bar{s}_2 + n - 1,\ s_1 - s \\ s_1 - \bar{s}_1,\ s_1 - \bar{s}_2 \end{array}\middle|\ 1\right) \, .$$

For the Hahn polynomials $h_n^{(\alpha,\beta)}(s, N)$ we have

$$\tilde{A} = -1 \, , \quad s_1 = 0 \, , \quad s_2 = N + \alpha \, , \quad \bar{s}_1 = -(\beta + 1) \, , \quad \bar{s}_2 = N - 1 \, ,$$

and we come to the expression (2.7.19) obtained earlier.

If degrees of $\sigma(s)$ and $\sigma(s) + \tau(s)$ are less than 2, the expressions for $y_n(s)$ may be obtained by taking the additional limit when $s_2 \to \infty$. We choose $s_1 = 0$, $\tilde{A} = -1/s_2$ and the constants $\bar{s}_1$, $\bar{s}_2$ such that the functions $\sigma(s)$ and $\sigma(s) + \tau(s)$ become the corresponding ones if $s_1 \to +\infty$ for the Meixner, Kravchuk and Charlier polynomials:

$y_n(s)$	$m_n^{(\gamma,\mu)}(s)$	$k_n^{(p)}(s)$	$c_n^{(\mu)}(s)$
$\bar{s}_1$	$-\gamma$	N	$\sqrt{s_2}$
$\bar{s}_2$	μs_2	$-\frac{p}{q}s_2$	$-\mu\sqrt{s_2}$

As a result we obtain the expressions coinciding with (2.7.12), (2.7.13) and (2.7.9) respectively.

3.11.5. In Sect. 3.11.2 by using generalized q-hypergeometric functions (3.11.17) we obtained general formulas (3.11.16) and (3.11.25) for the poylnomials of hypergeometric type

$$\tilde{y}_n[x(s)] \equiv y_n(s) \quad \text{with} \quad x(s) = \tilde{c}_1 q^s + \tilde{c}_2 q^{-s} + \tilde{c}_3 \ ,$$

$$\left(\tilde{c}_1/\tilde{c}_2 = q^\mu\right) \ , \qquad \sigma(s) = A \prod_{i=1}^{4} \psi_q(s - s_i)$$

in the case when values of μ and s_i $(i = 1, 2, 3, 4)$ are fixed numbers. By using the limiting process when either $q^{s_i} \to 0$ (or $q^{-s_i} \to 0$) with fixed μ we may also derive explicit expressions for $y_n(s)$ in the case when the function $\sigma(s)$ has less than four zeros. To do this it is sufficient to employ the simple relations

$$\lim_{q^t \to 0} \frac{\psi_q(a + t)}{\psi_q(t)} = q^{-a/2} \ , \tag{3.11.31}$$

$$\lim_{q^t \to 0} \frac{\psi_q(a - t)}{\psi_q(t)} = -q^{a/2} \ ; \tag{3.11.32}$$

$$\lim_{q^t \to 0} \frac{(a + t|q)_k}{[\psi_q(t)]^k} = q^{-k(a+(k-1)/2)/2} \ , \tag{3.11.33}$$

$$\lim_{q^t \to 0} \frac{(a - t|q)_k}{[\psi_q(t)]^k} = (-1)^k q^{k(a+(k-1)/2)/2} \ ; \tag{3.11.34}$$

$$(k = 0, 1, \ldots) \ .$$

From (3.11.12) and (3.11.25) when $A(q) = A(1/q)$ it is seen that the functions $\sigma(s)$ and $y_n(s)q^{-\mu n/2}$ do not vary if q is replaced by $1/q$. Therefore, the limiting expressions for $\sigma(s)$ and $y_n(s)q^{-\mu n/2}$ for $q^{-s_i} \to 0$ may be obtained from the respective limiting expressions for $q^{s_i} \to 0$ after replacing q by $1/q$.

3.11.5.1. We shall consider the limiting processes when $q^{s_i} \to 0$.

1) Let $q^{s_4} \to 0$. Then with $A = -1/\psi_q(s_4)$ from (3.11.12) by using (3.11.32) we obtain

$$\sigma(s) = q^{s/2} \prod_{i=1}^{3} \psi_q(s - s_i) . \tag{3.11.35}$$

For $y_n(s)$ we use (3.11.25). By means of (3.11.33) we obtain

$$\lim_{q^{s_4} \to 0} A^n (s_1 + s_4 + \mu | q)_n = (-1)^n q^{-n(s_1 + \mu + (n-1)/2)/2} ,$$

$$\lim_{q^{s_4} \to 0} \frac{\left(\sum_{i=1}^{4} s_i + 2\mu + n - 1 | q\right)_k}{(s_1 + s_4 + \mu | q)_k} = \lim_{q^{s_4} \to 0} \frac{\left(\sum_{i=1}^{4} s_i + 2\mu + n - 1 | q\right)_k}{[\psi_q(s_4)]^k}$$

$$\times \frac{[\psi_q(s_4)]^k}{(s_1 + s_4 + \mu | q)_k} = \frac{q^{-k\left(\sum_{i=1}^{3} s_i + 2\mu + n - 1\right)/2}}{q^{k(s_1 + \mu)/2}} ,$$

whence

$$y_n(s) = \frac{(-1)^n B_n}{(\tilde{c}_1 \kappa^2)^n} q^{-n(s_1 + (n-1)/2)/2}(s_1 + s_2 + \mu | q)_n (s_1 + s_3 + \mu | q)_n$$

$$\times {}_3F_2 \left(\begin{matrix} -n, \ s_1 - s, \ s_1 + s + \mu \\ s_1 + s_2 + \mu, \ s_1 + s_3 + \mu \end{matrix} \bigg| q, q^{-(s_2 + s_3 + \mu + n - 1)/2} \right) . \tag{3.11.36}$$

It is easy to verify that (3.11.36) for $q \to 1$, $\mu = 1$ coincides with formula (3.11.30) for the dual Hahn polynomials at $s_1 = a$, $s_2 = -b$, $s_3 = c$, $B_n = (-1)^n/n!$ and $\tilde{c}_1 q^{-\mu/2} \kappa^2 = C_1 = 1$. Therefore it is natural to call the polynomials (3.11.36) *the dual q-Hahn polynomials*.

When $q^{-s_4} \to 0$, by using (3.11.35) and (3.11.36) and replacing q by $1/q$ in the expressions for $\sigma(s)$ and $y_n(s)q^{-\mu n/2}$ we obtain

$$\sigma(s) = q^{-s/2} \prod_{i=1}^{3} \psi_q(s - s_i) , \tag{3.11.37}$$

$$y_n(s) = \frac{(-1)^n B_n}{(\tilde{c}_1 \kappa^2)^n} q^{n(s_1 + 2\mu + (n-1)/2)/2}(s_1 + s_2 + \mu | q)_n (s_1 + s_3 + \mu | q)_n$$

$$\times {}_3F_2 \left(\begin{matrix} -n, \ s_1 - s, \ s_1 + s + \mu \\ s_1 + s_2 + \mu, \ s_1 + s_3 + \mu \end{matrix} \bigg| q, q^{(s_2 + s_3 + \mu + n - 1)/2} \right) . \tag{3.11.38}$$

2) It is also convenient to use formula (3.11.25) when $q^{s_i} \to 0$ $(i = 2, 3, 4)$, $A = -1/\prod_{i=2}^{4} \psi_q(s_i)$. In this case

$$\sigma(s) =)q^{3s/2}\psi_q(s - s_1) . \tag{3.11.39}$$

Due to (3.11.33) and (3.11.34) we have

$$\lim_{\substack{q^{s_i}\to 0\\(i=2,3,4)}} A^n \prod_{i=2}^{4}(s_1+s_i+\mu|q)_n$$

$$=\lim_{\substack{q^{s_i}\to 0\\(i=2,3,4)}} (-1)^n \frac{(s_1+s_2+\mu|q)_n}{[\psi_q(s_2)]^n}\frac{(s_1+s_3+\mu|q)_n}{[\psi_q(s_3)]^n}\frac{(s_1+s_4+\mu|q)_n}{[\psi_q(s_4)]^n}$$

$$=(-1)^n q^{-3n(s_1+\mu+(n-1/2)/2} \; ;$$

$$\lim_{\substack{q^{s_i}\to 0\\(i=2,3,4)}} \frac{(\sum_{i=1}^{4} s_i+2\mu+n-1|q)_k}{(s_1+s_2+\mu|q)_k(s_1+s_3+\mu|q)_k(s_1+s_4+\mu|q)_k}$$

$$=\lim_{\substack{q^{s_i}\to 0\\(i=2,3,4)}} \frac{\dfrac{\left(\sum_{i=1}^{4} s_i+2\mu+n-1)|q\right)_k}{[\psi_q(s_2+s_3+s_4)]^k}}{\dfrac{(s_1+s_2+\mu|q)_k}{[\psi_q(s_2)]^k}\dfrac{(s_1+s_3+\mu|q)_k}{[\psi_q(s_3)]^k}\dfrac{(s_1+s_4+\mu|q)_k}{[\psi_q(s_4)]^k}}$$

$$\times\left(\frac{\psi_q(s_2+s_3+s_4)}{\prod_{i=2}^{4}\psi_q(s_i)}\right)^k = q^{-k(n-k-2s_1-\mu)/2}\kappa^{2k} \; ,$$

because

$$\lim_{\substack{q^{s_i}\to 0\\(i=2,3,4)}} \frac{\psi_q(s_2+s_3+s_4)}{\prod_{i=2}^{4}\psi_q(s_i)} = \kappa^2 \; .$$

Therefore it follows from (3.11.25) that in the case of (3.11.39)

$$y_n(s)=\frac{(-1)^n B_n}{(\tilde{c}_1\kappa^2)^n} q^{-3n(s_1+(n-1)/2)/2-\mu n}\sum_{k=0}^{n}\frac{(-n|q)_k(s_1-s|q)_k(s_1+s+\mu|q)_k}{(1|q)_k}$$

$$\times \kappa^{2k}q^{-k(n-k-2s_1-\mu)/2} \; . \tag{3.11.40}$$

When $q^{-s_i}\to 0$ $(i=2,3,4)$, after replacing q by $1/q$ in the expressions for $\sigma(s)$ and $y_n(s)q^{-\mu n/2}$ we obtain

$$\sigma(s)=q^{-3s/2}\psi_q(s-s_1) \; , \tag{3.11.41}$$

$$y_n(s)=\frac{(-1)^n B_n}{(\tilde{c}_1\kappa^2)^n} q^{3n(s_1+(n-1)/2)/2+2\mu n}\sum_{k=0}^{n}\frac{(-n|q)_k(s_1-s|q)_k(s_1+s+\mu|q)_k}{(1|q)_k}$$

$$\times \kappa^{2k}q^{k(n-k-2s_1-\mu)/2} \; . \tag{3.11.42}$$

When considering the other limit cases it is convenient to use for $y_n(s)$ the formula following from (3.11.16) and (3.11.23). It has the form

$$y_n(s)=(-1)^n\left(\frac{A}{\tilde{c}_1 q^{-\mu/2}\kappa^2}\right)^n B_n(s-s_3+1-n|q)_n(s-s_4+1-n|q)_n$$

$$\times (1-n-\mu-s_1-s_2|q)_n$$

$$\times {}_4F_3\left(\begin{matrix} -n,\; 1-n-\mu-s_3-s_4,\; s+s_2+\mu,\; s+s_1+\mu \\ s_1+s_2+\mu,\; s-s_3+1-n,\; s-s_4+1-n \end{matrix}\bigg|q,1\right) \; . \tag{3.11.43}$$

3) Let $q^{s_2} \to 0$, $q^{s_4} \to 0$, $A = 1/[\psi_q(s_2)\psi_q(s_4)]$. Then by using (3.11.32) we obtain from (3.11.12) that

$$\sigma(s) = q^s \psi_q(s - s_1)\psi_q(s - s_3) . \tag{3.11.44}$$

Due to (3.11.33) and (3.11.34) we have

$$\lim_{\substack{q^{s_i} \to 0 \\ (i=2,4)}} \frac{(s - s_4 + 1 - n|q)_n}{[\psi_q(s_4)]^n} \frac{(1 - n - \mu - s_1 - s_2|q)_n}{[\psi_q(s_2)]^n} = q^{n(s-s_1+1-n-\mu)/2} ;$$

$$\lim_{\substack{q^{s_i} \to 0 \\ (i=2,4)}} \frac{(1 - n - \mu - s_3 - s_4|q)_k (s + s_2 + \mu|q)_k}{(s_1 + s_2 + \mu|q)_k (s - s_4 + 1 - n|q)_k} = q^{k(s_1-s_3-2s-\mu)/2} .$$

Therefore, from (3.11.43) we obtain

$$y_n(s) = \frac{(-1)^n B_n}{(\tilde{c}_1 \kappa^2)^n} q^{n(s-s_1+1-n)/2}(s - s_3 + 1 - n|q)_n$$

$$\times {}_2F_1 \left(\begin{matrix} -n, \; s + s_1 + \mu \\ s - s_3 + 1 - n \end{matrix} \middle| q, q^{(s_1-s_3-2s-\mu)/2} \right) . \tag{3.11.45}$$

When $q^{-s_2} \to 0$ and $q^{-s_4} \to 0$, by replacing q by $1/q$ in the expressions for $\sigma(s)$ and $y_n(s)q^{-\mu n/2}$ we obtain, respectively,

$$\sigma(s) = q^{-s} \psi_q(s - s_1)\psi_q(s - s_3) ; \tag{3.11.46}$$

$$y_n(s) = \frac{(-1)^n B_n}{(\tilde{c}_1 \kappa^2)^n} q^{-n(s-s_1+1-n-2\mu)/2}(s - s_3 + 1 - n|q)_n$$

$$\times {}_2F_1 \left(\begin{matrix} -n, \; s + s_1 + \mu \\ s - s_3 + 1 - n \end{matrix} \middle| q, q^{-(s_1-s_3-2s-\mu)/2} \right) . \tag{3.11.47}$$

4) Let $q^{s_i} \to 0$ $(i = 1, 2, 3, 4)$, $A = 1/ \prod_{i=1}^4 \psi_q(s_i)$. Then by using (3.11.32) we obtain from (3.11.12) that

$$\sigma(s) = q^{2s} . \tag{3.11.48}$$

For $y_n(s)$ we shall proceed from formula (3.11.43). Due to (3.11.33) and (3.11.34) we have

$$\lim_{\substack{q^{s_i} \to 0 \\ (i=1,2,3,4)}} A^n(s - s_3 + 1 - n|q)_n(s - s_4 + 1 - n|q)_n(1 - n - \mu - s_1 - s_2|q)_n$$

$$= \lim_{\substack{q^{s_i} \to 0 \\ (i=1,2,3,4)}} \frac{s - s_3 + 1 - n|q)_n}{[\psi_q(s_3)]^n} \frac{(s - s_4 + 1 - n|q)_n}{[\psi_q(s_4)]^n} \frac{(1 - n - \mu - s_1 - s_2|q)_n}{[\psi_q(s_1 + s_2)]^n}$$

$$\times \left(\frac{\psi_q(s_1 + s_2)}{\psi_q(s_1)\psi_q(s_2)} \right)^n = \kappa^n q^{(n/2)[2s-\mu-3(n-1)/2]} ;$$

$$\lim_{\substack{q^{s_i} \to 0 \\ (i=1,2,3,4)}} \frac{(1 - n - \mu - s_3 - s_4|q)_k (s + s_2 + \mu|q)_k (s + s_1 + \mu|q)_k}{(s_1 + s_2 + \mu|q)_k (s - s_3 + 1 - n|q)_k (s - s_4 + 1 - n|q)_k}$$

$$
= \lim_{\substack{q^{s_i} \to 0 \\ (i=1,2,3,4)}} \frac{(1 - n - \mu - s_3 - s_4|q)_k}{[\psi_q(s_3 + s_4)]^k} \frac{(s + s_2 + \mu|q)_k}{[\psi_q(s_2)]^k} \frac{(s + s_1 + \mu|q)_k}{[\psi_q(s_1)]^k}
$$

$$
\times \frac{[\psi_q(s_1 + s_2)]^k}{(s_1 + s_2 + \mu|q)_k} \frac{[\psi_q(s_3)]^k}{(s - s_3 + 1 - n|q)_k} \frac{[\psi_q(s_4)]^k}{(s - s_4 + 1 - n|q)_k}
$$

$$
\times \left[\frac{\psi_q(s_3 + s_4)\psi_q(s_2)\psi_q(s_1)}{\psi_q(s_1 + s_2)\psi_q(s_3)\psi_q(s_4)} \right]^k = (-1)^k q^{k(n-k)/2 - k\mu - 2ks} .
$$

Thus, from (3.11.43) it follows that

$$
y_n(s) = \frac{(-1)^n B_n}{(\tilde{c}_1 \kappa)^n} q^{-3n(n-1)/4} \sum_{k=0}^{n} (-1)^k \frac{(-n|q)_k}{(1|q)_k}
$$

$$
\times q^{k(n-k-2\mu)/2 + (n-2k)s} . \tag{3.11.49}
$$

For $q^{-s_i} \to 0$ $(i = 1, 2, 3, 4)$ we obtain

$$
\sigma(s) = q^{-2s} , \tag{3.11.50}
$$

$$
y_n(s) = \frac{(-1)^n B_n}{(\tilde{c}_1 \kappa)^n} q^{\mu n + 3n(n-1)/4} \sum_{k=0}^{n} (-1)^k \frac{(-n|q)_k}{(1|q)_k}
$$

$$
\times q^{-k(n-k-2\mu)/2 - (n-2k)s} . \tag{3.11.51}
$$

The polynomials (3.11.49) for the case $x(s) = \cos\theta$ ($\tilde{c}_1 = 1/2$, $\mu = 0$, $\tilde{c}_3 = 0$, $q^s = e^{i\theta}$) correspond to the q-Hermite polynomials [R22].

3.11.5.2. We have considered various explicit representations of polynomials of hypergeometric type at $x(s) = \tilde{c}_1 q^s + \tilde{c}_2 q^{-s} + \tilde{c}_3$, $\tilde{c}_1/\tilde{c}_2 = q^\mu$ for fixed μ. Now we shall consider the cases when $x(s) = \tilde{c}_1 q^s + \tilde{c}_3$ and $x(s) = \tilde{c}_2 q^{-s} + \tilde{c}_3$, which can be derived from the general case by using the limiting process for $q^{-\mu} \to 0$ and fixed $\tilde{c}_1, \tilde{c}_3$ or for $q^\mu \to 0$ and fixed $\tilde{c}_2, \tilde{c}_3$.

First, we consider the case when $q^{-\mu} \to 0$. We show with the use of (3.11.12) and (3.11.25) that the functions $\sigma(s)$ and $y_n(s)$ have limits when $q^{-\mu} \to 0$ if we set $s_3 = -\mu + d_1$, $s_4 = -\mu + d_2$ and $A = [q^{(d_1+d_2)/2}/\psi_q^2(\mu)]\tilde{A}$, where s_1, s_2, d_1, d_2 and $\tilde{A}$ are constants independent of μ. From (3.11.12) and (3.11.32) it follows that when $q^{-\mu} \to 0$ we obtain in the limit that

$$
\sigma(s) = \tilde{A} q^s \psi_q(s - s_1)\psi_q(s - s_2) . \tag{3.11.52}
$$

In order to derive formulas for $y_n(s)$ we shall use (3.11.25). Due to (3.11.34) we have

$$
\lim_{q^{-\mu} \to 0} \left(A q^{\mu/2} \right)^n (s_1 + s_2 + \mu|q)_n
$$

$$
= \lim_{q^{-\mu} \to 0} \tilde{A}^n q^{n(d_1+d_2)/2} \frac{(s_1 + s_2 + \mu|q)_n}{[\psi_q(-\mu)]^n} \frac{(q^{\mu/2})^n}{[\psi_q(-\mu)]^n}
$$

$$
= (\tilde{A}\kappa)^n q^{n(s_1 + s_2 + d_1 + d_2 + (n-1)/2)/2} ;
$$

$$\lim_{q^{-\mu}\to 0} \frac{(s_1 + s + \mu|q)_k}{(s_1 + s_2 + \mu|q)_k}$$

$$= \lim_{q^{-\mu}\to 0} \frac{(s_1 + s + \mu|q)_k}{[\psi_q(-\mu)]^k} \frac{[\psi_q(-\mu)]^k}{(s_1 + s_2 + \mu|q)_k} = q^{k(s-s_2)/2} \ .$$

From this we obtain, when $q^{-\mu} \to 0$,

$$y_n(s) = \frac{\tilde{A}^n B_n}{(\tilde{c}_1 \kappa)^n} (s_1 + d_1|q)_n (s_1 + d_2|q)_n q^{n(s_1+s_2+d_1+d_2+(n-1)/2)/2}$$

$$\times {}_3F_2 \left(\begin{matrix} -n,\ s_1 + s_2 + d_1 + d_2 + n - 1,\ s_1 - s \\ s_1 + d_1,\ s_1 + d_2 \end{matrix} \middle| q, q^{(s-s_2)/2} \right). \tag{3.11.53}$$

If we choose $\tilde{c}_1 = 1/\kappa$ and $\tilde{c}_3 = -1/\kappa$ then for $q \to 1$ the function $x(s) = \tilde{c}_1 q^s + \tilde{c}_3$ transforms into $x(s) = s$. In this case, from (3.11.53) with $q \to 1$ by passing to the limit we may obtain formula (2.7.19) for the Hahn polynomials $h_n^{(\alpha,\beta)}(s)$ setting $s_1 = 0$, $s_2 = N + \alpha$, $d_1 = \beta + 1$, $d_2 = 1 - N$ and $\tilde{A} = -1$. Therefore, if $q \neq 1$ and for the above values of parameters s_1, s_2, d_1, d_2, $\tilde{A}$, $\tilde{c}_1$, $\tilde{c}_3$ the polynomials $y_n(s)$ obtained by formula (3.11.53) at $x(s) = \tilde{c}_1 q^s + \tilde{c}_3$ are naturally called the *q-analogs of the Hahn polynomials* $h_n^{(\alpha,\beta)}(s)$.

We may also consider the limit case of the polynomials (3.11.53) for $q \to 1$ when a value of $x = q^s (\tilde{c}_1 = 1,\ \tilde{c}_2 = \tilde{c}_3 = 0)$ is fixed. We show that in this case, with a certain choice of parameters s_1, s_2, d_1 and d_2, the polynomial

$$p_n(x, q) = {}_3F_2 \left(\begin{matrix} -n,\ s_1 + s_2 + d_1 + d_2 + n - 1,\ s_1 - s \\ s_1 + d_1,\ s_1 + d_2 \end{matrix} \middle| q, q^{(s-s_2)/2} \right)$$

transforms into the Jacobi polynomial, for which

$$\frac{P_n^{(\alpha,\beta)}(x)}{P_n^{(\alpha,\beta)}(1)} = {}_2F_1 \left(\begin{matrix} -n,\ \alpha + \beta + n + 1 \\ \alpha + 1 \end{matrix} \middle| \frac{1-x}{2} \right)$$

to within a constant multiplier while $q \to 1$. The formulas for $p_n(x, q)$ and $P_n^{(\alpha,\beta)}(x)$ being compared, it is natural to put

$$s_1 + s_2 + d_1 + d_2 + n - 1 = \alpha + \beta + n + 1 \ , \quad s_1 = 0 \ , \quad d_1 = \alpha + 1 \ .$$

For the limit relation

$$\lim_{q\to 1} p_n(x, q) = \frac{P_n^{(\alpha,\beta)}(x)}{P_n^{(\alpha,\beta)}(1)} \tag{3.11.53a}$$

to be valid the condition

$$\lim_{q\to 1} \frac{(-s|q)_k}{(d_2|q)_k} q^{k(s-s_2)/2} = \left(\frac{1-x}{2} \right)^k$$

must be fulfilled. We have

$$\frac{(-s|q)_k}{(d_2|q)_k} q^{k(s-s_2)/2} = \prod_{l=0}^{k-1} \frac{q^{(-s+l)/2} - q^{(s-l)/2}}{q^{(d_2+l)/2} - q^{-(d_2+l)/2}} q^{(s-s_2)/2}$$

$$= \prod_{l=0}^{k-1} \frac{q^{l/2} - q^{s-l/2}}{q^{(s_2+d_2+l)/2} - q^{(s_2-d_2-l)/2}}$$

$$= \prod_{l=0}^{k-1} \frac{q^{l/2} - xq^{-l/2}}{q^{(s_2+d_2+l)/2} - q^{(s_2-d_2-l)/2}} \ .$$

Since $s_2 + d_2 = \beta + 1$, the above condition will obviously be satisfied if

$$\lim_{q \to 1} q^{(s_2-d_2)/2} = -1 \ .$$

To achieve this it is sufficient to put

$$s_2 = \beta - \gamma + \frac{i\pi}{\ln q} \ , \qquad d_2 = \gamma + 1 - \frac{i\pi}{\ln q}$$

where γ is a constant independent of q.

Thus, the relation (3.11.53a) is really valid for the polynomial

$$p_n(x, q) = {}_3F_2 \left(\begin{matrix} -n, \ \alpha + \beta + n + 1, \ -s \\ \alpha + 1, \ \gamma + 1 - \dfrac{i\pi}{\ln q} \end{matrix} \ \middle| \ q, q^{(s+\gamma-\beta-(i\pi/\ln q))/2} \right) .$$

In the book [G7a] the so called *big q-Jacobi polynomials*

$$P_n(x; a, b, c; q) = {}_3\varphi_2 \left(\begin{matrix} q^{-n}, \ abq^{n+1}, \ x \\ aq, \ cq \end{matrix} \ \middle| \ q, q \right)$$

$$(a = q^\alpha, \ b = q^\beta, \ c = -q^\gamma)$$

are considered. They were introduced by Andrews and Askey [A17] and may be obtained from the polynomials $p_n(x, q)$ after replacement of s by $-s$ and q by $1/q$ (for such replacements $x = q^s$ does not change).

Now we shall consider various limiting processes in (3.11.52) and (3.11.53) with $x = \tilde{c}_1 q^s + \tilde{c}_3$ as $q^{s_2} \to 0$ or $q^{s_1} \to 0$, $q^{s_2} \to 0$.

1) Let $q^{s_2} \to 0$ and $s_1 + s_2 + d_2 = \delta$ (where δ is a constant). Then, by using (3.11.32) we obtain from (3.11.52) with $\tilde{A} = -1/\psi_q(s_2)$:

$$\sigma(s) = q^{3s/2}\psi_q(s - s_1) \ . \tag{3.11.54}$$

Utilizing (3.11.34) for (3.11.53) yields

$$\lim_{q^{s_2} \to 0} \tilde{A}^n (s_1 + d_2 | q)_n = \lim_{q^{s_2} \to 0} \frac{(\delta - s_2 | q)_n}{[\psi_q(s_2)]^n} = q^{n(\delta+(n-1)/2)/2} \ ;$$

$$\lim_{q^{s_2} \to 0} \frac{q^{k(s-s_2)/2}}{(s_1 + d_2 | q)_k} = \lim_{q^{s_2} \to 0} \frac{q^{k(s-s_2)/2}}{[\psi_q(s_2)]^k} \frac{[\psi_q(s_2)]^k}{(\delta - s_2 | q)_k} = \kappa^k q^{k(s-\delta-(k-1)/2)/2} \ .$$

Hence, as $q^{s_2} \to 0$ we obtain from (3.11.53)

$$y_n(s) = \frac{B_n}{(\tilde{c}_1 \kappa)^n} q^{n(d_1+2\delta+n-1)/2}(s_1 + d_1 | q)_n$$

$$\times \sum_{k=0}^{n} \frac{(-n|q)_k (d_1 + \delta + n - 1|q)_k (s_1 - s|q)_k}{(s_1 + d_1|q)_k (1|q)_k} \kappa^k q^{k(s - \delta - (k-1)/2)/2} . \qquad (3.11.55)$$

But if in (3.11.53) instead of passage to the limit $q^{s_2} \to 0$ we use $q^{s_1} \to 0$ it may be seen that there is the limit of $y_n(s)$ when $s_1 + d_1 = \delta$, where δ is a constant independent of s_1. Really, in this case, at $\tilde{A} = -1/\psi_q(s_1)$ by passing to a limit we obtain from (3.11.52)

$$\sigma(s) = q^{3s/2} \psi_q(s - s_2) . \qquad (3.11.56)$$

Furthermore,

$$\lim_{q^{s_1} \to 0} \tilde{A}^n (s_1 + d_2|q)_n = \lim_{q^{s_1} \to 0} \frac{(-1)^n (s_1 + d_2|q)_n}{[\psi_q(s_1)]^n} = (-1)^n q^{-n(d_2 + (n-1)/2)/2} ;$$

$$\lim_{q^{s_1} \to 0} \frac{(s_1 - s|q)_k}{(s_1 + d_2|q)_k} q^{k(s - s_2)/2} = q^{k(s + (d_2 - s_2)/2)} .$$

Hence, proceeding from (3.11.53) as $q^{s_1} \to 0$ we find

$$y_n(s) = \frac{(-1)^n B_n}{(\tilde{c}_1 \kappa)^n} q^{n(\delta + s_2)/2}$$

$$\times \, {}_2F_1 \left(\begin{array}{c} -n, \ \delta + s_2 + d_2 + n - 1 \\ \delta \end{array} \middle| q, q^{s + (d_2 - s_2)/2} \right) . \qquad (3.11.57)$$

If we choose the parameters in (3.11.57) from the conditions

$$\frac{(-1)^n B_n}{(\tilde{c}_1 \kappa)^n} q^{n(\delta + s_2)/2} = 1 , \quad \delta = \alpha + 1 , \quad d_2 = s_2 = (\beta + 1)/2 ,$$

then with fixed $x = q^s$ from (3.11.57) we can obtain the so called *little q-Jacobi polynomials* [A15, G7a]

$$p_n(x, q) = {}_2F_1 \left(\begin{array}{c} -n, \ \alpha + \beta + n + 1 \\ \alpha + 1 \end{array} \middle| q, x \right) ,$$

which have the property

$$\lim_{q \to 1} p_n \left(\frac{1 - x}{2}, q \right) = {}_2F_1 \left(\begin{array}{c} -n, \ \alpha + \beta + n + 1 \\ \alpha + 1 \end{array} \middle| \frac{1 - x}{2} \right) = \frac{P_n^{(\alpha, \beta)}(x)}{P_n^{(\alpha, \beta)}(1)}$$

analogous to (3.11.53a).

2) Let $q^{s_i} \to 0$ $(i = 1, 2)$ and $s_1 + s_2 + d_2 = \delta$ where δ is a constant independent of s_1 and s_2. By using (3.11.32) we obtain from (3.11.52) with $\tilde{A} = 1/[\psi_q(s_1)\psi_q(s_2)]$

$$\sigma(s) = q^{2s} . \qquad (3.11.58)$$

Utilization of (3.11.34) for (3.11.53) yields

$$\lim_{\substack{q^{s_i}\to 0 \\ (i=1,2)}} \tilde{A}^n (s_1 + d_1|q)_n (s_1 + d_2|q)_n$$

$$= \lim_{\substack{q^{s_i}\to 0 \\ (i=1,2)}} \frac{(s_1 + d_1|q)_n}{[\psi_q(s_1)]^n} \frac{(\delta - s_2|q)_n}{[\psi_q(s_2)]^n} = (-1)^n q^{n(\delta - d_1)/2} \; ;$$

$$\lim_{\substack{q^{s_i}\to 0 \\ (i=1,2)}} \frac{(s_1 - s|q)_k q^{k(s-s_2)/2}}{(s_1 + d_1|q)_k (s_1 + d_2|q)_k}$$

$$= \lim_{\substack{q^{s_i}\to 0 \\ (i=1,2)}} \frac{(-s + s_1|q)_k}{[\psi_q(s_1)]^k} \frac{[\psi_q(s_1)]^k}{(d_1 + s_1|q)_k} \frac{[\psi_q(s_2)]^k}{(\delta - s_2|q)_k} \frac{q^{k(s-s_2)/2}}{[\psi_q(s_2)]^k}$$

$$= \kappa^k q^{k(d_1 - \delta - (k-1)/2)/2 + ks} \; .$$

Since

$$s_1 + s_2 + d_1 + d_2 = d_1 + \delta$$

we obtain from (3.11.53)

$$y_n(s) = \frac{(-1)^n B_n}{(\tilde{c}_1 \kappa)^n} q^{n\delta + n(n-1)/4}$$

$$\times \sum_{k=0}^{n} \frac{(-n|q)_k}{(1|q)_k} (\delta + d_1 + n - 1|q)_k \kappa^k q^{k(d_1 - \delta - (k-1)/2)/2 + ks} . \quad (3.11.59)$$

Since the function $x(s) = \tilde{c}_1 q^s + \tilde{c}_3$ transforms into the function $x(s) = \tilde{c}_2 q^{-s} + \tilde{c}_3$ if q is replaced by $1/q$ and $\tilde{c}_1$ by $\tilde{c}_2$, formulas (3.11.52–59) for $\sigma(s)$ and $y_n(s)$, obtained for $x(s) = \tilde{c}_1 q^s + \tilde{c}_3$ after replacing q by $1/q$ and $\tilde{c}_1$ by $\tilde{c}_2$, can be extended to the case when $x(s) = \tilde{c}_2 q^{-s} + \tilde{c}_3$. In particular, formula (3.11.53) with $\tilde{c}_1 = 1$ gives after replacement of q by $1/q$ the polynomials corresponding to *the q-Hahn polynomials* at $x(s) = q^{-s}$ [H1–6, A17].

In order to define the weight function $\varrho(s)$ corresponding to the above $\sigma(s)$ and $y_n(s)$ for the cases when $x(s) = \tilde{c}_1 q^s + \tilde{c}_3$ and $x(s) = \tilde{c}_2 q^{-s} + \tilde{c}_3$ $(\mu = \pm\infty)$ we should use the original equation (3.2.9) instead of (3.11.26):

$$\Delta[\sigma(s)\varrho(s)] = \tau(s)\varrho(s)\Delta x(s - 1/2) \; .$$

Here the function $\tau(s)$ is connected with $y_1(s)$ by the relation

$$\tau(s) = \frac{1}{B_1} y_1(s) \; , \qquad (3.11.60)$$

which follows from the Rodrigues formula. For the above cases the function $\tau(s)$ depends on two parameters: for example, in the case (3.11.53) it depends on d_1 and d_2, while in the cases (3.11.55) and (3.11.59) it depends on d_1 and δ. Also of possible interest are the cases where some of these parameters tend to $\pm\infty$.

3) We consider one of the cases, corresponding to the Stieltjes-Wigert polynomials [S24, W6, C18], for which

$$x(s) = q^s \; , \quad \sigma(s) = q^{2s} \; , \quad \varrho(s) = q^{-s^2/2}$$

and, hence,

$$\tau(s) = \frac{\Delta[\sigma(s)\varrho(s)]}{\varrho(s)\Delta x(s - 1/2)} = \frac{1}{\kappa}\left(q^{3/2} - q^s\right) . \tag{3.11.61}$$

In order to derive the formula for the Stieltjes-Wigert polynomials we shall proceed from (3.11.59) for $\tilde{c}_1 = 1$. The parameters d_1 and δ are determined by comparing formulas (3.11.60) and (3.11.61) for $\tau(s)$. According to (3.11.60)

$$\tau(s) = -\frac{1}{\kappa}q^\delta\left[1 - \psi_q(\delta + d_1)\kappa q^{(d_1 - \delta)/2 + s}\right] = \frac{1}{\kappa}\left[-q^\delta + \left(q^{d_1 + \delta} - 1\right)q^s\right] .$$

It is seen that formula (3.11.61) may be obtained at $q^\delta = -q^{3/2}(\delta = 3/2 + i\pi/\ln q)$ and $q^{d_1} \to 0$. Since

$$\lim_{q^{d_1} \to 0}(\delta + d_1 + n - 1|q)_k q^{k(d_1 - \delta - (k-1)/2)/2}$$

$$= \lim_{q^{d_1} \to 0}\frac{(\delta + d_1 + n - 1|q)_k}{[\psi_q(d_1)]^k}\frac{[\psi_q(d_1)]^k}{q^{-k(d_1 - \delta - (k-1)/2)/2}}$$

$$= \frac{(-1)^k}{\kappa^k}\left(q^\delta\right)^{-k}q^{-k(n+k-2)/2} = \frac{1}{\kappa^k}q^{-k(n+k+1)/2} ,$$

we obtain from (3.11.59) with $q^\delta = -q^{3/2}$, $q^{d_1} \to 0$ for Stieltjes-Wigert polynomials

$$y_n(s) = \tilde{y}_n(x) = \frac{B_n}{\kappa^n}q^{n(n+5)/4}\sum_{k=0}^{n}\frac{(-n|q)_k}{(1|q)_k}q^{-k(n+k+1)/2}x^k . \tag{3.11.62}$$

4) In conclusion we shall establish the relation between the polynomials (3.11.55) at $x(s) = \tilde{c}_1 q^s + \tilde{c}_3$ and the Meixner, Kravchuk and Charlier polynomials considered in Chap. 2, for which $x(s) = s$ and $\sigma(s) = s$. It is natural to expect that the Meixner and Kravchuk polynomials [see (2.7.12) and (2.7.11a)] may be obtained from (3.11.55) at $s_1 = 0$, $\tilde{c}_1 = 1/\kappa$, $\tilde{c}_3 = -1/\kappa$ as $q \to 1$ if we choose the parameters δ and d_1 to be dependent on q in a certain way.

We shall show that in (3.11.55) there exists a limit for $y_n(s)$ as $q \to 1$ if d_1 is chosen to be independent of q, while the parameter $\delta = \delta(q)$ is chosen from the condition $q^{-\delta} = \alpha$, where α is a constant independent of q. Indeed, in this case

$$\lim_{q \to 1}(d_1 + \delta + n - 1|q)_k \kappa^k q^{-k\delta/2}$$

$$= \lim_{q \to 1}\prod_{l=0}^{k-1}\left[\left(q^{(d_1 + \delta + n - 1 + l)/2} - q^{-(d_1 + \delta + n - 1 + l)/2}\right)q^{-\delta/2}\right] = (1 - \alpha)^k .$$

Proceeding from the above we obtain from (3.11.55)

$$y_n(s) = \frac{B_n}{\alpha^n}(d_1)_n\sum_{k=0}^{n}\frac{(-n)_k(-s)_k(1 - \alpha)^k}{(d_1)_k k!}$$

154

$$= \frac{B_n}{\alpha^n}(d_1)_n \, {}_2F_1 \left(\begin{array}{c} -n, \ -s \\ d_1 \end{array} \middle| (1-\alpha) \right) \tag{3.11.63}$$

as $q \to 1$. It coincides with formula (2.7.12) for the Meixner polynomials $m_n^{(\gamma,\mu)}(s)$ at $d_1 = \gamma$ and $\alpha = 1/\mu$, and with formula (2.7.11a) for the Kravchuk polynomials $k_n^{(p)}(s,N)$ at $d_1 = -N$ and $\alpha = 1 - 1/p$.

Let us show that formula (2.7.9) for the Charlier polynomials $c_n^{(\mu)}(s)$ may be derived from (3.11.55) as $q \to 1$ ($s_1 = 0$, $\tilde{c}_1 = 1/\kappa$ and $\tilde{c}_3 = -1/\kappa$), if we choose $d_1 = d_1(q)$ and $\delta = \delta(q)$ from the conditions

$$q^{d_1} = 1/\mu\kappa \,, \quad q^{\delta} = (\mu\kappa)^2 \quad \left(\kappa = q^{1/2} - q^{-1/2} \right) \,.$$

Indeed, in this case we have

$$\lim_{q \to 1} q^{n(d_1 + 2\delta + n - 1)/2}(d_1|q)_n$$

$$= \lim_{q \to 1} \prod_{l=0}^{n-1} \left[q^{(d_1 + 2\delta + n - 1)/2} \left(\frac{q^{(d_1+l)/2} - q^{-(d_1+l)/2}}{\kappa} \right) \right]$$

$$= \lim_{q \to 1} \prod_{l=0}^{n-1} \left[(\mu\kappa)^{-1/2}(\mu\kappa)^2 q^{(n-1)/2} \left((\mu\kappa)^{-1/2}q^{l/2} - (\mu\kappa)^{1/2}q^{-l/2} \right) \Big/ \kappa \right]$$

$$= \mu^n \,.$$

$$\lim_{q \to 1} \frac{(d_1 + \delta + n - 1|q)_k}{(d_1|q)_k} \, \kappa^k q^{-k\delta/2}$$

$$= \lim_{q \to 1} \prod_{l=0}^{k-1} \left[\frac{q^{(d_1+\delta+n-1+l)/2} - q^{-(d_1+\delta+n-1+l)/2}}{q^{(d_1+l)/2} - q^{-(d_1+l)/2}} \, \kappa q^{-\delta/2} \right]$$

$$= \lim_{q \to 1} \prod_{l=0}^{k-1} \left[\frac{(\mu\kappa)^{1/2}q^{(n-1+l)/2} - (\mu\kappa)^{-1/2}q^{-(n-1+l)/2}}{(\mu\kappa)^{-1/2}q^{l/2} - (\mu\kappa)^{1/2}q^{-l/2}} \, \kappa \frac{1}{\mu\kappa} \right] = \left(-\frac{1}{\mu} \right)^k \,.$$

Therefore, at $B_n = 1/\mu^n$ and as $q \to 1$ we obtain from (3.11.55)

$$y_n(s) = \sum_{k=0}^{n-1} \frac{(-n)_k(-s)_k}{k!} \left(-\frac{1}{\mu} \right)^k = {}_2F_0 \left(-n, -s; -1/\mu \right)$$

which coincides with (2.7.9).

3.12 Particular Solutions
of the Hypergeometric Type Difference Equation

3.12.1. In Sect. 3.11.2 it was proved that the difference equation of hypergeometric type (3.11.26a), equivalent to (3.1.21), on the lattice (3.11.10) with $\sigma(s)$ given in the most general form (3.11.12) has the polynomial solutions (3.11.25) for $\lambda = \lambda_n$ determined by (3.2.6) (see also Sect. 3.7.1). Let us consider the function

$$y_\nu(s) = {}_4F_3\left(\begin{array}{c} -\nu, \sum_{i=1}^4 s_i + 2\mu + \nu - 1, s_1 - s, s_1 + s + \mu \\ s_1 + s_2 + \mu, s_1 + s_3 + \mu, s_1 + s_4 + \mu \end{array}\middle| q, 1\right), \qquad (3.12.1)$$

which is obtained from (3.11.25) to within a constant factor by replacing n by a parameter ν and trying to determine if this function is a solution of the equation (see Eq. (3.11.26a))

$$\sigma(-s - \mu)\frac{\Delta y(s)}{\Delta x(s)} - \sigma(s)\frac{\nabla y(s)}{\nabla x(s)} = -\lambda \Delta x(s - 1/2)y(s) \qquad (3.12.2)$$

for some value of λ that depends on a parameter ν.

First of all we must know if the generalized q-hypergeometric series (3.12.1) converge. We have

$$y_\nu(s) = \sum_{k=0}^\infty A_k(s_1 - s|q)_k(s_1 + s + \mu|q)_k , \qquad (3.12.3)$$

where

$$A_k = \frac{(-\nu|q)_k \left(\sum_{i=1}^4 s_i + 2\mu + \nu - 1|q\right)_k}{(s_1 + s_2 + \mu|q)_k(s_1 + s_3 + \mu|q)_k(s_1 + s_4 + \mu|q)_k(1|q)_k} . \qquad (3.12.4)$$

Let us show that the series (3.12.3) converges uniformly in the region $|s| \leq$ const. It is sufficient to verify if the ratio of two successive terms of the series for arbitrarily large k is less than some constant $c < 1$. In our case after using (3.11.1) and (3.11.31) with $t = k$ and $q < 1$ we obtain

$$\lim_{k\to\infty} \frac{A_{k+1}(s_1 - s|q)_{k+1}(s_1 + s + \mu|q)_{k+1}}{A_k(s_1 - s|q)_k(s_1 + s + \mu|q)_k}$$

$$= \lim_{k\to\infty} \frac{\psi_q(-\nu + k)\psi_q\left(\sum_{i=1}^4 s_i + 2\mu + \nu - 1 + k\right)\psi_q(s_1 - s + k)\psi_q(s_1 + s + \mu + k)}{\psi_q(s_1 + s_2 + \mu + k)\psi_q(s_1 + s_3 + \mu + k)\psi_q(s_1 + s_4 + \mu + k)\psi_q(k + 1)}$$

$$= q < 1$$

(for $q > 1$ this ratio tends to $1/q$ that is also less than unity). Consequently the series (3.12.3) for $y_\nu(s)$ converges uniformly for all s in the region $|s| \leq$ const.

Further,

$$\frac{\Delta y_\nu(s)}{\Delta x(s)} = \sum_{k=0}^\infty A_k \frac{\Delta}{\Delta x(s)}\left[(s_1 - s|q)_k(s_1 + s + \mu|q)_k\right] . \qquad (3.12.5)$$

In subsequent transformations we shall use the following identities, which can be easily verified:

$$\psi_q(a)\psi_q(b + c) - \psi_q(b)\psi_q(c + a) = \psi_q(a - b)\psi_q(c) , \qquad (3.12.6)$$

$$ab - cd = (a - c)(b + d)/2 + (b - d)(a + c)/2 . \qquad (3.12.7)$$

Since

$$\Delta x(s) = \tilde{c}_1 \left(q^{s+1} + q^{-s-\mu-1} - q^s - q^{-s-\mu} \right) = B\psi_q(2s + \mu + 1) , \qquad (3.12.8)$$

where

$$B = \tilde{c}_1 q^{-\mu/2} \kappa^2 \quad (\kappa = q^{1/2} - q^{-1/2}) ,$$

then by using (3.12.6) we obtain

$$\frac{\Delta}{\Delta x(s)} \left[(s_1 - s|q)_k (s_1 + s + \mu|q)_k \right]$$
$$= \frac{(s_1 - s - 1|q)_k (s_1 + s + \mu + 1|q)_k - (s_1 - s|q)_k (s_1 + s + \mu|q)_k}{B\psi_q(2s + \mu + 1)}$$
$$= (s_1 - s|q)_{k-1} (s_1 + s + \mu + 1|q)_{k-1}$$
$$\times \frac{\psi_q(s_1 - s - 1)\psi_q(s_1 + s + \mu + k) - \psi_q(s_1 - s + k - 1)\psi_q(s_1 + s + \mu)}{B\psi_q(2s + \mu + 1)}$$
$$= -(s_1 - s|q)_{k-1} (s_1 + s + \mu + 1|q)_{k-1} \frac{\psi_q(k)}{B} . \qquad (3.12.9)$$

Thus by using (3.11.12), (3.12.5) and the equality

$$\frac{\nabla f(s)}{\nabla x(s)} = \frac{\Delta f(t)}{\Delta x(t)}\bigg|_{t=s-1}$$

we obtain

$$\sigma(-s - \mu)\frac{\Delta y_\nu(s)}{\Delta x(s)} - \sigma(s)\frac{\nabla y_\nu(s)}{\nabla x(s)}$$
$$= -\frac{A}{B} \sum_{k=0}^{\infty} \mathcal{A}_k (s_1 - s|q)_{k-1} (s_1 + s + \mu|q)_{k-1} \psi_q(k) f_k(s) , \qquad (3.12.10)$$

where

$$f_k(s) = \psi_q(s_1 + s + \mu + k - 1) \prod_{i=2}^{4} \psi_q(s + s_i + \mu)$$
$$- \psi_q(s_1 - s + k - 1) \prod_{i=2}^{4} \psi_q(s_i - s) .$$

We transform the expression for $f_k(s)$ by using (3.12.7) with

$$a = a(s) = \psi_q(s_1 + s + \mu + k - 1)\psi_q(s + s_2 + \mu) ,$$
$$b = b(s) = \psi_q(s + s_3 + \mu)\psi_q(s + s_4 + \mu) ,$$
$$c = c(s) = a(-s - \mu) , \quad d = d(s) = b(-s - \mu) .$$

Owing to (3.12.6)

$$a(s) - c(s) = \psi_q(2s + \mu)\psi_q(s_1 + s_2 + \mu + k - 1) ,$$
$$b(s) - d(s) = \psi_q(2s + \mu)\psi_q(s_3 + s_4 + \mu) .$$

By using an explicit form of the function $\psi_q(s)$ it is easy to verify that the expressions $a(s) + c(s)$ and $b(s) + d(s)$ are polynomials of the first degree in $x(s)$. The product $\psi_q(s_1 - s + k - 1)\psi_q(s_1 + s + \mu + k - 1)$ is also a polynomial of the first degree in $x(s)$. Therefore

$$
\begin{aligned}
\frac{a(s) + c(s)}{2} &= M_1\psi_q(s_1 - s + k - 1)\psi_q(s_1 + s + \mu + k - 1) + M_2 \, , \\
\frac{b(s) + d(s)}{2} &= M_3\psi_q(s_1 - s + k - 1)\psi_q(s_1 + s + \mu + k - 1) + M_4 \, ,
\end{aligned}
\tag{3.12.11}
$$

where M_1, M_2, M_3 and M_4 are constants which can be found by putting $s = s_1 + k - 1$ and by equating the coefficients at q^s in (3.12.11):

$$
\begin{aligned}
M_1 &= -\tfrac{1}{2}\left(q^{(s_1+s_2+\mu+k-1)/2} + q^{-(s_1+s_2+\mu+k-1)/2}\right) \, , \\
M_2 &= \tfrac{1}{2}\psi_q(2s_1 + 2k - 2 + \mu)\psi_q(s_1 + s_2 + \mu + k - 1) \, , \\
M_3 &= -\tfrac{1}{2}\left(q^{(s_3+s_4+\mu)/2} + q^{-(s_3+s_4+\mu)/2}\right) \, , \\
M_4 &= \tfrac{1}{2}\left[\psi_q(s_1 + s_3 + \mu + k - 1)\psi_q(s_1 + s_4 + \mu + k - 1)\right. \\
&\quad \left. + \psi_q(s_1 - s_3 + k - 1)\psi_q(s_1 - s_4 + k - 1)\right] \, .
\end{aligned}
$$

As a result we obtain

$$
\begin{aligned}
f_k(s) &= \psi_q(2s + \mu)\big\{\psi_q(s_1 - s + k - 1)\psi_q(s_1 + s + \mu + k - 1) \\
&\quad \times \left[M_1\psi_q(s_3 + s_4 + \mu) + M_3\psi_q(s_1 + s_2 + \mu + k - 1)\right] \\
&\quad + M_2\psi_q(s_3 + s_4 + \mu) + M_4\psi_q(s_1 + s_2 + \mu + k - 1)\big\} \, .
\end{aligned}
\tag{3.12.12}
$$

The expression for $f_k(s)$ may be simplified by using (3.12.7) and (3.12.6). By putting

$$
\begin{aligned}
a &= \frac{1}{\kappa}\, q^{(s_3+s_4+\mu)/2} \, , & c &= \frac{1}{\kappa}\, q^{-(s_3+s_4+\mu)/2} \, , \\
b &= \frac{1}{\kappa}\, q^{(s_1+s_2+\mu+k-1)/2} \, , & d &= \frac{1}{\kappa}\, q^{-(s_1+s_2+\mu+k-1)/2}
\end{aligned}
$$

in (3.12.7) we obtain

$$
\begin{aligned}
M_1\psi_q(s_3 + s_4 + \mu) &+ M_3\psi_q(s_1 + s_2 + \mu + k - 1) \\
&= -\psi_q\left(\sum_{i=1}^{4} s_i + 2\mu + k - 1\right) \, .
\end{aligned}
\tag{3.12.13}
$$

Further,

$$
\begin{aligned}
M_2\psi_q(s_3 &+ s_4 + \mu) + M_4\psi_q(s_1 + s_2 + \mu + k - 1) \\
&= \tfrac{1}{2}\psi_q(s_1 + s_2 + \mu + k - 1)\big[\psi_q(2s_1 + 2k - 2 + \mu)\psi_q(s_3 + s_4 + \mu) \\
&\quad + \psi_q(s_1 - s_3 + k - 1)\psi_q(s_1 - s_4 + k - 1)\big] \\
&\quad + \tfrac{1}{2}\prod_{i=2}^{4}\psi_q(s_1 + s_i + \mu + k - 1) \, .
\end{aligned}
\tag{3.12.14}
$$

In order to simplify (3.12.14) we use Eq. (3.12.6) with

$$a = s_1 + s_3 + \mu + k - 1 , \quad b = s_3 + s_4 + \mu , \quad c = s_1 - s_3 + k - 1 .$$

Then we obtain

$$\psi_q(2s_1 + 2k - 2 + \mu)\psi_q(s_3 + s_4 + \mu) + \psi_q(s_1 - s_3 + k - 1)\psi_q(s_1 - s_4 + k - 1)$$
$$= \psi_q(s_1 + s_3 + \mu + k - 1)\psi_q(s_1 + s_4 + \mu + k - 1) . \tag{3.12.15}$$

As a result, owing to (3.12.13–15), the Eq. (3.12.12) for $f_k(s)$ will have the form

$$f_k(s) = -\psi_q(2s + \mu)\big[\psi_q(s_1 - s + k - 1)\psi_q(s_1 + s + \mu + k - 1)$$
$$\times \psi_q\left(\sum_{i=1}^{4} s_i + 2\mu + k - 1\right) + \psi_q(s_1 + s_2 + \mu + k - 1)$$
$$\times \psi_q(s_1 + s_3 + \mu + k - 1)\psi_q(s_1 + s_4 + \mu + k - 1)\big] .$$

If we substitute this expression into (3.12.10) and use the equality

$$\psi_q(k)\mathcal{A}_k\psi_q(s_1 + s_2 + \mu + k - 1)\psi_q(s_1 + s_3 + \mu + k - 1)$$
$$\times \psi_q(s_1 + s_4 + \mu + k - 1)$$
$$= \mathcal{A}_{k-1}\psi_q(-\nu + k - 1)\psi_q\left(\sum_{i=1}^{4} s_i + 2\mu + \nu + k - 2\right) ,$$

Eq. (3.12.10) may be rewritten in the form

$$\sigma(-s - \mu)\frac{\Delta y_\nu(s)}{\Delta x(s)} - \sigma(s)\frac{\nabla y_\nu(s)}{\nabla x(s)} = \lim_{N\to\infty} \frac{A}{B}\,\psi_q(2s + \mu)$$
$$\times \left[\sum_{k=0}^{N} \mathcal{A}_k(s_1 - s|q)_k(s_1 + s + \mu|q)_k\psi_q(k)\psi_q\left(\sum_{i=1}^{4} s_i + 2\mu + k - 1\right)\right.$$
$$- \sum_{k=1}^{N} \mathcal{A}_{k-1}(s_1 - s|q)_{k-1}(s_1 + s + \mu|q)_{k-1}\psi_q(-\nu + k - 1)$$
$$\left.\times \psi_q\left(\sum_{i=1}^{4} s_i + 2\mu + \nu + k - 2\right)\right] .$$

Replacing k by $k + 1$ in the second sum we obtain

$$\sigma(-s - \mu)\frac{\Delta y_\nu(s)}{\Delta x(s)} - \sigma(s)\frac{\nabla y_\nu(s)}{\nabla x(s)} = \frac{A}{B}\,\psi_q(2s + \mu)$$
$$\times \lim_{N\to\infty}\left\{\sum_{k=0}^{N-1} \mathcal{A}_k(s_1 - s|q)_k(s_1 + s + \mu|q)_k\right.$$
$$\times \left[\psi_q(k)\psi_q\left(\sum_{i=1}^{4} s_i + 2\mu + k - 1\right) - \psi_q(-\nu + k)\right.$$

$$\times \psi_q \left(\sum_{i=1}^{4} s_i + 2\mu + \nu + k - 1) \right) \Bigg] + g_{\nu N}(s) \Bigg\} , \qquad (3.12.16)$$

where

$$g_{\nu N}(s) = \mathcal{A}_N(s_1 - s|q)_N \psi_q(N)\psi_q(N)(s_1 + s + \mu|q)_N$$

$$\times \psi_q \left(\sum_{i=1}^{4} s_i + 2\mu + N - 1 \right) . \qquad (3.12.17)$$

According to (3.12.16) with $a = k$, $b = -\nu + k$, $c = \sum_{i=1}^{4} s_i + 2\mu + \nu - 1$ we have

$$\psi_q(k)\psi_q \left(\sum_{i=1}^{4} s_i + 2\mu + k - 1 \right) - \psi_q(-\nu + k)\psi_q \left(\sum_{i=1}^{4} s_i + 2\mu + \nu + k - 1 \right)$$

$$= \psi_q(\nu)\psi_q \left(\sum_{i=1}^{4} s_i + 2\mu + \nu - 1 \right) .$$

Therefore, owing to (3.12.3) and (3.12.8),

$$\sigma(-s - \mu) \frac{\Delta y_\nu(s)}{\Delta x(s)} - \sigma(s) \frac{\nabla y_\nu(s)}{\nabla x(s)}$$

$$= \frac{A}{B} \psi_q(2s + \mu)\psi_q(\nu)\psi_q \left(\sum_{i=1}^{4} s_i + 2\mu + \nu - 1 \right) y_\nu(s) + \lim_{N \to \infty} g_{\nu N}(s)$$

$$= -\lambda_\nu \Delta x(s - 1/2)y_\nu(s) + g_\nu(s) ,$$

where

$$\lambda_\nu = -\frac{A}{B^2} \psi_q(\nu)\psi_q \left(\sum_{i=1}^{4} s_i + 2\mu + \nu - 1 \right) , \qquad (3.12.18)$$

$$g_\nu(s) = \lim_{N \to \infty} g_{\nu N}(s) . \qquad (3.12.19)$$

Thus, we see that the function $y_\nu(s)$ defined by (3.12.1) is the solution of the non-homogeneous difference equation of hypergeometric type

$$\sigma(-s - \mu) \frac{\Delta y_\nu(s)}{\Delta x(s)} - \sigma(s) \frac{\nabla y_\nu(s)}{\nabla x(s)} + \lambda \Delta x(s - 1/2)y_\nu(s) = g_\nu(s) , \quad (3.12.20)$$

where the constant ν is connected with $\lambda \equiv \lambda_\nu$ by the relation (3.12.18). In order to find the parameter ν for any given λ it is necessary to solve a quadratic equation in q^ν. The function $g_\nu(s)$ is defined by formulas (3.12.19) and (3.12.17). We can find the explicit expression for $g_\nu(s)$ by using the limiting relation

$$\lim_{N \to \infty} q^{aN/2} \frac{(a|q)_N}{\tilde{\Gamma}_q(N)} = \frac{1}{\tilde{\Gamma}_q(a)} q^{-a(a-3)/4}(1 - q)^{-a} , \qquad (3.12.21)$$

which can be found by using (3.11.1) and the asymptotic behavior of the function $\tilde{\Gamma}_q(s)$ for $s \to +\infty$ (see Sect. 3.6.1.1). Hence by using (3.12.19), (3.12.18) and (3.12.4) for $0 < q < 1$ we obtain

$$g_\nu(s) = \frac{\alpha_\nu q^{-s(s+\mu)/2}}{\tilde{\Gamma}_q(s_1 - s)\tilde{\Gamma}_q(s_1 + s + \mu)} \,, \tag{3.12.22}$$

where

$$\alpha_\nu = \frac{\tilde{\Gamma}_q(s_1 + s_2 + \mu)\tilde{\Gamma}_q(s_1 + s_3 + \mu)\tilde{\Gamma}_q(s_1 + s_4 + \mu)}{\tilde{\Gamma}_q(-\nu)\tilde{\Gamma}_q\left(\sum_{i=1}^4 s_i + 2\mu + \nu - 1\right)} q^{\beta_\nu/4} \,, \tag{3.12.23}$$

$$\beta_\nu = -s_1(s_1 - 3) - (s_1 + \mu)(s_1 + \mu - 3) - \nu(\nu + 3) - 2\left(\sum_{i=1}^4 s_i + 2\mu\right) + 4$$

$$- \left(\sum_{i=1}^4 s_i + 2\mu + \nu - 1\right)\left(\sum_{i=1}^4 s_i + 2\mu + \nu - 4\right)$$

$$+ (s_1 + s_2 + \mu)(s_1 + s_2 + \mu - 3) + (s_1 + s_3 + \mu)(s_1 + s_3 + \mu - 3)$$

$$+ (s_1 + s_4 + \mu)(s_1 + s_4 + \mu - 3) \,. \tag{3.12.24}$$

The formula for $g_\nu(s)$ with $q > 1$ can be found from (3.12.22–24) by replacing q by $1/q$ since the function $g_\nu(s)$ does not vary after this change.

In order to find the solution of the homogeneous difference equation (3.12.2) we shall try to find, along with the particular solution $y_\nu(s)$ (see (3.12.1)) of the non-homogeneous equation (3.12.20), some other solution $\bar{y}_\nu(s)$ of the non-homogeneous equation. We shall seek this solution in the form

$$\bar{y}_\nu(s) = \varphi(s)u(s) \,, \tag{3.12.25}$$

where the formula for $u(s)$ is obtained by replacing $s_i\,(i = 1, 2, 3, 4)$ by $\bar{s}_i$ and ν by $\bar{\nu}$:

$$u(s) = {}_4F_3\left(\begin{array}{c} -\bar{\nu}, \sum_{i=1}^4 \bar{s}_i + 2\mu + \bar{\nu} - 1, \bar{s}_1 - s, \bar{s}_1 + s + \mu \\ \bar{s}_1 + \bar{s}_2 + \mu, \bar{s}_1 + \bar{s}_3 + \mu, \bar{s}_1 + \bar{s}_4 + \mu \end{array}\middle| q, 1\right) \,. \tag{3.12.26}$$

It is evident that the function $u(s)$ satisfies the non-homogeneous difference equation of hypergeometric type similar to (3.12.20)

$$\bar{\sigma}(-s - \mu)\frac{\Delta u(s)}{\Delta x(s)} - \bar{\sigma}(s)\frac{\nabla u(s)}{\nabla x(s)} + \bar{\lambda}_\nu \Delta x(s - 1/2)u(s) = \bar{g}_\nu(s) \,, \tag{3.12.27}$$

where $\bar{\sigma}(s)$, $\bar{g}_\nu(s)$ and $\bar{\lambda}_\nu$ are obtained from $\sigma(s)$, $g_\nu(s)$ and λ_ν after replacing s_i by $\bar{s}_i$ and ν by $\bar{\nu}$.

Owing to (3.12.25) we have the following formulas for $\Delta\bar{y}_\nu(s)$ and $\nabla\bar{y}_\nu(s)$:

$$\Delta\bar{y}_\nu(s) = \varphi(s + 1)\Delta u(s) + u(s)\Delta\varphi(s) \,,$$

$$\nabla\bar{y}_\nu(s) = \varphi(s - 1)\nabla u(s) + u(s)\nabla\varphi(s) \,,$$

whence

$$\sigma(-s-\mu)\frac{\Delta\bar{y}_\nu(s)}{\Delta x(s)} - \sigma(s)\frac{\nabla\bar{y}_\nu(s)}{\nabla x(s)} + \lambda_\nu\Delta x(s-1/2)\bar{y}_\nu(s)$$

$$= \varphi(s)\left[\sigma(-s-\mu)\frac{\varphi(s+1)}{\varphi(s)}\frac{\Delta u(s)}{\Delta x(s)} - \sigma(s)\frac{\varphi(s-1)}{\varphi(s)}\frac{\nabla u(s)}{\nabla x(s)}\right.$$

$$\left. + \bar{\lambda}(s)\Delta x(s-1/2)u(s)\right], \tag{3.12.28}$$

where

$$\bar{\lambda}(s) = \lambda_\nu + \frac{1}{\Delta x(s-1/2)}\left[\frac{\sigma(-s-\mu)}{\varphi(s)}\frac{\Delta\varphi(s)}{\Delta x(s)} - \frac{\sigma(s)}{\varphi(s)}\frac{\nabla\varphi(s)}{\nabla x(s)}\right]. \tag{3.12.29}$$

We choose the function $\varphi(s)$ and parameters $\bar{s}_i$ from the conditions

$$\begin{cases} \sigma(-s-\mu)\dfrac{\varphi(s+1)}{\varphi(s)} = \bar{\sigma}(-s-\mu), \\[2ex] \sigma(s)\dfrac{\varphi(s-1)}{\varphi(s)} = \bar{\sigma}(s). \end{cases} \tag{3.12.30}$$

In this case according to (3.12.27) we have

$$\sigma(-s-\mu)\frac{\varphi(s+1)}{\varphi(s)}\frac{\Delta u(s)}{\Delta x(s)} - \sigma(s)\frac{\varphi(s-1)}{\varphi(s)}\frac{\nabla u(s)}{\nabla x(s)}$$

$$= \bar{\sigma}(-s-\mu)\frac{\Delta u(s)}{\Delta x(s)} - \bar{\sigma}(s)\frac{\nabla u(s)}{\nabla x(s)} = -\bar{\lambda}_\nu\Delta x(s-1/2)u(s) + \bar{g}_\nu(s).$$

Therefore (3.12.28) may be rewritten in the form

$$\sigma(-s-\mu)\frac{\Delta\bar{y}_\nu(s)}{\Delta x(s)} - \sigma(s)\frac{\nabla\bar{y}_\nu(s)}{\nabla x(s)} + \lambda_\nu\Delta x(s-1/2)\bar{y}_\nu(s)$$

$$= \left[\bar{\lambda}(s) - \bar{\lambda}_\nu\right]\Delta x(s-1/2)\bar{y}_\nu(s) + \varphi(s)\bar{g}_\nu(s).$$

If we could choose the function $\varphi(s)$ and parameters $\bar{s}_i$ and $\bar{\nu}$ in such a form that the equations (3.12.30) were valid and $\bar{\lambda}(s) = \text{const}$ (and moreover, $\bar{\lambda}(s) = \bar{\lambda}_\nu$), then the function $\bar{y}_\nu(s)$ would satisfy the non-homogeneous difference equation similar to (3.12.20) for $y_\nu(s)$

$$\sigma(-s-\mu)\frac{\Delta\bar{y}_\nu}{\Delta x(s)} - \sigma(s)\frac{\nabla\bar{y}_\nu(s)}{\nabla x(s)} + \lambda_\nu\Delta x(s-1/2)\bar{y}_\nu(s) = \varphi(s)\bar{g}_\nu(s). \tag{3.12.31}$$

From (3.12.30) it follows that

$$\frac{\varphi(s+1)}{\varphi(s)} = \frac{\bar{\sigma}(-s-\mu)}{\sigma(-s-\mu)} = \frac{\sigma(s+1)}{\bar{\sigma}(s+1)}.$$

Therefore the function $\bar{\sigma}(s)$ must be connected with $\sigma(s)$ by the relation

$$\bar{\sigma}(s+1)\bar{\sigma}(-s-\mu) = \sigma(s+1)\sigma(-s-\mu), \tag{3.12.32}$$

i.e.

$$\prod_{i=1}^{4}\psi_q(s+1-\bar{s}_i)\psi_q(s+\mu+\bar{s}_i)=\prod_{i=1}^{4}\psi_q(s+1-s_i)\psi_q(s+\mu+s_i). \quad (3.12.33)$$

This equality can be satisfied by setting, for example,

$$\bar{s}_1=1-\mu-s_2\,,\quad \bar{s}_2=1-\mu-s_1\,,\quad \bar{s}_3=s_3\,,\quad \bar{s}_4=s_4\,. \quad (3.12.34)$$

In this case the function $\varphi(s)$ is the solution of the equation

$$\frac{\varphi(s+1)}{\varphi(s)}=\frac{\psi_q(s+1-s_1)\psi_q(s+1-s_2)}{\psi_q(s+s_1+\mu)\psi_q(s+s_2+\mu)}\,, \quad (3.12.35)$$

whence

$$\varphi(s)=\frac{\tilde{\Gamma}_q(s+1-s_1)\tilde{\Gamma}_q(s+1-s_2)}{\tilde{\Gamma}_q(s+s_1+\mu)\tilde{\Gamma}_q(s+s_2+\mu)}\,. \quad (3.12.36)$$

Let us show that $\bar{\lambda}(s)=\text{const}$. By virtue of $(3.12.29,35)$ and $(3.12.8)$ we have

$$\begin{aligned}
\left[\bar{\lambda}(s)-\lambda_\nu\right]\Delta x(s-1/2)&=\frac{\sigma(-s-\mu)}{\Delta x(s)}\left[\frac{\varphi(s+1)}{\varphi(s)}-1\right]\\
&-\frac{\sigma(s)}{\nabla x(s)}\left[1-\frac{\varphi(s-1)}{\varphi(s)}\right]=\frac{A}{B}\frac{\prod_{i=3}^{4}\psi_q(s+\mu+s_i)}{\psi_q(2s+\mu+1)}\\
&\times\left[\psi_q(s+1-s_1)\psi_q(s+1-s_2)-\psi_q(s+\mu+s_1)\psi_q(s+\mu+s_2)\right]\\
&-\frac{A}{B}\frac{\psi_q(s-s_3)\psi_q(s-s_4)}{\psi_q(2s+\mu-1)}\left[\psi_q(s-s_1)\psi_q(s-s_2)\right.\\
&\left.-\psi_q(s+\mu+s_1-1)\psi_q(s+\mu+s_2-1)\right]\,. \quad (3.12.37)
\end{aligned}$$

By using $(3.12.6)$ with $a=s-s_1$, $b+c=s-s_2$, $a-b=s+\mu+s_2-1$ and $c=s-s_3$ we obtain

$$\begin{aligned}
\psi_q(s-s_1)&\psi_q(s-s_2)-\psi_q(s+\mu+s_1-1)\psi_q(s+\mu+s_2-1)\\
&=\psi_q(1-\mu-s_1-s_2)\psi_q(2s+\mu-1)\,.
\end{aligned}$$

Moreover, by replacing s by $s+1$ in this relation we have

$$\begin{aligned}
\psi_q(s+1-s_1)&\psi_q(s+1-s_2)-\psi_q(s+\mu+s_1)\psi_q(s+\mu+s_2)\\
&=\psi_q(1-\mu-s_1-s_2)\psi_q(2s+\mu+1)\,.
\end{aligned}$$

Therefore $(3.12.37)$ may be rewritten in the form

$$\begin{aligned}
\bar{\lambda}(s)-\lambda_\nu&=\frac{A}{B^2}\frac{\psi_q(1-\mu-s_1-s_2)}{\psi_q(2s+\mu)}\\
&\times\left[\psi_q(s+\mu+s_3)\psi_q(s+\mu+s_4)-\psi_q(s-s_3)\psi_q(s-s_4)\right]\,.
\end{aligned}$$

By using $(3.12.6)$ with $a=s+\mu+s_3$, $b=s_3+s_4+\mu$, $c=s-s_3$ and $(3.12.18)$ we obtain

$$\bar{\lambda}(s) = \lambda_\nu + \frac{A}{B^2}\,\psi_q(1 - \mu - s_1 - s_2)\psi_q(s_3 + s_4 + \mu)$$

$$= -\frac{A}{B^2}\left[\psi_q(\nu)\psi_q\left(\sum_{i=1}^{4} s_i + 2\mu + \nu - 1\right)\right.$$

$$\left. -\psi_q(1 - \mu - s_1 - s_2)\psi_q(s_3 + s_4 + \mu)\right].$$

This expression may be simplified with the aid of (3.12.6) by setting $a = s_1 + s_2 + \mu - 1$, $b = -\nu$, $c = s_3 + s_4 + \mu + \nu$:

$$\bar{\lambda}(s) = -\frac{A}{B^2}\,\psi_q(\bar{\nu})\psi_q\left(\sum_{i=1}^{4} s_i + 2\mu + \bar{\nu} - 1\right), \tag{3.12.18a}$$

where $\bar{s}_i$ are defined by (3.12.34) and

$$\bar{\nu} = s_1 + s_2 + \mu + \nu - 1 . \tag{3.12.38}$$

Thus after comparing (3.12.18) and (3.12.18a) we see that we obtain $\bar{\lambda}(s) = \bar{\lambda}_\nu$ for the above choice of parameters $\bar{s}_i$ and $\bar{\nu}$ (see (3.12.34) and (3.12.38)) and the function

$$\bar{y}_\nu(s) = \varphi(s)\,{}_4F_3\left(\begin{array}{c} -\bar{\nu}, \sum_{i=1}^{4}\bar{s}_i + 2\mu + \bar{\nu} - 1, \bar{s}_1 - s, \bar{s}_1 + s + \mu \\ \bar{s}_1 + \bar{s}_2 + \mu, \bar{s}_1 + \bar{s}_3 + \mu, \bar{s}_1 + \bar{s}_4 + \mu \end{array}\middle|\, q, 1\right) \tag{3.12.39}$$

is the solution of Eq. (3.12.31) ($\varphi(s)$ is defined by (3.12.36)).

The function $\bar{g}_\nu(s)$ is obtained from $g_\nu(s)$ by replacing ν by $\bar{\nu}$ and s_i by $\bar{s}_i$. Therefore the non-homogeneity in the right-hand side of (3.12.31) takes the form

$$\varphi(s)\bar{g}_\nu(s) = \frac{\bar{\alpha}_\nu q^{-s(s+\mu)/2}\tilde{\Gamma}_q(s + 1 - s_1)}{\tilde{\Gamma}_q(s + s_1 + \mu)\tilde{\Gamma}_q(s + s_2 + \mu)\tilde{\Gamma}_q(1 - \mu - s_2 - s)} \tag{3.12.40}$$

($\bar{\alpha}_\nu$ is obtained from formula (3.12.31) for α_ν by change ν by $\bar{\nu}$, s_i by $\bar{s}_i$).

With the aid of particular solutions $y_\nu(s)$ and $\bar{y}_\nu(s)$ which satisfy the equations (3.12.20) and (3.12.31) we can find the solution of the homogeneous equation (3.12.2) in the following way. Let

$$y(s) = c_1(s)y_\nu(s) + c_2(s)\bar{y}_\nu(s) , \tag{3.12.41}$$

where $c_1(s)$ and $c_2(s)$ are arbitrary periodic functions with period 1:

$$c_1(s + 1) = c_1(s) , \quad c_2(s + 1) = c_2(s) .$$

Since

$$\frac{\Delta y(s)}{\Delta x(s)} = \frac{y(s+1) - y(s)}{x(s+1) - x(s)}$$

$$= c_1(s)\frac{y_\nu(s+1) - y_\nu(s)}{x(s+1) - x(s)} + c_2(s)\frac{\bar{y}_\nu(s+1) - \bar{y}_\nu(s)}{x(s+1) - x(s)}$$

$$= c_1(s)\frac{\Delta y_\nu(s)}{\Delta x(s)} + c_2(s)\frac{\Delta \bar{y}_\nu(s)}{\Delta x(s)}$$

and

$$\frac{\nabla y(s)}{\nabla x(s)} = c_1(s)\frac{\nabla y_\nu(s)}{\nabla x(s)} + c_2(s)\frac{\nabla y_\nu(s)}{\nabla x(s)} \, ,$$

we obtain that $y(s)$ satisfies the non-homogeneous difference equation

$$\sigma(-s-\mu)\frac{\Delta y(s)}{\Delta x(s)} - \sigma(s)\frac{\nabla y(s)}{\nabla x(s)} + \lambda \Delta x(s-1/2)y(s) = g(s) \, , \qquad (3.12.42)$$

where

$$g(s) = c_1(s)g_\nu(s) + c_2(s)\varphi(s)\bar{g}_\nu(s) = \frac{q^{-s(s+\mu)/2}}{\tilde{\Gamma}_q(s + s_1 + \mu)}$$

$$\times \left[c_1(s)\frac{\alpha_\nu}{\tilde{\Gamma}_q(s_1 - s)} + c_2(s)\frac{\bar{\alpha}_\nu \tilde{\Gamma}_q(s + 1 - s_1)}{\tilde{\Gamma}_q(s + s_2 + \mu)\tilde{\Gamma}_q(1 - \mu - s_2 - s)} \right] . \quad (3.12.43)$$

The function $y(s)$, defined by (3.12.41), would be the solution of (3.12.2), if we could find such periodic functions $c_1(s)$ and $c_2(s)$ for which $g(s) = 0$. This condition is fulfilled, for example, for functions

$$c_1(s) = \bar{\alpha}_\nu \, ,$$

$$c_2(s) = -\alpha_\nu \frac{\tilde{\Gamma}_q(s + s_2 + \mu)\tilde{\Gamma}_q(1 - \mu - s_2 - s)}{\tilde{\Gamma}_q(s_1 - s)\tilde{\Gamma}_q(s + 1 - s_1)} \, . \qquad (3.12.44)$$

The periodicity of $c_1(s)$ is evident. The condition $c_2(s + 1) = c_2(s)$ is fulfilled because for the functions $f(t)$ of the form

$$f(t) = \tilde{\Gamma}_q(t)\tilde{\Gamma}_q(1 - t)$$

we have the property $f(t + 1) = -f(t)$ since by virtue of (3.6.9)

$$\frac{f(t+1)}{f(t)} = \frac{\tilde{\Gamma}_q(t+1)}{\tilde{\Gamma}_q(t)}\frac{\tilde{\Gamma}_q(-t)}{\tilde{\Gamma}_q(-t+1)} = \frac{\psi_q(t)}{\psi_q(-t)} = -1 \, .$$

Thus, as a result we have the function

$$y(s) = \bar{\alpha}_\nu \, {}_4F_3\left(\begin{array}{c} -\nu, \sum_{i=1}^{4} s_i + 2\mu + \nu - 1, s_1 - s, s_1 + s + \mu \\ s_1 + s_2 + \mu, s_1 + s_3 + \mu, s_1 + s_4 + \mu \end{array} \middle| q, 1 \right)$$

$$- \alpha_\nu \frac{\tilde{\Gamma}_q(s + 1 - s_2)\tilde{\Gamma}_q(1 - \mu - s_2 - s)}{\tilde{\Gamma}_q(s + s_1 + \mu)\tilde{\Gamma}_q(s_1 - s)}$$

$$\times {}_4F_3\left(\begin{array}{c} -\bar{\nu}, \sum_{i=1}^{4} \bar{s}_i + 2\mu + \bar{\nu} - 1, \bar{s}_1 - s, \bar{s}_1 + s + \mu \\ \bar{s}_1 + \bar{s}_2 + \mu, \bar{s}_1 + \bar{s}_3 + \mu, \bar{s}_1 + \bar{s}_4 + \mu \end{array} \middle| q, 1 \right) , \quad (3.12.45)$$

where $\bar{s}_1 = 1 - \mu - s_2$, $\bar{s}_2 = 1 - \mu - s_1$, $\bar{s}_3 = s_3$, $\bar{s}_4 = s_4$, $\bar{\nu} = s_1 + s_2 + \mu + \nu - 1$, α_ν is defined by (3.12.23), and $\bar{\alpha}_\nu$ is defined by the same formula on replacing s_i by $\bar{s}_i$ and ν by $\bar{\nu}$. The Eq. (3.12.2) for $y(s)$ does not vary on permutation of constants s_1, s_2, s_3, s_4. Therefore along with the solution (3.12.45) this equation has other solutions obtained from (3.12.45) by replacing s_1, s_2, s_3, s_4 by s_{k_1}, s_{k_2}, s_{k_3}, s_{k_4}, where the indices (k_1, k_2, k_3, k_4) are obtained as a result of any permutation of the indices $(1, 2, 3, 4)$.

Particular solutions of (3.12.2) can also be obtained with the aid of transformations $y(s) = \varphi(s)u(s)$ if the function $\varphi(s)$ is chosen such that the equation for $u(s)$ is a difference equation of hypergeometric type but with different parameters s_i and ν.

Addendum to Chapter 3

Let us consider some polynomial systems introduced in works of other authors in terms of basic hypergeometric series. For this purpose we rewrite (3.11.25) using (3.11.18–20) in the form

$$y_n(s) = \frac{\tilde{A}^n B_n q^{-n(3n-5)/4}}{[(1-q)\tilde{c}_2\xi_1\tilde{\mu}^2]^n}(\tilde{\mu}\xi_1\xi_2,q)_n(\tilde{\mu}\xi_1\xi_3,q)_n(\tilde{\mu}\xi_1\xi_4,q)_n$$

$$\times {}_4\varphi_3\left(\begin{matrix} q^{-n}, \tilde{\mu}^2\xi_1\xi_2\xi_3\xi_4 q^{n-1}, \xi_1 q^{-s}, \tilde{\mu}\xi_1 q^s \\ \tilde{\mu}\xi_1\xi_2, \ \tilde{\mu}\xi_1\xi_3, \ \tilde{\mu}\xi_1\xi_4 \end{matrix}\,\bigg|\, q,q\right) \tag{A1}$$

with

$$\sigma(s) = \tilde{A}q^{-2s}(q^s - \xi_1)(q^s - \xi_2)(q^s - \xi_3)(q^s - \xi_4)$$

$$(\tilde{A} = \frac{1}{\kappa^4}Aq^{-1/2\sum_{i=1}^{4}s_i}, \quad \xi_i = q^{s_i}$$

where $(a,q)_n = (1-a)(1-aq)\ldots(1-aq^{n-1})$, $\tilde{\mu} = \tilde{c}_1/\tilde{c}_2 = q^\mu$.

Different parametrizations are used in the literature for the ${}_4\varphi_3$-polynomials. Discrete orthogonality for the *q-Racah polynomials*

$$y_n(s) = R_n(x(s); \alpha, \beta, \gamma, \delta : q)$$

introduced in [A27, A29] corresponds to the following choice of parameters:

$$\tilde{c}_1 = \tilde{\mu} = \gamma\delta q, \quad \tilde{c}_2 = 1, \quad \tilde{c}_3 = 0,$$
$$\xi_1 = 1, \quad \xi_2 = \alpha/\gamma\delta, \quad \xi_3 = \beta/\gamma, \quad \xi_4 = 1/\delta.$$

In this case we have

$$R_n(x) = {}_4\varphi_3\left(\begin{matrix} q^{-n}, \alpha\beta q^{n+1}, q^{-s}, \gamma\delta q^{s+1} \\ \alpha q, \beta\delta q, \gamma q \end{matrix}\,\bigg|\, q,q\right),$$

$$x = x(s) = q^{-s} + \gamma\delta q^{s+1} \tag{A2}$$

and

$$\sigma(s) = \frac{\tilde{A}}{(\gamma\delta)^2}q^{-2s}(q^s - 1)(\gamma\delta q^s - \alpha)(\gamma q^s - \beta)(\delta q^s - 1),$$

$$\sigma(s) + \tau(s)\Delta x \left(s - \tfrac{1}{2}\right) = \sigma(-s - 1 - \ln(\gamma\delta)/\ln q)$$

$$= \frac{\tilde{A}}{(\gamma\delta)^2}q^{-2s-2}(1 - \alpha q^{s+1})(1 - \beta\delta q^{s+1})(1 - \gamma q^{s+1})(1 - \gamma\delta q^{s+1}).$$

Since

$$\frac{(a,q)_{s+1}}{(a,q)_s} = 1 - aq^s \quad (s = 0, 1, \ldots) ,$$

one can find from (3.5.1) that

$$\varrho(s) = (\alpha\beta)^{-s} \frac{(\alpha q, q)_s (\beta\delta q, q)_s (\gamma q, q)_s (\gamma\delta q, q)_s}{(q, q)_s (\alpha^{-1}\gamma\delta q, q)_s (\beta^{-1}\gamma q, q)_s (\delta q, q)_s} .$$

According to the conditions

$$\sigma(a) = 0 , \quad \sigma(s) + \tau(s)\Delta x \left(s - \tfrac{1}{2}\right)\Big|_{s=b-1} = 0$$

orthogonality property (3.3.4) is valid for the q-Racah polynomials (A2), when $a = 0$, $b = N + 1$ and any of parameters αq, $\beta\delta q$, γq or $\gamma\delta q$ is equal to q^{-N}.

When $\delta = 0$, $\gamma q = q^{-N}$ we come from (A2) to the q-Hahn polynomials [H1–6, A17]

$$Q_n(x(s)) = {}_3\varphi_2 \left(\begin{matrix} q^{-n}, \ \alpha\beta q^{n+1}, \ q^{-s} \\ \alpha q, \ q^{-N} \end{matrix} \ \middle| \ q, q \right) ,$$

$$x(s) = q^{-s} ,$$

(A3)

for which in (3.3.4) $a = 0$, $b = N + 1$ and

$$\varrho(s) = (\alpha\beta)^{-s} \frac{\alpha q, q)_s (q^{-N}, q)_s}{(q, q)_s (\beta^{-1} q^{-N}, q)_s} .$$

In the case $\tilde{c}_1 = \tilde{c}_2 = 1/2$, $\xi_1 = a$, $\xi_2 = b$, $\xi_3 = c$, $\xi_4 = d$ from (A1) we obtain the *Askey-Wilson polynomials* [A29]:

$$y_n(x) = p_n(x, a, b, c, d)|q) = a^{-n}(ab, q)_n (ac, q)_n (ad, q)_n$$

$$\times {}_4\varphi_3 \left(\begin{matrix} q^{-n}, \ abcdq^{n-1}, \ aq^s, \ aq^{-s} \\ ab, \ ac, \ ad \end{matrix} \ \middle| \ q, q \right) ,$$

(A4)

$$x(s) = \tfrac{1}{2}(q^s + q^{-s}) = \cos\theta , \quad q^s = e^{i\theta} .$$

Part II

Applications

4. Classical Orthogonal Polynomials of a Discrete Variable in Applied Mathematics

4.1 Quadrature Formulas of Gaussian Type

In the approximate calculation of definite integrals and of sums of a large number of terms, numerical analysis makes extensive use of quadrature formulas of Gaussian type, which depend on properties of orthogonal polynomials.

Quadrature formulas of Gaussian type for definite integrals are formulas of the form

$$\int_a^b f(x)\varrho(x)dx \approx \sum_{j=1}^{n} \lambda_j f(x_j) , \tag{4.1.1}$$

where $\varrho(x) > 0$, and the coefficients λ_j and abscissas $x_j (j = 1, 2, \ldots, n)$ are chosen so that (4.1.1) is exact for all polynomials of degree $2n - 1$.

If the moments of the weight function

$$c_k = \int_a^b x^k \varrho(x)dx$$

are known, the numbers λ_j and x_j can be found from the system of equations

$$\sum_{j=1}^{n} \lambda_j x_j^k = c_k \quad (k = 0, 1, \ldots, 2n - 1) .$$

However, we usually apply a different method to construct quadrature formulas of form (4.1.1). It turns out that the $x_j (j = 1, 2, \ldots, n)$ are the zeros of a polynomial $p_n(x)$ that is orthogonal on (a, b) with weight function $\varrho(x)$. For the proof we consider the function

$$f(x) = x^k \tilde{p}_n(x) ,$$

where

$$\tilde{p}_n(x) = (x - x_1)(x - x_2)\ldots(x - x_n)$$

is a polynomial of degree n whose zeros are the abscissas of the quadrature formula. When $k = 0, 1, \ldots, n - 1$ the function $f(x)$ is a polynomial of at most degree $2n - 1$. Therefore, if $f(x)$ is substituted into (4.1.1), the quadrature formula ought to produce the exact value of the integral for every $k < n$. Hence for $k = 0, 1, \ldots, n - 1$ we have

$$\int_a^b f(x)\varrho(x)dx = \int_a^b x^k \tilde{p}_n(x)\varrho(x)dx$$

$$= \sum_{j=1}^n \lambda_j x^k (x - x_1)(x - x_2)\ldots(x - x_n)\Big|_{x=x_j} = 0 .$$

Therefore the polynomial $\tilde{p}_n(x)$ is orthogonal to every power of order less than n, and consequently must be the same, up to a constant factor, as the polynomial $p_n(x)$ of degree n that is orthogonal on (a, b) with weight function $\varrho(x)$. Hence we can conclude that to determine the abscissas x_j of the quadrature formula it is enough to construct $p_n(x)$ and find its zeros.

To find the coefficients λ_j, which are known as the Christoffel numbers, it is convenient to take $f(x)$ in (4.1.1) to be a polynomial of degree less than $2n$ that is zero at all the abscissas except $x = x_j$. As a result we obtain

$$\lambda_j = \frac{1}{f(x_j)} \int_a^b f(x)\varrho(x)dx .$$

If we take, for example, $f(x) = [p_n(x)/(x - x_j)]^2$ or $f(x) = p_n(x)p_{n-1}(x)/(x - x_j)$, we obtain the following expressions for the λ_j:

$$\lambda_j = \int_a^b \left[\frac{p_n(x)}{p_n'(x_j)(x - x_j)} \right]^2 \varrho(x)dx , \tag{4.1.2}$$

$$\lambda_j = \frac{1}{p_n'(x_j)p_{n-1}(x_j)} \int_a^b \frac{p_n(x)}{x - x_j} p_{n-1}(x)\varrho(x)dx . \tag{4.1.3}$$

It is clear from (4.1.2) that $\lambda_j > 0$. The integral on the right of (4.1.3) is easily evaluated. Since

$$\frac{p_n(x)}{x - x_j} = \frac{a_n}{a_{n-1}} p_{n-1}(x) + q_{n-2}(x) ,$$

where a_n is the leading coefficient of $p_n(x)$, and $q_{n-2}(x)$ is a polynomial of degree $n - 2$, we obtain, by the properties of the orthogonal polynomials $p_n(x)$,

$$\lambda_j = \frac{a_n}{a_{n-1}} \frac{d_{n-1}^2}{p_n'(x_j)p_{n-1}(x_j)} , \tag{4.1.4}$$

where $d_n^2 = \int_a^b p_n^2(x)\varrho(x)dx$ is the squared norm.

Let us note that all arguments connected with deriving the quadrature formulas of Gaussian type for evaluating the integrals remain valid if instead of the integral $\int_a^b f(x)\varrho(x)dx$ we consider the sum

$$S_N = \sum_{k=0}^{N-1} f(x_k)\varrho_k \ . \tag{4.1.5}$$

The quantity

$$\tilde{S}_N = \sum_{j=1}^{n} \lambda_j f(y_j) \tag{4.1.6}$$

is used as an approximate value of S_N for large N. The abscissas y_j and the coefficients λ_j for given N are uniquely determined from the condition $S_N = \tilde{S}_N$ for all polynomials of at most degree $2n-1$. In this case to obtain the abscissas of quadrature formula and the Christoffel numbers one should use the polynomials of a discrete variable, $p_n(x)$, that satisfy the orthogonality condition of the form

$$\sum_{k=0}^{N-1} p_n(x_k)x_k^m \varrho_k = 0$$

with $m < n$.

This method allows simplification of the calculation of the sums that contain the functions $f(x)$, which is very difficult to calculate, by using the sums with a much smaller number of summands.

Let us look at some typical examples of the use of Gaussian quadrature formulas.

Example 1. Let us consider the application of Gaussian integration formulas to the calculation of sums of the form

$$S_N = \sum_{k=0}^{N-1} f(k) \ . \tag{4.1.7}$$

The Gaussian quadrature formula lets us replace the sum by a sum of fewer terms:

$$S_N \approx \sum_{j=1}^{n} \lambda_j f(x_j) \ . \tag{4.1.8}$$

In this case the abscissas x_j are the zeros of polynomials $p_n(x)$ that have the orthogonality properties

$$\sum_{k=0}^{N-1} p_n(k)p_m(k) = 0 \quad (m \neq n) \ .$$

The corresponding orthogonal polynomials are the Chebyshev polynomials of a discrete variable $t_n(x)$. As an illustration we give a comparative table of the results of calculating the sums S_N for $f(k) = \sqrt{l+k}$ (with an integer l) for different numbers of quadrature points. Note that when $n = N$ the quadrature formula gives the exact value of the original sum.

$N - 1$	10		1000	
n \ l	1	10	1	10
1	26.944	42.603	22 405	22 606
3	25.808	42.360	21 207	21 453
5	25.786	42.360	21 148	21 405
N	25.785	42.360	21 129	21 395

Example 2. It is required to calculate the sum

$$S_N = \sum_{k=0}^{N} C_N^k p^k (1 - p)^{N-k} \frac{1}{1+k} = \frac{1}{p(1+N)}[1 - (1 - p)^{N+1}] . \qquad (4.1.9)$$

In this case the weight function

$$\varrho_k(p, N) = C_N^k p^k (1 - p)^{N-k} , \quad k = 0, 1, \ldots, N ; \quad 0 < p < 1 \qquad (4.1.10)$$

is determined by the binomial distribution that is well known from probability theory. Consequently, the polynomials $p_n(x)$ must satisfy the orthogonality relation

$$\sum_{k=0}^{N} \varrho_k p_n(k) p_m(k) = 0 \quad (m \neq n) ,$$

where ϱ_k is determined by (4.1.10).

The corresponding polynomials $p_n(x)$ are the Kravchuk polynomials $k_n^{(p)}(x, N)$. The results of approximate calculations of the sum for different numbers of quadrature points n are given in the Table below.

N	100		1000	
n \ p	0.1	0.01	0.1	0.01
1	9.09091–0.2	5.00000–0.1	9.90099–0.3	9.09091–0.2
3	9.86833–0.2	6.24372–0.1	9.98997–0.3	9.94731–0.2
5	9.89662–0.2	6.31113–0.1	9.99001–0.3	9.98352–0.2
10	9.90057–0.2	6.31315–0.1		9.98928–0.2
15	9.90074–0.2			9.98956–0.2
$N + 1$	9.90075–0.2	6.31315–0.1	9.99001–0.3	9.98958–0.2

Example 3. It is very convenient to use the quadrature formulas of Gaussian type for calculating sums in quantum mechanics when sums with a large number of terms have to be evaluated. Summing is carried out over different values of quantum numbers that describe states of quantum mechanical system. Each term in the sum is usually a rather complicated expression, and therefore it is quite desirable to reduce the number of terms by using the quadrature formulas for sums. As an illustration we consider the application of quadrature formulas for calculating an effective cross-section $\sigma_{\mathrm{ff}}(\omega)$ of the light bremsstrahlung ("free-free transitions") in a plasma with a given temperature (for example, in stellar matter):

$$\sigma_{\mathrm{ff}}(\omega) = A\omega \sum_{l=0}^{\infty} \sum_{l'=l\pm 1} (l + l' + 1)$$

$$\times \int_{\varepsilon_0}^{\infty} \frac{\left(R_{\varepsilon l}^{\varepsilon' l'}\right)^2 d\varepsilon}{\left[1 + \exp\left(\frac{\varepsilon - \mu}{\theta}\right)\right]\left[1 + \exp\left(-\frac{\varepsilon - \mu}{\theta}\right)\right]}. \qquad (4.1.11)$$

Here ω is the light frequency, θ is the temperature (we use the atomic unit system), A, ε_0 and μ are some constants, and $R_{\varepsilon l}^{\varepsilon' l'}$ is an integral over spatial variables which contains the product of radial electron wave functions for continuous spectrum states with energies ε and $\varepsilon' = \varepsilon + \omega$ for certain values of orbital quantum numbers l and l'. The radial electron wave functions are determined by solving a nonlinear system of differential equations with the use of the method of successive approximations in the modified Hartree-Fock-Slater model [F5, N3, N4].

In order to reduce the time of calculation of $\sigma_{\mathrm{ff}}(\omega)$ in integrating over the variable ε we use the quadrature formulas of Gaussian type

$$\int_0^{\infty} f(x)\exp(-x/\theta)dx \approx \theta \sum_{j=1}^{N} \lambda_j f(\theta y_j),$$

where $x = \varepsilon - \varepsilon_0$, y_j are the zeros of the Laguerre polynomial $L_n^0(y)$.

A main contribution to the sum in l with given energy $\varepsilon = \varepsilon_j = \theta y_j + \varepsilon_0$ is made by the terms for which $l \le l_{\max}(\varepsilon_j)$.

For summing over l it is convenient to use the quadrature formula of Gaussian type

$$\sum_{l=0}^{l_{\max}} f(l) = \sum_{i=1}^{i_{\max}} a_i f(l_i),$$

where l_j are the zeros of the Chebyshev polynomial of a discrete variable. In practical calculations using the quadrature formulas of Gaussian type for summing over l proves very efficient: the calculation time can be reduced by more than a factor of 10 with an accuracy to within 5%.

4.2 Compression of Information
by Means of the Hahn Polynomials

4.2.1 Problem Formulation

The development of many scientific fields is connected with using a vast amount
of numerical data. In view of this the problem of data storing, compressing and
processing is of imperative importance in modern science and technology. For
example, it concerns effective processing of information obtained from flight
vehicles, the compression of information for creating data banks of physical
and chemical properties of various matters as well as geophysical and seismic
observations. Besides, the necessity of efficient data storing is governed not only
by scientific but also by economic considerations.

Along with the ordinary compression of information for its subsequent storage other methods of information processing are frequently required to be used
simultaneously with compression. Among them are first of all damping and
smoothing of high frequency noises and revealing random errors in computations. In addition, sometimes an analytic representation of tabulated data in the
form of interpolation formulas, splines, and various approximation proves also
to be helpful. The necessity of conjoint solutions of these problems is obvious.

At present much use is being made of spectral methods of information processing. These methods are based on the best approximation on the average of
the function $f(x)$ by using the partial Fourier sums over an orthogonal system
of functions $y_n(x)$:

$$f(x) \approx \sum_{n=0}^{N} c_n y_n(x) \ . \tag{4.2.1}$$

The function $f(x)$ describes experimental data which are given in an analytic
or tabular form. As is known the coefficients c_n are calculated by using the
orthogonality relations for functions $y_n(x)$:

$$c_n = \frac{1}{d_n^2} \, (y_n, f) \ .$$

Here d_n^2 is the squared norm of function $y_n(x)$ while the parentheses mean a
respective scalar product.

The approximation (4.2.1) of the original function solves several problems.
First, by memorizing a number of initial Fourier coefficients instead of values of
$f(x)$ we are capable of restoring these values (though, only approximately). Thus,
at the expense of a loss in accuracy that depends on the number of coefficients
c_n used, we achieve the compression of information. Second, we select the
most important low frequency characteristics and reject, as a rule, harmful high
frequency characteristics (arising, possibly, from errors in experimental data or
calculations). Third, the method is convenient for analytical processing of results

since the mathematical description of data now uses the known properties of functions $y_n(x)$.

Following the above merits of spectral methods we may formulate the requirements to be satisfied by the orthogonal basis of functions $\{y_n(x)\}$:

(1) high accuracy of approximation of different functions;
(2) simplicity and unification of the methods for calculating the functions $y_n(x)$ for any n and x;
(3) well studied analytical properties of the basis, which may be used for an analysis of the function characteristics.

As a system of functions $\{y_n(x)\}$ it is convenient to use classical orthogonal polynomials – the Jacobi, Laguerre and Hermite polynomials. A variety of systems of classical orthogonal polynomials allows the choosing of the basis most appropriate to a specific case. Properties of the classical orthogonal polynomials are described in detail in the literature on special functions.

In this section the Hahn polynomials – the difference analogs of the Jacobi polynomials – are used as a system of functions $\{y_n(x)\}$. (The description of properties of the Hahn polynomials may be found in Sect. 2.4.2.)

Using the classical orthogonal polynomials of a discrete variable is justified if the function of interest is represented in the form of tabulated values. In such cases we may, of course, use also the classical orthogonal polynomials of a continuous argument, as well as other bases (for example, trigonometric), however, then calculating the coefficients c_n by means of the orthogonality relation will require rather laborious numerical integration. Using the polynomials of a discrete variable allows us to avoid this procedure and replace the integration by summation.

Similar methods of information compression evidently may be used also for the functions of several variables.

4.2.2 Description of the Method

We consider the method of numerical processing of data given by a table of values for the function of two variables $f(x, y)$, which is approximated on the average by linear combinations of products of orthogonal Hahn polynomials in each of variables:

$$\{h_{n_1}^{(\alpha_1,\beta_1)}(x, N_1) \times h_{n_2}^{(\alpha_2,\beta_2)}(y, N_2)\,,$$
$$n_1 = 0, 1, \ldots, M_1 - 1\,; \quad n_2 = 0, 1, \ldots, M_2 - 1\}\,, \tag{4.2.2}$$

where α_i and β_i are parameters of the Hahn polynomials. Original information, i.e. the tabulated values of functions of two variables

$$f_{k_1 k_2} = f(x_{k_1}, y_{k_2})\,; \quad k_1 = 0, 1, \ldots, N_1 - 1\,; \quad k_2 = 0, 1, \ldots, N_2 - 1\,,$$

is replaced by a set of the Fourier coefficients $c_{n_1 n_2}$ calculated by the formula

$$c_{n_1 n_2} = \frac{1}{d_{n_1}^2 d_{n_2}^2} \sum_{k_1=0}^{N_1-1} \sum_{k_2=0}^{N_2-1} \left[f_{k_1 k_2} \times h_{n_1}^{(\alpha_1,\beta_1)}(x_{k_1}; N_1) \right.$$

$$\left. \times h_{n_2}^{(\alpha_2,\beta_2)}(y_{k_2}; N_2) \times \varrho^{(\alpha_1,\beta_1)}(x_{k_1}, N_1) \times \varrho^{(\alpha_2,\beta_2)}(y_{k_2}, N_2) \right] \,, \quad (4.2.3)$$

where $\varrho^{(\alpha,\beta)}$ is the weight of the Hahn polynomials, i.e. this transformation may be schematically presented in the following way:

$$\{f_{k_1 k_2}\} \to \{c_{n_1,n_2}\} \quad k_i = 0, 1, \ldots, N_i - 1 \,; \quad n_i = 0, 1, \ldots, M_i - 1 \,.$$

The Hahn polynomials in (4.2.2) may be calculated by means of the recursion relations of the form

$$x h_n^{(\alpha,\beta)}(x, N) = \alpha_n h_{n+1}^{(\alpha,\beta)}(x, N) + \beta_n h_n^{(\alpha,\beta)}(x, N) + \gamma_n h_{n-1}^{(\alpha,\beta)}(x, N) \,. \quad (4.2.4)$$

We note that despite the fact that the Hahn polynomials are orthogonal on a discrete set of points the relation (4.2.4) is valid for any values of x.

By using the obtained set of coefficients $c_{n_1 n_2}$ ($n_i = 0, 1, \ldots, M_i - 1$) we can find an approximate value of the function $f(x, y)$ at an arbitrary point of the domain

$$D = \{(x; y) | x \in [0, N_1 - 1] \,, \quad y \in [0, N_2 - 1]\} \,.$$

To do this we should evaluate values of the Hahn polynomials $h_{n_1}^{(\alpha_1,\beta_1)}(x, N_1)$ and $h_{n_2}^{(\alpha_2,\beta_2)}(y, N_2)$ with the aid of recursion relation (4.2.4) for $n_i = 0, 1, \ldots, M_i - 1$ at points x and y, respectively. Then a value of $f(x, y)$ is determined by using the Fourier sums

$$f(x, y) \approx \sum_{n_1=0}^{M_1-1} \sum_{n_2=0}^{M_2-1} c_{n_1 n_2} h_{n_1}^{(\alpha_1,\beta_1)}(x, N_i) \times h_{n_2}^{(\alpha_2,\beta_2)}(y, N_2) \,. \quad (4.2.5)$$

Thus owing to the transformation $\{f_{k_1 k_2}\} \to \{c_{n_1 n_2}\}$, we compress the initial information by $N_1 \times N_2 / (M_1 \times M_2)$ times.

With the use of expansion (4.2.5) approximate values of the function at original points $(x_{k_1}; y_{k_2})$, where the table of $\{f_{k_1 k_2}\}$ was given, may be determined. We denote the restored values by $\tilde{f}_{k_1 k_2}$. Comparing the original table of $\{f_{k_1 k_2}\}$ with the obtained values $\{\tilde{f}_{k_1 k_2}\}$ enables us to evaluate a relative error

$$\varepsilon_{k_1 k_2} = \frac{\tilde{f}_{k_1 k_2} - f_{k_1 k_2}}{|f_{k_1 k_2}|} \,.$$

Integral criteria of the approximation accuracy are the maximal error E_0, the mean absolute error E_1 and the mean square error E_2:

$$E_0 = \max_{k_1, k_2} |\varepsilon_{k_1 k_2}| \,; \quad E_1 = \frac{\sum_{k_1 k_2} |\varepsilon_{k_1 k_2}|}{N_1 N_2} \,; \quad E_2 = \sqrt{\frac{\sum_{k_1 k_2} \varepsilon_{k_1 k_2}^2}{N_1 N_2}} \quad (4.2.6)$$

(summation is carried out over all the lattice points).

The expansion (4.2.5) allows one to obtain partial derivatives and integrals of function $f(x,y)$. In order to obtain the derivative $\partial f/\partial x$ or the integral $\int f(x,y)dx$, it is sufficient, owing to the linear nature of Eq.(4.2.5), to know the derivative and the integral of polynomials $h_n^{(\alpha,\beta)}(x,N)$, which may be obtained by expanding $h_n^{(\alpha,\beta)}(x,N)$ in powers of x. The expansion may be easily derived from the relation (2.7.19) between the Hahn polynomials and the generalized hypergeometric functions

$$h_n^{(\alpha,\beta)}(x,N) = \frac{(-1)^n(N-1)!(\beta+1)_n}{n!(N-n-1)!} \; {}_3F_2\left(\begin{matrix}-n;\alpha+\beta+n+1;-x\\\beta+1;1-N\end{matrix}\;\middle|\;1\right) \, ,$$

where

$$_pF_q\left(\begin{matrix}a_1,a_2,\ldots,a_p\\b_1,b_2,\ldots,b_q\end{matrix}\;\middle|\;1\right) = \sum_{k=0}^{\infty} \frac{(a_1)_k(a_2)_k\ldots(a_p)_k}{(b_1)_k(b_2)_k\ldots(b_q)_k} \frac{1}{k!} \, ,$$

$$(a)_0 = 1 \, , \quad (a)_k = a(a+1)\ldots(a+k-1) = \frac{\Gamma(a+k)}{\Gamma(a)} \, .$$

Hence we have

$$h_n^{(\alpha,\beta)}(x,N) = \frac{(-1)^n(N-1)!(\beta+1)_n}{n!(N-n-1)!} \sum_{k=0}^{n} \frac{(-n)_k(\alpha+\beta+n+1)_k(-x)_k}{(\beta+1)_k(1-N)_k k!} \, .$$

$$(4.2.7)$$

Since

$$(-x)_k = (-1)^k \sum_{m=0}^{k} s_k^{(m)} x^m \, , \tag{4.2.8}$$

where $s_k^{(m)}$ are the Stirling numbers of the first kind tabulated in [A1], a combination of Eqs. (4.2.5, 7, 8) lets us obtain the coefficients in the expansion of $f(x,y)$ in powers of x and hence to calculate the coefficients in expansion of $\partial f/\partial x$ and $\int f(x,y)dx$ in powers of x.

The most important parameters that characterize the method of data approximation with the use of the Hahn polynomials are an accuracy of approximation, the time required for data processing on a computer, the coefficient of information compression and a degree of smoothing.

Searching for optimal values of the parameters of the Hahn polynomials $\alpha_1,\beta_1,\alpha_2,\beta_2$ shows that for a wide class of functions the approximation accuracy weakly depends on the parameters α and β, when $-1/2 \leq \alpha_i \leq 1/2$, $-1/2 \leq \beta_i \leq 1/2$ (beyond the interval $(-1/2,1/2)$ the approximation error usually sharply increases). This situation may be explained by a behaviour of weight functions of the Hahn polynomials on the orthogonality interval. Application of the Hahn polynomials with α and β whose values are outside the interval $(-1/2,1/2)$ is expedient only when there is a need to increase the accuracy of approximation on a certain portion of the orthogonality interval.

The accuracy and the coefficient of information compression depend on parameters M_1 and M_2. For specific computations we should determine the most convenient relationship between these characteristics. An essential feature of the method is the high quality of data smoothing, which may be seen from the plots of approximation.

In the above discussion we have not shown all the advantages and disadvantages of the data processing method by using classical orthogonal polynomials of a discrete variable. However, by using information available we may draw the following conclusions:

- the method should be used for a class of functions of one and two variables which may be approximated with high accuracy by polynomials of moderate degrees, i.e. for the functions without high frequency oscillations or with harmful high frequency components;
- the method provides multiple compression of information (which is particularly important for large tables of function values) with relatively high accuracy of approximation;
- required computational time is very small;
- the functions in the form of the Fourier sums may be easily investigated analytically and then used in computations.

Hence we may assume that the method of data processing with the use of classical orthogonal polynomials of a discrete variable will be applied to many problems of computational mathematics, statistics, economics, geology, geophysics.

4.3 Spherical Harmonics Orthogonal on a Discrete Set of Points

4.3.1. A great variety of the mathematical physics problems are solved with the use of expansions with respect to spherical harmonics

$$Y_{lm}(\vartheta, \varphi) = \theta_{lm}(\vartheta)\phi_m(\varphi) , \tag{4.3.1}$$

where

$$\phi_m(\varphi) = \frac{1}{\sqrt{2\pi}}\, e^{im\varphi}\ (-l \leq m \leq l) ,$$

$$\theta_{lm}(\vartheta) = \frac{\sqrt{\varrho(x)}}{d_{l-m}} P_{l-m}^{(m,m)}(x)\quad (m \geq 0,\ x = \cos\vartheta)$$

$$(Y_{l,-m}(\vartheta, \varphi) = (-1)^m Y_{lm}^*(\vartheta, \varphi)) ,$$

$P_n^{(\alpha,\beta)}(x)$ is the Jacobi polynomial, $\varrho(x)$ and d_n are the weight and the norm for $P_n^{(m,m)}(x)$. It is easy to verify that the functions $\phi_m(\varphi)$ have the following property of orthogonality on a discrete set of points:

$$\sum_{j=0}^{M-1} \phi_m(\varphi_j)\phi_{m'}^*(\varphi_j)\Delta\varphi = \delta_{mm'}\,, \tag{4.3.2}$$

where $\varphi_j = -\pi + (2\pi/M)(j+1/2)$ $(\Delta\varphi = 2\pi/M$, $M-1$ is a positive integer).

By using classical orthogonal polynomials of a discrete variable that are orthogonal on a lattice uniform in $\cos\theta$ or θ we may obtain approximate expressions for the Jacobi polynomials. It enables us to construct two variants of spherical harmonics that are orthogonal to a discrete set of points:

1) the functions that are orthogonal on the lattice uniform in $x = \cos\vartheta$ and in φ;
2) the functions that are orthogonal on a lattice uniform in ϑ and φ.

4.3.2 The Spherical Harmonics that are Orthogonal on the Lattice Uniform in $x = \cos\vartheta$ and φ

In order to obtain expressions for the spherical harmonics $\tilde{Y}_{lm}(\vartheta,\varphi)$ that satisfy the orthogonality relations on the lattice uniform in $x = \cos\vartheta$ and φ we use the asymptotic relation (2.6.5) that connects the Hahn polynomials $h_n^{(\alpha,\beta)}(s,N)$ and the Jacobi polynomials $P_n^{(\alpha,\beta)}(x)$, where $s = \tilde{N}(1+x)/2 - (\beta+1)/2$, $\tilde{N} = N + (\alpha+\beta)/2$:

$$\frac{1}{\tilde{N}^n} h_n^{(\alpha,\beta)}(s,N) = P_n^{(\alpha,\beta)}(x) + \mathrm{O}\left(\frac{1}{\tilde{N}^2}\right)\,, \tag{4.3.3}$$

$$\left(\frac{2}{\tilde{N}}\right)^{\alpha+\beta} \bar{\varrho}(s) = \varrho(x)\left[1 + \mathrm{O}\left(\frac{1}{\tilde{N}^2}\right)\right]\,, \tag{4.3.4}$$

$$\frac{1}{2^{2n}}\left(\frac{2}{\tilde{N}}\right)^{\alpha+\beta+2n+1} \bar{d}_n^2 = d_n^2\left[1 + \mathrm{O}\left(\frac{1}{\tilde{N}^2}\right)\right]\,. \tag{4.3.5}$$

Here $\varrho(x)$ and d_n are the weight and the norm for the Jacobi polynomials $P_n^{(\alpha,\beta)}(x)$; $\bar{\varrho}(s)$ and $\bar{d}_n$ are the weight and the norm for the Hahn polynomials $h_n^{(\alpha,\beta)}(s,N)$. From the relations (4.3.3–5) it follows that when $\tilde{N} \to \infty$

$$\frac{\sqrt{\varrho(x)}}{d_n} P_n^{(m,m)}(x) = \frac{\sqrt{\bar{\varrho}(s)}}{\bar{d}_n\sqrt{\Delta x}} h_n^{(m,m)}(s,N) + \mathrm{O}\left(\frac{1}{\tilde{N}^2}\right)\,,$$

$$\left(s = \frac{\tilde{N}}{2}(1+x) - \frac{m+1}{2}\,, \quad \tilde{N} = N - m\,, \quad \Delta x = \frac{2}{\tilde{N}}\right)\,.$$

Hence we obtain the following approximate expressions for the spherical harmonics $Y_{lm}(\vartheta,\varphi)$:

$$\tilde{Y}_{lm}(\vartheta,\varphi) = \frac{\sqrt{\bar{\varrho}(s)}}{\bar{d}_{l-m}\sqrt{\Delta x}} h_{l-m}^{(m,m)}(s,N)\phi_m(\varphi) \quad (m \geq 0)\,, \tag{4.3.6}$$

$$\tilde{Y}_{l,-m}(\vartheta,\varphi) = (-1)^m \tilde{Y}_{lm}^*(\vartheta,\varphi)\,.$$

The functions $\tilde{Y}_{lm}(\vartheta,\varphi)$ satisfy the following orthogonality relations:

$$\sum_{i=0}^{\tilde{N}-m-1} \sum_{j=0}^{M-1} \tilde{Y}_{lm}(\vartheta_{im},\varphi_j)\tilde{Y}_{l'm'}^{*}(\vartheta_{im'},\varphi_j)\Delta x \Delta\varphi = \delta_{ll'}\delta_{mm'} , \qquad (4.3.7)$$

where

$$\cos\vartheta_{im} = x_{2i+m+1} , \quad x_k = -1 + \frac{k}{\tilde{N}} \quad (i = 0,1,\ldots,\tilde{N}-m-1) ,$$

$$\varphi_j = -\pi + \frac{\pi}{M}(2j+1) \quad (j = 0,1,\ldots,M-1) ,$$

$$\Delta x = \frac{2}{\tilde{N}} , \quad \Delta\varphi = \frac{2\pi}{M} .$$

For approximate representation of the functions on the sphere their expansions in the spherical harmonics $Y_{lm}(\vartheta,\varphi)$ are usually used with a finite number of terms. Replacing $Y_{lm}(\vartheta,\varphi)$ by $\tilde{Y}_{lm}(\vartheta,\varphi)$ we obtain an approximate representation of $f(\vartheta,\varphi)$ suitable for applications. It has the form

$$f(\vartheta,\varphi) = \sum_{l,m} a_{lm}\tilde{Y}_{lm}(\vartheta,\varphi) . \qquad (4.3.8)$$

For the coefficients a_{lm} according to the orthogonality property (4.3.7) we obtain

$$a_{lm} = \sum_{i=0}^{N-1} \sum_{j=0}^{M-1} f(\vartheta_{im},\varphi_j)\tilde{Y}_{lm}^{*}(\vartheta_{im},\varphi_j)\Delta x \Delta\varphi . \qquad (4.3.9)$$

Hence when using the expansion of $f(\vartheta,\varphi)$ in the functions $\tilde{Y}_{lm}(\vartheta,\varphi)$ we should know the values of $f(\theta_k,\varphi_j)$ for

$$\cos\vartheta_k = -1 + \frac{k}{\tilde{N}} \quad (k = 1,2,\ldots,2\tilde{N}-1) ,$$

$$\varphi_j = \left(-1 + \frac{2j+1}{M}\right)\pi \quad (j = 0,1,\ldots,M-1) .$$

For calculating the functions $\tilde{Y}_{lm}(\vartheta,\varphi)$ in (4.3.8) we may use the recursion relation that follows from the recursion relation for the Hahn polynomials (see Sect. 2.5.5 and Table 2.1):

$$sh_n^{(m,m)}(s) = \frac{(n+1)(2m+n+1)}{(2m+2n+1)(2m+2n+2)}h_{n+1}^{(m,m)}(s) + \frac{1}{2}(N-1)h_n^{(m,m)}(s)$$
$$+ \frac{(m+n)(2m+n+N)(N-n)}{2(2m+2n+1)}h_{n-1}^{(m,m)}(s) .$$

As a result, for $l \geq m \geq 0$ we obtain

$$\cos\vartheta\,\tilde{Y}_{lm}(\vartheta,\varphi) = \sqrt{\frac{(l+1)^2 - m^2}{4(l+1)^2 - 1}}\sqrt{1 - \left(\frac{l+1}{\tilde{N}}\right)^2}\,\tilde{Y}_{l+1,m}(\vartheta,\varphi)$$

$$+ \sqrt{\frac{l^2 - m^2}{4l^2 - 1}} \sqrt{1 - \left(\frac{l}{\tilde{N}}\right)^2} \, \tilde{Y}_{l-1,m}(\vartheta, \varphi) \, ,$$

$$\tilde{Y}_{m-1,m}(\vartheta, \varphi) = 0 \; ; \tag{4.3.10}$$

$$\tilde{Y}_{mm}(\vartheta, \varphi) = \frac{e^{im\varphi}}{\sqrt{2\pi}} \frac{\sqrt{(2m+1)!}}{m! \, 2^{m+1/2}} \sqrt{\prod_{k=0}^{m-1} \frac{1 - (x + (2k+1-m)/\tilde{N})^2}{1 - [(k+1)/\tilde{N}]^2}} \, , \tag{4.3.11}$$

$$x = \cos\vartheta, \quad \tilde{Y}_{00}(\vartheta, \varphi) = 1/\sqrt{4\pi} \, .$$

4.3.3 The Spherical Harmonics that are Orthogonal on the Lattices Uniform in Angles ϑ and φ

The polynomials orthogonal on the lattice $x(s) = \cos(2\omega s)$ were obtained in Sect. 3.6.2 from the Racah polynomials that are orthogonal on the quadratic lattice and tend to the Jacobi polynomials when $N \to \infty$. Therefore by analogy with the previous arguments for the lattice uniform in $x = \cos\vartheta$ we should expect that for large N rather good approximations can be obtained for the spherical harmonics by using expressions of the form

$$\tilde{Y}_{lm}(\vartheta, \varphi) = \frac{\sqrt{\bar{\varrho}(s)}}{\bar{d}_{l-m}} u_{l-m}^{(m,m)}(x(s), a, b, N) \phi_m(\varphi) \tag{4.3.12}$$

with

$$a = (m+1)/2 \, , \quad b = a + N \, , \quad N = \tilde{N} - m \, ,$$

$$x(s) = \cos\vartheta(s) \, , \quad \vartheta = \vartheta(s) = 2\omega s \, , \quad \omega = \pi/2\tilde{N} \, .$$

Here $\bar{\varrho}(s)$ and d_n are the weight and the norm of the polynomial $u_n^{(m,m)}[x(s), a, b, N] \equiv u_n^{(m,m)}[x(s)]$, which is the q-analog of the Racah polynomial $u_n^{(\alpha,\beta)}(s, a, b, N)$ on the lattice $x(s) = \cos(2\omega s)$ and which satisfies the orthogonality relation

$$-\sum_{i=0}^{N-1} u_n^{(m,m)}[x(s_i)] u_k^{(m,m)}[x(s_i)] \bar{\varrho}(s_i) \Delta x(s_i - 1/2) = \delta_{nk} \bar{d}_n^2 \, . \tag{4.3.13}$$

(In the orthogonality relation we replaced $\Delta x(s_i - 1/2)$ by $-\Delta x(s_i - 1/2)$, since $\Delta x(s_i - 1/2) = -2\sin\omega \sin(2\omega s) < 0$.) In this case (see (3.6.39))

$$\bar{\varrho}(s) = \left[\frac{\tilde{\Gamma}_q(s + (m+1)/2)}{\tilde{\Gamma}_q(s - (m-1)/2)} \right]^2$$

$$\times \frac{\tilde{\Gamma}_q(\tilde{N} + s + (m+1)/2)}{\tilde{\Gamma}_q(\tilde{N} + s - (m-1)/2)} \frac{\tilde{\Gamma}_q(\tilde{N} - s + (m+1)/2)}{\tilde{\Gamma}_q(\tilde{N} - s - (m-1)/2)} \, , \tag{4.3.14}$$

where $q = e^{i\omega}$ ($\bar{\varrho}(s) = 1$ for $m = 0$). The function $\tilde{\Gamma}_q(s)$ is a generalization of the gamma-function $\Gamma(s)$:

$$\lim_{q \to 1} \tilde{\Gamma}_q(s) = \Gamma(s) \, , \quad \tilde{\Gamma}_q(1) = 1 \, .$$

Since

$$\frac{\tilde{\Gamma}_q(s+1)}{\tilde{\Gamma}(s)} = \psi_q(s) , \quad \psi_q(s) = \frac{\sin \omega s}{\sin \omega} , \tag{4.3.15}$$

we have

$$\frac{\tilde{\Gamma}_q(s+p+1)}{\tilde{\Gamma}_q(s)} = \prod_{k=0}^{p} \psi_q(s+k) . \tag{4.3.16}$$

Hence, if we use additionally the relation

$$\psi_q(\tilde{N}+s) = \psi_q(\tilde{N}-s) , \tag{4.3.17}$$

the expression (4.3.14) for $\bar{\varrho}(s)$ can be rewritten in the form

$$\bar{\varrho}(s) = \left[\prod_{k=0}^{m-1} \psi_q \left(s - \frac{m-1}{2} + k \right) \psi_q \left(\tilde{N} - s + \frac{m-1}{2} - k \right) \right]^2 . \tag{4.3.18}$$

In order to find out the relationship between $\bar{\varrho}(s)$ and the weight $\varrho(x) = (1-x^2)^m$ for the Jacobi polynomials $P_n^{(m,m)}(x)$ we use the equalities

$$\psi_q(s+c) = \frac{\sin (\vartheta/2 + \omega c)}{\sin \omega} , \quad \psi_q(\tilde{N} - s - c) = \frac{\cos (\vartheta/2 + \omega c)}{\sin \omega} .$$

As a result we obtain

$$\bar{\varrho}(s) = \frac{1}{(\sin \pi/2N)^{4m}} \prod_{k=0}^{m-1} \sin^2 \left(\frac{\vartheta}{2} + \frac{2k+1-m}{4\tilde{N}}\pi \right) \cos^2 \left(\frac{\vartheta}{2} + \frac{2k+1-m}{4\tilde{N}}\pi \right)$$

$$= \frac{1}{(4\sin^4 \pi/2\tilde{N})^m} \prod_{k=0}^{m-1} \sin^2 \left(\vartheta + \frac{2k+1-m}{2\tilde{N}} \right) . \tag{4.3.19}$$

On replacing the index k by $m-1-k$ the product $\prod_{k=0}^{m-1} f(k)$ does not change. Therefore

$$\prod_{k=0}^{m-1} \sin^2 \left(\vartheta + \frac{2k+1-m}{2\tilde{N}}\pi \right)$$

$$= \prod_{k=0}^{m-1} \sin \left(\vartheta + \frac{2k+1-m}{2\tilde{N}}\pi \right) \sin \left(\vartheta - \frac{2k+1-m}{2\tilde{N}}\pi \right)$$

$$= \sin^{2m} \vartheta + O \left(\frac{1}{\tilde{N}^2} \right) ,$$

i.e.

$$4 \left(\sin \frac{\pi}{2\tilde{N}} \right)^{4m} \bar{\varrho}(s) = \varrho(x) + O \left(\frac{1}{\tilde{N}^2} \right) . \tag{4.3.20}$$

In order to prove the analogous relation

$$\tilde{Y}_{lm}(\vartheta, \varphi) = Y_{lm}(\vartheta, \varphi) + O \left(\frac{1}{\tilde{N}^2} \right) , \tag{4.3.21}$$

we obtain the recursion relations (in l) for the functions $\tilde{Y}_{lm}(\vartheta, \varphi)$. Since for the polynomials $u_n^{(m,m)}(x, a, b, N) \equiv u_n^{(m,m)}(x)$ we have the recursion relation

$$x u_n^{(m,m)}(x) = \frac{\bar{a}_n}{\bar{a}_{n+1}} u_{n+1}^{(m,m)}(x)$$
$$+ \left(\frac{\bar{b}_n}{\bar{a}_n} - \frac{\bar{b}_{n+1}}{\bar{a}_{n+1}} \right) u_n^{(m,m)}(x) + \frac{\bar{a}_{n-1}}{\bar{a}_n} \frac{\bar{d}_n^2}{\bar{d}_{n-1}^2} u_{n-1}^{(m,m)}(x) ,$$

where $\bar{a}_n$ and $\bar{b}_n$ are the coefficients of x^n and x^{n-1} in the polynomial $u_n^{(m,m)}(x)$, for the functions

$$v_n^{(m)}(x) = (1/\bar{d}_n)\sqrt{\bar{\varrho}(s)} u_n^{(m,m)}[x(s)]$$

that enter into the expression for $\tilde{Y}_{lm}(\vartheta, \varphi)$, we obtain the recursion relation

$$x v_n^{(m)}(x) = c_n v_{n+1}^{(m)}(x) + \left(\frac{\bar{b}_n}{\bar{a}_n} - \frac{\bar{b}_{n+1}}{\bar{a}_{n+1}} \right) v_n^{(m)}(x) + c_{n-1} v_{n-1}^{(m)}(x) , \qquad (4.3.22)$$

where

$$c_n = \frac{\bar{a}_n}{\bar{a}_{n+1}} \frac{\bar{d}_{n+1}}{\bar{d}_n} .$$

According to (3.7.2) and (3.7.15) we have

$$\frac{\bar{b}_n}{\bar{a}_n} = \frac{\alpha_n}{\alpha_{n-1}} \left[\frac{\tilde{\tau}_{n-1}(0)}{\tilde{\tau}'_{n-1}} - \beta_n \right] = \psi_q(n) \frac{\tilde{\tau}_{n-1}(0)}{\tilde{\tau}'_{n-1}} \qquad (4.3.23)$$
$$(\alpha_n = \tilde{\Gamma}_q(n+1) , \quad \beta_n = 0) ,$$

$$c_n = \frac{\alpha_{n+1}}{\alpha_n} \left(-\frac{A_{n-1,n}}{A_{n,n+1}} \right) \sqrt{ \left(-\frac{A_{n+1,n+1}}{A_{nn}} \right) \frac{S_{n+1}}{S_n} \frac{\tilde{\tau}'_{n-1}}{\tilde{\tau}'_n} } . \qquad (4.3.24)$$

Here

$$A_{in} = (-1)^i \prod_{k=0}^{i-1} (\lambda_n - \lambda_k) ; \quad A_{0n} = 1 ; \qquad (4.3.25)$$

$$\lambda_n = -\sum_{k=0}^{n-1} \tilde{\tau}'_k , \qquad (4.3.26)$$

$$S_n = -\sum_i \varrho_n(s_i) \Delta x_n \left(s_i - \tfrac{1}{2} \right) ,$$

$$\varrho_n(s) = \varrho(s+n) \prod_{k=1}^{n} \sigma(s+k)$$

$$= \left[\prod_{i=1}^{n+m} \psi_q \left(s + i - \frac{m+1}{2} \right) \psi_q \left(\tilde{N} - s - i + \frac{m+1}{2} \right) \right]^2$$

$$= \left[\frac{\tilde{\Gamma}_q(s + n + (m+1)/2)}{\tilde{\Gamma}_q(s - (m-1)/2)} \frac{\tilde{\Gamma}_q(\tilde{N} - s + (m+1)/2)}{\tilde{\Gamma}_q(\tilde{N} - s - n - (m-1)/2)} \right]^2 \qquad (4.3.27)$$

(when deriving (4.3.27) we used equality (4.3.16)).

Also, owing to (3.7.19),

$$\frac{S_{n+1}}{S_n} = \frac{\tilde{\sigma}_n(x_n^*)}{\cos\omega + \tilde{\sigma}_n''/2\tilde{\tau}_n'} , \qquad (4.3.28)$$

where x_n^* is a root of the equation $\tilde{\tau}_n(x_n) = 0$. Hence for calculating c_n we need explicit expressions for the functions $\tilde{\tau}_k(x_k)$, $\tilde{\sigma}_k(x_k)$. In this case

$$\tilde{\tau}_k[x_k(s)] = \frac{\sigma(-s-k) - \sigma(s)}{\Delta x_k(s-1/2)} ,$$

$$\tilde{\sigma}_k[x_k(s)] = \frac{\sigma(-s-k) + \sigma(s)}{2} ,$$

$$\sigma(s) = \psi_q^2\left(s - \frac{m+1}{2}\right) \psi_q\left(\tilde{N} + s - \frac{m+1}{2}\right) \psi_q\left(\tilde{N} - s + \frac{m+1}{2}\right)$$

$$= \left[\psi_q\left(s - \frac{m+1}{2}\right) \psi_q\left(\tilde{N} - s + \frac{m+1}{2}\right)\right]^2$$

$$= \frac{\sin^2 2\omega(s - (m+1)/2)}{4\sin^4\omega} . \qquad (4.3.29)$$

From this we obtain

$$\tilde{\tau}_k[x_k(s)] = -\frac{\sin 2\omega(k+m+1)}{4\sin^5\omega} x_k(s) , \qquad (4.3.30)$$

$$\tilde{\sigma}_k[x_k(s)] = \frac{1}{4\sin^4\omega}[\cos^2\omega(k+m+1) - \cos 2\omega(k+m+1)x_k^2(s)] . \quad (4.3.31)$$

Hence

$$\tilde{\tau}_k' = -\frac{\sin 2\omega(k+m+1)}{4\sin^5\omega} , \qquad x_n^* = 0 ;$$

$$\tilde{\sigma}_n(x_n^*) = \frac{\cos^2\omega(n+m+1)}{4\sin^4\omega} , \qquad \frac{\tilde{\sigma}_n''}{2} = -\frac{\cos 2\omega(n+m+1)}{4\sin^4\omega} ;$$

$$\lambda_n = \frac{1}{4\sin^5\omega} \sum_{k=0}^{n-1} \sin 2\omega(k+m+1) = \frac{\sin\omega n \sin\omega(n+k+2m+1)}{4\sin^6\omega} , \quad (4.3.32)$$

$$\lambda_n - \lambda_k = \frac{\sin\omega(n-k)\sin\omega(n+k+2m+1)}{4\sin^6\omega}$$

$$\frac{S_{n+1}}{S_n} = \frac{\cos^2\omega(n+m+1)\sin 2\omega(n+m+1)}{4\sin^4\omega \sin\omega(2n+2m+3)} .$$

As a result for the coefficients c_n and $\bar{b}_n/\bar{a}_n$ we find

$$\frac{\bar{b}_n}{\bar{a}_n} = 0 ,$$

$$c_n = \cos\omega(n+m+1)\sqrt{\frac{\sin\omega(n+1)\sin\omega(n+2m+1)}{\sin\omega(2n+2m+1)\sin\omega(2n+2m+3)}} . \qquad (4.3.33)$$

The recursion relation for $\tilde{Y}_{lm}(\vartheta,\varphi)$ takes the form

$$\cos\vartheta\,\tilde{Y}_{lm}(\vartheta,\varphi) = c_{l-m}\tilde{Y}_{l+1,m}(\vartheta,\varphi) + c_{l-m-1}\tilde{Y}_{l-1,m}(\vartheta,\varphi) \,,$$

$$c_{l-m} = \cos\omega(l+1)\sqrt{\frac{\sin\omega(l+1-m)\sin\omega(l+1+m)}{\sin\omega(2l+1)\sin\omega(2l+3)}} \,, \tag{4.3.34}$$

$$\tilde{Y}_{m-1,m}(\vartheta,\varphi) = 0 \,.$$

In order to use (4.3.34) we have to determine the function $\tilde{Y}_{mm}(\vartheta,\varphi)$. According to (4.3.12)

$$\tilde{Y}_{mm}(\vartheta,\varphi) = \frac{\sqrt{\bar{\varrho}(s)}}{\bar{d}_0}u_0^{(m,m)}[x(s)]\phi_m(\varphi) \,.$$

Since

$$u_0^{(m,m)}[x(s)] = 1 \,,$$

$$\bar{d}_0^2 = -\sum_i\varrho(s_i)\Delta x(s_i-1/2) = S_0 \,,$$

we have

$$\tilde{Y}_{mm}(\vartheta,\varphi) = \sqrt{\frac{\bar{\varrho}(s)}{S_0}}\phi_m(\varphi) \,. \tag{4.3.35}$$

For calculating S_0 we use (4.3.32) and the equality

$$S_{N-1} = -\varrho_{N-1}(a)\Delta x_{N-1}\left(a-\tfrac{1}{2}\right) \,. \tag{4.3.36}$$

From (4.3.32) it follows that

$$\begin{aligned}
\frac{S_{n+1}}{S_n} &= \frac{\sin^4\omega(\tilde{N}-n-m-1)\sin^2\omega(n+m+1)}{\sin^4\omega\,\sin\omega(2n+2m+2)\sin\omega(2n+2m+3)} \\
&= \frac{\psi_q^4(\tilde{N}-n-m-1)\psi_q^2(n+m+1)}{\psi_q(2n+2m+2)\psi_q(2n+2m+3)} \,,
\end{aligned}$$

and hence, because of (4.3.15),

$$S_n = C\frac{\tilde{\Gamma}_q^2(n+m+1)}{\tilde{\Gamma}_q^4(\tilde{N}-m-n)\tilde{\Gamma}_q(2n+2m+2)} \,,$$

where C is a constant that does not depend on n. Hence

$$\frac{S_0}{S_{N-1}} = \frac{\tilde{\Gamma}_q^2(m+1)}{\tilde{\Gamma}_q(2m+2)}\frac{\tilde{\Gamma}_q(2\tilde{N})}{\tilde{\Gamma}_q^2(\tilde{N})}\frac{1}{\tilde{\Gamma}_q^4(\tilde{N}-m)} \,. \tag{4.3.37}$$

Let us simplify the expression obtained. According to (4.3.16)

$$\frac{\tilde{\Gamma}_q(2\tilde{N})}{\tilde{\Gamma}_q(\tilde{N})} = \prod_{k=0}^{\tilde{N}-1}\psi_q(\tilde{N}+k) = \prod_{k=0}^{\tilde{N}-1}\psi_q(k+1) = \tilde{\Gamma}_q(N)\psi_q(\tilde{N}) \,,$$

whence it follows that

$$\frac{\tilde{\Gamma}_q(2\tilde{N})}{\tilde{\Gamma}_q^2(\tilde{N})} = \frac{1}{\sin\omega} \ .$$

By virtue of (4.3.27)

$$\varrho_{N-1}(a) = \varrho_{\tilde{N}-m-1}\left(\frac{m+1}{2}\right) = \tilde{\Gamma}_q^4(\tilde{N}) \ ,$$

and hence

$$S_{N-1} = 2\sin\omega\,\tilde{\Gamma}_q^4(\tilde{N}) \ .$$

Thus

$$S_0 = 2\left[\frac{\tilde{\Gamma}_q(\tilde{N})}{\tilde{\Gamma}_q(\tilde{N}-m)}\right]^4 \frac{\tilde{\Gamma}_q^2(m+1)}{\tilde{\Gamma}_q(2m+2)} \ .$$

Finally we obtain

$$\tilde{Y}_{mm}(\vartheta,\varphi) = \frac{1}{(2\sin^2\omega)^m} \frac{\sqrt{(1/2)\tilde{\Gamma}_q(2m+2)}}{\tilde{\Gamma}_q(m+1)} \frac{\tilde{\Gamma}_q^2(\tilde{N}-m)}{\tilde{\Gamma}_q^2(\tilde{N})}$$
$$\times \prod_{k=0}^{m-1} \sin\left[\vartheta + \frac{\pi}{2\tilde{N}}(2k-m-1)\right]\phi_m(\varphi) \ , \quad \tilde{Y}_{00}(\vartheta,\varphi) = \frac{1}{\sqrt{4\pi}} \ . \tag{4.3.38}$$

We note that

$$\frac{\tilde{\Gamma}_q(\tilde{N})}{\tilde{\Gamma}_q(\tilde{N}-m)} = \prod_{k=0}^{m-1} \psi_q(\tilde{N}-m-k) = \prod_{k=1}^{m} \frac{\cos\omega k}{\sin\omega} \ .$$

By using (4.3.34, 38) we may prove by induction that

$$\tilde{Y}_{lm}(\vartheta,\varphi) = Y_{lm}(\vartheta,\varphi) + O\left(\frac{1}{\tilde{N}^2}\right) \ , \tag{4.3.39}$$

when $\tilde{N} \to \infty$. For the functions $Y_{lm}(\vartheta,\varphi)$ we have the recursion relation

$$\cos\vartheta\, Y_{lm}(\vartheta,\varphi) = \sqrt{\frac{(l+1)^2 - m^2}{4(l+1)^2 - 1}}\, Y_{l+1,m}(\vartheta,\varphi)$$
$$+ \sqrt{\frac{l^2 - m^2}{4l^2 - 1}}\, Y_{l-1,m}(\vartheta,\varphi) \ , \tag{4.3.40}$$

$$Y_{m-1,m}(\vartheta,\varphi) = 0 \ , \quad Y_{00}(\vartheta,\varphi) = \frac{1}{\sqrt{4\pi}} \ ,$$

$$Y_{mm}(\vartheta,\varphi) = \frac{e^{im\varphi}}{\sqrt{4\pi}} \frac{\sqrt{(2m+1)!}}{2^m m!} \sin^m\vartheta \ . \tag{4.3.41}$$

Therefore from comparison of (4.3.34, 38) and (4.3.40, 41) it is seen that the relation (4.3.39) is valid if for $\tilde{N} \to \infty$:

$$\cos \omega \, (l + 1) = 1 + O\left(\frac{1}{\tilde{N}^2}\right) \quad \left(\omega = \frac{\pi}{2\tilde{N}}\right) ,$$

$$\frac{\sin \omega \, (l + 1 - m) \sin \omega \, (l + 1 + m)}{\sin \omega \, (2l + 1) \sin \omega \, (2l + 3)} = \frac{(l + 1)^2 - m^2}{4(l + 1)^2 - 1} + O\left(\frac{1}{\tilde{N}^2}\right) ; \tag{4.3.42}$$

$$\frac{1}{(\sin^2 \omega)^m} \frac{\sqrt{\tilde{\Gamma}_q(2m + 2)}}{\tilde{\Gamma}_q(m + 1)} \frac{\tilde{\Gamma}_q^2(\tilde{N} - m)}{\tilde{\Gamma}_q^2(\tilde{N})} = \frac{\sqrt{(2m + 1)!}}{m!} + O\left(\frac{1}{\tilde{N}^2}\right) ; \tag{4.3.43}$$

$$\prod_{k=0}^{m-1} \sin \left[\vartheta + \frac{\pi}{2\tilde{N}}(2k - m + 1)\right] = \sin^m \vartheta + O\left(\frac{1}{\tilde{N}^2}\right) . \tag{4.3.44}$$

The relation (4.3.42) follows from the fact that when $\omega \to 0$ ($\tilde{N} \to \infty$) we have

$$\frac{\sin \omega x}{\omega} = x + O(\omega^2) , \quad \cos \omega x = 1 + O(\omega^2) . \tag{4.3.45}$$

To prove (4.3.43) it is sufficient to use (4.3.35) and the equality

$$\frac{1}{(\sin^2 \omega)^m} \frac{\sqrt{\tilde{\Gamma}_q(2m + 2)}}{\tilde{\Gamma}_q(m + 1)} \frac{\tilde{\Gamma}_q^2(\tilde{N} - m)}{\tilde{\Gamma}_q^2(\tilde{N})}$$

$$= \frac{\sqrt{\prod_{k=1}^{2m+1}(\sin \omega k / \sin \omega)}}{\prod_{k=1}^{m}(\sin \omega k / \sin \omega)} \left[\prod_{k=1}^{m} \cos \omega k\right]^2 .$$

To prove (4.3.44) we use the fact that the replacement of k by $m - 1 - k$ does not change the product on the left-hand side of (4.3.44), i.e.

$$\prod_{k=0}^{m-1} \sin \left[\vartheta + \frac{\pi}{2\tilde{N}}(2k - m + 1)\right] = \prod_{k=0}^{m-1} \sin \left[\vartheta - \frac{\pi}{2\tilde{N}}(2k - m + 1)\right] ,$$

whence

$$\prod_{k=0}^{m-1} \sin \left[\vartheta + \frac{\pi}{2\tilde{N}}(2k - m + 1)\right]$$

$$= \sqrt{\prod_{k=0}^{m-1} \left\{\sin \left[\vartheta + \frac{\pi}{2\tilde{N}}(2k - m + 1)\right] \sin \left[\vartheta - \frac{\pi}{2\tilde{N}}(2k - m + 1)\right]\right\}} .$$

Since for $\tilde{N} \to \infty$

$$\sin \left[\vartheta + \frac{\pi}{2\tilde{N}}(2k - m + 1)\right] \sin \left[\vartheta - \frac{\pi}{2\tilde{N}}(2k - m + 1)\right] = \sin^2 \vartheta + O\left(\frac{1}{\tilde{N}^2}\right) ,$$

the relation (4.3.44) holds.

4.3.4 Conclusions

Thus we have obtained sets of functions $\tilde{Y}_{lm}(\theta, \varphi)$ in two forms, (4.3.6) and (4.3.12), that satisfy the orthogonality relations (4.3.7) and (4.3.2, 13), respectively. By using the recursion relations (4.3.10, 14) and (4.3.34) the functions $\tilde{Y}_{lm}(\theta, \varphi)$ may be evaluated at any point of the sphere and then employed in expansions of the form (4.3.8). In practice it is more preferable to make computations with the use of the recursion relations than the Rodrigues formula or the expansions of polynomials in powers of an independent variable. We note that the lattice, uniform in x, describes better a behavior of the function $f(\theta, \varphi)$ depending on angle θ near the equator $\left(\theta = \frac{\pi}{2}\right)$ than near the poles $(\theta = 0, \pi)$. The lattice, uniform in θ, is condensed in x near the poles, since at $\Delta\theta = \text{const}$ the step $\Delta x \approx \sin\theta\, \Delta\theta$ decreases on approaching the poles, when $\theta \sim 0$ and $\theta \sim \pi$.

In the both cases the step of the lattice is given by $\tilde{N}$. Depending on whether m is even or odd numbered, we obtain two noncoinciding lattices for each form of functions $\tilde{Y}_{lm}(\theta, \varphi)$. Since for the given $\tilde{N}$ we have $N = \tilde{N} - m$, the orthogonality interval decreases with a growth of m (as should be expected, only summands with a small m make a contribution to the expansion (4.3.8) near the poles $x = \pm 1$). In Fig. 4.1 the crosses and circles mark the points of the lattice, uniform in x, with $\tilde{N} = 6$ for different m.

The spherical harmonics $\tilde{Y}_{lm}(\theta, \varphi)$ are sufficiently convenient for practical applications on computers, since on expanding the functions given on a surface of the sphere to obtain the expansion coefficients the summation over a finite number of the points of a lattice in θ and φ is used instead of respective integration for the functions $Y_{lm}(\theta, \varphi)$ (cf. Sect. 4.2).

We note also that the difference analogs of spherical harmonics introduced in [R20] were obtained as a result of numerical solution of the eigenvalue problem for the Beltrami equation. The algorithm proposed in [R20] requires computations of eigenvalues and eigenvectors for each given number of lattice points.

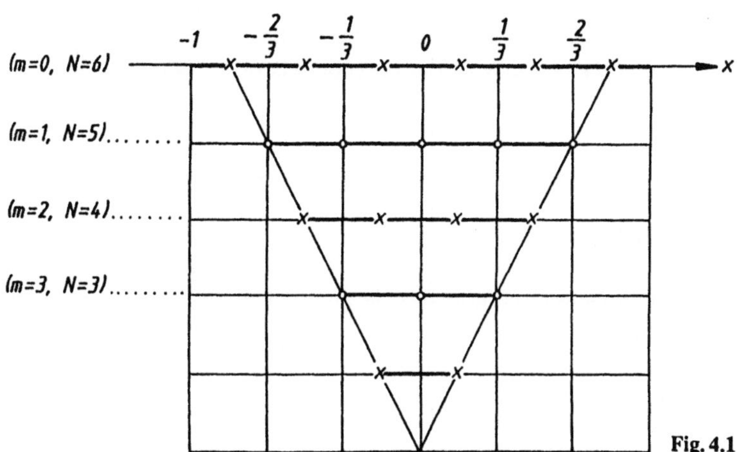

Fig. 4.1

4.4 Some Finite-Difference Methods of Solution of Partial Differential Equations

4.4.1. Classical orthogonal polynomials of a discrete variable are the solutions of difference equations that approximate differential equations arising in solving many problems of mathematical physics by the method of separation of variables. We consider some fields where the classical orthogonal polynomials of a discrete variable may be applied to solving partial differential equations.

4.4.1.1. Let the exact solution of an initial problem be given as a series in particular solutions that are obtained by the method of separation of variables. These solutions can be often expressed in a simple way through classical orthogonal polynomials. If the series quickly converges we may restrict ourselves to a small number of terms and instead of the classical orthogonal polynomials take their difference analogs – the classical polynomials of a discrete variable. In order to calculate the coefficients of expansion it is natural to use the orthogonality property of the polynomials of a discrete variable. This leads to the situation where in the approximate solution instead of the integrals being present in the exact solution the sums arise at points of the lattice on which the polynomials of a discrete variable are orthogonal. As a result we obtain a rather simple procedure for determining an approximate solution of the partial differential equation.

4.4.1.2. We now consider another possibility of using the difference equations for classical orthogonal polynomials. Let an initial partial differential equation be replaced by the difference equation so that in solving the difference equation by the method of separation of variables the difference equations should arise whose solutions can be expressed through the classical orthogonal polynomials of a discrete variable. As a result the solution of a difference equation corresponding to the partial differential equation is presented in the form of the sum (or the series) in functions that are connected with classical orthogonal polynomials of a discrete variable in a simple way. The coefficients of the expansion can be found by using the orthogonality property of the polynomials of a discrete variable.

4.4.2. As an example we shall find the bounded solution of the Laplace equation $\Delta u = 0$ in a ball when on the sphere $r = 1$ the boundary condition

$$u(r, \vartheta, \varphi)|_{r=1} = f(\vartheta) , \tag{4.4.1}$$

where r, ϑ, φ are the spherical coordinates, is given. The exact solution of the problem has the form

$$u(r, \vartheta, \varphi) = \sum_{l=0}^{\infty} a_l r^l P_l(\cos \vartheta) , \tag{4.4.2}$$

where $P_l(x)$ are the Legendre polynomials,

$$a_l = \frac{2l+1}{2} \int_0^\pi f(\vartheta) P_l(\cos\vartheta) \sin\vartheta\, d\vartheta \ . \tag{4.4.3}$$

If $f(\vartheta)$ is a sufficiently smooth function, the coefficients a_l quickly decrease with growing l.

4.4.2.1. If we use the first method of approximate solution, we should restrict ourselves in (4.4.2) by a finite number of summands ($l \leq \bar{l}$) and then replace the Legendre polynomial by its difference analog, for example the Chebyshev polynomial $t_l(s)$, where $s = (N(1 + x))/2 - 1/2$ (see (2.6.6)). As a result we obtain an approximate representation of the solution in the form of the sum

$$u(r,\vartheta,\varphi) \approx \sum_{l=0}^{\bar{l}} b_l r^l t_l(s) \ .$$

The coefficients b_l are determined by using the boundary condition (4.4.1).

$$f(\vartheta) \approx \sum_{l=0}^{\bar{b}} b_l t_l(s) \ .$$

If we use the orthogonality property of the Chebyshev polynomials, then for the coefficients b_l we obtain the expression

$$b_l = \frac{1}{d_l^2} \sum_{s_i=0}^{N-1} f(\vartheta_i) t_l(i) \ , \tag{4.4.4}$$

where $\cos\vartheta_i = (2i+1)/N - 1$, d_l^2 is the squared norm of polynomial $t_l(s)$. When the problem is solved by this method the N should be chosen from the condition $N \gg \bar{l}$, since for the fixed value of l the limit relation holds (see Eq. (2.6.6)):

$$\frac{1}{N^l} t_l \left[\frac{N}{2}(1 + x) - \frac{1}{2} \right] = P_l(x) + O\left(\frac{1}{N^2}\right) \ , \quad N \to \infty \ .$$

Remark. When solving the Laplace equation $\Delta u = 0$ with a more general boundary condition $u(r,\vartheta,\varphi)|_{r=1} = f(\vartheta,\varphi)$ we may represent the exact solution as a series in the functions $\sqrt{\varrho(x)} P_n^{(m,m)}(x) \cos m\varphi$, $\sqrt{\varrho(x)} P_n^{(m,m)}(x) \sin m\varphi$ ($x = \cos\vartheta$, $\varrho(x)$ is the weight for the Jacobi polynomials $P_n^{(m,m)}(x)$, $m = 0, 1, \ldots, n$; $n = 0, 1, \ldots$). In this case the difference analog of the function $\sqrt{\varrho(x)} P_n^{(m,m)}(x)$ is the function (see Sect. 4.3)

$$\sqrt{\tilde{\varrho}(s)} h_n^{(m,m)}(s) \ , \quad s = \frac{N+m}{2}(1 - x) - \frac{m+1}{2}$$

and $\tilde{\varrho}(s)$ is the weight for the Hahn polynomials $h_n^{(m,m)}(s)$. The functions $\sqrt{\tilde{\varrho}(s)} h_n^{(m,m)}(s)$ can be expressed through the Clebsch-Gordan coefficients for which detailed tables are available (see [V5] and Sect. 5.2).

4.4.2.2. We consider now the solution of the Laplace equation $\Delta u = 0$ with the boundary condition (4.4.1) by applying the second method. Since in this case the solution does not depend on the angle φ the equation $\Delta u = 0$ can be rewritten in the form

$$\frac{\partial}{\partial r}\left(r^2\frac{\partial u}{\partial r}\right) + \frac{\partial}{\partial x}\left[(1-x^2)\frac{\partial u}{\partial x}\right] = 0\,, \qquad x = \cos\vartheta\,.$$

We replace this equation by the second order difference equation on the uniform lattices:

$$\frac{1}{h_1^2}\Delta_1\left(r_{i-1/2}^2\nabla_1 u_{ik}\right) + \frac{1}{h_2^2}\Delta_2\left[(1-x_{k-1/2}^2)\nabla_2 u_{ik}\right] = 0\,. \tag{4.4.5}$$

Here $r_i = r_0 + ih_1$, $x_k = x_0 + kh_2$, $i = 0,1,\ldots,N_1$, $k = 0,1,\ldots,N_2$; $r_{i-1/2} = r_i - h_1/2$, $x_{k-1/2} = x_k - h_2/2$; $\Delta_1, \Delta_2, \nabla_1, \nabla_2$ are the corresponding difference operators in the variables i and k.

The differential operators $(\partial/\partial r)(r^2(\partial/\partial r))$ and $(\partial/\partial x)[(1-x^2)\partial/\partial x]$ have singularities at $r = 0$ and $x = \pm 1$ (the leading coefficient at these points is zero). We choose the values r_0, x_0, h_2 so that the corresponding difference operators have singularities at the boundary points of the lattice. This requirement will be satisfied if we put $r_0 = h_1/2$, $x_0 = -1 + h_2/2$, $x_{N_2-1/2} = 1$, $h_2 = 2/N_2$. Besides, the mesh h_1 is chosen from the condition that the value $r = 1$, on which the boundary condition is given, enters into the lattice in r. It yields $h_1 = 2/(2N_1 + 1)$.

We shall solve (4.4.5) by the method of separation of variables. The particular solution will be searched for in the form $u_{ik} = R_i y_k$. As a result we have

$$\frac{(1/h_1^2)\Delta_1(r_{i-1/2}^2\nabla_1 R_i)}{R_i} = -\frac{(1/h_2^2)\Delta_2[(1-x_{k-1/2}^2\nabla_2 y_k]}{y_k} = \lambda\,,$$

where λ is a constant. These equations are equivalent to the following system of equations:

$$i^2\Delta\nabla R_i + (2i+1)\Delta R_i - \lambda R_i = 0\,, \tag{4.4.6}$$

$$k(N_2 - k)\Delta\nabla y_k + (N_2 - 2k - 1)\Delta y_k + \lambda y_k = 0\,. \tag{4.4.7}$$

We shall first consider (4.4.7). At $k = 0$ this equation connects only the values at the points $k = 0$ and $k = 1$. Therefore the solution is uniquely determined by giving the value of y_0. By using analogous considerations we obtain that the solution is uniquely determined by giving the value of y_k at $k = N_2 - 1$. From these arguments it is seen that the values of y_0 and y_{N_2-1} must be compatible, i.e. nontrivial solutions of Eq. (4.4.7) exist only for certain values of λ. In this case nontrivial solutions of Eq. (4.4.7) exist only for $\lambda = \lambda_l = l(l+1)$, $l = 0,1,\ldots$, and $y_k(\lambda_l) = t_l(k,N_2)$, where $t_l(x,N)$ are the Chebyshev polynomials of a discrete variable at $N = N_2$.

Now we consider the solutions of Eq. (4.4.6) for $\lambda = \lambda_l$. The solution of this equation is uniquely determined by giving the value of R_i at $i = 0$. Equation

(4.4.6) is the difference equation of hypergeometric type and for $\lambda = \lambda_l = l(l+1)$ it has the solution in the form of the polynomial which is determined by the Rodrigues formula

$$R_i(\lambda_l) \equiv R_{il} = \Delta_1^l \left[\prod_{p=0}^{l-1} (i-p)^2 \right] = l!\, t_l(i,0) \, . \tag{4.4.8}$$

Thus, Eq. (4.4.5) has particular solutions of the form $u_{ik}(\lambda_l) = u_{ikl} = R_{il}\, t_l(k, N_2)$. The general solution of Eq. (4.4.6) is the superposition of these particular solutions:

$$u_{ik} = \sum_{l=0}^{N_2-1} C_l R_{il} t_l(k, N_2) \, . \tag{4.4.9}$$

To satisfy the boundary condition (4.4.1) we require that the equality

$$f(\vartheta_k) = \sum_{l=0}^{N_2-1} C_l R_{N_1 l} t_l(k, N_2) \tag{4.4.10}$$

is valid at points $\vartheta = \vartheta_k$. The coefficients C_l are calculated by using the orthogonality property of the Chebyshev polynomials $t_l(x)$:

$$C_l = \frac{1}{R_{N_1 l} d_l^2} \sum_{k=0}^{N_2-1} f(\vartheta_k) t_l(k) \, , \tag{4.4.11}$$

where d_l^2 is the squared norm of the polynomials $t_l(x, N_2)$.

For determining some specific values of u_{ik} by Eq. (4.4.9) for the polynomials R_{il} and $t_l(k)$ we may use the recursion formulas in l.

Remark. When the partial differential equations are solved by the method of separation of variables the generalized equations of hypergeometric type often arise in the form [N18]

$$u'' + \frac{\tilde{\tau}(x)}{\sigma(x)} u' + \frac{\tilde{\sigma}(x)}{\sigma^2(x)} u = 0 \, ,$$

where $\sigma(x)$, $\tilde{\sigma}(x)$ are polynomials of at most degree 2; $\tilde{\tau}(x)$ is the polynomials of at most degree 1. These equations are usually reduced to the equations of hypergeometric type

$$\sigma(x) y'' + \tau(x) y' + \lambda y = 0 \, ,$$

where $\tau(x)$ is the polynomial of at most degree one and λ is a constant, by using the replacement of the variable $u(x) = \varphi(x) y(x)$. Here $\varphi(x)$ is the solution of the differential equation $\varphi' = (\pi(x)/\sigma(x))\varphi$; $\pi(x)$ is the polynomial of at most degree one (see [N18]).

In this connection we are interested in the analogs of a generalized equation of hypergeometric type, which can be reduced to the difference equation of hypergeometric type by a simple replacement of the variable.

Let us consider the equation

$$\Delta\nabla u_k + \frac{\tilde{\tau}_k}{\sigma_k}\Delta u_k + \frac{\tilde{\sigma}_k}{\sigma_k\sigma_{k+1}}u_{k+1} = 0 , \qquad (4.4.12)$$

where σ_k and $\tilde{\sigma}_k$ are polynomials of at most degree two with respect to k; $\tilde{\tau}_k$ is the polynomial of at most degree one. We show that by using the replacement $u_k = \varphi_k y_k$ Eq. (4.4.12) can be reduced to the difference equation of hypergeometric type

$$w_k\Delta\nabla y_k + \tau_k\Delta y_k + \lambda y_k = 0 , \qquad (4.4.13)$$

where τ_k and w_k are polynomials of at most degree one and two, respectively, and λ is a constant, provided φ_k is the solution of the difference equation of the form

$$\nabla\varphi_k = \frac{\pi_k}{\sigma_k}\varphi_k .$$

Here π_k is an appropriately chosen polynomial of at most degree one. We have $\varphi_{k-1} = \varphi_k w_k/\sigma_k$, where $w_k = \sigma_k - \pi_k$. Therefore the equation (4.4.13) is equivalent to

$$w_k\Delta\nabla y_k + (\tilde{\tau}_k + 2\pi_k)\Delta y_k + \frac{\bar{\sigma}_k}{w_{k+1}}y_{k+1} = 0 ,$$

where

$$\bar{\sigma}_k = \tilde{\sigma}_k + \tilde{\tau}_k\pi_{k+1} + \pi_k\pi_{k+1} + \sigma_k\Delta\pi_k - \pi_k\Delta\sigma_k .$$

The coefficients of the polynomial π_k and the constant λ can be chosen so that the equality $\bar{\sigma}_k = \lambda w_{k+1}$ is satisfied. As a result the equation for y_k will take the form (4.4.13), where $\tau_k = \tilde{\tau}_k + 2\pi_k + \lambda$ is the polynomial of at most degree one.

4.5 Systems of Differential Equations with Constant Coefficients. The Genetic Model of Moran and Some Problems of the Queueing Theory

4.5.1. We consider the system of differential equations

$$\frac{du_n(t)}{dt} = \alpha_n u_{n+1}(t) + \beta_n u_n(t) + \gamma_n u_{n-1}(t)$$

$$(n = 0, 1, \ldots, N - 1 ; \quad u_{-1}(t) = u_N(t) = 0 , \quad \alpha_n \neq 0) \qquad (4.5.1)$$

with initial conditions

$$u_n(0) = u_n^0 , \qquad (4.5.2)$$

where u_n^0 are constants. As is known the system (4.5.1) has particular solutions of the form $u_n(t) = y_n(x)\exp(xt)$ for some values of $x = x_i (i = 0, 1, \ldots, N-1)$ with $y_0(x) = 1$. By substituting the supposed form of the solution into (4.5.1) we obtain the following recursion relation for $y_n(x)$:

$$xy_n(x) = \alpha_n y_{n+1}(x) + \beta_n y_n(x) + \gamma_n y_{n-1}(x)$$
$$(y_{-1}(x) = y_N(x) \equiv 0, \quad y_0(x) = 1) .$$

(4.5.3)

From (4.5.3) and the conditions $y_{-1}(x) = 0$, $y_0(x) = 1$ it follows that $y_n(x)$ is a polynomial of degree n. The values $x = x_i (i = 0, 1, \ldots, N-1)$ are the roots of the equation

$$y_N(x_i) = 0 .$$

(4.5.4)

If the roots x_i are simple, a general solution of the system (4.5.1) is a linear combination of the particular solutions obtained, i.e.

$$u_n(t) = \sum_{i=0}^{N-1} C_i y_n(x_i) e^{x_i t} .$$

(4.5.5)

In order to determine the solution of the system of Eqs. (4.5.1) with the initial conditions (4.5.2) it is convenient to obtain as a preliminary the particular solutions of (4.5.1) $u_n(t) = u_{nm}(t)$ that satisfy the simplest initial conditions

$$u_{nm}(0) = \delta_{nm}$$

(4.5.6)

with fixed values of $m = 0, 1, \ldots, N-1$. Then the solution of (4.5.1) with the initial conditions (4.5.2) can be written in the form

$$u_n(t) = \sum_{m=0}^{N-1} u_m^0 u_{nm}(t) .$$

(4.5.7)

In some cases the solution of the problem posed may be simplified by making use of the following considerations. Let the polynomials $y_n(x)$ satisfy the recursion relation (4.5.3) in which the coefficients α_n, β_n, γ_n are real and $\alpha_{n-1}\gamma_n > 0$ for $n \geq 1$. As shown in Sect. 3.9 under these conditions the roots of Eq. (4.5.4) are real and different. Moreover, there are the constants $\varrho_i > 0$, $d_n^2 > 0$ such that the polynomials $y_n(x)$ satisfy orthogonality relations of the form

$$\sum_{i=0}^{N-1} y_n(x_i)y_m(x_i)\varrho_i = \delta_{mn} d_n^2 .$$

(4.5.8)

It is easy to verify that in this case the solution of the system (4.5.1) $u_n(t) \equiv u_{nm}(t)$ with the initial conditions (4.5.6) can be obtained from (4.5.5) by assuming

$$C_i \equiv C_{im} = \frac{1}{d_m^2} y_m(x_i)\varrho_i ,$$

i.e.

$$u_{nm}(t) = \sum_{i=0}^{N-1} e^{x_i t} \frac{y_n(x_i)y_m(x_i)}{d_m^2} \varrho_i . \tag{4.5.9}$$

According to (4.5.7) the solution of the system (4.5.1) with the initial conditions (4.5.2) has the form

$$u_n(t) = \sum_{m=0}^{N-1} \sum_{i=0}^{N-1} u_m^0 e^{x_i t} \frac{y_n(x_i)y_m(x_i)}{d_m^2} \varrho_i . \tag{4.5.10}$$

In some particular cases by comparing the recursion relation (4.5.3) and the recursion relations for known systems of orthogonal polynomials of a discrete variable the polynomials $y_n(x)$ can be expressed through the known polynomials of a discrete variable, and the orthogonality relation (4.5.8) can be obtained from the orthogonality relations for these polynomials with the use of a linear change of variable and multiplication by the normalization factors.

4.5.2. As an example we consider the system of differential equations arising in *the genetic model of Moran* [M16]. The problem is formulated as follows.

Let there exist N gametes that belong to two types a and A. The number of gametes of each type is varied as a result of their coupling. After the coupling one of the gametes (the fertilizer) does not change its type while the other gamete after the impregnation (posterity) becomes the gamete of the same type as the fertilizer and, moreover, it can then mutate into the other type. Let the probability of mutating the gamete-posterity of type a into the gamete of type A be equal to γ_1, and the probability of mutating to gamete-posterity of type A into the gamete of type a to γ_2. We assume also that the probability of gamete coupling during the time interval dt is $\lambda dt + \mathrm{o}(dt)$ with $\lambda = \text{const}$, while the probability of two or more couplings is $\mathrm{o}(dt)$. At the gamete coupling each gamete is chosen randomly, and the probability of choosing the gamete of a certain type is proportional to the number of gametes of this type. For example, if at the time t there are n gametes of type a the probability of choosing one gamete of type a or A is equal to n/N or $(N-n)/N$, respectively, while the probabilities of choosing the pairs aa, aA, Aa or AA (the first gamete in the pair is a fertilizer) are equal to $(n/N)^2$, $n(N-n)/N^2$, $(N-n)n/N^2$ or $(N-n)^2/N^2$, respectively. Hence, if we assume that at the time t there are n gametes of type a, the probability that in time dt the gamete couplings yield $n+1$ gametes of type a is equal to

$$\lambda dt \left[\frac{n(N-n)}{N^2}(1-\gamma_1) + \left(\frac{N-n}{N}\right)^2 \gamma_2 \right] + \mathrm{o}(dt) = a_{n+1,n} dt + \mathrm{o}(dt) ,$$

where the multiplier $(1-\gamma_1)$ in the first summand is the probability that, owing to the couplings of gametes of type aA, the posterior gamete will not mutate into the type A; the second summand corresponds to the couplings of gametes of type AA and the subsequent mutation of the gamete-posterity into type a. Analogously we may obtain that the probability that after coupling we shall have

$n - 1$ gametes of type a is equal to

$$\lambda dt \left[\left(\frac{n}{N} \right)^2 \gamma_1 + \frac{(N-n)n}{N^2} (1 - \gamma_2) \right] + \mathrm{o}(dt) = a_{n-1,n} dt + \mathrm{o}(dt) \ .$$

Let $u_{nm}(t)$ be the probability of the fact that at the time t there are n gametes of type a provided that at the time $t = 0$ there were m gametes of type a. Then for the functions $u_{nm}(t)$ we obtain the relations

$$u_{nm}(t + dt) - u_{nm}(t) = -[a_{n-1,n} dt + a_{n+1,n} dt] u_{nm}(t) + a_{n,n+1} dt u_{n+1,m}(t)$$
$$+ a_{n,n-1} dt \, u_{n-1,m}(t) + \mathrm{o}(dt)$$
$$(u_{-1,m}(t) = 0 \ , \quad u_{N+1,m}(t) = 0) \ .$$

As a result we obtain the following system of differential equations

$$\frac{du_{nm}(t)}{dt} = \alpha_n u_{n+1,m}(t) + \beta_n u_{nm}(t) + \gamma_n u_{n-1,m}(t) \tag{4.5.11}$$
$$(n = 0, 1, \ldots, N \ ; \quad u_{-1,m}(t) = 0 \ , \quad u_{N+1,m}(t) = 0)$$

with the initial conditions

$$u_{nm}(0) = \delta_{nm} \ , \tag{4.5.12}$$

where

$$\alpha_n = a_{n,n+1} \ , \quad \beta_n = -(a_{n-1,n} + a_{n+1,n}) \ , \quad \gamma_n = a_{n,n-1} \ ;$$
$$a_{n+1,n} = \frac{\lambda}{N^2} (N - n)[n(1 - \gamma_1) + (N - n)\gamma_2] \ , \tag{4.5.13}$$
$$a_{n-1,n} = \frac{\lambda n}{N^2} [n\gamma_1 + (N - n)(1 - \gamma_2)] \ .$$

For the functions $u_{nm}(t)$ the system of Eqs. (4.5.11) belongs to the type (4.5.1) with $N - 1$ replaced by N. Hence according to (4.5.9) the solution of these equations may be given in the form

$$u_{nm}(t) = \sum_{i=0}^{N} e^{x_i t} \frac{y_n(x_i) y_m(x_i)}{d_m^2} \varrho_i \ , \tag{4.5.14}$$

where $y_n(x)$ are polynomials satisfying the recursion relation

$$x y_n(x) = \alpha_n y_{n+1}(x) + \beta_n y_n(x) + \gamma_n y_{n-1}(x) \tag{4.5.15}$$

with $y_{-1}(x) = 0$ and $y_0(x) = 1$, and the orthogonality relations of the form

$$\sum_{i=0}^{N} y_n(x_i) y_m(x_i) \varrho_i = d_n^2 \delta_{nm} \ . \tag{4.5.16}$$

Here x_i are roots of the equation

$$y_{N+1}(x) = 0 \ . \tag{4.5.17}$$

In this case the recursion relation (4.5.15) can be written in the form

$$
-\frac{N^2}{(1-\gamma_1-\gamma_2)\lambda}xy_n(x) = -(n+1)\left(\frac{N(1-\gamma_2)}{1-\gamma_1-\gamma_2}-1-n\right)y_{n+1}(x)
$$

$$
+\left(\frac{\gamma_2}{1-\gamma_1-\gamma_2}N^2+\frac{2-\gamma_1-3\gamma_2}{1-\gamma_1-\gamma_2}Nn-2n^2\right)y_n(x)
$$

$$
-(N+1-n)\left(\frac{N\gamma_2}{1-\gamma_1-\gamma_2}-1+n\right)y_{n-1}(x)\,. \tag{4.5.18}
$$

By using a linear change of independent variable and introducing normalization factors, the relation (4.5.18) may be reduced to the recursion relation for the dual Hahn polynomials (see Table 3.7). As a result we obtain

$$
y_n(x) = A_n w_n^{(c)}(\xi,a,b)\,, \tag{4.5.19}
$$

where

$$
a = \frac{\gamma_1+\gamma_2}{1-\gamma_1-\gamma_2}\frac{N}{2}-1\,,\quad b = a+N+1\,,\quad c = \frac{\gamma_2-\gamma_1}{1-\gamma_1-\gamma_2}\frac{N}{2}\,;\tag{4.5.20}
$$

$$
-\frac{N^2}{(1-\gamma_1-\gamma_2)\lambda}x = \xi - a(a+1)\,;\tag{4.5.21}
$$

$$
A_n = (-1)^n\frac{\Gamma(N+a+1-c-n)}{\Gamma(N+a+1-c)}\,.\tag{4.5.22}
$$

For the functions $u_{nm}(t)$ we can use Eq. (4.5.14) in which the constants x_i, ϱ_i and d_n^2 can be determined with the aid of the orthogonality relation for the dual Hahn polynomials:

$$
\varrho_i = (2a+2i+1)\frac{\Gamma(2a+i+1)\Gamma(a+c+i+1)}{i!(N-i)!\Gamma(2a+N+2+i)\Gamma(a-c+i+1)}\,,\tag{4.5.23}
$$

$$
d_n^2 = A_n^2\frac{\Gamma(a+c+n+1)}{n!(N-n)!\Gamma(a-c+N-n+1)}\,,\tag{4.5.24}
$$

$$
x_i = -\frac{\lambda}{N^2}(1-\gamma_1-\gamma_2)i(i+2a+1)\quad(\xi_i = s_i(s_i+1),\ s_i = a+i)\,.\tag{4.5.25}
$$

From (4.5.25) it follows that $x_0 = 0$, and $x_i < 0$ for $1 \le i \le N$. Hence there is a limit

$$
\lim_{t\to\infty} u_{nm}(t) = \frac{y_n(0)y_m(0)}{d_m^2}\varrho_0\,.
$$

By using (4.5.19–25) and (3.11.30) we obtain

$$
\lim_{t\to\infty} u_{nm}(t) = \frac{N!\Gamma(2a+2)}{\Gamma(a+c+1)\Gamma(2a+N+2)\Gamma(a-c+1)}
$$

$$
\times\frac{\Gamma(N+a+1-c-n)\Gamma(a+1+c+n)}{(N-n)!n!} = U_n\,,
$$

which coincides with the *Polya distribution* to within the notations (see Table 4.1).

The solution of the system (4.5.11) with initial conditions $u_n(0) = u_n^0 (\sum_n u_n^0 = 1)$ has the form

$$u_n(t) = \sum_m u_m^0 u_{nm}(t) \, .$$

Hence for $t \to \infty$

$$\lim_{t \to \infty} u_n(t) = U_n \sum_m u_m^0 = U_n \, ,$$

i.e. the limiting distribution does not depend on initial conditions.

4.5.3. As another example we consider *the linear growth birth and death processes* [F1a, K5a, I4]. The problem is formulated as follows. We assume that elements of a population may split or die. For any living element the probability of splitting into two in a small time interval $\Delta t = h$ is equal to $\lambda h + o(h)$, while the corresponding probability of death is equal to $\mu h + o(h)$, where λ and μ are constants. It is also assumed that there is no interaction among elements of the population.

If the state n describes the current population size, then the average instantaneous rate of growth is $\lambda n + a$. The factor λn represents the natural growth of the population due to its current size, while the second factor a may be interpreted as the infinitesimal rate of increase of the population due to an external source such as immigration.

Similarly the probability of the state of the process decreasing by one after a small duration of time $\Delta t = h$ is $\mu n h + o(h)$. Therefore by following the arguments given in the Moran model for the probability $u_n(t)$ that at the time t the population contains n elements we shall obtain a system of equations in the form of (4.5.11), i.e.

$$\frac{du_n}{dt} = \mu(n+1)u_{n+1} - [(\lambda + \mu)n + a]u_n + [\lambda(n-1) + a]u_{n-1} \, , \tag{4.5.26}$$

$$(u_{-1}(t) = 0) \, .$$

This system belongs to the above considered type except that the number of its equations is infinite. If a particular solution of the system (4.5.26) is searched for in the form $u_n(t) = y_n(x)\exp(xt)$, then for $y_n(x)$ we obtain a recursion relation that for $\mu < \lambda$ and $a > 0$ coincides with the recursion relation for the polynomials $(1/n!)m_n^{(a/\lambda, \mu/\lambda)}(x)$, where $m_n(x)$ is the Meixner polynomial (see Table 2.3). Therefore the particular solution of equation (4.5.26) has the form

$$u_n(t) \equiv u_n(t, x) = \frac{1}{n!} e^{t(\mu - \lambda)(x + a/\lambda)} m_n^{(a/\lambda, \mu/\lambda)}(x) \, .$$

Since the Meixner polynomials satisfy the orthogonality relation

$$\sum_{k=0}^{\infty} \frac{m_n^{(a/\lambda,\mu/\lambda)}(k)}{n!} m_l^{(a/\lambda,\mu/\lambda)}(k) \left(\frac{\mu}{\lambda}\right)^k \frac{(a/\lambda)_k}{k!} = \delta_{nl} \frac{d_l^2}{l!} \, ,$$

where d_l^2 is the squared norm of the polynomial $m_l^{(a/\lambda,\mu/\lambda)}(x)$, the solution of system (4.5.26) $u_n(t) = u_{nl}(t)$ corresponding to the initial condition $u_{nl}(0) = \delta_{nl}$ has the form

$$
\begin{aligned}
u_{nl}(t) &= \sum_{l=0}^{\infty} e^{t(\mu - \lambda)(k + a/\lambda)} \frac{m_n^{(a/\lambda,\mu/\lambda)}(k)}{n!} m_l^{(a/\lambda,\mu/\lambda)}(k) \frac{l!}{d_l^2} \frac{(\mu/\lambda)^k (a/\lambda)_k}{k!} \\
&= \frac{(\mu/\lambda)^l (1 - \mu/\lambda)^{a/\lambda}}{n!(a/\lambda)_l} e^{-t(\lambda - \mu)a/\lambda} \\
&\quad \times \sum_{k=0}^{\infty} \left[\frac{\mu}{\lambda} e^{-t(\lambda - \mu)}\right]^k \frac{(a/\lambda)_k}{k!} m_n^{(a/\lambda,\mu/\lambda)}(k) m_l^{(a/\lambda,\mu/\lambda)}(k) \, . \quad (4.5.28)
\end{aligned}
$$

The expression (4.5.28) may be further transformed by using the relation obtained in [M9a], namely

$$
\begin{aligned}
\sum_{k=0}^{\infty} &\frac{(r - k + 1)_k}{k!} s^k F(-k, -p, -r, u) F(-k, -q, -r, v) \\
&= (1 + s)^{r-p-q}(1 + s - su)^p(1 + s - sv)^q \\
&\quad \times F\left(-p, -q, -r, -\frac{suv}{(1 + s - su)(1 + s - sv)}\right) \, , \quad (4.5.29)
\end{aligned}
$$

where $F(\alpha, \beta, \gamma, z)$ is a hypergeometric function.
Since

$$m_n^{(\gamma,\mu)}(x) = (\gamma)_n F(-n, -x, \gamma, 1 - (1/\mu))$$

(see Table 2.4), we have as a result that

$$
\begin{aligned}
u_{nl}(t) &= \frac{(a/\lambda)_n (1 - \mu/\lambda)^{a/\lambda}}{n!} \left(\frac{\mu}{\lambda}\right)^l e^{-t(\lambda - \mu)a/\lambda} \frac{[1 - e^{-t(\lambda - \mu)}]^{n+l}}{[1 - (\mu/\lambda)e^{-t(\lambda - \mu)}]^{n+l+a/\lambda}} \\
&\quad \times F\left(-n, -l, \frac{a}{\lambda}, \frac{(\mu - \lambda)^2 e^{-t(\lambda - \mu)}}{\mu\lambda[1 - e^{-t(\lambda - \mu)}]^2}\right) \, . \quad (4.5.30)
\end{aligned}
$$

In the explicit form the above expression determines the probability that at time t the considered population consists of n elements if at the initial time $t = 0$ it consisted of l elements. It is easy to see that $u_{nl}(t) \geq 0$ when $t \geq 0$.

Equation (4.5.30) is obtained under the assumption that $a > 0$, $\mu < \lambda$. However, it remains valid for any positive λ and μ. The case of interest $a = 0$ [F1a] arises from (4.5.30) if we take the limit $a \to 0$:

$$u_{nl}(t) = \left(\frac{\mu}{\lambda}\right)^l \left[\frac{1 - e^{-t(\lambda-\mu)}}{1 - (\mu/\lambda)e^{-t(\lambda-\mu)}}\right]^{n+l} l\frac{(\lambda-\mu)^2 e^{-t(\lambda-\mu)}}{\mu\lambda[1 - e^{-t(\lambda-\mu)}]^2}$$

$$\times F\left(-n+1, -l+1, 2, \frac{(\lambda-\mu)^2 e^{-t(\lambda-\mu)}}{\mu\lambda[1 - e^{-t(\lambda-\mu)}]^2}\right), \quad n \geq 1; \quad (4.5.31)$$

$$u_{0l}(t) = \left[\mu\frac{1 - e^{-t(\lambda-\mu)}}{\lambda - \mu e^{-t(\lambda-\mu)}}\right]^l .$$

The solution of (4.5.26) with $\mu = \lambda$ may be obtained from (4.5.30) and (4.5.31) by taking the limit:

$$u_{nl}(t) = \frac{(\lambda t)^{n+l}}{(1 + \lambda t)^{n+l+a/\lambda}} \frac{(a/\lambda)_n}{n!} F\left(-n, -l, \frac{a}{\lambda}, \frac{1}{\lambda^2 t^2}\right), \quad a > 0; \quad (4.5.32)$$

$$u_{nl}(t) = \frac{l}{\lambda^2 t^2}\left(\frac{\lambda t}{1 + \lambda t}\right)^{n+l} F\left(-n+1, -l+1, 2, \frac{1}{\lambda^2 t^2}\right),$$
$$\tag{4.5.33}$$

$$u_{0l}(t) = \left(\frac{\lambda t}{1 + \lambda t}\right)^l, \quad a = 0 .$$

By taking the limit $t \to \infty$ for $\mu > \lambda$ and $a > 0$ we obtain from (4.5.30) that

$$U_n = \lim_{t\to\infty} u_{nl}(t) = \left(1 - \frac{\lambda}{\mu}\right)^{a/\lambda} \frac{(a/\lambda)_n}{n!} \left(\frac{\lambda}{\mu}\right)^n . \tag{4.5.34}$$

Hence in this case the limiting distribution is the *negative binomial distribution* with parameters $p = 1 - \lambda/\mu$, $r = a/\lambda$ (see Table 4.1).

The solution $u_n(t)$ of (4.5.26) with arbitrary initial conditions $u_n(0) = u_n^0$ $\left(\sum_n u_n^0 = 1\right)$ has the form

$$u_n(t) = \sum_{l=0}^{\infty} u_l^0 u_{nl}(t) .$$

According to (4.5.34) we obtain, taking the limit $t \to \infty$, that

$$\lim_{t\to\infty} u_n(t) = U_n \sum_l u_l^0 = U_n ,$$

i.e. the limiting distribution (4.5.34) does not depend on initial conditions.

In conclusion we shall discuss some limiting cases of the linear growth process. At $a = \mu = 0$ the system of differential equations arising from (4.5.26) is

$$\frac{du_n}{dt} = -\lambda n u_n + \lambda(n-1)u_{n-1}, \quad n \geq 1, \quad \frac{du_0}{dt} = 0 . \tag{4.5.35}$$

This corresponds to the so called Yule process [F1a, K5a]. This process is an example of the pure birth process that arises in physics and biology.

Solutions of system (4.5.35) may be obtained from (4.5.31) by taking the limit $\mu \to 0$. They have the form

$$u_{nl}(t) = \frac{(n-1)!}{(l-1)!(n-l)!}e^{-\lambda lt}(1-e^{-\lambda t})^{n-l}\,, \quad n \geq l > 0\,; \tag{4.5.36}$$

$$u_{nl}(t) = 0\,, \quad n < l\,.$$

Now we consider the pure death process for which $a = \lambda = 0$. In this case according to (4.5.31) we have

$$u_{nl}(t) = \frac{l!}{n!(l-n)!}e^{-n\mu t}(1-e^{-\mu t})^{l-n}\,. \tag{4.5.37}$$

4.5.4. Let us consider *the power-supply problem* [F1a]. One electric circuit supplies N welders who use the current only intermittenly. If at time t a welder uses current, the probability that he ceases using before time $t + h$ is $\mu h + o(h)$; if at time t he requires no current, the probability that he calls for current before $t + h$ is $\lambda h + o(h)$. The welders work independently of each other.

If n welders use current, then $N - n$ welders do not use current and the probability of a new call for current within a time interval $\Delta t = h$ is $\lambda(N - n)h + o(h)$; on the other hand, the probability that one of the n welder ceases using current is $n\mu h + o(h)$. Therefore for the probability $u_n(t)$ that at the time t exactly n welders are working the system of Eqs. (4.5.11) takes the form

$$\frac{du_n}{dt} = \mu(n+1)u_{n+1} - [\mu n + \lambda(N-n)]u_n + \lambda(N-n+1)u_{n-1} \tag{4.5.38}$$

$$(u_{-1}(t) = 0\,, \quad u_{N+1}(t) = 0)\,.$$

Particular solutions of the system (4.5.38) are

$$u_n(t) = e^{-(\lambda+\mu)xt}\left(-\frac{\lambda+\mu}{\mu}\right)^n k_n^{(p)}(x, N)\,, \tag{4.5.39}$$

where $k_n^{(p)}(x, N)$ are the Kravchuk polynomials at $p = \lambda/(\lambda+\mu)$. Hence according to (4.5.9) the solution $u_n = u_{nm}(t)$ of system (4.5.38), which satisfies the initial condition $u_{nm}(0) = \delta_{nm}$, has the form

$$u_{nm}(t) = (-1)^{n+m}m!(N-m)! \sum_{l=0}^{N} e^{-t(\lambda+\mu)l}$$

$$\times k_n^{(p)}(l, N)k_m^{(p)}(l, N)\frac{p^{l-m}q^{N-n-l}}{l!(N-l)!}\,, \tag{4.5.40}$$

where $p = \lambda/(\lambda+\mu)$ and $q = \mu/(\lambda+\mu)$. The right-hand side in the equality may be simplified if we present the Kravchuk polynomials in terms of the hypergeometric functions (see Table 2.4) and use Eq. (4.5.29) in the form

$$\sum_{l=0}^{N} C_N^l s^l F\left(-l, -n, -N, \frac{1}{p}\right) F\left(-l, -m, -N, \frac{1}{p}\right)$$

202

$$= (1+s)^{N-m-n} \left(1 - s\frac{q}{p}\right)^{n+m} F\left(-n, -m, -N, -\frac{s}{(p-sq)^2}\right), \quad (4.5.41)$$

where $C_N^l = N!/[l!(N-l)!]$, $q = 1 - p$.

With the involvement of (2.7.11a) the resulting expression is

$$u_{nm}(t) = C_N^n p^n q^{N-n} \left[1 - e^{-(\lambda+\mu)t}\right]^{m+n} \left[1 + \frac{p}{q}e^{-(\lambda+\mu)t}\right]^{N-n-m}$$

$$\times F\left(-n, -m, -N, -\frac{e^{-(\lambda+\mu)t}}{pq[1 - e^{-(\lambda+\mu)t}]^2}\right), \quad (4.5.42)$$

where $p = \lambda/(\lambda + \mu)$, $q = \mu/(\lambda + \mu)$.

It is easy to see that $u_{nm}(t) \geq 0$ when $t \geq 0$. By taking the limit $t \to \infty$ we obtain the *binomial distribution*

$$U_n = \lim_{t\to\infty} u_{nm}(t) = C_N^n p^n q^{N-n}, \quad (4.5.43)$$

where $p = \lambda/(\lambda + \mu)$, $q = \mu/(\lambda + \mu)$.

4.5.5. Further, we now consider *the simplest trunking problem*. Suppose that infinitely many telephone trunklines or channels are available, and that the probability of a conversation ending between t and $t + h$ is $\mu h + o(h)$. It is, of course, assumed that the durations of the conversations are mutually independent.

If at the time t n lines are busy, the probability that one of them will be freed within time interval $\Delta t = h$ is then $n\mu h + o(h)$. The probability that within this time two or more conversations terminate is obviously of the order of magnitude h^2 and therefore negligible. The probability of a new call arriving is $\lambda h + o(h)$. The probability of a combination of several calls, or of a call arriving and a conversation ending is again $o(h)$. Then for the probability $u_n(t)$ of the event that at the time t there are n busy lines we obtain the system of differential equations

$$\frac{du_n}{dt} = \lambda u_{n-1} - (\lambda + \mu n)u_n + \mu(n+1)u_{n+1} \quad (4.5.44)$$

$$(u_{-1}(t) = 0) .$$

The particular solution of the above system has the form

$$u_n(t) = \frac{e^{-\mu x t}}{n!} \left(\frac{\lambda}{\mu}\right)^n c_n^{(\lambda/\mu)}(x) , \quad (4.5.45)$$

where $c_n(x)$ are the Charlier polynomials. The solution $u_n = u_{nm}(t)$ of (4.5.44), which satisfies the initial condition $u_{nm}(0) = \delta_{nm}$, has the form

$$u_{nm}(t) = \frac{(\lambda/\mu)^n e^{-\lambda/\mu}}{n!} \sum_{k=0}^{\infty} \frac{e^{-t\mu k}}{k!} \left(\frac{\lambda}{\mu}\right)^k c_n^{(\lambda/\mu)}(k) c_m^{(\lambda/\mu)}(k) . \quad (4.5.46)$$

By using the generating function for the Charlier polynomials [M9, G2] the right-hand side of (4.5.46) may be summed as

$$\sum_{k=0}^{\infty} \frac{(\mu_1\mu_2 s)^k}{k!} c_n^{(\mu_1)}(k) c_m^{(\mu_2)}(k) = e^{\mu_1\mu_2 s}(1-\mu_1 s)^m (1-\mu_2 s)^n$$

$$\times \, _2F_0\left(-m,-n,\frac{s}{(1-\mu_1 s)(1-\mu_2 s)}\right). \tag{4.5.47}$$

As a result for the simplest trunking problem by putting $\mu_1 = \mu_2 = \lambda/\mu$, $s = \mu e^{-\mu t}/\lambda$ we obtain the solution of system (4.5.44) in the form

$$u_{nm}(t) = \frac{e^{-\lambda/\mu}}{n!}\left(\frac{\lambda}{\mu}\right)^n \exp\left(\frac{\lambda}{\mu}e^{-\mu t}\right)(1-e^{-\mu t})^{m+n}$$

$$\times \, _2F_0\left(-m,-n,\frac{\mu e^{-\mu t}}{\lambda(1-e^{-\mu t})^2}\right). \tag{4.5.48}$$

It is easy to see that $u_{nm}(t) \geq 0$ when $t \geq 0$. Taking the limit $t \to \infty$ we obtain according to (4.5.48) that

$$U_n = \lim_{t\to\infty} u_{nm}(t) = \frac{e^{-\lambda/\mu}}{n!}\left(\frac{\lambda}{\mu}\right)^n. \tag{4.5.49}$$

Thus in the simplest trunking problem the limiting distribution is *the Poisson distribution* with parameter λ/μ. It does not depend on initial conditions.

In the limit $\mu \to 0$ for the Poisson process we may deduce from (4.5.44) the system of differential equations

$$\frac{du_n}{dt} = \lambda u_{n-1} - \lambda u_n, \quad (u_{-1}(t) = 0). \tag{4.5.50}$$

The solutions of this system may be obtained from (4.5.48) by taking the limit $\mu \to 0$. They are

$$u_{nm}(t) = \delta_{m0}\frac{(\lambda t)^n}{n!}e^{-\lambda t}. \tag{4.5.51}$$

4.5.6. In conclusion let us discuss the application of Charlier polynomials to the *servicing of machines*. The problem is as follows [F1a]. We consider automatic machines which normally require no human care except that they may break down and need servicing. The machine is characterized by two constants λ and μ with the following properties. When the machine is being serviced at time t the probability that the servicing time terminates before $t+h$ and the machine reverts to the working state equals $\lambda h + o(h)$. If at time t the machine is in working order, the probability that it will need servicing before $t + h$ equals $\mu h + o(h)$. For an efficient machine μ should be relatively small and λ relatively large.

We suppose that N machines with the same parameters λ and μ and working independently are serviced by a single repairman. A machine which breaks down is serviced immediately unless the repairman is servicing another machine, in which case a waiting line is formed. We say that the system is in state E_n if n machines are working. For $0 \leq n \leq N - 1$ this means that one machine is being serviced and $N - n - 1$ are in the queue; in the state E_N all machines work and the repairman is idle.

A transition $E_n \to E_{n+1}$ occurs if the machine being serviced reverts to the working state, whereas a transition $E_n \to E_{n-1}$ is caused by a breakdown of one among n working machines.

If $u_n(t)$ is the probability that at time t there are n working machines, we may obtain for functions $u_n(t)$ with $0 \leq n \leq N$ the following system of equations:

$$\frac{du_n}{dt} = \lambda u_{n-1} - (\lambda + \mu n)u_n + \mu(n+1)u_{n+1} \tag{4.5.52}$$

$$(0 \leq n \leq N - 1 , \quad u_{-1}(t) = 0) ,$$

$$\frac{du_N}{dt} = \lambda u_{N-1} - \mu N u_N . \tag{4.5.53}$$

Equations (4.5.52) coincide with (4.5.44) for $0 \leq n \leq N - 1$. Therefore a particular solution of this system for $0 \leq n \leq N - 1$ has the form of (4.5.45), i.e.

$$u_n(t) = \frac{e^{-\mu x t}}{n!} \left(\frac{\lambda}{\mu}\right)^n c_n^{(\lambda/\mu)}(x) . \tag{4.5.54}$$

In order that Eq. (4.5.53) be satisfied for $u_n(t)$ in such a form it is sufficient to require that the values of $x = x_i$ $(i = 0, 1, \ldots, N)$ are the roots of the equation

$$c_N^{(\lambda/\mu)}(x_i) = c_{N+1}^{(\lambda/\mu)}(x_i) . \tag{4.5.55}$$

Specifically, owing to the equality $c_n^{(\lambda/\mu)}(0) = 1$ we have $x_0 = 0$.

By using the Darboux-Christoffel formula (1.4.18) we may easily show that if the condition (4.5.55) is satisfied the orthogonality relation

$$\sum_{i=0}^{N} c_n(x_i)c_m(x_i)f_i = d_n^2 \delta_{nm}$$

is valid for the Charlier polynomials $c_n(x) = c_n^{(\lambda/\mu)}(x)$. Here

$$f_i^{-1} = \frac{\alpha_N}{d_N^2} c_N(x_i)[c_{N+1}'(x_i) - c_N'(x_i)]$$

(cf. with Sect. 3.9). Therefore the solution $u_n(t) = u_{nm}(t)$ of the system (4.5.52–53) that satisfies the initial conditions $u_{nm}(0) = \delta_{mn}$ has the form

$$u_{nm}(t) = \frac{(\lambda/\mu)^n}{n!} \sum_{i=0}^{N} f_i e^{-\mu t x_i} c_n^{(\lambda/\mu)}(x_i)c_m^{(\lambda/\mu)}(x_i) . \tag{4.5.56}$$

This relation determines the probability of the event that n machines will be working at epoch t if m machines were working at epoch $t = 0$.

Since $x_0 = 0$ we take the limit $t \to \infty$ and obtain from (4.5.56) that

$$\lim_{t \to \infty} u_{nm}(t) = \frac{(\lambda/\mu)^n}{n!} f_0 = U_n .\tag{4.5.57}$$

The value of constant f_0 may be obtained most easily from the condition $\sum_n U_n = 1$, which yields

$$f_0^{-1} = \sum_{k=0}^{N} \frac{(\lambda/\mu)^k}{k!} .$$

This result is known as Erlang's loss formula [F1a].

4.6 Elementary Applications to Probability Theory

As already been noted in the previous chapters the weight functions of the considered orthogonal polynomials coincide in some cases with the well-known distributions of probability theory. In this connection we may propose a simple way of calculating the basic parameters that characterize these distributions – expectation value, variance (standard deviation), skewness and excess, moments, etc. by building on the properties of orthogonal polynomials studied above. To do this we only need knowledge of the main polynomial characteristics: leading coefficients a_n and b_n, squared norms d_n^2 or coefficients of recursion relations

$$\alpha_n = \frac{a_n}{a_{n+1}} , \qquad \beta_n = \frac{b_n}{a_n} - \frac{b_{n+1}}{a_{n+1}} , \qquad \gamma_n = \alpha_{n-1} \frac{d_n^2}{d_{n-1}^2}$$

at $n = 1, 2$ (see Eq. (1.4.17) and Tables 4.1–3).

Let x be a random variable which takes discrete values x_i with probabilities f_i (f_i is a discrete distribution function, $\sum_i f_i = 1$). We consider a system of orthogonal polynomials of a discrete variable, for which

$$(y_m, y_n) = \sum_i y_m(x_i) y_n(x_i) f_i = d_n^2 \delta_{mn} .\tag{4.6.1}$$

As usual we shall assume that $y_0 = 1$ ($a_0 = 1$, $b_0 = 0$), which corresponds to existing tables. By following the normalization of the distribution function f_i accepted in probability theory we choose $d_0^2 = 1$. We introduce the moments of the distribution function in the form

$$C_p = \sum_i x_i^p f_i .\tag{4.6.2}$$

The expectation value of a random variable x^p (i.e. $\bar{x}^p = M x^p$) is equal to C_p.

In order to calculate the expectation value of the random variable x,

$$\bar{x} = M x = C_1 = \sum_i x_i f_i ,\tag{4.6.3}$$

it is convenient to use the relation (4.6.1) at $m = 0$, $n = 1$. We have

$$(y_0, y_1) = a_1 \bar{x} + b_1 = y_1(\bar{x}) = 0 \;,$$

whence

$$\bar{x} = -\frac{b_1}{a_1} = \beta_0 \;. \tag{4.6.4}$$

For the classical orthogonal polynomials of a discrete variable we have $x_i = x(s_i)$, $f_i = \varrho(s_i)\Delta x(s_i - 1/2)$ and it follows from the Rodrigues formulas (2.2.8) and (3.2.19) that

$$y_1(x) = B_1 \tau(x) \;.$$

Therefore the expectation value $\bar{x}$ may be found also from the equation $\tau(\bar{x}) = 0$.
For calculating the standard deviation

$$\sigma^2 = Dx = \overline{(x - \bar{x})^2} = M(x - Mx)^2 = Mx^2 - (Mx)^2 = C_2 - C_1^2 \;, \tag{4.6.5}$$

where σ is a variance, we use the equality $(y_1, y_1) = d_1^2$. Since $y_1(x) = a_1(x - \bar{x})$, we have

$$a_1^2(x - \bar{x}, x - \bar{x}) = d_1^2 \;,$$

whence for σ^2 we obtain the simple expression

$$\sigma^2 = \frac{d_1^2}{a_1^2} = \alpha_0 \gamma_1 = \frac{\gamma_1^2}{d_1^2} \;. \tag{4.6.6}$$

As characteristics of the distribution function the skewness Γ_1 and the excess Γ_2 are frequently used. They are determined by

$$\Gamma_1 = \frac{\overline{(x - \bar{x})^3}}{\sigma^3} = \frac{M(x - Mx)^3}{(Dx)^{3/2}} \;, \tag{4.6.7}$$

$$\Gamma_2 = \frac{\overline{(x - \bar{x})^4}}{\sigma^4} - 3 = \frac{M(x - Mx)^4}{(Dx)^2} - 3 \;. \tag{4.6.8}$$

We denote them by Γ_1 and Γ_2 instead of the usual γ_1 and γ_2 because the coefficients of recursion relations (1.4.14) given above are designated by γ_n. The quantities Γ_1 and Γ_2 allow us to judge how much a given distribution differs from normal, for which $\Gamma_1 = \Gamma_2 = 0$.
We first determine the skewness Γ_1. According to the definition we have

$$(x - \bar{x})^3 = C_3 - 3C_2C_1 + 2C_1^3 \;.$$

For evaluating the right-hand side it is convenient to use the relation $(y_1, y_2) = 0$. By means of the recursion relation

$$xy_1 = \alpha_1 y_2 + \beta_1 y_1 + \gamma_1 y_0 \tag{4.6.9}$$

we obtain

$$\alpha_1(y_1, y_2) = (y_1, x y_1 - \beta_1 y_1 - \gamma_1 y_0) = (y_1, x y_1) - \beta_1 d_1^2$$
$$= a_1^2(x - \bar{x}, x(x - \bar{x})) - \beta_1 d_1^2$$
$$= a_1^2(C_3 - 2C_2 C_1 + C_1^3) - \beta_1 d_1^2 = 0 \, ,$$

whence

$$C_3 - 2C_2 C_1 + C_1^3 = \beta_1 \frac{d_1^2}{a_1^2} = \beta_1 \sigma^2 \, . \tag{4.6.10}$$

By subtracting the equality (4.6.5) multiplied by C_1 from (4.6.10) we find

$$(x - \bar{x})^3 = \sigma^2(\beta_1 - \bar{x}) \, .$$

Therefore

$$\Gamma_1 = \frac{\beta_1 - \bar{x}}{\sigma} = \frac{\beta_1 - \beta_0}{\sigma} \, . \tag{4.6.11}$$

Now let us evaluate the excess Γ_2 in (4.6.8). By definition

$$\overline{(x - \bar{x})^4} = C_4 - 4C_3 C_1 + 6C_2 C_1^2 - 3C_1^4 \, . \tag{4.6.12}$$

By using (4.6.9) and (4.6.1) we obtain

$$(x y_1, x y_1) = \alpha_1 d_2^2 + \beta_1 d_1^2 + \gamma_1^2 \, .$$

Since $y_1 = a_1(x - \bar{x})$, we have

$$C_4 - 2C_3 C_1 + C_2 C_1^2 = \frac{\sigma^2}{d_1^2}(\alpha_1^2 d_2^2 + \beta_1 d_1 \quad \gamma_1^2) \, . \tag{4.6.13}$$

We subtract the relation (4.6.10) multiplied by $2C_1$ from (4.6.13) and add (4.6.5) multiplied by C_1^2. As a result according to (4.6.12) we obtain

$$\overline{(x - \bar{x})^4} = \sigma^2 \left[\alpha_1^2 \frac{d_2^2}{d_1^2} + \frac{\gamma_1^2}{d_1^2} + (\beta_1 - \bar{x})^2 \right] \, ,$$

whence

$$\Gamma_2 = \Gamma_1^2 + \frac{\alpha_1 \gamma_2}{\sigma^2} - 2 \tag{4.6.14}$$

in accordance with

$$\alpha_1 \frac{d_2^2}{d_1^2} = \gamma_2 \, , \quad \frac{\gamma_1^2}{d_1^2} = \sigma^2 \, , \quad \beta_1 - \bar{x} = \sigma \Gamma_1 \, .$$

Formulas (4.6.4, 6, 11) and (4.6.14) for $\bar{x}$, σ^2, Γ_1 and Γ_2 are derived under the assumption that the random variable x takes discrete values. It is obvious that all these considerations are valid for continuous random variables too, since we used only general properties of orthogonal polynomials.

The weight functions of polynomials considered in this book are solutions of Eq. (3.2.9); they form a class of distributions (discrete and continuous), which

Table 4.1. Discrete distribution functions related to the Hahn, Meixner, Kravchuk and Charlier polynomials

Distribution	Distribution function $\varrho(x_i),\ x_i = i$	Expectation value $\bar{x}$	Standard deviation σ^2	Skewness Γ_1	Excess Γ_2	Related polynomials $y_n(x)$
Polya distribution	$C_N^i\,\dfrac{(b/s)_i(c/s)_{N-i}}{(M/N)_N}$ $b = pM,\ c = (1-p)M$ $(i = 0, 1, \ldots, N)$	pN	$\dfrac{p(1-p)N(N+M/s)}{1+M/s}$	$\dfrac{(1-2p)(2N+M/s)}{(2+M/s)\sigma}$	$-2 + [\sigma(2+M/s)]^{-2}\Big\{(1-2p)^2(2N+M/s)^2 + 2M(N-1)$ $\times\ \dfrac{[1+(1-p)M/s](1+pM/s)(1+M/s+N)}{s(1+M/s)(3+M/s)}\Big\}$	$h_n^{(\alpha,\beta)}(x, N+1),$ $\alpha = c/s - 1,$ $\beta = b/s - 1$
Hypergeometric distribution	$\dfrac{C_{Mp}^i C_{M-Mp}^{N-i}}{C_M^N}$ $(i = 0, 1, \ldots, N)$	pN	$\dfrac{p(1-p)(M-N)N}{M-1}$	$\dfrac{(1-2p)(M-2N)}{(M-2)\sigma}$	$-2 + [\sigma(M-2)]^{-2}\Big[(1-2p)^2(M-2N)^2 + 2M(M-N-1)$ $\times\ \dfrac{(N-1)(M-Mp-1)(Mp-1)}{(M-1)(M-3)}\Big]$	$\tilde{h}_n^{(\mu,\nu)}(x, N+1),$ $\mu = (1-p) - N,$ $\nu = p - N.$
Geometric distribution	$p(1-p)^i,\ 0 < p < 1$ $(i = 0, 1, 2, \ldots)$	$\dfrac{1-p}{p}$	$\dfrac{1-p}{p^2}$	$\dfrac{2-p}{\sqrt{1-p}}$	$6 + \dfrac{p^2}{1-p}$	$m_n^{(1,\,1-p)}(x)$
Pascal's distribution (negative binomial distribution)	$C_{r+i-1}^i p^r(1-p)^i$ $\Gamma > 0,\ 0 < p < 1$ $(i = 0, 1, 2, \ldots)$	$r\,\dfrac{1-p}{p}$	$r\,\dfrac{1-p}{p^2}$	$\dfrac{2-p}{\sqrt{r(1-p)}}$	$\dfrac{6}{r} + \dfrac{p^2}{r(1-p)}$	$m_n^{(r,\,1-p)}(x)$
Binomial distribution	$C_N^i p^i q^{N-i},\ q = 1-p$ $0 < p < 1$ $(i = 0, 1, \ldots, N)$	pN	pqN	$\dfrac{q-p}{\sqrt{pqN}}$	$\dfrac{1-6pq}{pqN}$	$k_n^{(p)}(x, N)$
Poisson distribution	$e^{-\mu}\mu^i/i!,\ \mu > 0$ $(i = 0, 1, 2, \ldots)$	μ	μ	$\dfrac{1}{\sqrt{\mu}}$	$\dfrac{1}{\mu}$	$c_n^{(\mu)}(x)$

Table 4.2. Some continuous distribution functions related to the Jacobi, Laguerre and Hermite polynomials

Distribution	Probability density $\varrho(x)$	Expectation value $\bar{x}$	Standard deviation σ^2	Skewness Γ_1	Excess Γ_2	Related polynomials $y_n(x)$
Uniform distribution	$\dfrac{1}{b-a}, b>a$ $(a \leqslant x \leqslant b)$	$\dfrac{a+b}{2}$	$\dfrac{(b-a)^2}{12}$	0	$-\dfrac{6}{5}$	Legendre polynomials $P_n\left(\dfrac{2x-a-b}{b-a}\right)$
Beta-distribution	$\dfrac{\Gamma(\alpha+\beta)}{\Gamma(\alpha)\Gamma(\beta)} x^{\alpha-1}(1-x)^{\beta-1},$ $\alpha>0, \beta>0$ $(0 \leqslant x \leqslant 1)$	$\dfrac{\alpha}{\alpha+\beta}$	$\dfrac{\alpha\beta}{(\alpha+\beta)^2(\alpha+\beta+1)}$	$\dfrac{2(\beta-\alpha)(\alpha+\beta+1)^{1/2}}{(\alpha+\beta+2)(\alpha\beta)^{1/2}}$	$\dfrac{6}{\alpha\beta(\alpha+\beta+2)(\alpha+\beta+3)}$ $\times [(\alpha-\beta)^2(\alpha+\beta+1)$ $- \alpha\beta(\alpha+\beta+2)]$	Jacobi polynomials $P_n^{(\beta-1,\alpha-1)}(2x-1)$
Exponential distribution	$\lambda e^{-\lambda x}, \lambda>0$ $(0 \leqslant x < \infty)$	$\dfrac{1}{\lambda}$	$\dfrac{1}{\lambda^2}$	2	6	Laguerre polynomials $L_n^0(\lambda x)$
Gamma-distribution	$\dfrac{\lambda^\alpha}{\Gamma(\alpha)} x^{\alpha-1}e^{-\lambda x}, \alpha>0, \lambda>0$ $(0 < x < \infty)$	$\dfrac{\alpha}{\lambda}$	$\dfrac{\alpha}{\lambda^2}$	$\dfrac{2}{\alpha^{1/2}}$	$\dfrac{6}{\alpha}$	Laguerre polynomials $L_n^{\alpha-1}(\lambda x)$
Normal distribution (Gauss distribution)	$\dfrac{1}{\sigma\sqrt{2\pi}} e^{-(x-m)^2/2\sigma^2}, \sigma>0$ $(-\infty < x < \infty)$	m	σ^2	0	0	Hermite polynomials $H_n\left(\dfrac{x-m}{\sigma\sqrt{2}}\right)$

Table 4.3. Some continuous distribution functions related to the polynomial solutions of the difference equation of hypergeometric type

Distribution	Probability density $\varrho(x)$	Expectation value $\bar{x}$	Standard deviation σ^2	Skewness Γ_1	Excess Γ_2	Related polynomials $y_n(x)$
Logistic distribution	$\dfrac{\pi/(4\sigma\sqrt{3})}{\cosh^2[\pi(x-m)/(2\sigma\sqrt{3})]},\ \sigma>0$ $(-\infty<x<\infty)$	m	σ^2	0	$\dfrac{6}{5}$	Hahn polynomials of an imaginary argument $p_n^{(0,0)}(t,0),\ t=\dfrac{x-m}{\sigma\sqrt{3}}$
Hyperbolic secant distribution [L2]	$\dfrac{1}{\cosh\pi x},\ -\infty<x<\infty$	0	$\dfrac{1}{4}$	0	2	Pollaczek polynomials $P_n^{(1/2)}(x,\pi/2)$
Hyperbolic cosecant distribution [L2]	$\dfrac{2x}{\sinh\pi x},\ -\infty<x<\infty$	0	$\dfrac{1}{2}$	0	1	Pollaczek polynomials $P_n^{(1)}(x,\pi/2)$
Generalized hyperbolic secant distribution [L2]	$\dfrac{2^{a-1}}{2\pi c\Gamma(a)}\left|\Gamma\left(\dfrac{a}{2}+\dfrac{ix}{2c}\right)\right|^2$ $(a>0,c>0,$ $-\infty<x<\infty)$	0	ac^2	0	$\dfrac{2}{a}$	Pollaczek polynomials $P_n^{(a/2)}(x/2c,\pi/2)$
The Meixner–Pollaczek distribution [M8, P10, L2] (continuous binomial distribution [A17]	$\dfrac{(2\sin\varphi)^{2\lambda}}{2\pi\,\Gamma(2\lambda)}\,e^{(2\varphi-\pi)x}$ $\times|\Gamma(\lambda+ix)|^2,\lambda>0,0<\varphi<\pi$ $(-\infty<x<\infty)$	$-\lambda\,\dfrac{\cos\varphi}{\sin\varphi}$	$\dfrac{\lambda}{2\sin^2\varphi}$	$-\sqrt{\dfrac{2}{\lambda}}\cos\varphi$	$\dfrac{1+2\cos^2\varphi}{\lambda}$	Pollaczek polynomials $P_n^{(\lambda)}(x,\varphi)$
Logarithmic normal (lognormal) distribution	$\dfrac{1}{xa\sqrt{2\pi}}\,e^{-(\ln x-m)^2/2a^2}$ $(a>0,x>0)$	$e^{m+a^2/2}$	$e^{a^2+2m}(e^{a^2}-1)$	$(e^{a^2}+2)$ $\times(e^{a^2}-1)^{1/2}$	$e^{4a^2}+2e^{3a^2}$ $+3e^{2a^2}-6$	Stieltjes–Wigert distribution $p_n(t,k),\ t=xe^{a^2-m}$ $k=1/(a\sqrt{2})$

contain the Pearson distributions (see, for example, [K26]) as limiting cases. Values of the constants $\bar{x}$, σ^2, Γ_1 and Γ_2 calculated on the basis of the discussed method for some well-known distributions (see, for example, [A1, K26]) are presented in Tables 4.1–3.

Remark. Considering the normalization accepted in probability theory it is necessary to make the transformations $\varrho(x) \to \varrho(x)/d_0^2$, $d_n^2 \to d_n^2/d_0^2$ and to take into account a possible linear transformation of an independent variable by using the tables of main data for the systems of polynomials.

4.7 Estimation of the Packaging Capacity of Metric Spaces

When considering some questions in geometry, coding theory and other fields of mathematics one needs accurate estimates of maximal packaging capacities of compact metric spaces filled by nonintersecting balls of fixed radius. The problem is posed in the following way. Let $\mathcal{M}$ be a metric space, i.e. the set any two elements x and y of which have definite numbers $d(x, y)$ called the distance between x and y with the following properties:

$$d(y, x) = d(x, y) \geq 0 \,,$$

where $d(x, y) = 0$ only if $x = y$ and $d(x, y) \leq d(x, z) + d(z, y)$ $(x, y, z \in \mathcal{M})$. The ball of radius $\mathcal{D}/2$ with its center at point $x \in \mathcal{M}$ is called a set of elements $y \in \mathcal{M}$, for which $d(x, y) \leq \mathcal{D}/2$. The balls are called nonintersecting if they have no common elements. It is required to estimate from above the number $\nu(\mathcal{D})$ of nonintersecting balls of radius $\mathcal{D}/2$ for the set $\mathcal{M}$ under the assumption that the set is compact.

Since the balls of radius $\mathcal{D}/2$ are nonintersecting if the distance between their centers is not less than $\mathcal{D}$, in our estimation of the number of centers of nonintersecting balls we come to the other formulation of the above problem: it is required to estimate from above the number of elements, $\nu(\mathcal{D})$ in the set $\mathcal{M}$ if the distance between elements is not less than $\mathcal{D}$.

4.7.1. We consider a solution of this problem by using the method proposed in [L6] for the simplest case when the set $\mathcal{M}$ is a sphere S of unit radius in three-dimensional space with the center in the origin. For points on sphere S we shall use spherical coordinates (ϑ, φ). By the distance between points on the sphere we shall mean the length of an arc drawn through these points, $(d(x, y) \leq \pi)$. Obviously this distance numerically coincides with a value of the angle between the sphere radii crossing these points. Let $W = W(\mathcal{D})$ be a set of points on sphere S, the distances between which are not less than $\mathcal{D}$, and the number of points of the set $W(\mathcal{D})$ be equal to $\nu(\mathcal{D})$. To estimate $\nu(\mathcal{D})$ it is convenient to use the addition theorem for spherical harmonics $Y_{lm}(\vartheta, \varphi)$ [E7, N18]. Let the points $x \in W(\mathcal{D})$ and $y \in W(\mathcal{D})$ have spherical coordinates (ϑ, φ) and $(\tilde{\vartheta}, \tilde{\varphi})$, while the distance between them is $d(x,y)$. Then by the addition theorem we have

$$\frac{4\pi}{2l+1} \sum_{m=-l}^{l} Y_{lm}(\vartheta,\varphi) Y_{lm}^{*}(\tilde{\vartheta},\tilde{\varphi}) = P_l(t) \, , \qquad (4.7.1)$$

where $P_l(t)$ is the Legendre polynomial, $t = t(d(x,y)) = \cos(d(x,y))$.

Let $f(t)$ be an arbitrary polynomial of degree n:

$$f(t) = \sum_{l=0}^{n} f_l P_l(t) \, , \quad t = t(d(x,y)) \, , \quad x \in W, y \in W \, . \qquad (4.7.2)$$

We sum both sides of (4.7.2) over all points $x \in W(\mathcal{D})$ and $y \in W(\mathcal{D})$, including $x = y$. If $x = y$, we have $d(x,y) = 0$, $t(0) = 1$. Therefore the sum of the left-hand side of (4.7.2) over $x \in W(\mathcal{D})$, $y \in W(\mathcal{D})$ is equal to

$$\nu(\mathcal{D}) f(1) + \sum_{x \neq y} f(t) \, .$$

Note that $d(x,y) \geq \mathcal{D}$ for $x \neq y$, and consequently $t = \cos(d(x,y)) \leq \cos \mathcal{D}$, i.e. $-1 \leq t \leq t(\mathcal{D})$ for $x \neq y$. While summing up the right-hand side of (4.7.2) we obtain by virtue of (4.7.1)

$$\sum_{x,y} P_l(t) = \frac{4\pi}{2l+1} \sum_{m=-l}^{l} \left[\sum_{x \in W} Y_{lm}(\vartheta,\varphi) \right] \left[\sum_{y \in W} Y_{lm}(\tilde{\vartheta},\tilde{\varphi}) \right]^{*}$$

$$= \frac{4\pi}{2l+1} \sum_{m} \left| \sum_{x \in W} Y_{lm}(\vartheta,\varphi) \right|^{2} \, , \qquad (4.7.3)$$

since

$$\sum_{x \in W} Y_{lm}(\vartheta,\varphi) = \sum_{y \in W} Y_{lm}(\tilde{\vartheta},\tilde{\varphi}) \, .$$

On the other hand, $P_0(t) = 1$, and hence

$$\sum_{x,y} P_0(t) = \nu^{2}(\mathcal{D}) \, . \qquad (4.7.4)$$

As a result we obtain the equality

$$\nu(\mathcal{D}) f(1) + \sum_{x \neq y} f(t) = \nu^{2}(\mathcal{D}) f_0 + \sum_{l=1}^{n} \frac{4\pi}{2l+1} f_l \left[\sum_{m} \left| \sum_{x} Y_{lm}(\vartheta,\varphi) \right|^{2} \right] . \qquad (4.7.5)$$

If the polynomial $f(t)$ satisfies the conditions

$$f_l \geq 0 \quad \text{for} \quad 0 \leq l \leq n \, , \qquad (4.7.6)$$

$$f(t) \leq 0 \quad \text{for} \quad -1 \leq t \leq t(\mathcal{D}) \qquad (4.7.7)$$

from (4.7.5) it follows that

$$\nu(\mathcal{D}) \le \frac{f(1)}{f_0} \; . \tag{4.7.8}$$

For constructing the polynomial $f(t)$ with the properties (4.7.6) and (4.7.7) we first prove that in the decomposition

$$P_{l_1}(t)P_{l_2}(t) = \sum_{l=0}^{l_1+l_2} C_l^{(l_1,l_2)} P_l(t) \tag{4.7.9}$$

all the coefficients $C_l^{(l_1,l_2)}$ are non-negative for any l_1 and l_2. For the proof we put in (4.7.9) $t = t(d(x,y))$, $x \in W$, $y \in W$, where x and y are points on the sphere S with spherical coordinates (ϑ,φ) and $(\tilde{\vartheta},\tilde{\varphi})$. If we use the addition theorem for $P_{l_1}(t)$, $P_{l_2}(t)$, $P_l(t)$ as well as the orthogonalitiy relations for the spherical harmonics

$$\int_S Y_{lm}(\vartheta,\varphi)Y^*_{l'm'}(\vartheta,\varphi)d\Omega = \delta_{ll'}\delta_{mm'} \quad (d\Omega = \sin\vartheta\, d\vartheta\, d\varphi) , \tag{4.7.10}$$

after multiplication of (4.7.9) by $Y^*_{l'm'}(\vartheta,\varphi)Y_{l'm'}(\tilde{\vartheta},\tilde{\varphi})$ and integration over all possible values (ϑ,φ) and $(\tilde{\vartheta},\tilde{\varphi})$ the equality

$$C_{l'}^{(l_1,l_2)} = \frac{4\pi(2l^1+1)}{(2l_1+1)(2l_2+1)}$$
$$\times \sum_{m_1=-l_1}^{l_1} \sum_{m_2=-l_2}^{l_2} \left| \int_S Y_{l_1 m_1}(\vartheta,\varphi)Y_{l_2 m_2}(\vartheta,\varphi)Y^*_{l'm'}(\vartheta,\varphi)d\Omega \right|^2 \tag{4.7.11}$$

is obtained. From this equality it follows that in (4.7.9) the coefficients $C_l^{(l_1,l_2)}$ are non-negative for any l_1, l_2.

By means of the property considered it is easy to prove that the polynomial of degree $n = 2k+1$

$$f(t) = (t-s)\left[\frac{P_{k+1}(t)P_k(s) - P_{k+1}(s)P_k(t)}{t-s} \right]^2 \tag{4.7.12}$$

satisfies the conditions (4.7.6) and (4.7.7) at $s = t(\mathcal{D})$ and $t \le s$ if the condition

$$t_k < s < t_{k+1} \tag{4.7.13}$$

is fulfilled, where t_k is the greatest zero of the polynomial $P_k(t)$. Fulfillment of the condition (4.7.7) may be directly verified. We prove that the condition (4.7.6) is also satisfied by using the Darboux-Christoffel formula:

$$\frac{P_{k+1}(t)P_k(s) - P_{k+1}(s)P_k(t)}{t-s} = d_k^2 \frac{a_{k+1}}{a_k} \sum_{l=0}^{k} \frac{P_l(t)P_l(s)}{d_l^2} , \tag{4.7.14}$$

where a_l and d_l^2 are, respectively, the leading coefficient and the squared norm of polynomials $P_l(t)$. By means of this formula the expression (4.7.12) may be transformed into the form

$$f(t) = d_k^2 \frac{a_{k+1}}{a_k} [P_{k+1}(t)P_k(s) - P_{k+1}(s)P_k(t)] \sum_{l=0}^{k} \frac{P_l(t)P_l(s)}{d_l^2} . \qquad (4.7.15)$$

From properties of zeros of the orthogonal polynomials and the equality $P_l(1) = 1$ it follows that for the values of s satisfying (4.7.13) we have

$$P_l(s) > 0 \qquad \text{for} \quad 0 \le l \le k \quad (\text{since } s > t_k) , \quad P_{k+1}(s) < 0 ,$$
$$a_k > 0 , \qquad a_{k+1} > 0 . \qquad (4.7.16)$$

Therefore, $f(t)$ is represented as the sums of products $P_l(t)P_{k+1}(t)$ and $P_l(t)P_k(t)$ with positive coefficients. Thus by virtue of the above considered property of the coefficients in the expansion (4.7.9) the condition (4.7.6) is satisfied for the polynomials $f(t)$ with the values of s that satisfy (4.7.13). By using this polynomials we may employ the inequality (4.7.8) to estimate the value of $\nu(\mathcal{D})$. The value of f_0 may be calculated by means of (4.7.9–11), (4.7.15) if we use the equalities

$$Y_{00}(\vartheta, \varphi) = \frac{1}{\sqrt{4\pi}} , \qquad \frac{a_{k+1}}{a_k} = \frac{2k+1}{k+1} ,$$
$$Y_{lm}(\vartheta, \varphi) = (-1)^m Y_{l,-m}^*(\vartheta, \varphi) .$$

As a result we obtain

$$f_0 = -\frac{P_k(s)P_{k+1}(s)}{k+1} , \qquad (4.7.17)$$

whence

$$\nu(\mathcal{D}) \le -\frac{(k+1)[P_k(s) - P_{k+1}(s)]^2}{(1-s)P_k(s)P_{k+1}(s)} . \qquad (4.7.18)$$

4.7.2. Now we consider applying a similar method for estimating the value of $\nu(\mathcal{D})$ in compact metric spaces $\mathcal{M}$ of a more general form. At first we obtain the addition theorem analogous to (4.7.1) for the spaces $\mathcal{M}$ by assuming that these spaces have a certain symmetry. We note that the addition theorem for spherical harmonics may be obtained from the symmetry property of the sphere S and spherical harmonics: after the rotation of the sphere S around its axis, going through the centre, a set of elements of S goes into the same set, while the spherical harmonics in new spherical coordinates (ϑ', φ') are connected with the spherical harmonics $Y_{lm}(\vartheta, \varphi)$ by the linear transformation

$$Y_{lm}(\vartheta', \varphi') = \sum_{m'=-l}^{l} a_{lm'} Y_{lm'}(\vartheta, \varphi) , \qquad (4.7.19)$$

where the matrix A_l with elements a_{lm} is unitary. Thus an irreducible representation of the rotation group of sphere S around the axis going through the centre occurs for the spherical harmonics. Therefore to obtain a relation similar to (4.7.1) we assume additionally that for the set $\mathcal{M}$ there is a group of motions $\mathcal{G}$, i.e. a group of one-to-one maps $\mathcal{M}$ onto $\mathcal{M}$ that preserve the value of $d(x, y)$ for any elements $x \in \mathcal{M}$, $y \in \mathcal{M}$. Moreover, we restrict ourselves by considering only the spaces $\mathcal{M}$ which are measurable and for which it follows from the equality $d(x, y) = d(x', y')$ that there is a motion $g \in \mathcal{G}$ such that $x' = gx$, $y' = gy$ (such spaces are called strongly homogeneous).

On the set $\mathcal{M}$ we consider the functions $v(x)$ (generally, complex) that belong to the space $\mathcal{L}_2(\mathcal{M}, \mu)$, i.e. the functions for which there is a convergence of integrals $\int_{\mathcal{M}} |v(x)|^2 d\mu(x)$ in measure μ of space $\mathcal{M}$ (the measure μ is invariant with respect to motions).

Let $\mathcal{L}(g)$ be a representation of the motion group, given by the formula $\mathcal{L}(g)v(x) = v(g^{-1}x)$. The representation may be expanded into a sum (finite or countable) of mutually nonequivalent irreducible representations $\mathcal{L}_n(g)$ [G13] which act on finite-dimensional spaces V_n ($n = 0, 1, \ldots$) of continuous functions. Let $v_{nm}(x)$ ($m = 1, 2, \ldots, r_n$) be an arbitrary orthonormalized basis of space V_n of dimension $r_n (r_0 = 1)$:

$$\int_{\mathcal{M}} v_{n_1, m_1}(x) v^*_{n_2 m_2}(x) d\mu(x) = \delta_{n_1 n_2} \delta_{m_1 m_2} . \tag{4.7.20}$$

The functions $v_{nm}(x)$ are analogues of the spherical harmonics $Y_{lm}(\vartheta, \varphi)$ on sphere S, because for them the relation analogous to (4.7.19)

$$\mathcal{L}(g)v_{nm}(x) = \sum_{m'=1}^{r_n} a^{(n)}_{mm'}(g) v_{nm'}(x) , \tag{4.7.21}$$

is valid. Here the matrix A_n with elements $a^{(n)}_{mm'}(g)$ is unitary. We consider the function analogous to the left-hand side of equality (4.7.1)

$$\frac{1}{r_n} \sum_{m=1}^{r_n} v_{nm}(x) v^*_{nm}(y) , \tag{4.7.22}$$

which is called a zonal spherical harmonic. It may be shown that this function depends only on $d(x, y)$. In fact if $d(x', y') = d(x, y)$, then $x' = gx$, $y' = gy$ (where g is an element of the motion group). By using (4.7.21) it is easy to verify that

$$\sum_{m=1}^{r_n} \mathcal{L}(g)v_{nm}(x)[\mathcal{L}(g)v_{nm}(y)]^* = \sum_{m=1}^{r_n} v_{nm}(x) v^*_{nm}(y) .$$

Therefore the function (4.7.22) may be designated by $\phi_n(d(x, y))$:

$$\phi_n(d(x, y)) = \frac{1}{r_n} \sum_{m=1}^{r_n} v_{nm}(x) v^*_{nm}(y) . \tag{4.7.23}$$

There is an analogue of relation (4.7.1) for the so called polynomial spaces $\mathcal{M}$ in which

$$\phi_n(d) = p_n(t(d)) \,, \tag{4.7.24}$$

where $p_n(t)$ is a polynomial of degree n, and $t(d)$ is a decreasing function that satisfies the conditions

$$t(0) = 1 \,, \quad t\left(\max_{x,y \in \mathcal{M}} d(x,y)\right) = -1 \,. \tag{4.7.25}$$

If we integrate (4.7.23) at $y = x$ and use (4.7.20), it may be shown that

$$p_n(1) = 1 \,. \tag{4.7.26}$$

For the function $\phi_n(d(x,y))$ by using (4.7.20), (4.7.23) we obtain the relation

$$r_{n_1} \int_{\mathcal{M}} \phi_{n_1}(d(x,z))\phi_{n_2}(d(z,y))d\mu(z) = \delta_{n_1,n_2}\phi_{n_1}(d(x,y)) \,, \tag{4.7.27}$$

which for $x = y$ by virtue of (4.7.24), (4.7.26) is equivalent to the orthogonality relation of the form

$$r_n \int_{-1}^{1} p_n(t)p_{n'}(t)dw(t) = \delta_{nn'} \tag{4.7.28}$$

for the polynomials $p_n(t)$.

Thus for the spaces satisfying the above mentioned requirements we obtained the relation analogous to (4.7.1)

$$\frac{1}{r_n} \sum_{m=0}^{r_n} v_{nm}(x)v_{nm}^*(y) = p_n(t) \,, \tag{4.7.29}$$

where $t = t(d(x,y))$ and $p_n(t)$ are polynomials that satisfy the condition $p_n(1) = 1$ and the orthogonality relations of the form (4.7.28). By applying a method similar to the one considered above it may be shown that the coefficients $C_n^{(n_1,n_2)}$ in the expansion

$$p_{n_1}(t)p_{n_2}(t) = \sum_{n=0}^{n_1+n_2} C_n^{(n_1,n_2)}p_n(t)$$

are non-negative, and for any polynomial

$$f(t) = \sum_{n=0}^{k} f_n p_n(t) \tag{4.7.30}$$

the equality similar to (4.7.5)

$$\nu(\mathcal{D})f(1) + \sum_{\substack{x,y \in W(\mathcal{D}) \\ (x \neq y)}} f(t)$$

$$= \nu^2(\mathcal{D})f_0 + \sum_{n=1}^{k} \frac{f_n}{r_n} \sum_{m=1}^{r_n} \left| \sum_{x \in W(\mathcal{D})} v_{nm}(x) \right|^2 \,. \tag{4.7.31}$$

is valid.

If the polynomial $f(t)$ satisfies the conditions

$$f(t) \leq 0 \qquad \text{for} \quad -1 \leq t \leq t(\mathcal{D}) , \tag{4.7.32}$$

$$f_n \geq 0 , \tag{4.7.33}$$

it follows from (4.7.31) that

$$\nu(\mathcal{D}) \leq \frac{f(1)}{f_0} . \tag{4.7.34}$$

By using the above considered method it may be shown that the polynomial

$$f(t) = (t - s) \left[\frac{p_{k+1}(t)p_k(s) - p_{k+1}(s)p_k(t)}{t - s} \right]^2 \tag{4.7.35}$$

with $s = t(\mathcal{D})$ satisfies the conditions (4.7.33) if the inequality

$$t_k < s < t_{k+1} , \tag{4.7.36}$$

where t_k is the largest zero of polynomial $p_k(t)$, is valid.

The estimate (4.7.34) for $\nu(\mathcal{D})$ may be improved if we use the polynomials $f(t)$ of a more complex form than (4.7.35). We note that in the case when the set $\mathcal{M}$ has a finite number of elements a finite sum instead of integration appears in the orthogonality relations (4.7.20) and (4.7.28). For many spaces $\mathcal{M}$ of interest the polynomials $p_n(t)$ are reduced to the classical orthogonal polynomials of a discrete variable. Specifically it occurs for the Hamming space B^N, whose elements are the vectors $x = (x_1, x_2, \ldots, x_N)$ with $x_i = \pm 1$. In this space the distance $d(x, y)$ may be introduced in the following way: $d(x, y)$ is equal to the number of noncoinciding components x_i and y_i belonging to the vectors $x = (x_1, x_2, \ldots, x_N)$ and $y = (y_1, y_2, \ldots, y_N)$:

$$d(x, y) = N - \sum_{i=1}^{N} \delta_{x_i, y_i} .$$

The space B^N contains 2^N elements, and the motion group in this space is the permutation group

$$(x_1, x_2, \ldots, x_N) \to (x_{i_1}, x_{i_2}, \ldots, x_{i_N}) ,$$

where $(i_1, i_2, \ldots, i_N)$ is the permutation of indices $1, 2, \ldots, N$. Moreover, the motion group contains also the inversion transformation in which the components $x_{i_1}, x_{i_2}, \ldots, x_{i_p}$ are replaced by $-x_{i_1}, -x_{i_2}, \ldots, -x_{i_p}$, respectively.

It is easy to verify that under such transformations the distance $d(x, y)$ does not change. Let us show now that for $d(x, y) = d(x', y')$ there is such a transformation g that $x' = gx$, $y' = gy$. In fact let $d(x, y) = N - p$, and for elements $x = (x_1, x_2, \ldots, x_N)$ and $y = (y_1, y_2, \ldots, y_N)$ the components with numbers $i_1, i_2, \ldots, i_p$ coincide, while for vectors x' and y' the components with numbers $i'_1, i'_2, \ldots, i'_p$ coincide. Then $y_i = -x_i$ for any i that are not equal to $i_1, i_2, \ldots, i_p$, and accordingly $y_i' = -x_i'$ for any i that are not equal to $i_1', i_2', \ldots, i_p'$. If g is a transformation connected with both the permutation of the indices and inversion, which transfers x into x', while the indices $i_1, i_2, \ldots, i_p$ transform into $i_1', i_2', \ldots, i_p'$. Then $x' = gx$, $y' = gy$, $d(x,y) = d(x',y')$, which was to be proved.

We shall find a zonal spherical harmonic for the space B^N. Since the arbitrary function $v(x)$ has only 2^N values when $x \in B^N$, the function space in B^N is a vector space of dimension 2^N. The scalar product (v,w) for functions $v(x)$ and $w(x)$ may be determined in the usual way:

$$(v,w) = \frac{1}{2^N} \sum_{x \in B^N} v(x) w^*(x) . \tag{4.7.37}$$

In this case the functions $v_{nm}(x)$ may be taken in the form of products

$$v_{nm}(x) = x_{i_1} \cdot x_{i_2} \cdot \ldots \cdot x_{i_N} , \tag{4.7.38}$$

where m is the number of permutation, $(1,2,\ldots,n) \rightarrow (i_1, i_2, \ldots, i_N)$, $m = 1, 2, \ldots, C_N^n$ with $C_N^n = N(N-1)\ldots(N-n+1)/n!$. For the considered motion g in the space B^N we have

$$\mathcal{L}(g) v_{nm}(x) = v_{nm'}(x) \quad (m' = m'(g)) ,$$

i.e. an irreducible representation of the motion group occurs on the functions $v_{nm}(x)$. We show that the functions $v_{nm}(x)$ are orthogonal. In fact, since

$$\sum_{x \in B^N} x_i = 0 , \qquad \sum_{x \in B^N} x_i^2 = 2^N , \tag{4.7.39}$$

we have

$$\sum_{x \in B^N} v_{nm}(x) v_{n'm'}(x) = 0$$

if $n \neq n'$, or $n = n'$ but $m \neq m'$. Therefore

$$(v_{nm}(x), v_{n'm'}(x)) = \frac{1}{2^N} \sum_{x \in B^N} v_{nm}(x) v_{n'm'}(x) = \delta_{nn'} \delta_{mm'} . \tag{4.7.40}$$

In this case a zonal spherical harmonic may be derived by using (4.7.23), (4.7.38):

$$\phi_n(d(x,y)) = \frac{1}{r_n} \sum_{i_1, i_2, \ldots, i_n} \left(x_{i_1} y_{i_1}\right) \left(x_{i_2} y_{i_2}\right) \ldots \left(x_{i_n} y_{i_n}\right) , \tag{4.7.41}$$

where $r_n = C_N^n$.

Summation is carried out over all permutations of indices $(1,2,\ldots,n)$. If among pairs $(x_{i_1}, y_{i_1}), \ldots, (x_{i_n}, y_{i_n})$ there are j pairs with different values of x_i and y_i, the contribution into the sum (4.7.41) from such summands is $(-1)^j$ and the number of summands is equal to $C_p^j C_{N-p}^{n-j}$. Hence when $d(x,y) = N - p$ we have

$$\phi_n(d(x,y)) = \frac{1}{C_N^n} \sum_{j=0}^{p} (-1)^j C_p^j C_{N-p}^{n-j}$$

$$= \sum_{j=0}^{N} (-1)^j \frac{\prod_{k=0}^{j-1}(p-k) \prod_{k=0}^{n-j-1}(N-p-k)}{C_N^n j!(n-j)!} \qquad (4.7.42)$$

(the right-hand side of (4.7.42) is complemented with the summands equal to zero when $p < j \leq N$). From (4.7.42) it follows that the function $\phi_n(d)$ is a polynomial of degree n in d. By using the relation analogous to (4.7.27) for the polynomials $\phi_n(d)$ with $x = y$ we may obtain the orthogonality relation

$$\sum_{d=0}^{N} \phi_{n_1}(d)\phi_{n_2}(d)\frac{C_N^d}{2^N} = C_N^{n_1}\delta_{n_1,n_2} \ . \qquad (4.7.43)$$

From this relation it follows that the function $\phi_n(d)$ is connected with the Kravchuk polynomial $k_n^{(1/2)}(d)$ by the relation

$$\phi_n(d) = \frac{k_n^{(1/2)}(d)}{k_n^{(1/2)}(0)} \quad (\phi_n(0) = 1) \ . \qquad (4.7.44)$$

In this case the polynomials $p_n(t)$, which may be used for estimating the values of $\nu(\mathcal{D})$, are obviously connected with the polynomial $\phi_n(d)$ as

$$\phi_n(d) = p_n(t) \ , \qquad (4.7.45)$$

where $t = 1 - 2d/N$.

5. Classical Orthogonal Polynomials of a Discrete Variable and the Representations of the Rotation Group

The representations of the three-dimensional rotation group are closely related to many mathematical disciplines and enjoy numerous physical applications. Suffice it to mention the quantum theory of angular momentum that plays an essential role in the mathematics of modern physics. This theory formulates a fundamental concept of the invariance of a quantum system under rotations, thus reflecting the isotropy of a real physical space. It deduces some practical corollaries of this invariance and uses all the information due to the symmetry of the system to advantage.

The quantum theory of angular momentum and the associated aspects of the rotation group representations have been fairly well developed. The basic quantities of the theory, such as the generalized spherical functions, Clebsch-Gordan coefficients, and Wigner's $6j$-symbols (or Racah's coefficients proportional to these Wigner symbols), have been the key subjects of numerous review papers and books (see, e.g. [B7, B14, B24–B27, C22, E4, G13, H7, L4, L18, L22, M10, N1, S6, S14, S18, S35, V4, V5, V9, W4, W7, Y1, Y2]).

By the early sixties the theory had its principal statements united in a single formalism. The analytical relations were derived, extensive numerical tables compiled for the basic quantities, and close connections established with other mathematical disciplines. These achievements have facilitated practical computations in atomic and nuclear spectroscopy, quantum radiation, nuclear decay investigations, and, last but not least, chemical and nuclear reactions. The theory is intensively used in solving the problems of quantum chemistry, plasma physics, quantum optics, and astrophysics.

Among recent additions to the theory one may find close connections between representations of the rotation group SO(3) and the classical orthogonal polynomials of a discrete variable. It has been found that the generalized spherical functions – the matrix elements of irreducible representations of SO(3) – are expressible in terms of Kravchuk polynomials [K21]. The Clebsch-Gordan coefficients arising in the decomposition of a tensor product of two irreducible representations of SO(3) into irreducible components are closely related with the Hahn polynomials [G13, R26, K5, K20, S16]. In turn, the $6j$-symbols which occur in the decomposition of a tensor product of three irreducible representations of SO(3) can be identified with Racah's polynomials accurate to a normalizing factor [W8, S17]. As a result of these findings the theory of angular momentum has solidified its logical foundation. The Clebsch-Gordan coefficients and

$6j$-symbols have found their place in the theory of special functions as discrete analogues of Jacobi polynomials on linear and squares meshes, respectively.

The ensuing chapter treats the aforementioned findings in sufficient depth. The treatment is based on the theory of classical orthogonal polynomials of a discrete variable. Because of the close linkage of the theory of angular momentum and rotation group representations, the presentation that follows draws upon the more rigorous mathematical language of group representation theory. The necessary mathematical preliminaries are fairly minimal.

5.1 Generalized Spherical Functions and Their Relations with Jacobi and Kravchuk Polynomials

Generalized spherical functions often occur in quantum mechanical applications. They effect the transformations of a wavefunction of a quantum system characterized by a certain angular momentum in rotations of the system of coordinates; accurate to a normalizing factor these functions coincide with the wavefunctions of symmetric tops and so on. Therefore a thoroughful study into the properties of these quantities is most desirable. In this section we wish to elucidate the spherical functions from the standpoint of their relations with the Jacobi and Kravchuk polynomials.

5.1.1 The Three-Dimensional Rotation Group and Its Irreducible Representations

To begin with we recall the fundamental properties of the rotation group representations which will be useful in our subsequent considerations.

5.1.1.1 The Rotation Group. Suppose that we have a Cartesian coordinate system specified in a Euclidean space by three mutually orthogonal unit vectors e_x, e_y, and e_z originating at a point 0. Any rotations g of the coordinate system about 0 will be completely defined by giving three real-valued parameters. Indeed any rotation of the system can be specified by indicating the direction of rotation axis (two parameters) and the magnitude of the angle (one parameter).

More often than not rotations of coordinate systems are defined in terms of the *Euler angles* α, β, and γ which are introduced as follows. With reference to Fig. 5.1 a rotation g carrying the axes x, y, z to new positions x', y', z' can be effected by three successive rotations around the coordinate axes, namely (1) a rotation about the z axis through an angle α, (2) a rotation about the new direction of the y axis, i.e. about y'', through an angle β, and (3) a rotation about the new direction of the z axis, i.e. about z', through an angle γ. Thus $g = g(\alpha, \beta, \gamma)$.

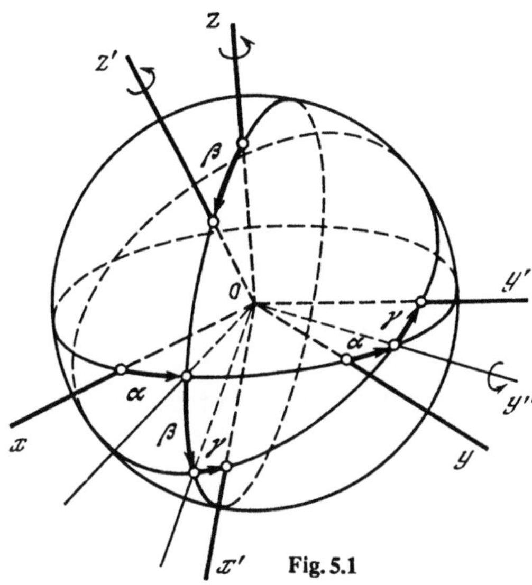

Fig. 5.1

The set of all rotations of the coordinate system about 0 forms the *three-dimensional group of rotations*[1] conventionally denoted by SO(3).

A group is a nonempty set G equipped with an operation of multiplication associating with any two elements g_1, $g_2 \in G$ a third element of G written as $g_1 g_2$ and called their product. It is required that this multiplication satisfy three conditions:

(1) For all g_1, g_2, g_3 in G, $g_1(g_2 g_3) = (g_1 g_2)g_3$.
(2) There exists a *unit element* e in G such that, for all x in G, $eg = ge = g$.
(3) For all g in G there is an *inverse element* g^{-1} in G such that $gg^{-1} = g^{-1}g = e$.

In the case of the SO(3) group, the product $g_2 g_1$ of two rotations g_1 and g_2 is understood as a rotation consisting in rotation g_1 followed by g_2; the unit element e is rotation through a zero angle; and an inverse g^{-1} to rotation g returns the coordinate system to the original position. Generally speaking, rotations g_1 and g_2 do not commute, that is, rotation $g_2 g_1$ does not always coincide with $g_1 g_2$. (To illustrate this assertion it suffices to evaluate rotations through $\frac{\pi}{2}$ about two mutually orthogonal directions.)

Any rotation is uniquely defined by Euler angles so long as $0 \leq \alpha < 2\pi$, $0 \leq \beta \leq \pi$, $0 \leq \gamma < 2\pi$. Unless Euler angles are within these limits, one should observe that for integer n_1, n_2, and n_3 a rotation $(\alpha + 2\pi n_1, \beta + 2\pi n_2, \gamma + 2\pi n_3)$ coincides with (α, β, γ), i.e.,

---

[1] In physics one usually deals with rotations of coordinate systems that leave unmoved all points of three-dimensional Euclidean space. In mathematical literature, conversely, one often is faced with rotations of the space proper, so that in place of the rotation group of the coordinate system one considers the isomorphic rotation group of the three-dimensional space (see, e.g. [G13, N1]).

$$g(\alpha + 2\pi n_1,\ \beta + 2\pi n_2,\ \gamma + 2\pi n_3) = g(\alpha,\ \beta,\ \gamma)\ .$$

We note also that a rotation $(\alpha,\ \beta,\ \gamma)$ is equivalent to $(\pi + \alpha,\ -\beta,\ -\pi + \gamma)$. The inverse rotation will be defined by the angles $(-\gamma,\ -\beta,\ -\alpha)$, which is equivalent to a rotation $(\pi - \gamma,\ \beta,\ -\pi - \alpha)$.

5.1.1.2 Representations of the Rotation Group.

Theoretical physics and especially quantum mechanics and its multiple applications make wide use of *representations of the rotation group.*

A finite dimensional representation $g \to T(g)$ of group G is said to be given if a linear operator $T(g)$ acting in some finite-dimensional linear space R is associated with every element g in G so that the product of elements of the group is identified with the product of operators

$$T(g_1)T(g_2) = T(g_1 g_2)\ , \tag{5.1.1}$$

and the unit element of the group is identified with the identity transformation

$$T(e) = 1\ . \tag{5.1.2}$$

The space R in which the representation $g \to T(g)$ acts is normally assumed to be a complex linear space[2] called the *space of the representation.* The basis of R is referred to as *the basis of the representation* with the dimensionality being equal to the number of linearly independent vectors of the basis.

In the finite-dimensional space of the representation, every operator $T(g)$ can be given by a matrix satisfying the conditions (5.1.1) and (5.1.2). If the elements of this matrix are continuous functions of elements in G, then the representation $g \to T(g)$ is said to be continuous.

Suppose that we have a representation $g \to T(g)$ of group G in a linear space R. A subspace $R' \subset R$ is called invariant under this representation if $T(g)\psi \subset R'$ for all g in G and ψ in R'. A representation $g \to R(g)$ is called *irreducible* if the space R has no other invariant subspace except zero and R itself.

The notion of *unitary* representation of a group G may be introduced also in a Euclidean space. We define the *scalar*, or *inner, product* in a complex linear space R as the one associating with any pair of vectors ψ_1 and ψ_2 in R the complex value $(\psi_1|\psi_2)$ by the following rules:

(1) $(\psi_1|\psi_2) = (\psi_2|\psi_1)^*$,
   where the asterisk stands for the complex conjugate;
(2) $(\psi_1|\alpha\psi_2 + \beta\psi_3) = \alpha(\psi_1|\psi_2) + \beta(\psi_1|\psi_3)$,
   implying that the inner product is linear in the second multiplicand; and
(3) $(\psi|\psi) \geq 0$, where $(\psi|\psi) = 0$ only for $\psi = 0$.

---

[2] For a definition of linear space, linear operator, and other necessary definitions of linear algebra, see, e.g. [G8, K27].

A representation $g \to T(g)$ acting in a Euclidean space R is called *unitary* if the operators $T(g)$ preserve the scalar product

$$(T(g)\psi_1|T(g)\psi_2) = (\psi_1|\psi_2)$$

for all g in G and ψ_1 and ψ_2 in R.

An operator $U^\dagger$ is the *conjugate* of U if, for all ψ_1 and ψ_2 in R, $(U^\dagger\psi_1|\psi_2) = (\psi_1|U\psi_2)$. For any unitary operator, $U^\dagger = U^{-1}$, where U^{-1} is an inverse operator: $UU^{-1} = U^{-1}U = 1$. Therefore for unitary representations $g \to T(g)$ we have $T^\dagger(g) = T^{-1}(g) = T(g^{-1})$. An operator A is said to be *self-adjoint* or *Hermitian* if $A^\dagger = A$.

For the case of the rotation group the examination of all continuous representations can be reduced to evaluating the finite-dimensional unitary irreducible representations [G13, N1] for which the following assertions hold.

Theorem 5.1. *Let a rotation g be given by the direction of the rotation axis $n = (n_x, n_y, n_z)$, $n^2 = 1$ and the magnitude of the angle of rotation φ, i.e.,*

$$g = g(n, \varphi) \, .$$

Then any representation of the rotation group, $g \to T(g)$, valid in some finite-dimensional linear space R may be written in the form

$$T(g) = \exp\left(-i\varphi n J\right) , \tag{5.1.3}$$

where $J = (J_x, J_y, J_z)$ is constituted by certain operators obeying the following commutation rules

$$[J_x, J_y] = iJ_z \, , \quad [J_y, J_z] = iJ_x \, , \quad [J_z, J_x] = iJ_y \, , \tag{5.1.4}$$

where $[A, B] = AB - BA$. For unitary representations, the operators J are Hermitian, i.e. $J^\dagger = J$.

Proof. Let us derive (5.1.3). We substitute the new variables $\xi = (\xi_x, \xi_y, \xi_z)$ for n and φ and rewrite $g = g(n, \varphi)$ in the form $g = g(\xi)$, $\xi = n\varphi$. Accordingly for the operators of the representation $g \to T(g)$ in R we get $T(g) = T(\xi)$. It is apparent that two rotations about a fixed axis n through the angles φt and $\varphi t'$ following one another are equivalent to one rotation about the same axis through the angle $\varphi(t + t')$,

$$g((t + t')\xi) = g(t\xi)g(t'\xi) \, .$$

Therefore in view of (5.1.1) we have

$$T((t + t')\xi) = T(t\xi)T(t'\xi) \, .$$

We differentiate both sides of this equation with respect to t' and let $t' = 0$.[3]

---

[3] A proof of $T(\xi)$ being differentiable can be found in [G13, N1].

This leads us to

$$\frac{d}{dt}T(t\boldsymbol{\xi}) = T(t\boldsymbol{\xi})A \;, \tag{5.1.5}$$

where

$$A = \frac{d}{dt'}T(t'\boldsymbol{\xi})|_{t'=0} \;.$$

By the chain rule of differentiation we find

$$A = \boldsymbol{\xi}\boldsymbol{A} = \xi_x A_x + \xi_y A_y + \xi_z A_z \;,$$

where

$$\boldsymbol{A} = \left.\frac{\partial T(\boldsymbol{\xi})}{\partial\boldsymbol{\xi}}\right|_{\boldsymbol{\xi}=0} . \tag{5.1.6}$$

Under the initial condition $T(0) = 1$ (in view of (5.1.2)) Eq. (5.1.5) has only one solution

$$T(t\boldsymbol{\xi}) = \mathrm{e}^{t\boldsymbol{\xi}\boldsymbol{A}} = 1 + t\boldsymbol{x}\boldsymbol{A} + \ldots + \frac{(t\boldsymbol{x}\boldsymbol{A})^k}{k!} + \ldots \;.$$

By letting $t = 1$, and observing that $\boldsymbol{\xi} = \boldsymbol{n}\varphi$ and $\boldsymbol{A} = -\mathrm{i}\boldsymbol{J}$ we arrive at (5.1.3).

For unitary representations, $T(g)T^{\dagger}(g) = 1$, which implies by virtue of (5.1.3) that $\boldsymbol{J}^{\dagger} = \boldsymbol{J}$.

Now we prove the commutation rules (5.1.4). Let $g = g(\boldsymbol{n}, \varphi)$ be a rotation about $\boldsymbol{n}$ through an angle φ, and $g' = g(\boldsymbol{n}', \varphi)$ a rotation through the same angle about another axis $\boldsymbol{n}'$. It is quite obvious that the rotation $g' = g(\boldsymbol{n}', \varphi)$ is equivalent to the sequence of three rotations: (1) sending $\boldsymbol{n}'$ to $\boldsymbol{n}$, (2) $g = g(\boldsymbol{n}, \varphi)$, and (3) carrying $\boldsymbol{n}$ to $\boldsymbol{n}'$.

If we denote by $\tilde{g}$ the rotation carrying $\boldsymbol{n}$ to $\boldsymbol{n}'$, then

$$g' = \tilde{g}g\tilde{g}^{-1} \;. \tag{5.1.7}$$

Correspondingly for operators $T(g)$ in R we have by definition

$$T(g') = T(\tilde{g})T(g)T(\tilde{g}^{-1}) \;. \tag{5.1.8}$$

Let $\tilde{g}$ be a rotation of the coordinate system about the y axis through an angle β carrying the axes x and z to new positions x' and z'. In view of (5.1.3) and (5.1.8) the operators $T(g) = \exp(-\mathrm{i}\varphi J_z)$ and $T(g') = \exp(-\mathrm{i}\varphi J_{z'})$, corresponding to rotations through φ about z and z', respectively, are related as

$$\exp(-\mathrm{i}\varphi J_{z'}) = \exp(-\mathrm{i}\beta J_y)\exp(-\mathrm{i}\varphi J_z)\exp(\mathrm{i}\beta J_y) \;.$$

Differentiating this relation with respect to φ and putting $\varphi = 0$ yields

$$J_{z'} = \exp(-\mathrm{i}\beta J_y)J_z\exp(\mathrm{i}\beta J_y) \;.$$

On the other hand, to rotate the system about z' we write the operator $\boldsymbol{n}\cdot\boldsymbol{J}$ entering (5.1.3) in two coordinate systems, x, y, z and x', y', z',

$$J_{z'} = \boldsymbol{n} \cdot \boldsymbol{J} = \sin\beta J_x + \cos\beta J_z \ .$$

Now

$$\sin\beta J_x + \cos\beta J_z = \exp\left(-i\beta J_y\right) J_z \exp\left(i\beta J_y\right) \ .$$

Differentiating this equation with respect to β and letting $\beta = 0$ we get $[J_y, J_z] = iJ_x$. The remaning two relations in (5.1.4) can be deduced in the same manner. $\qquad\square$

From (5.1.3) it is apparent that the operators J_x, J_y and J_z correspond to infinitesimal rotations about the axes x, y, and z, respectively. Quite appropriately they are referred to as *infinitesimal operators*.

By Theorem 5.1 the investigation of all continuous finite-dimensional unitary representations of the rotation group reduces to the evaluation of triples of Hermitian matrices J_x, J_y, J_z which obey the commutation rules (5.1.4). Then the operators of the representation $T(g)$ are constructed in accordance with (5.1.3).

Let us derive the matrices of J_x, J_y, and J_z in the basis of the eigenvectors of J_z. For this purpose we introduce the operators $J_\pm = J_x \pm iJ_y$ for which $[J_z, J_\pm] = \pm J_\pm$ and $[J_+, J_-] = 2J_z$. In our consideration we shall make use of the following lemma.

Lemma 5.1. *Let ψ be the eigenvector of the operator J_z corresponding to an eigenvalue λ, that is,*

$$J_z\psi = \lambda\psi \ .$$

Then the vectors $\psi_\pm = J_\pm\psi$ are either zero or the eigenvectors of J_z corresponding to the eigenvalues $\lambda \pm 1$ respectively.

Indeed

$$\begin{aligned}
J_z\psi_\pm &= J_z J_\pm\psi \\
&= J_\pm(J_z \pm 1)\psi \\
&= (\lambda \pm 1)J_\pm\psi \\
&= (\lambda \pm 1)\psi_\pm \ .
\end{aligned}$$

With this lemma at hand we can construct a system of eigenvectors ψ of operator J_z. The matrix of J_z is Hermitian and has in a finite-dimensional case only a finite number of real eigenvalues. Let j denote the largest eigenvalue of this matrix, $J_z\psi_j = j\psi_j$, and ψ_j denote a corresponding normalized eigenvector, $(\psi_j|\psi_j) = 1$.

If $J_-\psi_j \neq 0$, we let $J_-\psi_j = \alpha_j\psi_{j-1}$ and select the constant $\alpha_j > 0$ subject to the normalizing condition $(\psi_{j-1}|\psi_{j-1}) = 1$. By Lemma 5.1 ψ_{j-1} is the eigenvector of J_z for the eigenvalue $j - 1$. Likewise for $J_-\psi_{j-1} \neq 0$ we introduce the vector ψ_{j-2},

$$J_-\psi_{j-1} = \alpha_{j-1}\psi_{j-2} \ , \quad \alpha_{j-1} > 0 \ , \quad (\psi_{j-2}|\psi_{j-2}) = 1 \ .$$

Continuing this process we obtain for J_z a system of eigenvectors

$$\psi_j, \psi_{j-1}, \ldots, \psi_{j-k} \tag{5.1.9}$$

corresponding to the eigenvalues $j, j-1, \ldots, j-k$, respectively. Since the number of distinct eigenvalues of J_z is finite, the constructed sequence of vectors (5.1.9) has to terminate at some k, that is, we get $J_-\psi_{j-k} = 0$.

As a result we will have a finite system of orthogonal normalized eigenvectors of operator J_z,

$$J_z\psi_m = m\psi_m \;,$$

for which $J_-\psi_m = \alpha_m\psi_{m-1}$. For the last vector in the series (5.1.9) we have $\alpha_{j-k} = 0$.

Let us look now at how operator J_+ acts upon the vectors ψ_m constructed above. By Lemma 5.1, $J_+\psi_m$ is either zero or the eigenvector of J_z corresponding to the eigenvalue $m + 1$. Because j is the largest eigenvalue of J_z, $J_+\psi_j = 0$.

For $J_+\psi_{j-1}$ we have

$$J_+\psi_{j-1} = \frac{1}{\alpha_j}J_+J_-\psi_j = \frac{1}{\alpha_j}(J_-J_+ + 2J_z)\psi_j = \frac{2j}{\alpha_j}\psi_j \;,$$

that is

$$J_+\psi_{j-1} = \beta_{j-1}\psi_j \;, \quad \beta_{j-1} = 2j/\alpha_j > 0 \;.$$

We prove by induction that $J_+\psi_m$ is proportional to ψ_{m+1}, i.e.

$$J_+\psi_m = \beta_m\psi_{m+1} \;.$$

Assume that this equation holds for all vectors $\psi_j, \psi_{j-1}, \ldots, \psi_m$. Applying operator J_+ to vector ψ_{m-1} yields

$$
\begin{aligned}
J_+\psi_{m-1} &= \frac{1}{\alpha_m}J_+J_-\psi_m = \frac{1}{\alpha_m}(J_-J_+ + 2J_z)\psi_m \\
&= \frac{1}{\alpha_m}(J_-J_+\psi_m + 2m\psi_m) = \frac{1}{\alpha_m}(\beta_m J_-\psi_{m+1} + 2m\psi_m) \\
&= \frac{1}{\alpha_m}(\alpha_{m+1}\beta_m + 2m)\psi_m \;,
\end{aligned}
$$

whence $J_+\psi_{m-1} = \beta_{m-1}\psi_m$, where β_{m-1} is a constant defined by the equation

$$\alpha_m\beta_{m-1} - \alpha_{m+1}\beta_m = 2m \;. \tag{5.1.10}$$

This proves the assertion. It follows that the formula $J_+\psi_m = \beta_m\psi_{m+1}$ holds for all vectors in (5.1.9) where $\beta_j = 0$.

To complete the evaluation of matrices of $J_\pm$ in the basis (5.1.9) we are left to determine constants α_m and β_m. Since $J_+^+ = (J_x + iJ_y)^+ = J_x - iJ_y = J_-$, we have

$$\beta_m = (\psi_{m+1}|J_+\psi_m) = (\psi_m|J_-\psi_{m+1})^* = \alpha_{m+1} \;.$$

Now (5.1.10) takes the form $\alpha_m^2 - \alpha_{m+1}^2 = 2m$, whence

$$\alpha_m^2 - \alpha_{j+1}^2 = (\alpha_m^2 - \alpha_{m+1}^2) + (\alpha_{m+1}^2 - \alpha_{m+2}^2) + \ldots$$

$$\ldots + (\alpha_j^2 - \alpha_{j+1}^2) = 2m + 2(m+1) + \ldots + 2j \ .$$

Because $\alpha_{j+1} = \beta_j = 0 \, (J_+\psi_j = 0)$,

$$\alpha_m^2 = (j+m)(j-m+1) \ .$$

This formula indicates how many vectors there are in the series (5.1.9). For the last of them we have $J_-\psi_{j-k} = 0$, $\alpha_{j-k} = 0$, which does not hold unless $m = j - k = -j$. Consequently the number $j = k/2$ assumes only integer or half-integer values on the positive semiaxis. For a given j the series (5.1.9) consists of $2j + 1$ vectors ψ_m, $m = j, j - 1, \ldots, -j + 1, -j$.

The outlined argument is valid for any finite-dimensional unitary representation $T(g)$ of the group of rotations (both reducible and irreducible). For an irreducible representation $T(g)$ in a finite-dimensional space R, the constructed sequence of vectors $\psi_m \equiv \psi_{jm}, m = -j, -j+1, \ldots, j-1, j$, forms an orthonormal basis in R. Indeed the subspace R' formed by all feasible combinations of the orthonormal vectors ψ_m is invariant under the operators $J_\pm = J_x \pm iJ_y$, J_z and hence under J_x, J_y, J_z. In view of (5.1.3) this subspace will be invariant also under $T(g)$. Since the representation $T(g)$ is irreducible, the subspace R' coincides with the entire space R. Thus we have proved the following theorem.

Theorem 5.2. *Any finite-dimensional irreducible representation of the rotation group acts in some complex Euclidean space of dimension $2j + 1$, where j is a positive integer or half-integer, $j = 0, 1/2, 1$, etc. In the space of the irreducible representation, there exists an orthonormal basis ψ_{jm}, $m = -j, -j + 1, \ldots, j - 1, j$, transformed by the operators $J_\pm = J_x \pm iJ_y$, J_z as follows*

$$J_\pm\psi_{jm} = \sqrt{(j \mp m)(j \pm m + 1)}\psi_{j,m\pm 1} \ , \quad J_z\psi_{jm} = m\psi_{jm} \ . \tag{5.1.11}$$

The number j is referred to as the *weight* of the irreducible representation, and the basis ψ_{jm} as a *canonical basis*. An irreducible representation of the rotation group having a weight of j is frequently denoted by D^j. Various realizations of D^j for specific representation spaces R and operators $T(g)$ given on them can be found in [G13, N1].

5.1.1.3 Generalized Spherical Functions.

For an irreducible representation D^j, in the canonical basis ψ_{jm}, we know the matrix elements of operators J_x, J_y, J_z corresponding to infinitesimal rotations about the coordinate axes (see (5.1.11)). Now we wish to seek in the same basis the matrix elements of the operator $T(g)$ for an arbitrary rotation g.

The rotation operator $T(g)$ transforms the basis $\{\psi_{jm}\}$ into a new canonical basis $\{\psi'_{jm'}\}$

$$\psi'_{jm'} = T(g)\psi_{jm'}\ ,$$

where

$$\psi'_{jm'} = \sum_m D^j_{mm'}(g)\psi_{jm} \qquad (5.1.12)$$

and

$$D^j_{mm'}(g) = (\psi_{jm}|T(g)\psi_{jm'})\ .$$

For applications it is worthwhile to derive the matrix elements $D^j_{mm'}(g)$ for rotations g characterized by Euler's angles $g = g(\alpha,\beta,\gamma)$. Recalling the definitions of Euler angles α, β, and γ from (5.1.1) and (5.1.3), we obtain

$$T(g) = \exp(-i\gamma J_{z'})\exp(-i\beta J_{y''})\exp(-i\alpha J_z)\ ,$$

where J_z, $J_{y''}$, and $J_{z'}$ are the infinitesimal operators of rotation about the axes z, y'', and z', respectively (see Fig. 5.1).

Let us rearrange the expression for the operator $T(g)$. By virtue of (5.1.3) and (5.1.8) we have

$$\exp(-i\gamma J_{z'}) = \exp(-i\beta J_{y''})\exp(-i\gamma J_z)\exp(i\beta J_{y''})\ ,$$

therefore

$$T(g) = \exp(-i\beta J_{y''})\exp(-i\alpha J_z)\exp(-i\gamma J_z)\ .$$

Since in view of (5.1.8)

$$\exp(-i\beta J_{y''}) = \exp(-i\alpha J_z)\exp(-i\beta J_y)\exp(i\alpha J_z)\ ,$$

we finally get

$$T(g) = \exp(-i\alpha J_z)\exp(-i\beta J_y)\exp(-i\gamma J_z)\ . \qquad (5.1.13)$$

Let us derive the dependence of the matrix elements

$$D^j_{mm'}(\alpha,\beta,\gamma) = (\psi_{jm}|T(g)\psi_{jm'})$$

on the Euler angles α, β, γ. Now

$$D^j_{mm'}(\alpha,\beta,\gamma) = (\psi_{jm}|\exp(-i\alpha J_z)\exp(-i\beta J_y)\exp(-i\gamma J_z)\psi_{jm'})$$

$$= (\exp(i\alpha J_z)\psi_{jm}|\exp(-i\beta J_y)\exp(-i\gamma J_z)\psi_{jm'})\ .$$

Observing that in the basis ψ_{jm} operator J_z is diagonal, we have

$$D^j_{mm'}(\alpha,\beta,\gamma) = \exp(-im\alpha)d^j_{mm'}(\beta)\exp(-im'\gamma)\ ,$$

$$d^j_{mm'}(\beta) = (\psi_{jm}|\exp(-i\beta J_y)\psi_{jm'})\ . \qquad (5.1.14)$$

Because the matrix elements of $iJ_y = (J_+ - J_-)/2$ are real-valued in the basis ψ_{jm} (see (5.1.11)), $d^j_{mm'}(\beta)$ is a real function of angle β.

The functions $D^j_{mm'}(\alpha, \beta, \gamma)$ are known as *generalized spherical functions* or *Wigner's functions.*[4]

The representations of the group of rotations are closely connected with the momentum theory in quantum mechanics. The momentum operator $J = (J_x, J_y, J_z)$ obeys the commutation rules (5.1.4). Since the operator $J^2 = J_x^2 + J_y^2 + J_z^2$ commutes with J_z, i.e. $[J^2, J_z] = 0$, they have an eigenfunction Ψ_{jm} in common (see, e.g. [M10, L4]),

$$J^2 \Psi_{jm} = j(j+1)\Psi_{jm} , \quad J_z \Psi_{jm} = m\Psi_{jm} .$$

This eigenfunction Ψ_{jm} is the basis of the irreducible representation D^j of the group SO(3).

Let Ψ_{jm} be the wave function of a quantum system of momentum j, and m be its projection on the z axis. Let also $\Psi'_{jm'}$ be the wave function of the same system and m' be the projection on axis z'. Then

$$\Psi'_{jm'} = \sum_m D^j_{mm'}(\alpha, \beta, \gamma)\Psi_{jm} .$$

This leads us to a *quantum-mechanical interpretation* of the generalized spherical functions $D^j_{mm'}(\alpha, \beta, \gamma)$. According to the axiom of superposition the quantity $|D^j_{mm'}|^2 = [d^j_{mm'}(\beta)]^2$ is the probability of observing the state Ψ_{jm} in measuring a system which lies in the state $\Psi'_{jm'}$. This specifically suggests that $|d^j_{mm'}(\beta)| < 1$.

5.1.1.4 Some Properties of Generalized Spherical Functions

(1) The fact that the irreducible representations D^j are unitary implies that the matrix $D(\alpha, \gamma, \gamma) = D^j_{mm'}, (\alpha, \beta, \gamma)$ is unitary,

$$\sum_{m''} D^j_{mm''}(\alpha, \beta, \gamma)D^{*j}_{m'm''}(\alpha, \beta, \gamma) = \delta_{mm'} ,$$

that is, the matrix $D^\dagger(\alpha, \beta, \gamma)$, transposed and complex-conjugate to $D(\alpha, \beta, \gamma)$, coincides with its inverse $D^{-1}(\alpha, \beta, \gamma)$. Since $D^{-1}(\alpha, \beta, \gamma) = D(\pi - \gamma, \beta, -\pi - \alpha)$, then $D^j_{mm'}(\pi - \gamma, \beta, -\pi - \alpha) = (D^j_{m'm}(\alpha, \beta, \gamma))^*$, whence

$$d^j_{mm'}(\beta) = (-1)^{m-m'} d^j_{m'm}(\beta) . \tag{5.1.15}$$

Using the fact that functions $d^j_{mm'}(\beta)$ are real-valued it is not hard to deduce that

$$d^j_{mm'}(\beta) = d^j_{m'm}(-\beta) .$$

---

[4] Euler's angles and D functions allow for various definitions. In this text we adhere to the definition adopted in [V5] where page 102 lists relations between D functions used by different workers.

(2) Let us perform one after another two rotations of the coordinate system, $g_1 = g_1(\alpha_1, \beta_1, \gamma_1)$ and $g_2 = g_2(\alpha_2, \beta_2, \gamma_2)$, which in effect are equivalent to one rotation $g_2 g_1 = g = g(\alpha, \beta, \gamma)$. In rotation g_1 the operator $T(g_1)$ carries the basis ψ_{jm} of the irreducible representation D^j into a new basis $\psi_{jm''}$, which g_2 sends to $\psi'_{jm'}$

$$\psi''_{jm''} = T(g_1)\psi_{jm''} = \sum_m D^j_{mm''}(\alpha_1, \beta_1, \gamma_1)\psi_{jm} \ ,$$

$$\psi'_{jm'} = T(g_2)\psi''_{jm'} = \sum_{m''} D^j_{m''m'}(\alpha_2, \beta_2, \gamma_2)\psi''_{jm''}$$

$$= \sum_m \left[\sum_{m''} D^j_{mm''}(\alpha_1, \beta_1, \gamma_1) D^j_{m''m'}(\alpha_2, \beta_2, \gamma_2) \right] \psi_{jm} \ .$$

On the other hand, for rotation $g = g(\alpha, \beta, \gamma) = g_2 g_1$ we have

$$\psi'_{jm'} = T(g)\psi_{jm'} = \sum_m D^j_{mm'}(\alpha, \beta, \gamma)\psi_{jm} \ .$$

This leads us to the addition formula of generalized spherical functions

$$D^j_{mm'}(\alpha, \beta, \gamma) = \sum_{m''} D^j_{mm''}(\alpha_1, \beta_1, \gamma_1) D^j_{m''m'}(\alpha_2, \beta_2, \gamma_2) \ .$$

Thus when two rotations are multiplied, $g = g_2 g_1$, their matrices are multiplied in reverse order

$$D(\alpha, \beta, \gamma) = D(\alpha_1, \beta_1, \gamma_1) D(\alpha_2, \beta_2, \gamma_2) \ .$$

A similar rule holds for several successive rotations of a coordinate system.

5.1.2 Expressing the Generalized Spherical Functions in Terms of the Jacobi and Kravchuk Polynomials

In the preceding section we found the dependence of generalized spherical functions $D^j_{mm'}(\alpha, \beta, \gamma)$ on Euler's angles (5.1.14). Below we wish to demonstrate that $d^j_{mm'}(\beta)$ can be expressed in terms of the Jacobi and Kravchuk polynomials. For this purpose we first deduce differential relations for $d^j_{mm'}(\beta)$.

5.1.2.1 Differentiation Formulae For Functions $d^j_{mm'}(\beta)$. Let us rotate the coordinate system about the y axis through an angle β to a new orientation x', y, z'. Then in the space or irreducible representation D^j operator $T(g) = \exp(-\mathrm{i}\beta J_y)$ transforms the basis ψ_{jm} into a new one

$$\psi'_{jm'} = \exp(-\mathrm{i}\beta J_y)\psi_{jm'} \ , \tag{5.1.16}$$

and, because the matrix $D^j_{mm'}(\alpha, \beta, \gamma)$ is unitary and the functions $d^j_{mm'}(\beta)$ are real-valued, the Eq. (5.1.12) yields

$$\psi_{jm} = \sum_{m'} d^j_{mm'}(\beta)\psi'_{jm'} \, . \tag{5.1.17}$$

Let us put down the infinitesimal operator $\boldsymbol{n} \cdot \boldsymbol{J}$, responsible for rotation about z', in "old" and "new" coordinates

$$J_{z'} = J_x \sin\beta + J_z \cos\beta \, ,$$

i.e.,

$$J_x = J_{z'} \frac{1}{\sin\beta} - J_z \cot\beta \, . \tag{5.1.18}$$

From (5.1.17) and (5.1.18) we get

$$J_x \psi_{jm} = \sum_{m'} \frac{m' - m\cos\beta}{\sin\beta} d^j_{mm'}(\beta)\psi'_{jm'} \, .$$

Now we put (5.1.17) with the help of (5.1.16) in the form

$$\psi_{jm} = \exp(\mathrm{i}\beta J_y)\psi'_{jm} = \sum_{m} d^j_{mm'}(\beta)\psi'_{jm'} \, ,$$

differentiate this result with respect to β and by observing that $d\psi'_{jm}/d\beta = -\mathrm{i}J_y\psi'_{jm}$ in view of (5.1.16) we arrive at

$$\mathrm{i}J_y\psi_{jm} = \sum_{m'} \left[\frac{d}{d\beta} d^j_{mm'}(\beta) \right] \psi'_{jm'} \, .$$

Operators $J_{\pm}$ applied to the basis ψ_{jm} yield

$$\begin{aligned}
J_{\pm}\psi_{jm} &= (J_x \pm \mathrm{i}J_y)\psi_{jm} \\
&= \sum_{m'} \left[\pm\frac{d}{d\beta} d^j_{mm'}(\beta) + \frac{m' - m\cos\beta}{\sin\beta} d^j_{mm'}(\beta) \right] \psi'_{jm'} \, .
\end{aligned}$$

On the other hand,

$$\begin{aligned}
J_{\pm}\psi_{jm} &= \sqrt{(j \mp m)(j \pm m + 1)}\psi_{j,m\pm1} \\
&= \sqrt{(j \mp m)(j \pm m + 1)} \sum_{m'} d^j_{m\pm1,m'}(\beta)\psi'_{jm'} \, .
\end{aligned}$$

Thus we arrive at the following differentiation formulae for functions $d^j_{mm'}(\beta)$

$$\begin{aligned}
\pm\frac{d}{d\beta}d^j_{mm'}(\beta) &+ \frac{m' - m\cos\beta}{\sin\beta}d^j_{mm'}(\beta) \\
&= \sqrt{(j \mp m)(j \pm m + 1)}d^j_{m\pm1,m'}(\beta) \, . \tag{5.1.19}
\end{aligned}$$

5.1.2.2 An Explicit Expression for $d^j_{mm'}(\beta)$. To evaluate $d^j_{mm'}(\beta)$ we multiply both sides of (5.1.19) by a function $f^{\pm}(\beta)$ that satisfies the equality

$$\pm\frac{d}{d\beta}[f^{\pm}(\beta)d^j_{mm'}(\beta)] = \sqrt{(j \mp m)(j \pm m + 1)}f^{\pm}(\beta)d^j_{m\pm1,m'}(\beta) . \tag{5.1.20}$$

This will hold if

$$\pm\frac{d}{d\beta}f^{\pm}(\beta) = \frac{m' - m\cos\beta}{\sin\beta}f^{\pm}(\beta) ,$$

whence

$$f^{\pm}(\beta) = (1 - \cos\beta)^{\mp(m-m')/2}(1 + \cos\beta)^{\mp(m+m')/2} .$$

By substitution

$$f^{\pm}(\beta)d^j_{mm'}(\beta) = \left[\frac{(j - m)!}{(j + m)!}\right]^{\pm1/2} v^{\pm}_{jm}(s) , \quad s = \cos\beta , \tag{5.1.21}$$

into (5.1.20) we get

$$\mp\frac{d}{ds}v^{\pm}_{jm}(s) = v^{\pm}_{j,m\pm1}(s) , \tag{5.1.22}$$

where $v^{\pm}_{j,\pm(j+1)}(s) = 0$.

For $m = j$, (5.1.22), for the upper sign, takes the form

$$\frac{d}{d\cos\beta}\left[(1 - \cos\beta)^{-(j-m')/2}(1 + \cos\beta)^{-(j+m')/2}d^j_{jm'}(\beta)\right] = 0 ,$$

that is,

$$d^j_{jm'}(\beta) = C_{jm'}(1 - \cos\beta)^{(j-m')/2}(1 + \cos\beta)^{(j+m')/2} ,$$

where $C_{jm'}$ is a constant dependent on j and m'.

Consecutive differentiation of $v^{-}_{jj}(s)$ with respect to s yields by virtue of (5.1.22)

$$v^{-}_{jm} = \frac{d^{j-m}}{ds^{j-m}}v^{-}_{jj} .$$

In view of (5.1.21) this leads us to the explicit expression for $d^j_{mm'}(\beta)$, namely

$$d^j_{mm'}(\beta) = C_{jm'}\sqrt{\frac{(j + m)!}{(j - m)!(2j)!}}(1 - s)^{-(m-m')/2}(1 + s)^{-(m+m')/2}$$

$$\times \frac{d^{j-m}}{ds^{j-m}}\left[(1 - s)^{j-m'}(1 + s)^{j+m'}\right] , \quad s = \cos\beta . \tag{5.1.23}$$

The constant $C_{jm'}$ can be determined subject to the condition $d^j_{m'm'}(0) = 1$ as

$$C_{jm'} = (-1)^{j-m'} 2^{-j} \sqrt{\frac{(2j)!}{(j+m')!(j-m')!}} \; . \tag{5.1.24}$$

The two last expressions yield $d^j_{mm'}(\pi) = (-1)^{j-m'} \delta_{m,-m'}$. By the addition formula of generalized spherical functions we have

$$d^j_{mm'}(\pi + \beta) = \sum_{m''} d^j_{mm''}(\beta) d^j_{m''m'}(\pi) = (-1)^{j-m'} d^j_{m,-m'}(\beta) \; .$$

Likewise we find

$$d^j_{mm'}(\beta + \pi) = (-1)^{j+m} d^j_{-m,m'}(\beta) \; .$$

Since $d^j_{mm'}(\pi + \beta) = d^j_{mm'}(\beta + \pi)$,

$$d^j_{mm'}(\beta) = (-1)^{m-m'} d^j_{-m,-m'}(\beta) \; . \tag{5.1.25}$$

5.1.2.3 Relating $d^j_{mm'}(\beta)$ with Jacobi and Kravchuk Polynomials. Let us express $d^j_{mm'}$ in terms of Jacobi polynomials $P_n^{(\alpha,\beta)}(s)$. Comparing (5.1.23) with the Rodrigues formula for these polynomials, we obtain

$$d^j_{mm'}(\beta) = \frac{(-1)^{m-m'}}{2^m} \sqrt{\frac{(j+m)!(j-m)!}{(j+m')!(j-m')!}} (1-s)^{(m-m')/2}(1+s)^{(m+m')/2}$$

$$\times P_{j-m}^{(m-m',m+m')}(s) \; , \quad s = \cos\beta \; . \tag{5.1.26}$$

The Jacobi polynomials $P_n^{(\alpha,\beta)}(s)$ are related to Kravchuk polynomials $k_n^{(p)}(x, N)$ by (2.7.14). Therefore the functions $d^j_{mm'}(\beta)$ are expressable in terms of Kravchuk polynomials as

$$(-1)^{m-m'} d^j_{mm'}(\beta) = \frac{\sqrt{\varrho(x)}}{d_n} k_n^{(p)}(x, N) \; . \tag{5.1.27}$$

Here, $\varrho(x)$ and d_n are the weight and norm of the polynomials $k_n^{(p)}(x, N)$ (see Table 2.3), $n = j - m$, $x = j - m'$, $N = 2j$, and $p = \sin^2(\beta/2)$.

The Eq. (5.1.26) is convenient to use under the following parametrical constraints: $m + m' \geq 0$ and $m - m' \geq 0$. These inequalities may always be satisfied with the aid of the symmetry relations (5.1.25) and (5.1.15) that take place for the functions $d^j_{mm'}(\beta)$. Indeed by (5.1.25) one can first ensure that $m + m' \geq 0$. The Eq. (5.1.15) preserves this inequality and changes the sign in the second inequality. As a result an arbitrary function $d^j_{mm'}$ can always be driven to the domain $m + m' \geq 0$, $m - m' \geq 0$, where (5.1.26) holds true.

According to the noted symmetry properties the functions $d^j_{mm'}(\beta)$ depend in fact on $|m + m'|$ and $|m - m'|$. For the general case (5.1.26) should be rewritten in the form

$$\sqrt{\frac{2j+1}{2}} d^j_{mm'}(\beta) = (-1)^{\varphi} \frac{\sqrt{\varrho(s)}}{d_n} P_n^{(\mu,\nu)}(s) \; , \tag{5.1.26a}$$

where $\mu = |m - m'|$, $\nu = |m + m'|$, $n = j - (\mu + \nu)/2$, $s = \cos \beta$, $\varphi = (m - m' + |m - m'|)/2$, $\varrho(s)$ is the weight and d_n is the norm of Jacobi polynomials $P_n^{(\mu,\nu)}(s)$.

The Eq. (5.1.27) may be also rewritten as

$$d^j_{mm'}(\beta) = (-1)^\varphi \frac{\sqrt{\varrho(x)}}{d_n} k_n^{(p)}(x, N) , \qquad (5.1.27a)$$

where $n = j - (|m+m'| + |m - m'|)/2$, $x = j - (|m+m'| - |m - m'|)/2$, $N = 2j$, $p = \sin^2(\beta/2)$, $\varphi = (m - m' + |m - m'|)/2$, and $\varrho(x)$ and d_n are the weight and norm of Kravchuk polynomials. With the aid of the relations

$$k_n^{(p)}(m, N) = (-p)^{n-m} \frac{m!(N - m)!}{n!(N - n)!} k_m^{(p)}(n, N)$$

$$= q^{m+n-N} \frac{m!(N - m)!}{n!(N - n)!} k_{N-m}^{(p)}(N - n, N)$$

$$(m, n = 0, 1, \ldots, N)$$

all four expressions in (5.1.27a) can be readily transformed into one another. Any of the Eqs. (5.1.27a) may be used to derive the equivalent representations of $d^j_{mm'}$ through Kravchuk polynomials, each of which will be valid in the entire domain of variation of m and m'.

5.1.3 Major Properties of Generalized Spherical Functions

We deduce the basic properties of generalized spherical functions by invoking the derived properties of Jacobi and Kravchuk polynomials and with the aid of (5.1.14, 26) and (5.1.27).

5.1.3.1. By the symmetry relations (5.1.15) and (5.1.25) we can always ensure $m - m' \geq 0$ and $m + m' \geq 0$ in (5.1.26). Observing the orthogonality of Jacobi polynomials we find the orthogonality of $d^j_{mm'}(\beta)$, viz.,

$$\int_0^\pi d^j_{mm'}(\beta) d^{j'}_{mm'}(\beta) \sin \beta d\beta = \frac{2}{2j + 1} \delta_{jj'} . \qquad (5.1.28)$$

Likewise from (5.1.27) and the orthogonality of Kravchuk polynomials we arrive at the *unitary relation* for $d^j_{mm'}(\beta)$, viz.,

$$\sum_{m''=-j}^{j} d^j_{mm''}(\beta) d^j_{m'm''}(\beta) = \delta_{mm'} . \qquad (5.1.29)$$

The symmetry (5.1.15) corresponds to the self-duality of Kravchuk polynomials.

For generalized spherical functions $D^j_{mm'}(\alpha, \beta, \gamma)$ the analogues of (5.1.28) and (5.1.29) have the form

$$\int_0^{2\pi} d\alpha \int_0^\pi d\beta \sin \beta \int_0^{2\pi} d\gamma D^{*j}_{m_1 m_2} D^{j'}_{m'_1 m'_2} = \frac{8\pi^2}{2j + 1} \delta_{jj'} \delta_{m_1 m'_1} \delta_{m_2 m'_2} ,$$

$$\sum_{m''} D^{*j}_{mm''} D^{j}_{m'm''} = \delta_{mm'} \, ,$$

where both j and j' are simultaneously integers or half-integers. If one of them is an integer while the other a half-integer, then the right-hand side of the first equation should be multiplied by a factor of two thus doubling the volume of integration[5], which is achieved by doubling the upper limit of integration with respect to angle α or γ.

We also give the *condition of completeness*

$$\sum_{j=0,1/2,1,\dots} \sum_{m,m'=-j}^{j} \frac{2j+1}{16\pi^2} D^{*j}_{mm'}(\alpha,\beta,\gamma) D^{j}_{mm'}(\alpha',\beta',\gamma')$$
$$= \delta(\alpha - \alpha')\delta(\cos\beta - \cos\beta')\delta(\gamma - \gamma') \, ,$$

which follows from the completeness of the trigonometric system and the set of Jacobi polynomials.

5.1.3.2. From the differential equation for Jacobi polynomials we derive the differential equation for $d^{j}_{mm'}(\beta)$, viz.,

$$\frac{1}{\sin\beta}\frac{d}{d\beta}\left(\sin\beta\frac{d}{d\beta}d^{j}_{mm'}\right)$$
$$- \frac{m^2 - 2mm'\cos\beta + m'^2}{\sin^2\beta}d^{j}_{mm'} + j(j+1)d^{j}_{mm'} = 0 \, . \tag{5.1.30}$$

Because from (5.1.14) we have

$$\left(\frac{\partial}{\partial\alpha} + im\right) D^{j}_{mm'} = 0 \, , \qquad \left(\frac{\partial}{\partial\gamma} + im'\right) D^{j}_{mm'} = 0 \, ,$$

then using (5.1.30) we can derive the partial differential equation for $D^{j}_{mm'}(\alpha,\beta,\gamma)$ as follows:

$$\left[\frac{\partial^2}{\partial\beta^2} + \cot\beta\frac{\partial}{\partial\beta} + \frac{1}{\sin^2\beta}\left(\frac{\partial^2}{\partial\alpha^2} - 2\cos\beta\frac{\partial^2}{\partial\alpha\partial\gamma} + \frac{\partial^2}{\partial\gamma^2}\right)\right] D^{j}_{mm'}$$
$$+ j(j+1)D^{j}_{mm'} = 0 \, . \tag{5.1.31}$$

The boundary conditions have the form

$$D^{j}_{mm'}(\alpha \pm 2\pi k, \beta, \gamma) = D^{j}_{mm'}(\alpha, \beta \pm 2\pi k, \gamma)$$
$$= D^{j}_{mm'}(\alpha, \beta, \gamma \pm 2\pi k)$$
$$= D^{j}_{mm'}(\alpha, \beta, \gamma) \, ,$$

where $k = 0, 1, 2, \dots$ for integer j's and $k = 0, 2, 4, \dots$ for half-integer j's.

---

[5] In doing so we go over from the rotation group SO(3) to its two-fold covering counterpart SU(2) [G13]. We note also that $dg = (1/8\pi^2)\sin\beta\,d\alpha\,d\beta\,d\gamma$ is an invariant measure on the group SO(3).

In Sect. 6.2 we shall demonstrate that (5.1.31) coincides with Laplace's equation on a sphere in a four-dimensional Euclidean space subject to a suitable choice of the set of coordinates.

The difference equation for Kravchuk polynomials leads to the recursion relation in m' for $d^j_{mm'}(\beta)$, namely

$$2\frac{m'\cos\beta - m}{\sin\beta}d^j_{mm'} = \sqrt{(j-m')(j+m'+1)}d^j_{m,m'+1}$$
$$+ \sqrt{(j+m')(j-m'+1)}d^j_{m,m'-1}\,. \tag{5.1.32}$$

5.1.3.3. The recursion relation for Jacobi polynomials is equivalent to the *recursion relation in j* for $D^j_{mm'}(\alpha,\beta,\gamma)$:

$$\cos\beta D^j_{mm'} = \frac{\sqrt{(j^2-m^2)(j^2-m'^2)}}{j(2j+1)}D^{j-1}_{mm'} + \frac{mm'}{j(j+1)}D^j_{mm'}$$
$$+ \frac{\sqrt{[(j+1)^2-m^2][(j+1)^2-m'^2]}}{(j+1)(2j+1)}D^{j+1}_{mm'}\,. \tag{5.1.33}$$

5.1.3.4. As $n \to \infty$ Jacobi polynomials $P_n^{(\alpha,\beta)}(s)$ exhibit an asymptotic behavior as follows [N18]:

$$P_n^{(\alpha,\beta)}(\cos\theta) = \left\{\cos\left[\left(n+\frac{\alpha+\beta+1}{2}\right)\theta - (2\alpha+1)\tfrac{\pi}{4}\right]\middle/\sqrt{\pi n'}\right.$$
$$\times\left.\left(\sin\frac{\theta}{2}\right)^{\alpha+1/2}\left(\cos\frac{\theta}{2}\right)^{\beta+1/2}\right\}$$
$$+ O(n^{-3/2})\,, \quad 0 < \varepsilon \le \theta \le \pi - \varepsilon\,.$$

Therefore for $d^j_{mm'}(\beta)$ at $m \sim m' \sim 1$ and $j \gg 1$ we have the following *asymptotic expression*

$$d^j_{mm'}(\beta) \approx (-1)^{m-m'}\sqrt{\frac{2}{\pi(j-m)}}\left(\frac{2j+m-m'+1}{2j-m+m'+1}\right)^{(m+m')/2}$$
$$\times \frac{\cos[(j+1/2)\beta - (m-m'+1/2)\pi/2]}{\sqrt{\sin\beta}}\,, \quad 0 < \varepsilon \le \beta \le \pi - \varepsilon\,.$$

In the limit of $N \to \infty$ the Kravchuk polynomials $k_n^{(p)}(x,N)$ become Hermite polynomials $H_n(s)$ (see (2.6.7)). Therefore for $m \sim j \gg 1$ we get

$$d^j_{mm'}(\beta) \approx (-1)^{m-m'}[\sqrt{\pi j}2^{j-m}(j-m)!\sin\beta]^{-1/2}e^{-s^2/2}H_{j-m}(s)\,,$$

where $s = (j\cos\beta - m')/(\sqrt{j}\sin\beta)$.

5.1.3.5. For reference purposes we give the formulae for $d^j_{mm'}(\beta)$ at small values of j.

For $j = 0$

$$d_{00}^0(\beta) = 1 \; .$$

For $j = \frac{1}{2}$

$$d_{1/2,1/2}^{1/2}(\beta) = d_{-1/2,-1/2}^{1/2}(\beta) = \cos(\beta/2) \; ,$$

$$d_{-1/2,1/2}^{1/2}(\beta) = -d_{1/2,-1/2}^{1/2}(\beta) = \sin(\beta/2) \; .$$

For $j = 1$

$$d_{11}^1(\beta) = d_{-1,-1}^1(\beta) = \tfrac{1}{2}(1 + \cos \beta) \; ,$$

$$d_{10}^1(\beta) = -d_{01}^1(\beta) = d_{0,-1}^1 = -d_{-1,0}^1 = -\frac{1}{\sqrt{2}} \sin \beta \; ,$$

$$d_{1,-1}^1(\beta) = d_{-1,1}^1(\beta) = \tfrac{1}{2}(1 - \cos \beta) \; ,$$

$$d_{00}^1(\beta) = \cos \beta \; .$$

We recall also that

$$d_{jm}^j(\beta) = \sqrt{\frac{(2j)!}{(j+m)!(j-m)!}} \left(\cos \frac{\beta}{2}\right)^{j+m} \left(-\sin \frac{\beta}{2}\right)^{j-m} \; .$$

A more general relation evolves from (5.1.23) on calculating the derivative, viz.,

$$d_{mm'}^j(\beta) = (-1)^{j-m'} \sqrt{(j+m)!(j-m)!(j+m')!(j-m')!}$$

$$\times \sum_k (-1)^k \frac{(\cos \beta/2)^{m+m'+2k}(\sin \beta/2)^{2j-m-m'-2k}}{k!(j-m-k)!(j-m'-k)!(m+m'+k)!} \; .$$

Here the summation is carried out over all values of k producing positive integers under the factorial symbol.

For integer $j = l$ and $m = 0$, the generalized spherical functions reduce to spherical harmonics [V5]

$$D_{m0}^l(\alpha, \beta, \gamma) = \sqrt{\frac{4\pi}{2l+1}} Y_{lm}^*(\beta, \alpha) \; .$$

From the known properties of Jacobi and Kravchuk polynomials one can likewise examine other properties of generalized spherical functions, say, series expansion in these functions, the properties of zeros, differentiation formulae, etc. On the other hand, the addition formula for generalized spherical functions leads to the addition formulae for Jacobi and Kravchuk polynomials. Sequential multiplication of rotations yields

$$d_{mm'}^j(\beta) = i^{m-m'} \sum_{m''=-j}^{j} d_{m''m}^j(\tfrac{\pi}{2}) e^{-im''\beta} d_{m''m'}^j(\tfrac{\pi}{2}) \; .$$

For a more extended exposition of generalized spherical functions, the reader is referred to other works [Y1, V5, B24].

5.2 Clebsch-Gordan Coefficients and Hahn Polynomials

The addition of two momenta is a problem central to the quantum theory of angular momentum. It is solvable with the use of the Clebsch-Gordan coefficients (CGC) widely used in quantum-mechanical computations. Specifically by Wigner-Eckart's theorem they define the relative dependence of the matrix elements of transitions between states characterized by certain values of momenta and their projections. In this section we wish to express the Clebsch-Gordan coefficients through Hahn polynomials examined in Chap. 2. The properties of CGC will be evaluated step by step on the basis of this analogy.

From a mathematical standpoint the addition of two momenta in quantum mechanics consists in decomposing the tensor product of two irreducible representations of the rotation group into irreducible components. Therefore prior to beginning our evaluation we briefly survey the necessary concepts.

5.2.1 The Tensor Product of the Rotation Group Representations

5.2.1.1. Consider two representations of the rotation group, $g \to T_1(g)$ and $g \to T_2(g)$, acting in linear spaces R_1 and R_2, respectively. The linear combinations $\Psi = \sum_{m_1 m_2} c_{m_1 m_2} \psi_{m_1} \psi_{m_2}$ with arbitrary complex-valued coefficients $c_{m_1 m_2}$ belong to the space $R = R_1 \times R_2$, whose basis consists of all sorts of pairs $\psi_{m_1} \psi_{m_2}$. If $\psi_1 = \sum_{m_1} c_{m_1} \psi_{m_1}$ is a vector of space R_1 and $\psi_2 = \sum_{m_2} c_{m_2} \psi_{m_2}$, a vector of R_2, then their product is an element of $R = R_1 \times R_2$ equal to $\psi_1 \psi_2 = \sum_{m_1 m_2} c_{m_1} c_{m_2} \psi_{m_1} \psi_{m_2}$. The space $R = R_1 \times R_2$ is referred to as the *tensor product* of R_1 and R_2.

We construct a representation $g \to T(g)$ of the rotation group in $R = R_1 \times R_2$ as follows. A rotation g carries the basis ψ_{m_1} of R_1 in a new basis $T_1(g)\psi_{m_1}$, and the basis ψ_{m_2} of R_2 in a basis $T_2(g)\psi_{m_2}$. Assume that $T(g) = T_1(g) \times T_2(g)$ are linear operators which act on the basis vectors $\psi_{m_1} \psi_{m_2}$ of $R = R_1 \times R_2$ according to the rule

$$T(g)(\psi_{m_1} \psi_{m_2}) = (T_1(g)\psi_{m_1})(T_2(g)\psi_{m_2}) \, . \tag{5.2.1}$$

Because $\psi_{m_1} \psi_{m_2}$ is the basis in $R = R_1 \times R_2$, then (5.2.1) defines the action of linear operators $T(g)$ for all elements Ψ in R. In addition this expression indicates that operators $T(g)$ do specify a representation of the rotation group, that is, to the product of g_1 and g_2 there corresponds the product of operators $T(g_1)$ and $T(g_2)$.

A representation $g \to T(g)$ acting in space $R = R_1 \times R_2$ so that (5.2.1) is obeyed will be called the *tensor product of representations* $g \to T_1(g)$ and $g \to T_2(g)$.

The tensor product of three, four, etc., representations of the rotation group can be defined in a similar manner. In the following, unless confusion is likely to occur, we shall conventionally say a *product of representations* of the rotation group instead of the full name – tensor product of representations.

Let representations $g \to T_1(g)$ and $g \to T_2(g)$ be unitary, i.e., operators $T_1(g)$ and $T_2(g)$ acting in complex Euclidean spaces R_1 and R_2, respectively, preserve there the scalar products $(\psi_1|\psi_1')_1$ and $(\psi_2|\psi_2')_2$. We introduce the scalar product $(\Psi|\Psi')$ in space $R = R_1 \times R_2$ assuming that for the basis vectors $\Psi = \psi_{m_1}\psi_{m_2}$ of this space

$$(\psi_{m_1}\psi_{m_2}|\psi_{m_1'}\psi_{m_2'}) = (\psi_{m_1}|\psi_{m_1'})_1(\psi_{m_2}|\psi_{m_2'})_2 \,. \tag{5.2.2}$$

Now it is not hard to verify that from (5.2.1) and (5.2.2) it follows that the product of unitary representations $T(g) = T_1(g) \times T_2(g)$ is also a *unitary representation*

$$(T(g)\Psi|T(g)\Psi') = (\Psi|\Psi')$$

for all Ψ, Ψ' in $R = R_1 \times R_2$.

5.2.1.2. Let us derive the infinitesimal operators for the representation $T(g) = T_1(g) \times T_2(g)$. Let g be a rotation of the coordinate system about the axis n through an angle φ. By virtue of (5.1.3) and (5.2.1) we have

$$\exp(-i\varphi n \cdot J)(\psi_{m_1}\psi_{m_2}) = [\exp(-i\varphi n \cdot J_1)\psi_{m_1}][\exp(-i\varphi n \cdot J_2)\psi_{m_2}] \,.$$

Differentiating this expression with respect to φ and letting $\varphi = 0$ yields

$$J(\psi_{m_1}\psi_{m_2}) = (J_1\psi_{m_1})\psi_{m_2} + \psi_{m_1}(J_2\psi_{m_2}) \,.$$

By definition

$$J_1(\psi_{m_1}\psi_{m_2}) = (J_1\psi_{m_1})\psi_{m_2} \qquad J_2(\psi_{m_1}\psi_{m_2}) = \psi_{m_1}(J_2\psi_{m_2}) \,.$$

Hence

$$J = J_1 + J_2 \tag{5.2.3}$$

and $[J_1, J_2] = 0$.

5.2.1.3. Consider now the tensor product $D^{j_1} \times D^{j_2}$ of two irreducible representations D^{j_1} and D^{j_2} of the rotation group, which act in spaces R_1 and R_2, respectively. As a rule this representation is reducible. Its infinitesimal operators $J = J_1 + J_2$ obey the commutation rules (5.1.4). Therefore following along the lines of Theorem 5.2 we can construct a set of vectors Ψ_m, $-j \le m \le j$, which, in general, owing to the reducibility of the representation $D^{j_1} \times D^{j_2}$, serves as a basis for only some subspace $R_0 \subset R = R_1 \times R_2$ invariant under the action of operators J_x, J_y, and J_z.

Let us look at an orthogonal complement R' of R_0, i.e., at the set of all vectors Ψ' in R for which $(\Psi'|\Psi_0) = 0$ for any Ψ_0 in R_0. The subspace R' is also invariant with respect to operators J_x, J_y, J_z becvause $(J\Psi'|\Psi_0) = (\Psi'|J\Psi_0) = 0$, owing to the fact that the representation is unitary. Accordingly we may construct in R' a series of vectors Ψ_m', $-j' \le m \le j$, then again take the orthogonal complement and keep continuing in this way until the entire space $R = R_1 \times R_2$ is exhausted.

As a result we arrive at an assertion that in the space of the representation $D^{j_1} \times D^{j_2}$ there exists an orthonormal basis Ψ_{jm},

$$\Psi_{jm} = \sum_{m_1 m_2} \langle j_1 m_1 j_2 m_2 | jm \rangle \psi_{j_1 m_1} \psi_{j_2 m_2} , \qquad (5.2.4)$$

in which an irreducible representation D^j acts at some values of j which occur in the construction. The matrix elements of operators $J_\pm$, J_z are given in this basis by (5.1.11).

The coefficients of the decomposition (5.2.4) are known as the *Clebsch-Gordan coefficients* (CGC) and denoted also as $C^{jm}_{j_1 m_1 j_2 m_2}$. Below we express the Clebsch-Gordan coefficients in terms of Hahn polynomials and demonstrate that the possible values of weight j in (5.2.4) are as follows:

$$j = |j_1 - j_2|, \ |j_1 - j_2| + 1, \ \ldots, \ j_1 + j_2 - 1, \ j_1 + j_2 . \qquad (5.2.5)$$

Thus *the space of the representation $D^{j_1} \times D^{j_2}$ decomposes into the direct sum of the subspaces each having a basis Ψ_{jm}, $j = |j_1 - j_2|, \ldots, j_1 + j_2$, and an acting irreducible subrepresentation D^j*. Accordingly the decomposition (5.2.4) is often put down symbolically as

$$D^{j_1} \otimes D^{j_2} = \sum_{j=|j_1-j_2|}^{j_1+j_2} \oplus D^j .$$

This relation is referred to as the *Clebsch-Gordan series for the group* SO(3).

5.2.1.4. In quantum mechanical applications one can often encounter situations when the system under study may be regarded as consisting of two subsystems whose interaction is only weak compared with the total energy. To a zero approximation of perturbation theory one may then assume the Hamiltonian of the whole system equal to the sum of the Hamiltonians of individual subsystems, and construct the wave function as a product of the wave functions of subsystems. If in addition the Hamiltonian of each subsystem is invariant under rotations, then the respective wavefunctions $\psi_{j_1 m_1}$ and $\psi_{j_2 m_2}$ are characterized by the eigenvalues of operators J_1^2, J_{1z}, and J_2^2, J_{2z}, viz.,

$$J_i^2 \psi_{j_i m_i} = j_i(j_i + 1)\psi_{j_i m_i} ,$$
$$J_{iz} \psi_{j_i m_i} = m_i \psi_{j_i m_i} , \quad i = 1, 2 .$$

In order to evaluate a correction for the unperturbed energy level one needs to construct correct wavefunctions of the zeroth approximation, Ψ_{jm}, which are the eigenfunctions of operators $J^2 = (J_1 + J_2)^2$ and $J_z = J_{1z} + J_{2z}$, viz.,

$$J^2 \Psi_{jm} = j(j + 1)\Psi_{jm} , \quad J_z \Psi_{jm} = m \Psi_{jm} .$$

This gives rise to the problem of evaluating the eigenfunction of the total momentum $J = J_1 + J_2$ by the known eigenfunctions $\psi_{j_1 m_1}$ and $\psi_{j_2 m_2}$ of subsystem momenta J_1 and J_2, which is known as the *momentum addition problem* in quantum mechanics.

The eigenfunction of total angular momentum, Ψ_{jm}, is built by (5.2.4) with the help of the Clebsch-Gordan coefficients. The summation law (5.2.5) may be reformulated as a *vector model*. Let us construct the vector triangle $J = J_1 + J_2$, the length of the sides being respectively equal to integer or half-integer values, j, j_1, and j_2. Then j varies from the largest value $j_1 + j_2$ (when J_1 and J_2 are parallel) to the smallest value $j_1 - j_2$ (when J_1 and J_2 are antiparallel). Quite aptly the Clebsch-Gordan coefficients are often called *vector addition coefficients*, and the summation law (5.2.5) is called the *triangle condition*.

5.2.2 Expressing the Clebsch-Gordan Coefficients in Terms of Hahn Polynomials

In order to realize a decomposition of the representation $D^{j_1} \times D^{j_2}$ into irreducible components D^j, thus constructing a canonical basis Ψ_{jm} in the form of a linear combination of the initial basis vectors $\psi_{j_1 m_1} \psi_{j_2 m_2}$ in the space of the representation, one needs, by (5.2.4), to compute the Clebsch-Gordan coefficients. For this purpose we deduce recurrence relations between them.

5.2.2.1 The Formulae of Difference Differentiation. Let us apply operators $J_{\pm} = J_x \pm i J_y$ to both sides of (5.2.4). Because $J_{\pm} = J_{1\pm} + J_{2\pm}$ and $J_z = J_{1z} + J_{2z}$, where $J_{p\pm} = J_{px} \pm i J_{py}$, $p = 1, 2$, in agreement with (5.1.11), CGC are other than zero only when $m_1 + m_2 = m$ and the following recursion relations are satisfied

$$\sqrt{(j \mp m)(j \pm m + 1)} \langle j_1 m_1 j_2 m_2 | j, m \pm 1 \rangle$$
$$= \sqrt{(j_1 \pm m_1)(j_1 \mp m_1 + 1)} \langle j_1, m_1 \mp 1, j_2 m_2 | jm \rangle$$
$$+ \sqrt{(j_2 \pm m_2)(j_2 \mp m_2 + 1)} \langle j_1 m_1 j_2, m_2 \mp 1 | jm \rangle . \tag{5.2.6}$$

We shall seek the solution to these recursion relations in the form

$$\langle j_1 m_1 j_2, m - m_1 | jm \rangle = C^{\pm}_{m_1, m-m_1, m} u^{\pm}_{jm}(m_1) , \tag{5.2.7}$$

where

$$C^{\pm}_{m_1 m_2 m} = (-1)^{j_1 - m_1} \left[\frac{(j_1 + m_1)!(j_2 + m_2)!(j - m)!}{(j_1 - m_1)!(j_2 - m_2)!(j + m)!} \right]^{\pm 1/2}$$

[for $m = \pm(j+1)$, we have to assume $u^{\pm}_{j, \pm(j+1)} = 0$]. Then we arrive at the simple formulae for difference differentiation with respect to m_1, namely

$$u^{+}_{j, m+1}(m_1) = \nabla u^{+}_{jm}(m_1) , \quad u^{-}_{j, m-1}(m_1) = -\Delta u^{-}_{jm}(m_1) , \tag{5.2.8}$$

where Δ and ∇ are the difference operators,

$$\Delta u(m_1) = u(m_1 + 1) - u(m_1) , \quad \nabla u(m_1) = u(m_1) - u(m_1 - 1) .$$

5.2.2.2 An Explicit Expression for CGC. The relations (5.2.8) yield all CGC in complete analogy with the derivation of the general expression for $d^j_{mm'}(\beta)$ in Sect. 5.1.

At $m = j$ the first line of (5.2.8) gives

$$\nabla\left[C^-_{m_1,j-m_1,j}\langle j_1 m_1 j_2, j - m_1 | jj\rangle\right] = 0 \,,$$

i.e. accurate to a constant C

$$\langle j_1 m_1 j_2, j - m_1 | jj\rangle = C C^+_{m_1,j-m_1,j} \,, \tag{5.2.9}$$

whence

$$u^-_{jj} = C[C^+_{m_1,j-m_1,j}]^2 \,. \tag{5.2.10}$$

Having performed sequentially the difference differentiation of $u^-_{jj}(m_1)$ with respect to m_1, we turn to the second line of (5.2.8) to find $u^-_{jm} = (-\Delta)^{j-m} u^-_{jj}$ which in view of (5.2.7) and (5.2.10) is equivalent to the following formula for CGC

$$\langle j_1 m_1 j_2 m_2 | jm\rangle = (-1)^{j-m}\frac{C}{(2j)!}(-1)^{j_1-m_1}$$

$$\times \left[\frac{(j_1 - m_1)!(j_2 - m_2)!(j + m)!}{(j_1 + m_1)!(j_2 + m_2)!(j - m)!}\right]^{1/2}$$

$$\times \Delta^{j-m}\left[\frac{(j_1 + m_1)!(j_2 + j - m_1)!}{(j_1 - m_1)!(j_2 - j + m_1)!}\right] \,, \tag{5.2.11}$$

where $m_1 + m_2 = m$.

To determine C we compute the squared norm of Ψ_{jj}, $\|\Psi_{jj}\|^2 = (\Psi_{jj}|\Psi_{jj})$. From (5.2.4) and (5.2.9) we have

$$\|\Psi_{jj}\|^2 = \sum_{m_1=-j_1}^{j_1} |\langle j_1 m_1 j_2, j - m_1 | jj\rangle|^2$$

$$= \frac{|C|^2}{(2j)!}\sum_{m_1=j-j_2}^{j_1} \frac{(j_1 + m_1)!(j_2 + j - m_1)!}{(j_1 - m_1)!(j_2 - j + m_1)!} \,.$$

To compute the sum we refer to the familiar relation

$$\frac{\Gamma(x)\Gamma(y)}{\Gamma(x+y)} = \int_0^1 t^{x-1}(1 - t)^{y-1} dt \,, \quad x, y > 0 \,,$$

whence

$$\sum_{m_1=j-j_2}^{j_1} \frac{(j_1 + m_1)!(j_2 + j - m_1)!}{(j_1 - m_1)!(j_2 - j + m_1)!}$$

$$= \sum_{m_1=j-j_2}^{j_1} \frac{(j_1+j_2+j+1)!}{(j_1-m_1)!(j_2-j+m_1)!} \int_0^1 t^{j_1+m_1}(1-t)^{j_2+j-m_1}dt$$

$$= (j_1+j_2+j+1)! \int_0^1 dt\, t^{j+j_1-j_2}(1-t)^{j-j_1+j_2}$$

$$\times \sum_{m_1=j-j_2}^{j_1} \frac{t^{j_2-j+m_1}(1-t)^{j_1-m_1}}{(j_2-j+m_2)!(j_1-m_1)!}$$

$$= \frac{(j_1+j_2+j+1)!}{(j_1+j_2-j)!} \int_0^1 t^{j+j_1-j_2}(1-t)^{j-j_1+j_2}dt$$

$$= \frac{(j_1+j_2+j+1)!(j+j_1-j_2)!(j-j_1+j_2)!}{(2j+1)!(j_1+j_2-j)!}.$$

Hence

$$\|\Psi_{jj}\|^2 = \frac{|C|^2}{[(2j)!]^2} \frac{(j_1+j_2+j+1)!(j+j_1-j_2)!(j-j_1+j_2)!}{(2j+1)(j_1+j_2-j)!},$$

i.e. the vector Ψ_{jj} possesses a finite positive norm only for those weights j which satisfy the triangle condition (5.2.5). Letting in this case $\|\Psi_{jj}\| = 1$ and $C > 0$, we obtain

$$\frac{C}{(2j)!} = \left[\frac{(2j+1)(j_1+j_2-j)!}{(j+j_1-j_2)!(j-j_1+j_2)!(j_1+j_2+j+1)!} \right]^{1/2}. \tag{5.2.12}$$

Equations (5.2.11) and (5.2.12) uniquely define all the Clebsch-Gordan coefficients which are real-valued at a given choice of phases ($C > 0$).

Consider now vectors Ψ_{jm}, $|j_1-j_2| \le j \le j_1+j_2$, $-j \le m \le j$, constructed in the space of the representation $D^{j_1} \times D^{j_2}$ by the Eqs. (5.2.4, 11) and (5.2.12). The number of these vectors

$$\sum_{j=j_1-j_2}^{j_1+j_2} (2j+1) = (2j_1+1)(2j_2+1)$$

coincides with that in the initial basis $\psi_{j_1 m_1}\psi_{j_2 m_2}$, $-j_1 \le m_1 \le j_1$, $-j_2 \le m_2 \le j_2$. The vectors Ψ_{jm} are orthonormal since by construction they form an orthonormal set of eigenvectors of the Hermitian operators J^2 and J_z.

These vectors therefore form an orthonormal basis in the space of the representation $D^{j_1} \times D^{j_2}$ and so that in each subspace Ψ_{jm}, $-j \le m \le j$, at a given weight $j = |j_1-j_2|,\ldots,j_1+j_2$ there acts one irreducible subrepresentation D^j. Thus we have decomposed the representation $D^{j_1} \times D^{j_2}$ into irreducible components D^j.

5.2.2.3 Relating the Clebsch-Gordan Coefficients with the Hahn Polynomials.

Comparison of the Rodrigues formula for Hahn polynomials

$$h_n^{(\alpha,\beta)}(x,N) = \frac{(-1)^n \Gamma(N-x)\Gamma(x+1)}{n!\,\Gamma(\alpha+N-x)\Gamma(\beta+x+1)}$$
$$\times \nabla^n \left[\frac{\Gamma(\alpha+N-x)\Gamma(\beta+n+x+1)}{\Gamma(N-n-x)\Gamma(x+1)} \right]$$

with the Eqs. (5.2.11) and (5.2.12) enables the Clebsch-Gordan coefficients to be expressed through Hahn polynomials $h_n^{(\alpha,\beta)}(x,N)$ on account of $\Delta^n f(m_1) = \nabla^n f(m_1+n)$ as follows:

$$(-1)^{j_1-m_1} \langle j_1 m_1 j_2 m_2 | jm \rangle = \frac{\sqrt{\varrho(x)}}{d_n} h_n^{(\alpha,\beta)}(x,N) . \tag{5.2.13}$$

Here $\varrho(x)$ and d_n are the weight and norm of the polynomials $h_n^{(\alpha,\beta)}(x,N)$, $n = j-m$, $x = j_2-m_2$, $N = j_1+j_2-m+1$, $\alpha = m-m'$, $\beta = m+m'$, and $m' = j_1 - j_2$.

Because the Hahn polynomials $h_n^{(\alpha,\beta)}(s)$ are the difference analogues of Jacobi polynomials $P_n^{(\alpha,\beta)}(s)$ on a linear mesh, comparison of (5.1.26) and (5.2.13) indicates that CGC are the difference analogues of the functions $d^j_{mm'}(\beta)$ at $s = \cos\beta$. Equation (5.2.13) thus naturally explains the analogy between CGC and Jacobi polynomials [G13] (see also [R26, V9]). A particular case of (5.2.13) leads to Chebyshev polynomials of a discrete variable [M7, K12].

The Hahn polynomials $h_n^{(\alpha,\beta)}(x,N)$ are connected with the dual Hahn polynomials $w_k^{(c)}(x,a,b)$, $x = x(s) = s(s+1)$ (see (3.5.14)) therefore CGC can also be expressed in terms of the dual Hahn polynomials, viz.,

$$(-1)^{j_1+j_2-j} \langle j_1 m_1 j_2 m_2 | jm \rangle = \frac{\sqrt{\varrho(s)(2s+1)}}{d_k} w_k^{(c)}[x(s); a, b] , \quad m \geq |m'| . \tag{5.2.14}$$

Here $\varrho(s)$ and d_k are the weight and norm of $w_k^{(c)}(x,a,b)$, $k = j_2-m_2$, $x(s) = s(s+1)$, $s = j$, $a = m$, $b = j_1+j_2+1$, $c = m' = j_1 - j_2$.

An arbitrary CGC can always be reduced to the form with the conditions $m - m' \geq 0$ and $m + m' \geq 0$ satisfied by the symmetry relations[6]

$$\langle j_1 m_1 j_2 m_2 | jm \rangle = (-1)^{j_1+j_2-j} \langle j_1, -m_1, j_2, -m_2 | j, -m \rangle ,$$
$$\langle j_1 m_1 j_2 m_2 | jm \rangle = (-1)^{j_1+j_2-j} \langle j_2 m_2 j_1 m_1 | jm \rangle , \tag{5.2.15}$$
$$\langle j_1 m_1 j_2 m_2 | jm \rangle = \langle j_1' m_1' j_2' m_2' | jm' \rangle ,$$

where

$$j_1' = \tfrac{1}{2}(j_1+m_1) + \tfrac{1}{2}(j_2+m_2) ,$$
$$m_1' = \tfrac{1}{2}(j_1+m_1) - \tfrac{1}{2}(j_2+m_2) ,$$

---

[6] These symmetry relations can be derived as follows. The first relation occurs if in the outlined derivation of the general formula for CGC we start with the state $m = -j$, carry out the necessary consideration and then compare the resultant expression with (5.2.11). The second relation is readily derivable from (5.2.13) by (2.4.18). The last relation, referred to as Regge's symmetry, follows from (5.2.14) on account of $w_n^{(c)}(x,a,b) = w_n^{(a)}(x,c,b)$ [R18] (see also [B26]).

$$j_2' = \tfrac{1}{2}(j_1 - m_1) + \tfrac{1}{2}(j_2 - m_2) \,,$$
$$m_2' = \tfrac{1}{2}(j_1 - m_1) - \tfrac{1}{2}(j_2 - m_2) \,,$$
$$m' = j_1 - j_2 \,.$$

Indeed, observing the relation

$$\langle j_1 m_1 j_2 m_2 | j m \rangle = \langle j_2, -m_2, j_1, -m_1 | j, -m \rangle$$

for an arbitrary Clebsch-Gordan coefficient, one may secure $m + m' \geq 0$ if this inequality did not take place. Then the relation (5.2.15) that preserves the first inequality and changes the sign in the second is invoked finally to reach the domain of parameter variation where both inequalities are met.

The Eqs. (5.2.13) and (5.2.14) have been discussed above under the assumption that $m + m' \geq 0$ and $m - m' \geq 0$. In the general case, it is not hard to verify, by the symmetry properties noted earlier, that the representations of Clebsch-Gordan coefficients through Hahn polynomials have the form

$$\langle j_1 m_1 j_2 m_2 | j m \rangle = (-1)^\varphi \frac{\sqrt{\varrho(x)}}{d_n} h_n^{(\alpha,\beta)}(x, N) \,, \tag{5.2.13a}$$

where $\alpha = |m - m'|$, $\beta = |m + m'|$, $n = j - (\alpha + \beta)/2$, $N = j_1 + j_2 - (\alpha + \beta)/2 + 1$, $x = (j_1 + j_2 + m_1 - m_2 - \beta)/2$, $\varphi = (j_1 + j_2 - m_1 + m_2 - \alpha)/2$, $m' = j_1 - j_2$; and

$$\langle j_1 m_1 j_2 m_2 | j m \rangle = (-1)^\varphi \frac{\sqrt{\varrho(s)(2s+1)}}{d_k} w_k^{(c)}(x, a, b) \,, \tag{5.2.14a}$$

where $x = s(s+1)$, $s = j$, $a = (|m - m'| + |m + m'|)/2$, $b = j_1 + j_2 + 1$, $c = (|m+m'| - |m-m'|)/2$, $\varphi = j_1 + j_2 - j$, and $k = (j_1 + j_2 + m_1 - m_2 - |m+m'|)/2$.

The second relation with the aid of (5.2.15) may be rewritten in a convenient form with other values of the parameters as follows: $s = j$, $a = (|m - m'| + |m + m'|)/2$, $b = j_1 + j_2 + 1$, $c = (|m - m'| - |m + m'|)/2$, $\varphi = 0$, and $k = (j_1 + j_2 - m_1 + m_2 - |m - m'|)/2$.

From the last two expressions and the orthogonality of Hahn polynomials $h_n^{(\alpha,\beta)}(x, N)$ and $w_k^{(c)}(x, a, b)$ we obtain the orthogonality of Clebsch-Gordan coefficients

$$\sum_{m_1 m_2} \langle j_1 m_1 j_2 m_2 | j m \rangle \langle j_1 m_1 j_2 m_2 | j' m' \rangle = \delta_{jj'} \delta_{mm'} \,,$$

$$\sum_{jm} \langle j_1 m_1 j_2 m_2 | j m \rangle \langle j_1 m_1' j_2 m_2' | j m \rangle = \delta_{m_1 m_1'} \delta_{m_2 m_2'} \,.$$

Here the selection rule $m_1 + m_2 = m$ must be taken into account. These formulae are valid also because the transformation (5.2.4) is unitary. The reverse transformation is now straightforward

$$\psi_{j_1 m_1} \psi_{j_2 m_2} = \sum_{jm} \langle j_1 m_1 j_2 m_2 | j m \rangle \Psi_{jm} \,.$$

Although the literature devoted to the Clebsch-Gordan coefficients (see, e.g., [B24, V5, Y1]) is abundant, the connection of CGC with the Hahn polynomials is poorly documented. Until recently these quantities have been treated in isolation. In particular, the representations of CGC through the hypergeometric function $_3F_2(z)$ of $z = 1$ have been covered (see, e.g., [Y1, S14, V5]). The symmetry properties of $_3F_2(1)$ [B1] have been used even to derive a polynomial expression for these coefficients [B8]. However, only the recent publications [K5, K20] have filled the completely missed links of the hypergeometric functions arising in CGC theory with the Hahn polynomials for which representations through $_3F_2(1)$ have been known for a long time [K9] (see Table 2.4).

5.2.3 Basic Properties of the Clebsch-Gordan Coefficients

Equations (5.2.13) and (5.2.14) enable us to draw an analogy between the basic properties of the Clebsch-Gordan coefficients and the Hahn polynomials. The coefficients will be studied on the basis of the theory of orthogonal polynomials; a group-theoretical interpretation arises for the basic properties of Hahn polynomials.

5.2.3.1. Application of operators $J_\pm$ to both sides of (5.2.4) results in the recursion relations (5.2.6) for CGC, which corresponds to the difference differentiation formulae of Hahn polynomials

$$\Delta h_n^{(\alpha,\beta)}(x, N) = (\alpha + \beta + n + 1)h_{n-1}^{(\alpha+1,\beta+1)}(x, N - 1) \,,$$

$$[\sigma(x)\nabla + \tau(x)]h_{n-1}^{(\alpha+1,\beta+1)}(x, N - 1) + nh_n^{(\alpha,\beta)}(x, N) = 0 \,,$$

where the coefficients $\sigma(x)$ and $\tau(x)$ are listed in Table 2.1.

Likewise applying the operator

$$J^2 = (J_1 + J_2)^2 = J_1^2 + J_2^2 + J_{1+}J_{2-} + J_{1-}J_{2+} + 2J_{1z}J_{2z}$$

$(J_{p\pm} = J_{px} \pm iJ_{py}, \ p = 1,2)$ to both sides of (5.2.4) results in the *recursive relation* in m_1 and m_2

$$
\begin{aligned}
&\sqrt{(j_1 + m_1)(j_1 - m_1)(j_2 - m_2)(j_2 + m_2 + 1)} \\
&\quad \times \langle j_1, m_1 - 1, j_2, m_2 + 1|jm\rangle \\
&+ \sqrt{(j_1 - m_1)(j_1 + m_1 + 1)(j_2 + m_2)(j_2 - m_2 + 1)} \\
&\quad \times \langle j_1, m_1 + 1, j_2, m_2 - 1|jm\rangle \\
&= [j(j + 1) - j_1(j_1 + 1) - j_2(j_2 + 1) - 2m_1m_2]\langle j_1 m_1 j_2 m_2|jm\rangle \,, \quad (5.2.16)
\end{aligned}
$$

which is equivalent, on the one hand, to the difference equation for Hahn polynomials $h_n^{(\alpha,\beta)}(x, N)$ and, on the other hand, to the recursive relation for the dual Hahn polynomials $w_n^{(c)}(x, a, b)$. The recursive Eq. (5.2.16) is a convenient means for numerical evaluation of CGC [G1].

5.2.3.2. The recursive relation for Hahn polynomials $h_n^{(\alpha,\beta)}(x, N)$ leads to the recursive expression in j for CGC

$$a_{jm}\langle j_1 m_1 j_2 m_2 | j-1, m\rangle + (b_{jm} - m_1 + m_2)$$
$$\times \langle j_1 m_1 j_2 m_2 | jm\rangle + a_{j+1,m}\langle j_1 m_1 j_2 m_2 | j+1, m\rangle = 0 , \qquad (5.2.17)$$

where

$$a_{jm} = \left[\frac{(j^2 - m^2)(j^2 - m'^2)[(q+1)^2 - j^2]}{(2j-1)j^2(2j+1)}\right]^{1/2} ,$$

$$b_{jm} = \frac{mm'(q+1)}{j(j+1)} , \quad m' = j_1 - j_2 ; \quad q = j_1 + j_2 .$$

In view of (5.2.14) the equality (5.2.17) corresponds also to the difference equation for dual Hahn polynomials $w_k^{(c)}(x, a, b)$. The recursion relation (5.2.17) is closely connected with the representations of the four-dimensional rotation group SO(4) [B8] (see also Sect. 5.5.1).

5.2.3.3. A representation of CGC in terms of the Hahn polynomials $\tilde{h}_n^{(\mu,\nu)}(x, N)$ also takes place, viz.,

$$\langle j_1 m_1 j_2 m_2 | jm\rangle = \frac{\sqrt{\varrho(x)}}{d_n}\tilde{h}_n^{(\mu,\nu)}(x, N) , \qquad (5.2.18)$$

where $\varrho(x)$ and d_n are the weight and norm of $\tilde{h}_n^{(\mu,\nu)}(x, N)$ listed in Table 2.2, $n = j_1 + j_2 - j$, $x = j_1 - m_1$, $N = j_1 + j_2 - m + 1$, $\mu = m - m'$, $\nu = m + m'$, $m' = j_1 - j_2$. This equation can be verified by virtue of (5.2.17).

The formulae of difference differentiation,

$$\Delta\tilde{h}_n^{(\mu,\nu)}(x, N) = -(\mu + \nu + 2N - n - 1)\tilde{h}_{n-1}^{(\mu,\nu)}(x, N - 1) ,$$

$$[\sigma(x)\nabla + \tau(x)]\tilde{h}_{n-1}^{(\mu,\nu)}(x, N - 1) = n\tilde{h}_n^{(\mu,\nu)}(x, N) ,$$

are equivalent to the recursion relations for CGC:

$$\sqrt{(j_1 + j_2 - j)(j_1 + j_2 + j + 1)}\langle j_1 - \tfrac{1}{2}, m_1 - \tfrac{1}{2}, j_2 - \tfrac{1}{2}, m_2 + \tfrac{1}{2} | jm\rangle$$
$$= \sqrt{(j_1 + m_1)(j_2 - m_2)}\langle j_1 m_1 j_2 m_2 | jm\rangle$$
$$- \sqrt{(j_1 - m_1 + 1)(j_2 + m_2 + 1)}\langle j_1, m_1 - 1, j_2, m_2 + 1 | jm\rangle ,$$
$$\sqrt{(j_1 + j_2 - j)(j_1 + j_2 + j + 1)}\langle j_1 m_1 j_2 m_2 | jm\rangle$$
$$= \sqrt{(j_1 + m_1)(j_2 - m_2)}\langle j_1 - \tfrac{1}{2}, m_1 - \tfrac{1}{2}, j_2 - \tfrac{1}{2}, m_2 + \tfrac{1}{2} | jm\rangle$$
$$- \sqrt{(j_1 - m_1)(j_2 + m_2)}\langle j_1 - \tfrac{1}{2}, m_1 + \tfrac{1}{2}, j_2 - \tfrac{1}{2}, m_2 - \tfrac{1}{2}, | jm\rangle .$$

5.2.3.4. In view of (5.2.11) for $m_1 + m_2 = j$ we have

$$\langle j_1 m_1 j_2 m_2 | jj\rangle = \sqrt{\frac{(2j+1)!(j_1 + j_2 - j)!}{(j + j_1 - j_2)!(j - j_1 + j_2)!(j_1 + j_2 + j + 1)!}}$$

$$\times (-1)^{j_1-m_1}\sqrt{\frac{(j_1+m_1)!(j_2+m_2)!}{(j_1-m_1)!(j_2-m_2)!}} \ .$$

The representation of any other CGC as a sum of a finite number of terms results from (5.2.11) at $m = m_1 + m_2$, taking account of the formula

$$\Delta^n f(x) = \sum_{k=0}^{n}(-1)^{n+k}\frac{n!}{k!(n-k)!}f(x+k) \ ,$$

namely

$$\langle j_1 m_1 j_2 m_2 | j m \rangle = \sqrt{\frac{(j_1-m_1)!(j_2-m_2)!}{(j_1+m_1)!(j_2+m_2)!}}$$

$$\times \sqrt{\frac{(2j+1)(j_1+j_2-1)!(j+m)!(j-m)!}{(j+j_1-j_2)!(j-j_1+j_2)!(j_1+j_2+j+1)!}}$$

$$\times \sum_k (-1)^{j_1-m_1+k}$$

$$\times \frac{(j_1+m_1+k)!(j+j_2-m_1-k)!}{k!(j-m-k)!(j_1-m_1-k)!(j_2-j+m_1+k)!} \ . \tag{5.2.19}$$

Likewise from (5.2.18) we find with the aid of the Rodrigues formula for the polynomials $\tilde{h}_n^{(\mu,\nu)}(x,N)$ at $m_1 + m_2 = m$

$$\langle j_1 m_1 j_2 m_2 | j_1 + j_2, m \rangle$$

$$= \sqrt{\frac{(2j_1)!(2j_2)!(j_1+j_2+m)!(j_1+j_2-m)!}{(2j_1+2j_2)!(j_1+m_1)!(j_1-m_1)!(j_2+m_2)!(j_2-m_2)!}} \ ,$$

$$\langle j_1 m_1 j_2 m_2 | j m \rangle$$

$$= \sqrt{\frac{(2j+1)(j_1+j_2-j)!(j+j_1-j_2)!(j-j_1+j_2)!}{(j_1+j_2+j+1)!}} \sum_k (-1)^k \tag{5.2.20}$$

$$\times \frac{\sqrt{(j_1+m_1)!(j_1-m_1)!(j_2+m_2)!(j_2-m_2)!(j+m)!(j-m)!}}{k!(j_1+j_2-j-k)!(j_1-m_1-k)!(j_2+m_2-k)!(j-j_2+m_1+k)!(j-j_1-m_2+k)!}$$

The summation in the Eqs. (5.2.19) and (5.2.20) is carried out over all values of k producing integers under the factorial symbol. These expressions can be readily rewritten as CGC representations in terms of generalized hypergeometrical functions $_3F_2(1)$. Various equivalent forms can be produced for these representations [S14, V5, B8] by making use of the symmetry properties of these functions [B1]. Specifically in view of (5.2.13) and (2.7.19) we have

$$(-1)^{j_1-m_1}\langle j_1 m_1 j_2 m_2 | j m \rangle = \sqrt{\frac{(j_1+m_1)!(j_2+m_2)!(j+m)!}{(j_1-m_1)!(j_2-m_2)!(j-m)!}}$$

$$\times \sqrt{\frac{(2j+1)(j+j_1-j_2)!}{(j-j_1+j_2)!(j_1+j_2-j)!(j+j_1+j_2+1)!}}$$

$$\times (-1)^{j-m}\frac{(j_1+j_2-m)!}{(j_1-j_2+m)!}\,{}_3F_2\left(\begin{array}{c}-j+m,\ j+m+1,\ -j_2+m_2\\ j_1-j_2+m+1,\ -j_1-j_2+m\end{array}\bigg|\,1\right).$$

A more detailed examination of the CGC relation to the generalized hypergeometrical functions is given elsewhere [Y1, S14, V5, B8].

5.2.3.5. Let us look at the asymptotic behavior of the Clebsch-Gordan coefficients. For Hahn polynomials $h_n^{(\alpha,\beta)}(x,N)$ we have the asymptotic formula (see (2.6.5))

$$h_n^{(\alpha,\beta)}\left[\frac{\tilde{N}}{2}(1+s)-\frac{\beta+1}{2};N\right]=\tilde{N}^n[P_n^{(\alpha,\beta)}(s)+\mathrm{O}(\tilde{N}^{-2})]\,,$$

$$\tilde{N}=N+\frac{\alpha+\beta}{2}\,,\quad N\to\infty\,.$$

For the weight $\varrho(x)$ and the squared norm d_n^2 of the polynomials $h_n^{(\alpha,\beta)}(x,N)$ we have

$$\varrho(x)=\left(\frac{\tilde{N}}{2}\right)^{\alpha+\beta}(1-s)^\alpha(1+s)^\beta[1+\mathrm{O}(\tilde{N}^{-2})]\,,$$

$$d_n^2=\tilde{N}^{\alpha+\beta+2n+1}\frac{\Gamma(\alpha+n+1)\Gamma(\beta+n+1)}{(\alpha+\beta+2n+1)n!\Gamma(\alpha+\beta+n+1)}[1+\mathrm{O}(\tilde{N}^{-2})]\,,$$

$$\tilde{N}=N+\frac{\alpha+\beta}{2}\,,\quad N\to\infty\,.$$

Therefore, in accordance with (5.1.26) and (5.2.13), for $m\sim m'\sim j\ll j_1+j_2+1$ the Clebsch-Gordan coefficients of second order of accuracy become the Wigner d-functions

$$\langle j_1m_1j_2m_2|jm\rangle\approx(-1)^{j_2+m_2}\sqrt{\frac{2j+1}{j_1+j_2+1}}\,d^j_{mm'}(\beta)\,. \tag{5.2.21}$$

Here $\cos\beta=(m_1-m_2)/(j_1+j_2+1)$ and $m'=j_1-j_2$.

Equation (5.2.21) has a higher order of accuracy than the similar asymptotic formulae in use [E4, B40].

Equations (5.2.14) and (3.8.5) suggest that an asymptotic representation in terms of Laguerre polynomials can also be derived for CGC.

5.2.3.6. The Clebsch-Gordan coefficients $\langle j_1m_1j_2m_2|jm\rangle$ are zero if the triangle condition (5.2.5) or the condition $m_1+m_2=m$ is broken down. Besides, cases are known of CGC vanishing at specific values of angular momenta j_1, j_2, j and their projections m_1, m_2, m, despite the conditions $|j_1-j_2|\le j\le j_1+j_2$ and $m_1+m_2=m$ being satisfied. These situations are referred to as the *roots of the vector addition coefficients* [V5, B25]. The presence of a CGC root means that vector Ψ_{jm} in the decomposition (5.2.4) is orthogonal to some vectors $\psi_{j_1 m_1} \psi_{j_2 m_2}$ of the original basis in the space of the representation $D^{j_1} \times D^{j_2}$, namely

$$(\psi_{j_1 m_1} \psi_{j_2 m_2} | \Psi_{jm}) = \langle j_1 m_1 j_2 m_2 | jm \rangle = 0 \,.$$

In quantum mechanics these roots give rise to the additional selection rules forbidding the transitions whose amplitudes are proportional to zero CGC. Grechukhin [G21–23] suggests a critical experiment to verify physical models of the nucleus on the basis of the root $\langle 3220|32 \rangle = 0$.

In agreement with (5.2.13) and (5.2.14) the roots of vector addition coefficients are integer-valued roots of Hahn polynomials [S16]. The simplest condition for a root to exist has the form

$$h_1^{(\alpha,\beta)}(k, N) = (\alpha + \beta + 2)k - (\beta + 1)(N - 1) = 0 \,,$$
$$k = 0, 1, \ldots, N - 1 \,.$$

With reference to (5.2.13) it is not hard to verify that this condition will be satisfied, for example, by the coefficients $\langle 1010|10 \rangle = \langle 3220|32 \rangle = 0$.

5.2.3.7. For reference we note also the more complete *symmetry properties* of GCG which will be conveniently described in terms of the *Wigner 3jm-symbols*

$$\begin{pmatrix} j_1 & j_2 & j \\ m_1 & m_2 & -m \end{pmatrix} = (-1)^{j_1 - j_2 + m}(2j + 1)^{-1/2} C^{jm}_{j_1 m_1 j_2 m_2} \,,$$

which remains invariable at any even permutation of columns, and acquire a factor of $\varepsilon = (-1)^{j_1 + j_2 + j}$ at odd permutations and at a simultaneous reversal of sign of all projections, viz.,

$$\begin{pmatrix} j_1 & j_2 & j \\ m_1 & m_2 & -m \end{pmatrix} = \varepsilon \begin{pmatrix} j_2 & j_1 & j \\ m_2 & m_1 & -m \end{pmatrix} = \varepsilon \begin{pmatrix} j_1 & j_1 & j \\ -m_1 & -m_2 & m \end{pmatrix}$$
$$= \varepsilon \begin{pmatrix} j_1 & j & j_2 \\ m_1 & -m & m_2 \end{pmatrix} \,.$$

For CGC these equations take the form

$$C^{jm}_{j_1 m_1 j_2 m_2} = (-1)^{j_1 + j_2 - j} C^{jm}_{j_2 m_2 j_1 m_1} = (-1)^{j_1 + j_2 - j} C^{j,-m}_{j_1,-m_1,j_2,-m_2}$$
$$= (-1)^{j_1 - m_1} \sqrt{(2j + 1)/(2j_2 + 1)} C^{j_2,-m_2}_{j_1 m_1,j,-m} \,.$$

There exist also the Regge symmetry relations, which fail to reduce to simple changes in angular momenta and projections (see, e.g., the last relation in (5.2.15)). These properties are conveniently described in terms of the *Regge symbol* [R18, B26].

$$\begin{bmatrix} -j + j_2 + j & j_1 - j_2 + j & j_1 + j_2 - j \\ j_1 + m_1 & j_2 + m_2 & j - m \\ j_1 - m_1 & j_2 - m_2 & j + m \end{bmatrix} = \begin{pmatrix} j_1 & j_2 & j \\ m_1 & m_2 & -m \end{pmatrix} \,,$$

which does not change at even permutations of columns or rows and in transposing, and acquires the factor $\varepsilon = (-1)^{j_1+j_2+j}$ at odd permutations.

5.2.3.8. In conclusion we note also that the Darboux-Christoffel formula for Hahn polynomials leads to the following expression for CGC summation:

$$\sum_{j=m}^{j'} C^{jm}_{j_1 m_1 j_2 m_2} C^{jm}_{j_1 m'_1 j_2 m'_2} = \frac{a_{j'+1,m}}{2(m_2 - m'_2)}$$
$$\times \left(C^{j'm}_{j_1 m_1 j_2 m_2} C^{j'+1,m}_{j_1 m'_1 j_2 m'_2} - C^{j'+1,m}_{j_1 m_1 j_2 m_2} C^{j'm}_{j_1 m'_1 j_2 m'_2} \right) ,$$

where $m_1 + m_2 = m'_1 + m'_2 = m$, $m_2 \neq m'_2$, $m \geq |m'|$, $m' = j_1 - j_2$, and the values of a_{jm} are defined by (5.2.17). At $j' = j_1 + j_2$ this equation becomes the demonstration of CGC orthogonality.

Another CGC summation formula can be readily derived by (5.2.14) and the Darboux-Christoffel formula for dual Hahn polynomials, namely

$$\sum_{m_2=j_2}^{m'_2} C^{jm}_{j_1 m_1 j_2 m_2} C^{j'm}_{j_1 m_1 j_2 m_2} = \frac{\sqrt{(j_1 + m'_1 + 1)(j - m'_1)(j_2 + m'_2)(j_2 - m'_2 + 1)}}{(j - j')(j + j' + 1)}$$
$$\times \left(C^{jm}_{j_1, m'_1+1, j_2, m'_2-1} C^{j'm}_{j_1 m'_1 j_2 m'_2} - C^{jm}_{j_1 m'_1 j_2 m'_2} C^{j'm}_{j_1, m'_1+1, j_2, m'_2-1} \right) ,$$

where $m_1 + m_2 = m'_1 + m'_2 = m$, $j \neq j'$.

To close this section we considered some basic properties of the Clebsch-Gordan coefficients using their analogy with the properties of Hahn polynomials. Likewise one can deduce a number of other properties of applicational significance. A more expanded coverage of CGC theory can be found in [B8, B24–26, C22, E4, L18, L22, S6, S14, S35, V5, V9, Y1, Y1].

5.2.4 Irreducible Tensor Operators. The Wigner-Eckart Theorem

5.2.4.1. Phyiscal problem solving often produces matrix elements of the form

$$\langle j_1 m_1 | O_{jm} | j_2 m_2 \rangle = (\psi_{j_1 m_1} | O_{jm} \psi_{j_2 m_2}) , \tag{5.2.22}$$

where O_{jm} are the *irreducible tensor operators*, which on rotation of the coordinate system are transformed like the vectors of state ψ_{jm}, viz.

$$O'_{jm} = U O_{jm'} U^{-1} = \sum_m D^j_{mm'} O_{jm} ,$$
$$\psi'_{jm'} = U \psi_{jm'} = \sum_m D^j_{mm'} \psi_{jm} . \tag{5.2.23}$$

Here $U = T(g)$, $D^j_{mm'} = D^j_{mm'}(\alpha, \beta, \gamma)$ are the generalized spherical functions, and α, β and γ are the Euler angles specifiying the rotation of the coordinate set.

Examples of irreducible tensor operators O_{jm} are the operator of angular momentum $\boldsymbol{J}$, spin operator $\boldsymbol{S}$, coordinate operator $\boldsymbol{r}$, and momentum operator $\boldsymbol{p}$ of a quantum system. In many important applications, the operator of perturbation imposed on the system also possesses the properties of irreducible tensor operator. The respective matrix elements (5.2.22) enter the expression for the probability of transitions between the states with the given values of angular momenta and projections.

We examine the general method of computing the matrix elements (5.2.22), which is independent of the nature of operator O_{jm} and uses only its transformation properties.

5.2.4.2. In a rotation of the coordinate set, specified by Euler's angles α, β, and γ, Eq. (5.2.4), in view of (5.1.12), yields

$$C^{jm'}_{j_1 m'_1 j_2 m'_2} = \sum_{m\, m_1 m_2} D^{*j}_{mm'} C^{jm}_{j_1 m_1 j_2 m_2} D^{j_1}_{m_1 m'_1} D^{j_2}_{m_2 m'_2} , \tag{5.2.24}$$

where $D^{\lambda}_{\mu\mu'} = D^{\lambda}_{\mu\mu'}(\alpha, \beta, \gamma)$ are the generalized spherical functions, and $C^{j\mu}_{j_1 \mu_1 j_2 \mu_2}$ are the Clebsch-Gordan coefficients. This equation defines the law of CGC transformation under rotations.

Equation (5.2.24) may be rewritten in other equivalent forms by making use of the orthogonality of CGC and D-functions. Specifically from (5.2.24) one can readily obtain a decomposition of the product $D^{j_1}_{m_1 m'_1}(\alpha, \beta, \gamma)\, D^{j_2}_{m_2 m'_2}(\alpha, \beta, \gamma)$ in the generalized spherical functions $D^j_{mm'}(\alpha, \beta, \gamma)$, viz.

$$D^{j_1}_{m_1 m'_1} D^{j_2}_{m_2 m'_2} = \sum_j C^{j, m_1 + m_2}_{j_1 m_1 j_2 m_2} C^{j, m'_1 + m'_2}_{j_1 m'_1 j_2 m'_2} D^j_{m_1 + m_2, m'_1 + m'_2} ,$$

and the integral of the product of three D-functions, which will be useful in our later evaluations, viz.

$$\int D^{*j}_{mm'} D^{j_1}_{m_1 m'_1} D^{j_2}_{m_2 m'_2}\, dg = \frac{1}{2j+1} C^{jm}_{j_1 m_1 j_2 m_2} C^{jm'}_{j_1 m'_1 j_2 m'_2} , \tag{5.2.25}$$

where $dg = (1/V)\sin\beta\, d\alpha\, d\beta\, d\gamma$, $\int dg = 1$, and the domain of integration coincides with that in the property of orthogonality of generalized spherical functions (see Sect. 5.1).

5.2.4.3. By the transformation laws (5.2.23)

$$\langle j_1 m_1 | O_{jm} | j_2 m_2 \rangle = (\psi'_{j_1 m_1} | O'_{jm} \psi'_{j_2 m_2})$$

$$= \sum_{m'_1 m'_2 m'} D^{*j_1}_{m'_1 m_1} D^j_{m' m} D^{j_2}_{m'_2 m_2} (\psi_{j_1 m'_1} | O_{jm'} \psi_{j_2 m'_2}) .$$

Integrating both sides of this identity with respect to the measure dg, we see that it is not hard to obtain by virtue of (5.2.25)

$$\langle j_1 m_1 | O_{jm} | j_2 m_2 \rangle = C^{jm}_{j_1 m_1 j_2 m_2} I(j_1 j j_2) \, , \tag{5.2.26}$$

where

$$I(j_1 j j_2) = \frac{1}{2j_1 + 1} \sum_{m_1' m_2' m'} C^{jm'}_{j_1 m_1' j_2 m_2'} \langle j_1 m_1' | O_{jm'} | j_2 m_2' \rangle$$

is the reduced matrix element. The function $I(j_1 j j_2)$ remains invariant in rotations of the coordinate set and is independent of the projections m_1, m_2, and m.

The Eq. (5.2.26) indicates that the dependence of the matrix elements of the irreducible tensor operator (5.2.22) on the projections of momentum is completely determined by the Clebsch-Gordan coefficients. This is the main assertion of the Wigner-Eckart theorem [W7].

Many applications are interested in the ratios of matrix elements of the irreducible tensor operator, which in view of (5.2.26) take the form

$$\frac{\langle j_1 m_1 | O_{jm} | j_2 m_2 \rangle}{\langle j_1 m_1' | O_{jm'} | j_2 m_2' \rangle} = \frac{C^{j_1 m_1}_{jm j_2 m_2}}{C^{j_1 m_1'}_{jm' j_2 m_2'}} \, .$$

Equations (5.2.13) and (5.2.14) show that these rations can be expressed also as the ratios of Hahn polynomials.

Equation (5.2.26) enables one to calculate an arbitrary matrix element of the irreducible tensor operator for given values of angular momenta if at least one matrix element of this operator for specific values of projections is known, say the one corresponding to the simplest wave function of the system.

Numerous phyiscal applications of the Wigner-Eckart theorem have been reported, see, e.g. [B14, B24–28, L18].

5.3 The Wigner $6j$-Symbols and the Racah Polynomials

Many physical applications, primarily atomic and nuclear spectroscopy, widely invoke along with the Clebsch-Gordan coefficients the Wigner $6j$-symbols and Racah's coefficients proportional to these symbols. These quantities arise in decomposing the product of three irreducible representations of the rotating group into irreducible components or, what is the same, in combining three angular momenta in quantum mechanics.

5.3.1 The Racah Coefficients and Wigner $6j$-Symbols

The product of three irreducible representations of the rotating group $D^{j_1} \times D^{j_2} \times D^{j_3}$ is decomposable onto the irreducible components D^j in at least two ways in accordance with the following coupling schemes:

$$J_1 + J_2 = J_{12} \, , \quad J_{12} + J_3 = J \, , \tag{5.3.1}$$

or

$$J_2 + J_3 = J_{23} , \quad J_1 + J_{23} = J . \tag{5.3.2}$$

Here J_i are the infinitesimal operators of the representations D^{j_i}, $i = 1, 2, 3$.

The first way consists in firstly decomposing the product $D^{j_1} \times D^{j_2}$ in the sum of $D^{j_{12}}$ ($|j_1 - j_2| \le j_{12} \le j_1 + j_2$) and then decomposing each representation $D^{j_{12}} \times D^{j_3}$ in D^j. In the second way one is to decompose firstly $D^{j_2} \times D^{j_3}$ in the sum of $D^{j_{23}}$ ($|j_2 - j_3| \le j_{23} \le j_2 + j_3$), and then each $D^{j_1} \times D^{j_{23}}$ in D^j. This brings about, generally speaking, two distinct bases of the representation D^j. In quantum mechanics relations (5.3.1) and (5.3.2) define the addition of angular momenta J_1, J_2, and J_3 for three independent quantum subsystems into the total momentum J of the system.

Transformation between the bases $\Psi_{jm}^{j_{12}}$ and $\Psi_{jm}^{j_{23}}$ of the irreducible representation D^j, constructed by the relations (5.3.1) and (5.3.2), respectively, is effected by the unitary matrix $U(j_{12}, j_{23})$, viz.

$$\Psi_{jm}^{j_{23}} = \sum_{j_{12}} U(j_{12}, j_{23}) \Psi_{jm}^{j_{12}} , \tag{5.3.3}$$

$$\Psi_{jm}^{j_{12}} = \sum_{j_{23}} U^*(j_{12}, j_{23}) \Psi_{jm}^{j_{23}} . \tag{5.3.4}$$

The matrix $U(j_{12}, j_{23})$ is proportional to the $6j$-symbol

$$U(j_{12}, j_{23}) = (-1)^{j_1 + j_2 + j_3 + j} \sqrt{(2j_{12} + 1)(2j_{23} + 1)} \begin{Bmatrix} j_1 & j_2 & j_{12} \\ j_3 & j & j_{23} \end{Bmatrix} , \tag{5.3.5}$$

which in turn equals, accurate to the phase multiplier introduced to simplify the symmetry properties (see, e.g., [V5]), the Racah coefficient $W(abcd; ef)$, viz.

$$\begin{Bmatrix} a & b & e \\ d & c & f \end{Bmatrix} = (-1)^{a+b+c+d} W(abcd; ef) .$$

By invoking the relations (5.3.1) and (5.3.2) the functions $\Psi_{jm}^{j_{12}}$ and $\Psi_{jm}^{j_{23}}$ can be expressed in terms of linear combinations of the products $\psi_{j_1 m_1} \psi_{j_2 m_2} \psi_{j_3 m_3}$ with the help of the Eqs. (5.2.4). Equating the coefficients of $\psi_{j_1 m_1} \psi_{j_2 m_2} \psi_{j_3 m_3}$ on the left and right sides of (5.3.4) yields the following relation for the matrix $U(j_{12}, j_{23})$ with the Clebsch-Gordan coefficients:

$$\langle j_1 m_1 j_2 m_2 | j_{12} m_{12} \rangle \langle j_{12} m_{12} j_3 m_3 | j m \rangle$$
$$= \sum_{j_{23}} \langle j_2 m_2 j_3 m_3 | j_{23} m_{23} \rangle \langle j_1 m_1 j_{23} m_{23} | j m \rangle U^*(j_{12}, j_{23}) .$$

From this formula one can derive an explicit expression for $U(j_{12}, j_{23})$ – by multiplying both sides by $\langle j_2 m_2 j_3 m_3 | j'_{23} m_{23} \rangle$ and summing in m_2. In view of the orthogonality relation

$$\sum_{m_2} \langle j_2 m_2 j_3 m_3 | j_{23} m_{23} \rangle \langle j_2 m_2 j_3 m_3 | j'_{23} m_{23} \rangle = \delta_{j_{23} j'_{23}} ,$$

this leads to

$$\langle j_1 m_1 j_{23} m_{23} | j m \rangle U^*(j_{12}, j_{23}) = \sum_{m_2} \langle j_{12} m_{12} j_3, m - m_{12} | j m \rangle$$

$$\times \langle j_1 m_1 j_2 m_2 | j_{12} m_{12} \rangle \langle j_2 m_2 j_3, m_{23} - m_2 | j_{23} m_{23} \rangle , \qquad (5.3.6)$$

where $m_1 + m_2 = m_{12}$. It is apparent that the $6j$-symbols are real-valued, owing to the real valuedness of CGC. The property of unitarity of the transformation matrix (5.3.3) leads to the orthogonality of $6j$-symbols, namely

$$\sum_{j_{12}} (2j_{12} + 1)(2j_{23} + 1) \left\{ \begin{matrix} j_1 & j_2 & j_{12} \\ j_3 & j & j_{23} \end{matrix} \right\} \left\{ \begin{matrix} j_1 & j_2 & j_{12} \\ j_3 & j & j'_{23} \end{matrix} \right\} = \delta_{j_{23} j'_{23}} ,$$

$$\sum_{j_{23}} (2j_{12} + 1)(2j_{23} + 1) \left\{ \begin{matrix} j_1 & j_2 & j_{12} \\ j_3 & j & j_{23} \end{matrix} \right\} \left\{ \begin{matrix} j_1 & j_2 & j'_{12} \\ j_3 & j & j_{23} \end{matrix} \right\} = \delta_{j_{12} j'_{12}} . \qquad (5.3.7)$$

Racah [R1] has obtained a single sum expression for the matrix $U(j_{12}, j_{23})$. The elements of this matrix have recently been demonstrated to form a new system of orthogonal polynomials in a discrete variable, called Racah polynomials by Askey and Wilson [A27, W8]. The Racah polynomials $u_n^{(\alpha,\beta)}(x, a, b)$ are the difference analogues of the Jacobi polynomials $P_n^{(\alpha,\beta)}(s)$ on the square mesh; these were elucidated in Chap. 3 on the basis of a general approach different from those employed in [W8].

5.3.2 Expressing the $6j$-Symbols Through the Racah Polynomials

In order to establish a relation between the Wigner $6j$-symbols and the Racah polynomials we examine the behaviour of the right-hand side of (5.3.6) as a function of j_{23}. For $j_3 - j_2 \geq |m_{23}|$, from (5.2.14a) we obtain

$$\langle j_2 m_2 j_3 m_3 | j_{23} m_{23} \rangle$$

$$= \frac{\sqrt{\bar{\varrho}(j_{23})(2j_{23} + 1)}}{\bar{d}_{j_2 - m_2}} w_{j_2 - m_2}^{(m_{23})} [j_{23}(j_{23} + 1), j_3 - j_2, j_2 + j_3 + 1] ,$$

where $\bar{\varrho}(s)$ and $\bar{d}_k$ are the weight and norm of the dual Hahn polynomial $w_k^{(c)}(x, a, b)$. Therefore the right-hand side of (5.3.6) represents, accurate to the known factor $[\bar{\varrho}(j_{23})(2j_{23} + 1)]^{1/2}$, some polynomial $y_n(x)$ in $x = j_{23}(j_{23} + 1)$, viz.

$$y_n(x) = \sum_{m_2} \langle j_{12} m_{12} j_3, m - m_{12} | j m \rangle \langle j_1 m_1 j_2 m_2 | j_{12} m_{12} \rangle \bar{w}_{j_2 - m_2}(x) ,$$

where

$$\bar{w}_{j_2 - m_2}(x) = \frac{1}{\bar{d}_{j_2 - m_2}} w_{j_2 - m_2}^{(m_{23})} (x, j_3 - j_2, j_2 + j_3 + 1) .$$

In agreement with the section rules $m_1 + m_2 = m_{12}, -j_{12} \leq m_{12} \leq j_{12}$ the degree of $y_n(x)$ is $n = \max(j_2 - m_2) = j_{12} + m_1 + j_2$. We select the minimal degree by letting $m_1 = -j_1$ in (5.3.6); then

$$\langle j_1, -j_1, j_{23} m_{23} | j m \rangle U(j_{12}, j_{23}) = \sqrt{\bar{\varrho}\,(j_{23}(2j_{23} + 1)}\, y_{j_{12}-j_1+j_2}(x)\,, \qquad (5.3.8)$$

where

$$y_{j_{12}-j_1+j_2}(x) = \langle j_{12}, -j_{12}, j_3, m + j_{12} | j m \rangle$$
$$\times \langle j_1, -j_1, j_2, -j_{12} + j_1 | j_{12}, -j_{12} \rangle \bar{w}_{j_{12}-j_1+j_2}(x) + \dots .$$

Substituting in (5.3.8) specific values of CGC, derivable from (5.2.14) by virtue of the symmetry properties (5.2.15),

$$\langle j_1, -j_1, j_{23} m_{23} | j m \rangle = (-1)^{j_1+j_{23}-j}$$
$$\times \sqrt{\frac{(j_{23} + m_{23})!(j - m)!(j - j_1 + j_{23})!(2j_1)!(2j + 1)}{(j_{23} - m_{23})!(j + m)!(j + j_1 - j_{23})!(j_1 + j_{23} - j)!(j_1 + j_{23} + j + 1)}}\,,$$

and, making use of the unitarity properties of the transformation matrix (5.3.3)

$$\sum_{j_{23}} U(j_{12}, j_{23}) U(j'_{12}, j_{23}) = \delta_{j_{12} j'_{12}}\,,$$

we obtain that the polynomials $y_n(x)$ are orthogonal on the interval $[j_3 - j_2, j_2 + j_3 + 1]$ with the weight of Racah polynomials $u_n(x) = u_n^{(\alpha,\beta)}(x, a, b)$ at $\alpha = j_1 - j_2 - j_3 + j$, $\beta = j_1 - j_2 + j_3 - j$. Therefore the polynomials $y_n(x)$ and $u_n(x)$ differ only in a factor which can be readily determined by comparing the coefficients of the leading terms. As a result we arrive at the *expression of the Wigner 6j-symbols in terms of the Racah polynomials* $u_n^{(\alpha,\beta)}(x, a, b)$

$$(-1)^{j_1+j+j_{23}} \sqrt{2j_{12} + 1} \left\{ \begin{matrix} j_1 & j_2 & j_{12} \\ j_3 & j & j_{23} \end{matrix} \right\} = \frac{\sqrt{\varrho(s)}}{d_n} u_n^{(\alpha,\beta)}(x, a, b)\,. \qquad (5.3.9)$$

Here $\varrho(s)$ and d_n are the weight and norm of the polynomials $u_n^{(\alpha,\beta)}(x)$, $n = j_{12}-j_1+j_2$, $x = s(s+1)$, $s = j_{23}$, $a = j_3-j_2$, $b = j_2+j_3+1$, $\alpha = j_1-j_2-j_3+j$, $\beta = j_1 - j_2 + j_3 - j$. This expression has been derived by noting the orthogonality of Racah polynomials $u_n^{(\alpha,\beta)}(x, a, b)$ for which $\alpha > -1$, $2a + 1 > \beta > -1$, that is, on taking account of the parameters being integer valued, under the assumption that $j_1 - j_2 \geq |j_3 - j|$, $j_3 - j_2 \geq j_1 - j$.

By observing the familiar properties of symmetry of the $6j$-symbols

$$\left\{ \begin{matrix} j_1 & j_2 & j_{12} \\ j_3 & j & j_{23} \end{matrix} \right\} = \left\{ \begin{matrix} j_2 & j_1 & j_{12} \\ j & j_3 & j_{23} \end{matrix} \right\} \qquad (5.3.10)$$

$$\left\{ \begin{matrix} j_1 & j_2 & j_{12} \\ j_3 & j & j_{23} \end{matrix} \right\} = \left\{ \begin{matrix} j_3 & j & j_{12} \\ j_1 & j_2 & j_{23} \end{matrix} \right\}\,, \qquad (5.3.11)$$

and

$$\begin{Bmatrix} j_1 & j_2 & j_{12} \\ j_3 & j & j_{23} \end{Bmatrix} = \begin{Bmatrix} \dfrac{j_1 - j_2 + j_3 + j}{2} & \dfrac{-j_1 + j_2 + j_3 + j}{2} & j_{12} \\ \dfrac{j_1 + j_2 + j_3 - j}{2} & \dfrac{j_1 + j_2 - j_3 + j}{2} & j_{23} \end{Bmatrix} , \qquad (5.3.12)$$

this formula can be extended over arbitrary values of the parameters admissible by the angular momentum addition law. (We recall that the symmetry properties of the $6j$-symbols follow from the CGC symmetry properties [see (5.3.6)]. The most extensive coverage of the $6j$-symbol symmetry properties have been given by Regge [R19] (see also [B26]).) To demonstrate this we rewrite the constraints arising in the derivation of (5.3.9) in the form

$$j_1 - j_2 + j_3 - j \geq 0 ,$$
$$j_1 - j_2 - j_3 + j \geq 0 ,$$
$$-j_1 - j_2 + j_3 + j \geq 0 .$$

Then by virtue of (5.3.10) we can first of all satisfy the first inequality. Then by invoking (5.3.11), which retains the first constraint and changes the sign of the second, we meet the second line, unless it has been valid. The property (5.3.12) retains the first two inequalities and reverses the sign of the last one. Therfore by observing (5.3.12) we finally drive the said $6j$-symbol to the form when formula (5.3.9) is valid.

On the basis of the above considerations the most general case receives the representation of the $6j$-symbols in terms of the Racah polynomials in the form

$$\sqrt{2j_{12} + 1} \begin{Bmatrix} j_1 & j_2 & j_{12} \\ j_3 & j & j_{23} \end{Bmatrix} = (-1)^\varphi \frac{\sqrt{\varrho(s)}}{d_n} u_n^{(\alpha,\beta)}(x, a, b) , \qquad (5.3.13)$$

where $x = s(s+1)$, $s = j_{23}$, $\alpha = |M - M'|$, $\beta = |M + M'|$, $\gamma = |N|$, $n = j_{12} - (\alpha + \beta)/2$, $a = (\beta + \gamma)/2$, $b - 1 = (j_1 + j_2 + j_3 + j - \alpha)/2$, $\varphi = (j_1 + j_2 + j_3 + j + \alpha)/2 + j_{23}$, $M = j_1 - j_2$, $M' = j_3 - j$, and $N = -j_1 - j_2 + j_3 + j$. The last three designations have been utilized to emphasize the similarity with Eqs. (5.1.26a) and (5.2.13a).

The Hahn and Racah polynomials $h_n^{(\alpha,\beta)}(x, N)$ and $u_n^{(\alpha,\beta)}(x, a, b)$, are the difference analogues of Jacobi polynomials $P_n^{(\alpha,\beta)}(s)$ on linear and square meshes, respectively. Therefore by virtue of (5.1.26), (5.2.13) and (5.3.9) both the Clebsch-Gordan coefficents and $6j$-symbols are the difference analogues of Wigner's d-functions.

For the simplest particular case leading to analogues of the Legendre polynomials, a polynomial expression for the Racah coefficients has been noted by Biedenharn et al. [B22] with respect to the compilation of tables. Karasev and Shelepin have also touched upon this subject [K5]. A connection of Racah's coefficients with some relevant orthogonal polynomials has been recognized by Wilson [W8] by virtue of generalized hypergeometrical functions. These topics have been independently studied in [S17] on the basis of the quantum theory of angular momentum. It has been established later [N9] that the orthogonal polynomials encountered in [B22, K5, S17] (and earlier also in [L3, R25]) and named

Racah polynomials by Askey and Wilson [A27] are essentially the difference analogues of the Jacobi polynomials on a square mesh. A generalization of the theory has led to the classical orthogonal polynomials of a discrete variable on nonuniform meshes [N17] (see also [N18] and Chap. 3). The Eq. (5.3.9) stressing the analogy between the $6j$-symbols and Jacobi polynomials has been deduced in [N10].

5.3.3 Main Properties of the $6j$-Symbols

Equation (5.3.9) paves the way for the evaluation of the $6j$-symbols on the basis of the known properties of Racah polynomials.

5.3.3.1. For the Racah polynomials $u_n^{(\alpha,\beta)}[x(s), a, b]$, $x(s) = s(s+1)$, we have the formula of difference differentiation

$$\frac{\Delta u_n^{(\alpha,\beta)}[x(s), a, b]}{\Delta x(s)} = (\alpha + \beta + n + 1)u_{n-1}^{(\alpha+1,\beta+1)}\left[x\left(s + \tfrac{1}{2}\right), a + \tfrac{1}{2}, b - \tfrac{1}{2}\right] .$$

Another formula for difference differentiation results from substituting this relation into the difference equation for the Racah polynomials written in self-adjoint form

$$\frac{\Delta}{\Delta x(s - 1/2)}\left\{\sigma(s)\varrho(s)u_{n-1}^{(\alpha+1,\beta+1)}\left[x\left(s - \tfrac{1}{2}\right), a + \tfrac{1}{2}, b - \tfrac{1}{2}\right]\right\}$$
$$+ n\varrho(s)u_n^{(\alpha,\beta)}[x(s), a, b] = 0 ,$$

where the functions $\sigma(s)$ and $\varrho(s)$ are given in Table 3.6.

By virtue of (5.3.9) the above formulae for difference differentiation of Racah polynomials lead to the following recursion relations for the $6j$-symbols

$$\sqrt{\sigma(j_{23} + 1)}\begin{Bmatrix} j_1 & j_2 & j_{12} \\ j_3 & j & j_{23} + 1 \end{Bmatrix} + \sqrt{\sigma(-j_{23} - 1)}\begin{Bmatrix} j_1 & j_2 & j_{12} \\ j_3 & j & j_{23} \end{Bmatrix}$$
$$= (2j_{23} + 2)\sqrt{(j_{12} - j_1 + j_2)(j_{12} + j_1 - j_2 + 1)}$$
$$\times \begin{Bmatrix} j_1 + \tfrac{1}{2} & j_2 - \tfrac{1}{2} & j_{12} \\ j_3 & j & j_{23} + \tfrac{1}{2} \end{Bmatrix} , \tag{5.3.14}$$

$$\sqrt{\sigma(-j_{23})}\begin{Bmatrix} j_1 & j_2 & j_{12} \\ j_3 & j & j_{23} \end{Bmatrix} + \sqrt{\sigma(j_{23})}\begin{Bmatrix} j_1 & j_2 & j_{12} \\ j_3 & j & j_{23} - 1 \end{Bmatrix}$$
$$= 2j_{23}\sqrt{(j_{12} + j_1 - j_2)(j_{12} - j_1 + j_2 + 1)}$$
$$\times \begin{Bmatrix} j_1 - \tfrac{1}{2} & j_2 + \tfrac{1}{2} & j_{12} \\ j_3 & j & j_{23} - \tfrac{1}{2} \end{Bmatrix} , \tag{5.3.15}$$

where $\sigma(s) = (j - j_1 + s)(j_1 + j - s)(j_2 - j_3 + s)(j_2 + j_3 + s + 1)$. These relations may also be obtained on the basis of the Biedenharn-Elliott identity [B22, B20, E5]; they are closely connected with the irreducible representations of the group SU(3) (see Sect. 5.5.3).

260

Applying the operators $J_{12}^2 = (J_1 + J_2)^2$ and $J_{23}^2 = (J_2 + J_3)^2$ to both sides of (5.3.4) one can obtain for the $6j$-symbols the familiar recurrence relations [V5, S17] which correspond to the difference equation and the recurrence relation for Racah polynomials. Equation (3.2.19) leads to the Rodrigues' type formula for the $6j$-symbols [S27].

5.3.3.2. In view of (5.3.13) and the symmetry property of the $6j$-symbols

$$\begin{Bmatrix} j_1 & j_2 & j_{12} \\ j_3 & j & j_{23} \end{Bmatrix} = \begin{Bmatrix} j_3 & j_2 & j_{23} \\ j_1 & j & j_{12} \end{Bmatrix}$$

the relations (5.3.7) are equivalent to the discrete property of orthogonality of Racah polynomials $u_n^{(\alpha,\beta)}(x, a, b)$ subject to integer parameters.

Noting the property of orthogonality (5.3.7) we may also obtain the $6j$-symbol representation in terms of the Racah polynomials $\tilde{u}_n^{(c,d)}(x, a, b)$ [N13], viz.

$$\begin{Bmatrix} j_1 & j_2 & j_{12} \\ j_3 & j & j_{23} \end{Bmatrix} = \frac{(-1)^{j_{12}+j_3+j}}{(2j_{12}+1)^{1/2} d_{j_1+j_2-j_{12}}} \sqrt{\varrho(j_{23})}$$
$$\times \tilde{u}_{j_1+j_2-j_{12}}^{(j-j_1, j+j_1+1)}[j_{23}(j_{23}+1), j_3 - j_2, j_2 + j_3 + 1],$$

where $\varrho(s)$ and d_n are the weight and norm of the polynomials $u_n^{(c,d)}(x, a, b)$, $j_1 - j_2 \geq |j_3 - j|$, and $j_3 - j_2 \geq j_1 - j$.

5.3.3.3. Representing the Racah polynomials in (5.3.9) through $_4F_3(1)$ (see (3.11.28)) leads to the relation of the $6j$-symbols with the generalized hypergeometrical functions

$$\begin{Bmatrix} j_1 & j_2 & j_{12} \\ j_3 & j & j_{23} \end{Bmatrix} = (-1)^{j+j_2+j_{12}+j_{23}} \sqrt{C_{j_{12}} \varrho(j_{23})}$$

$$\times (2j_2)! [(j_1 - j_2 + j_3 - j)! (j_1 - j_2 + j_3 + j + 1)]^{-1}$$

$$\times {}_4F_3 \left(\begin{matrix} j_1 - j_2 - j_{12},\ j_1 - j_2 + j_{12} + 1,\ j_3 - j_2 - j_{23},\ j_3 - j_2 + j_{23} + 1 \\ j_1 - j_2 + j_3 - j + 1,\ -2j_2,\ j_1 - j + j_3 + j + 2 \end{matrix} \middle| 1 \right),$$

where

$$C_{j_{12}} = \frac{(j - j_2 + j_{12})! (j - j_{12} + j_3)! (j_3 - j + j_{12})! (j + j_3 + j_{12} + 1)!}{(j_{12} + j - j_3)! (j_1 + j_2 + j_{12} + 1)! (j_{12} - j + j_2)! (j_1 + j_2 - j_{12})!}$$

and

$$\varrho(j_{23}) = \frac{(j_{23} - j_2 + j_3)! (j_1 - j + j_{23})! (j_1 + j - j_{23})! (j_1 + j + j_{23} + 1)!}{(j - j_1 - j_{23})! (j_2 - j_3 + j_{23})! (j_2 + j_3 - j_{23})! (j_2 + j_3 + j_{23} + 1)!} .$$

Observing the symmetry properties, we find this representation can be rewritten in other equivalent forms (see, e.g., [S14, V5]).

At $n = j_{12} - j_1 + j_2 = 0$ we find for specific $6j$-symbols

$$\left\{ \begin{matrix} j_1 & j_2 & j_1 - j_2 \\ j_3 & j & j_{23} \end{matrix} \right\} = (-1)^{j_1+j+j_{23}} \frac{(2j_2)!}{(j_1 - j_2 + j_3 - j)!(j_1 - j_2 + j_3 + j + 1)!}$$
$$\times \sqrt{C_{j_1 - j_2} \varrho (j_{23})} \; .$$

5.3.3.4. Now we derive the *second-order-accuracy asymptotic formula* for the $6j$-symbols. From (3.8.2) we recall that for the Racah polynomials $u_n^{(\alpha,\beta)}(x, a, b)$ we have the following asymptotic equation

$$u_n^{(\alpha,\beta)}(x, a, b) = (\tilde{N}^2)^n [P_n^{(\alpha,\beta)}(z) + O(\tilde{N}^{-2})] \; ,$$

$$\tilde{N}^2 = \left(b + \frac{\alpha}{2} \right)^2 - \left(a - \frac{\beta}{2} \right)^2 \; , \quad b \to \infty \; , \quad z = \text{const} \; ,$$

where

$$x = s(s + 1) = -\frac{1}{4} + \left(a - \frac{\beta}{2} \right)^2 \frac{1-z}{2} + \left(b + \frac{\alpha}{2} \right)^2 \frac{1+z}{2} \; .$$

Under the same conditions we have for the weight $\varrho(s)$ and the squared norm d_n^2 of $u_n^{(\alpha,\beta)}(x, a, b)$

$$\varrho(s) = \left(\frac{\tilde{N}^2}{2} \right)^{\alpha+\beta} (1 - z)^\alpha (1 + z)^\beta [1 + O(\tilde{N}^{-2})] \; ,$$

$$d_n^2 = (\tilde{N}^2)^{\alpha+\beta+2n+1} \frac{\Gamma(\alpha + n + 1)\Gamma(\beta + n + 1)}{(\alpha + \beta + 2n + 1)n!\Gamma(\alpha + \beta + n + 1)}$$
$$\times [1 + O(\tilde{N}^{-2})] \; , \quad b \to \infty \; .$$

Therefore from (5.3.13) we find for $j_1 \sim j_2 \sim j_3 \sim j \gg j_{12}$

$$\left\{ \begin{matrix} j_1 & j_2 & j_{12} \\ j_3 & j & j_{23} \end{matrix} \right\} \approx \frac{(-1)^{j_2+j_3+j_{23}}}{\sqrt{(j_1 + j_2 + 1)(j_3 + j + 1)}} d_{j_1-j_2, j_3-j}^{j_{12}}(\theta) \; , \tag{5.3.16}$$

where

$$\cos \theta = \frac{(2j_{23} + 1)^2 - (j_1 + j_2 + 1)^2 - (j_3 + j + 1)^2}{2(j_1 + j_2 + 1)(j_3 + j + 1)} \; .$$

According to this formula the angle embraced by the sides of lenghts $j_1 + j_2 + 1$ and $j_3 + j + 1$ of the triangle in Fig. 5.2 is equal to $\pi - \theta$.

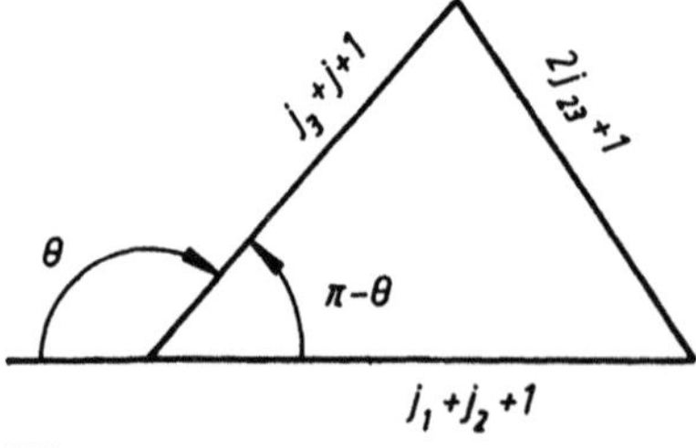

Fig. 5.2

The Eq. (5.3.16) has a higher order of accuracy than the similar asymptotic formulae [E4, V5].

In a very similar manner we may recall the asymptotic relation between the polynomials of Racah $u_n^{(\alpha,\beta)}(x, a, b)$ and Hahn $h_n^{(\alpha,\beta)}(z, N)$:

$$u_n^{(\alpha,\beta)}(x, a, b) = R^n \left[h_n^{(\alpha,\beta)}(z, N) + O(R^{-1}) \right] ,$$

$$\varrho(s) = R^{\alpha+\beta} \frac{\Gamma(\alpha + N - z)\Gamma(\beta + z + 1)}{\Gamma(N - z)\Gamma(z + 1)}[1 + O(R^{-1})] ,$$

$$d_n^2 = R^{\alpha+\beta+2n+1} \frac{\Gamma(\alpha + n + 1)\Gamma(\beta + n + 1)\Gamma(\alpha + \beta + n + N + 1)}{(\alpha + \beta + 2n + 1)n!\,\Gamma(\alpha + \beta + n + 1)(N - n - 1)!}$$
$$\times [1 + O(R^{-1})] ,$$

$$a \to \infty ,$$

where $x = s(s + 1)$, $s = a + z$, $b = a + N$, $R = a + b + (\alpha - \beta)/2$, and $n, z, \alpha, \beta,$ and N are some constants, to obtain with the help of (5.3.13) and (5.3.10) the asymptotic representation of the $6j$-symbols in terms of the Clebsch-Gordan coefficients [V5, B24]

$$(-1)^{j_1+j_2+j_3+j} \sqrt{(2j_{12} + 1)(j_3 + j + 1)} \begin{Bmatrix} j_1 & j_2 & j_{12} \\ j_3 & j & j_{23} \end{Bmatrix}$$
$$\approx \langle j_1, j_{23} - j, j_2, j_3 - j_{23} | j_{12}, j_3 - j \rangle$$

having the first order of accuracy.

5.3.3.5. Equation (5.3.6) suggests that the $6j$-symbols may be other than zero only when the angular momenta J_1, J_2, and J_3 are combined according to the relations (5.3.1) and (5.3.2) in agreement with the triangle condition (5.2.5). In addition there exist $6j$-symbols (called non-trivial zeros) which are equal to zero irrespective of whether or not all the rules of vector addition have been obeyed. To demonstrate,

$$\begin{Bmatrix} 2 & 2 & 2 \\ \frac{3}{2} & \frac{3}{2} & \frac{3}{2} \end{Bmatrix} = \begin{Bmatrix} 3 & 2 & 2 \\ 1 & 2 & 2 \end{Bmatrix} = \begin{Bmatrix} 6 & 4 & 9 \\ 5 & 5 & 2 \end{Bmatrix} = \begin{Bmatrix} 4 & 4 & 5 \\ 4 & 2 & 3 \end{Bmatrix}$$
$$= \begin{Bmatrix} 5 & 3 & 5 \\ 3 & 3 & 3 \end{Bmatrix} = 0 . \tag{5.3.17}$$

The fact that the last two $6j$-symbols become zero gives rise to a special G_2 group in the analysis of f-shells in atomic spectroscopy [S35, J5].

Equation (5.3.13) implies that all the roots of $6j$-symbols coincide with those of Racah's polynomials $u_n^{(\alpha,\beta)}(x, a, b)$ at points $x = s(s+1)$, $s = a, a+1, \ldots, b-1$. The simplest condition for a root to exist has the form

$$u_1^{(\alpha,\beta)}(x) = (\alpha + \beta + 2)x - (\alpha + 1)a(a - \beta) - (\beta + 1)b(b + \alpha)$$
$$+ (\alpha + 1)(\beta + 1) = 0 ,$$

$$x = s(s + 1), \quad s = a, a + 1, \ldots, b - 1 .$$

By invoking the symmetry relations of $6j$-symbols it is not hard to verify that this condition is valid for all cases listed in (5.3.17).

The non-trivial zeros of the $6j$-symbols have been discussed by other workers elsewhere [V5, B25, S17, V3, D8, V2, B36–39, R16, L19, B32].

Equation (5.3.13) may be used in a similar manner to derive a number of other attractive properties of the $6j$-symbols.

5.4 The Wigner $9j$-Symbols as Orthogonal Polynomials in Two Discrete Variables

In the preceding section of this chapter we demonstrated that the Clebsch-Gordan coefficients and Wigner's $6j$-symbols, arising in decompositions of the product of two and three irreducible representations of the rotation group, can be expressed, respectively, through the Hahn and Racah polynomials. The decomposition of a product of four irreducible representation gives rise to the $9j$-symbols (Fano coefficients), which are already orthogonal in two independent discrete variables. These quantities in turn are closely connected with a certain new set of orthogonal polynomials in two variables [S28]. The theory of angular momentum enables us to establish the main properties of such polynomials.

5.4.1 The $9j$-Symbols and the Relation with the Clebsch-Gordan Coefficients

The Wigner $9j$-symbols occur in applications alongside with the Clebsch-Gordan coefficients and $6j$-symbols. Specifically these quantities arise in decomposing the product of four rotation-group representations $D^{j_1} \times D^{j_2} \times D^{j_3} \times D^{j_4}$ to the irreducible components D^j. This decomposition can be effected in a few different ways. In particular, the decomposition into D^j proceeds according to these two coupling schemes

$$J_1 + J_2 = J_{12} , \quad J_3 + J_4 = J_{34} , \quad J_{12} + J_{34} = J \tag{5.4.1}$$

or

$$J_1 + J_3 = J_{13} , \quad J_2 + J_4 = J_{24} , \quad J_{13} + J_{24} = J . \tag{5.4.2}$$

Here J_i are the infinitesimal operators of the representations D^{j_i} (angular momenta to be added in quantum mechanics), $i = 1, 2, 3, 4$.

The transformation between the bases $\Psi_{jm}^{j_{12}j_{34}}$ and $\Psi_{jm}^{j_{13}j_{24}}$ constructed by the coupling schemes (5.4.1) and (5.4.2), respectively, is carried out by the unitary matrix $U_{j_{12}j_{34},j_{13}j_{24}}$, viz.

$$\Psi_{jm}^{j_{13}j_{24}} = \sum_{j_{12}j_{34}} U_{j_{12}j_{34},j_{13}j_{24}} \Psi_{jm}^{j_{12}j_{34}} ,$$

$$\Psi_{jm}^{j_{12}j_{34}} = \sum_{j_{13}j_{24}} U_{j_{12}j_{34},j_{13}j_{24}}^* \Psi_{jm}^{j_{13}j_{24}} . \tag{5.4.3}$$

This matrix is proportional to the $9j$-symbol

$$U_{j_{12}j_{34},j_{13}j_{24}} = \sqrt{(2j_{12}+1)(2j_{34}+1)(2j_{13}+1)(2j_{24}+1)}$$
$$\times \begin{Bmatrix} j_1 & j_2 & j_{12} \\ j_3 & j_4 & j_{34} \\ j_{13} & j_{24} & j \end{Bmatrix} . \tag{5.4.4}$$

These quantities are invariant under rotations.

Like the $6j$-symbols, the $9j$-symbol can be expressed through the Clebsch-Gordan coefficients. The functions $\Psi_{jm}^{j_{12}j_{34}}$ and $\Psi_{jm}^{j_{13}j_{24}}$ are not hard to represent as linear combinations of the functions $\psi_{j_1 m_1} \psi_{j_2 m_2} \psi_{j_3 m_3} \psi_{j_4 m_4}$ by observing the coupling schemes (5.4.1) and (5.4.2) and applying in succession the addition law (5.2.4). Equating the coefficients of $\psi_{j_1 m_1}\psi_{j_2 m_2}\psi_{j_3 m_3}\psi_{j_4 m_4}$ on both sides of (5.4.3) gives

$$C_{j_{12}m_{12}j_{34}m_{34}}^{jm} C_{j_1 m_1 j_2 m_2}^{j_{12}m_{12}} C_{j_3 m_3 j_4 m_4}^{j_{34}m_{34}}$$
$$= \sum_{j_{13}j_{24}} U_{j_{12}j_{34},j_{13}j_{24}}^* C_{j_{13}m_{13}j_{24}m_{24}}^{jm} C_{j_1 m_1 j_3 m_3}^{j_{13}m_{13}} C_{j_2 m_2 j_4 m_4}^{j_{24}m_{24}} .$$

To deduce the expression for the $9j$-symbol in terms of CGC from this equality it should be multiplied by $C_{j_1 m_1 j_3 m_3}^{j'_{13}m_{13}} C_{j_2 m_2 j_4 m_4}^{j'_{24}m_{24}}$ and summed over the values of the projections $m_1 + m_3 = m_{13}$ and $m_2 + m_4 = m_{24}$. Noting the property of orthogonality of CGC, we find

$$C_{j_{13}m_{13}j_{24}m_{24}}^{jm} U_{j_{12}j_{34},j_{13}j_{24}}^* = \sum_{\substack{m_1+m_3=m_{13} \\ m_2+m_4=m_{24}}} C_{j_{12}m_{12}j_{34}m_{34}}^{jm}$$
$$\times C_{j_1 m_1 j_2 m_2}^{j_{12}m_{12}} C_{j_3 m_3 j_4 m_4}^{j_{34}m_{34}} C_{j_1 m_1 j_3 m_3}^{j_{13}m_{13}} C_{j_2 m_2 j_4 m_4}^{j_{24}m_{24}} . \tag{5.4.5}$$

It is apparent that the $9j$-symbols are real-valued, owing to the fact that CGC are real-valued.

Equations (5.4.4) and (5.4.5) uniquely define all the phase and normalizing factors of the $9j$-symbols. The unitary property of the transform matrix (5.4.3) leads to the properties of orthogonality

$$\sum_{j_{12}j_{34}} (2j_{12}+1)(2j_{34}+1) \begin{Bmatrix} j_1 & j_2 & j_{12} \\ j_3 & j_4 & j_{34} \\ j_{13} & j_{24} & j \end{Bmatrix} \begin{Bmatrix} j_1 & j_2 & j_{12} \\ j_3 & j_4 & j_{34} \\ j'_{13} & j'_{24} & j \end{Bmatrix}$$
$$= \frac{\delta_{j_{13}j'_{13}} \delta_{j_{24}j'_{24}}}{(2j_{13}+1)(2j_{24}+1)} ,$$
$$\sum_{j_{13}j_{24}} (2j_{13}+1)(2j_{24}+1) \begin{Bmatrix} j_1 & j_2 & j_{12} \\ j_3 & j_4 & j_{34} \\ j_{13} & j_{24} & j \end{Bmatrix} \begin{Bmatrix} j_1 & j_2 & j'_{12} \\ j_3 & j_4 & j'_{34} \\ j_{13} & j_{24} & j \end{Bmatrix}$$
$$= \frac{\delta_{j_{12}j'_{12}} \delta_{j_{34}j'_{34}}}{(2j_{12}+1)(2j_{34}+1)} . \tag{5.4.6}$$

Although the $9j$-symbols can be expressed in terms of CGC by (5.4.4) and (5.4.5), these quantities are of independent value. Their properties have been studied to a far lesser extent than those of CGC and the $6j$-symbols. Specifically, up to recent times, for the $9j$-symbols no representation has been established in terms of generalized hypergemometric functions, nor have symmetry properties been found of the type of the Regge symmetry for CGC and $6j$-symbols, and continuous analogues are absent, etc. In what follows we wish to demonstrate that the $9j$-symbols are closely connected with a certain set of orthogonal polynomials in two discrete variables.

5.4.2 The Polynomial Expression for the $9j$-Symbols

Let us look at the behavior of the right-hand side of (5.4.5) as a function of j_{13} and j_{24}. Using (5.2.14a) we write the two last CGC subject to $|m_{ik}| + m'_{ik} \leq 0$ or $m_{ik} - |m'_{ik}| \geq 0$ in terms of the dual Hahn polynomials $w_n^{(c)}(x, a, b)$

$$C^{j_{ik} m_{ik}}_{j_i m_i j_k m_k} = \sqrt{\varrho(j_{ik})(2j_{ik} + 1)/d^2_{j_i - m_i}}\, w^{(c)}_{j_i - m_i}[j_{ik}(j_{ik} + 1), a, j_i + j_k + 1] \,,$$

where $\varrho(s)$ and d_n^2 are the weight and the square norm of the polynomials $w_n(x)$, $a = (|m_{ik} - m'_{ik}| + |m_{ik} + m'_{ik}|)/2$, $c = (|m_{ik} - m'_{ik}| - |m_{ik} + m'_{ik}|)/2$, $m'_{ik} = j_i - j_k$, and

$$\varrho(j_{ik}) = \frac{(j_{ik} + m'_{ki})!(j_{ik} + m_{ik})!}{(j_{ik} - m'_{ki})!(j_i + j_k - j_{ik})!(j_i + j_k + j_{ik} + 1)!(j_{ik} - m_{ik})!} \,.$$

It is apparent that the right-hand side of (5.4.5) is a polynomial in two variables, $x = j_{13}(j_{13} + 1)$ and $y = j_{24}(j_{24} + 1)$, multiplied by the known function $[\varrho(j_{13})\varrho(j_{24})(2j_{13} + 1)(2j_{24} + 1]^{1/2}$; the use of identical notation for the weights of these two dual Hahn polynomials with different parameters should not be a source of confusion.

Assume now in (5.4.5) that $m = j$, $m_{13} = j_3 - j_1$, and $m_{24} = j + j_1 - j_3$ (one can easily verify that this assumption satisfies one of the constraints imposed above) and substitute the specific CGC (see Sect. 5.2) in this expression. This leads us to the following *polynomial representation* for the $9j$-symbol [S28]:

$$\left\{ \begin{matrix} j_1 & j_2 & j_{12} \\ j_3 & j_4 & j_{34} \\ j_{13} & j_{24} & j \end{matrix} \right\} = (-1)^{j_{13} + j_1 - j_3} \sqrt{\varrho(j_{13}, j_{24})}\, \tilde{u}_{j_{12} j_{34}}(x, y) \,. \tag{5.4.7}$$

Here

$$\varrho(j_{13}, j_{24}) = \frac{(j + j_{13} - j_{24})!(j - j_{13} + j_{24})!(j_{13} + j_{24} + j + 1)!}{(j_{13} + j_{24} - j)!(j_{13} + j_1 - j_3)!(j_1 + j_3 - j_{13})!}$$
$$\times \frac{(j_3 - j_1 + j_{13})!(j_4 - j_2 + j_{24})!}{(j_1 + j_3 + j_{13} + 1)!(j_{24} + j_2 - j_4)!(j_2 + j_4 - j_{24})!(j_2 + j_4 + j_{24} + 1)!} \,,$$

and for the polynomial $\tilde{u}_{j_{12} j_{34}}(x, y)$ in $x = j_{13}(j_{13} + 1)$ and $y = j_{24}(j_{24} + 1)$ there holds the decomposition

$$\tilde{u}_{j_{12}j_{34}}(x,y) = \frac{1}{\sqrt{(2j_{12}+1)(2j_{34}+1)(2j+1)!}} \sum_{m_1 m_2} C^{jj}_{j_{12},m_1+m_2,j_{34},j-m_1-m_2}$$
$$\times \, C^{j_{12},m_1+m_2}_{j_1 m_1 j_2 m_2} C^{j_{34},j-m_1-m_2}_{j_3,j_3-j_1-m_1,j_4,j+j_1-j_3-m_2}$$
$$\times \, \bar{w}_{j_1-m_1}(x)\bar{w}_{j_2-m_2}(y) , \qquad\qquad (5.4.7a)$$

where

$$\bar{w}_{j_1-m_1}(x) = \frac{1}{d_{j_1-m_1}} w^{(j_3-j_1)}_{j_1-m_1}(x, j_3-j_1, j_1+j_3+1) ,$$

$$\bar{w}_{j_2-m_2} = \frac{1}{d_{j_2-m_2}} w^{(j+j_1-j_3)}_{j_2-m_2}(y, j_4-j_2, j_2+j_4+1) .$$

The orthogonality relations (5.4.6) for the $9j$-symbols imply that the polynomials $\tilde{u}_{j_{12}j_{34}}(x,y)$ will be orthogonal on a rectangular nonuniform mesh with the weight $\varrho(j_{13}, j_{24})$, viz.

$$\sum_{j_{13}j_{24}} \tilde{u}_{j_{12}j_{34}}(x_{j_{13}}, y_{j_{24}})\tilde{u}_{j'_{12}j'_{34}}(x_{j_{13}}, y_{j_{24}})\varrho(j_{13}, j_{24})\Delta x_{j_{13}-1/2}\Delta y_{j_{24}-1/2}$$

$$= \delta_{j_{12}j'_{12}}\delta_{j_{34}j'_{34}}/(2j_{12}+1)(2j_{34}+1) ,$$

where $x_{j_{13}} = j_{13}(j_{13}+1)$, $y_{j_{24}} = j_{24}(j_{24}+1)$, $\Delta x_{j_{13}-1/2} = 2j_{13}+1$, $\Delta y_{j_{24}-1/2} = 2j_{24}+1$.

A number of publications elucidate the properties of the $9j$-symbols in more detail, see e.g. [Y1, V5, B24, B25, L18].

5.4.3 Basic Properties of the Polynomials Related to the $9j$-Symbols

Using the quantum theory of angular momentum, we can establish for the polynomials $\tilde{u}(x,y)$ in (5.4.7) the formulae of difference differentiation, a difference equation, an analogue of the Rodrigues' type formula, a dual property of orthogonality, etc. We discuss some of these properties below using a simplified notation for the $9j$-symbol as follows

$$\begin{Bmatrix} j_1 & j_2 & j_{12} \\ j_3 & j_4 & j_{34} \\ j_{13} & j_{24} & j \end{Bmatrix} = \begin{Bmatrix} a & b & c \\ d & e & f \\ g & h & j \end{Bmatrix}$$

5.4.3.1. In order to deduce the formulae of difference differentiation for the polynomials $\tilde{u}(x,y)$ with $x = g(g+1)$ and $y = h(h+1)$ the following recurrence relations for the $9j$-symbols [J3, V5] will be of use

$$2g2h\sqrt{(a+b-c)(a+b+c+1)}\begin{Bmatrix} a-\frac{1}{2} & b-\frac{1}{2} & c \\ d & e & f \\ g-\frac{1}{2} & h-\frac{1}{2} & j \end{Bmatrix}$$

$$= m(h-g)\sqrt{\sigma(g,-h)}\begin{Bmatrix} a & b & c \\ d & e & f \\ g & h-1 & j \end{Bmatrix} - m(-g-h)\sqrt{\sigma(g,h)}$$

$$\times \begin{Bmatrix} a & b & c \\ d & e & f \\ g & h & j \end{Bmatrix} - m(g-h)\sqrt{\sigma(-g,h)} \begin{Bmatrix} a & b & c \\ d & e & f \\ g-1 & h & j \end{Bmatrix}$$

$$+ m(g+h)\sqrt{\sigma(-g,-h)} \begin{Bmatrix} a & b & c \\ d & e & f \\ g-1 & h-1 & j \end{Bmatrix} , \tag{5.4.8}$$

$$(2g+1)(2h+1)\sqrt{(a+b-c)(a+b+c+1)} \begin{Bmatrix} a & b & c \\ d & e & f \\ g & h & j \end{Bmatrix}$$

$$= l(g+h+1)\sqrt{\sigma(-g-1,-h-1)} \begin{Bmatrix} a-\tfrac{1}{2} & b-\tfrac{1}{2} & c \\ d & e & f \\ g+\tfrac{1}{2} & h+\tfrac{1}{2} & j \end{Bmatrix}$$

$$- l(g-h)\sqrt{\sigma(-g-1,h)} \begin{Bmatrix} a-\tfrac{1}{2} & b-\tfrac{1}{2} & c \\ d & e & f \\ g+\tfrac{1}{2} & h-\tfrac{1}{2} & j \end{Bmatrix}$$

$$+ l(h-g)\sqrt{\sigma(g,-h-1)} \begin{Bmatrix} a-\tfrac{1}{2} & b-\tfrac{1}{2} & c \\ d & e & f \\ g-\tfrac{1}{2} & h+\tfrac{1}{2} & j \end{Bmatrix}$$

$$- l(-g-h-1)\sqrt{\sigma(g,h)} \begin{Bmatrix} a-\tfrac{1}{2} & b-\tfrac{1}{2} & c \\ d & e & f \\ g-\tfrac{1}{2} & h-\tfrac{1}{2} & j \end{Bmatrix} , \tag{5.4.9}$$

where $m(t) = l(t-1) = t - j - 1$, and

$$\sigma(g,h) = \frac{(g+h-j)(a+d+g+1)(a-d+g)(b+e+h+1)(b-e+h)}{g+h+j+1} .$$

Indeed, starting from (5.4.8) and (5.4.7) and observing the identities

$$\sigma(g,h)\varrho(g,h) = \sigma(-g,h)\varrho(g-1,h)$$
$$= \sigma(g,-h)\varrho(g,h-1)$$
$$= \sigma(-g,-h)\varrho(g-1,h-1)$$
$$= \varrho^{(a-1/2,b-1/2)}(g-\tfrac{1}{2},h-\tfrac{1}{2}) \tag{5.4.10}$$

(we usually write only variable parameters), we arrive at the following *formula of difference differentiation* for the polynomials $u(g,h) = \tilde{u}(x,y)/\sqrt{\alpha}$

$$Mu^{(a,b)}(g,h) = u^{(a-1/2,b-1/2)}(g-\tfrac{1}{2},h-\tfrac{1}{2}) . \tag{5.4.11}$$

where M is the divided difference operator

$$M = M(g,h) = m(g+h)\frac{\nabla_g \nabla_h}{\nabla_g x(g) \nabla_h y(h)} - \frac{\nabla_g}{\nabla_g x(g)} - \frac{\nabla_h}{\nabla_h y(h)} .$$

Similarly with the substitution

$$\begin{Bmatrix} a & b & c \\ d & e & f \\ q & h & j \end{Bmatrix} = \frac{(-1)^{a-d+g}}{\sqrt{\alpha \varrho(g,h)}} v^{(a,b)}(g,h) , \tag{5.4.12}$$

where $\alpha = (a + b - c)!(a + b + c + 1)!$, and the identities (5.4.10), which will be rewritten for convenience in the form

$$\frac{\sigma(-g - 1, -h - 1)}{\varrho^{(a-1/2,b-1/2)}(g + 1/2, h + 1/2)} = \frac{\sigma(-g - 1, h)}{\varrho^{(a-1/2,b-1/2)}(g + 1/2, h - 1/2)}$$

$$= \frac{\sigma(g, -h - 1)}{\varrho^{(a-1/2,b-1/2)}(g - 1/2, h + 1/2)}$$

$$= \frac{\sigma(g, h)}{\varrho^{(a-1/2,b-1/2)}(g - 1/2, h - 1/2)}$$

$$= 1/\varrho(g, h) \, ,$$

we find from (5.4.9)

$$Lv^{(a-1/2,b-1/2)}(g - 1/2, h - 1/2) = v^{(a,b)}(g, h) \, , \tag{5.4.13}$$

where

$$L = L(g, h) = l(g + h + 1)\frac{\Delta_g \Delta_h}{\Delta_g x(g - 1/2)\Delta_h y(h - 1/2)}$$

$$+ \frac{\Delta_g}{\Delta_g x(g - 1/2)} + \frac{\Delta_h}{\Delta_h y(h - 1/2)} \, .$$

Since

$$v^{(a,b)}(g, h) = \alpha^{(a,b)}\varrho^{(a,b)}(g, h)u^{(a,b)}(g, h) \, ,$$

and

$$\sigma(g, h)\varrho(g, h) = \varrho^{(a-1/2,b-1/2)}(g - 1/2, h - 1/2) \, ,$$

we obtain another *formula of difference differentiation*:

$$L[\sigma(g, h)\varrho(g, h)u^{(a-1/2,b-1/2)}(g - 1/2, h - 1/2)] + \lambda\varrho(g, h)u^{(a,b)}(g, h) = 0 \, ,$$
$$\tag{5.4.14}$$

where $\lambda = (a + b - c)(a + b + c + 1)$.

5.4.3.2. To obtain a difference equation for the polynomials $u(g, h)$ it will be sufficient to resort to the Eqs. (5.4.11) and (5.4.14), namely

$$L(\sigma\varrho M u) + \lambda\varrho u = 0 \, . \tag{5.4.15}$$

The quantum theory of angular momentum affords polynomial solutions to this equation in the form (5.4.7a).

5.4.3.3. Now we deduce an analogue of the Rodrigues' type formula for the polynomials related to the $9j$-symbols. By virtue of (5.4.14)

$$\varrho(g, h)u(g, h) = - \frac{1}{\lambda^{(a,b)}}L[\varrho^{(a-1/2,b-1/2)}$$

$$\times (g - 1/2, h - 1/2)u^{(a-1/2,b-1/2)}(g - 1/2, h - 1/2)]$$

$$= \frac{(-1)^2}{\lambda^{(a,b)}\lambda^{(a-1/2,b-1/2)}}$$

$$= L_0 L_1 [\varrho^{(a-1,b-1)}(g-1,h-1)u^{(a-1,b-1)}(g-1,h-1)]$$

$$= \frac{(-1)^n}{\lambda^{(a,b)}\lambda^{(a-1/2,b-1/2)}\ldots\lambda^{(a-(n-1)/2,b-(n-1)/2)}}$$
$$\times L_0 L_1 \ldots L_n [\varrho^{(a-n/2,b-n/2)}(g-n/2,h-n/2)$$
$$\times u^{(a-n/2,b-n/2)}(g-n/2,h-n/2) \, .$$

As a result for $n = a + b - c$ we get

$$u(g,h) = [B_n/\varrho(g,h)]L^{(n)}[\varrho^{(a-n/2,b-n/2)}(g-n/2,h-n/2)$$
$$\times u^{(a-n/2,b-n/2)}(g-n/2,h-n/2)] \, , \qquad (5.4.16)$$

where $L^{(n)} = L_1 L_1 \ldots L_{n-1}$, $L_p = (g-p/2, h-p/2)$, and

$$B_n = (-1)^n \frac{(a+b-c-n)!(a+b-n+c+1)!}{(a+b-c)!(a+b+c+1)!} \, .$$

We have also

$$M u^{(a-n/2,b-n/2)}(g,h) = 0$$

and

$$\lambda^{(a-n/2,b-n/2)} = 0 \, .$$

Equation (5.4.16) is the analogue of the Rodrigues' type formula we sought after for the considered polynomials in two discrete variables.

Proceeding in a similar manner on the base of the quantum theory of angular momentum, one may establish a number of other valuable properties for the polynomials (5.4.7a). The construction of a systematic mathematical theory of these polynomials would lead us astray from the main purpose of the book.

A portion of the findings evaluated above for the $9j$-symbols seem to afford a further generalization over arbitrary $3nj$-symbols which arise in combining $n+1$ angular momenta. Physical considerations suggest a number of new sets of polynomials orthogonal with respect to many discrete variables. Construction of their mathematical theory is awaited with interest.

5.5 The Classical Orthogonal Polynomials
of a Discrete Variable
in Some Problems of Group Representation Theory

The theory of special functions of mathematical physics penetrates deeply into analysis, the theory of functions of a complex variable, theoretical and mathematical physics, and has numerous applications. Special functions are a well studied field (see the Introduction) and their close relationship with group representation theory has been extensively documented [W7, V9, M13]. Until recently, however, only little attention has been attracted to group-theoretical interpretation of the properties of the classical orthogonal polynomials in a discrete variable.

In the preceding sections of this chapter we elucidated the close relations existent between the classical orthogonal polynomials in a discrete variable and the representations of the three-dimensional rotation group. In this section we shall briefly outline some generalizations. More specifically we shall examine the relationship between the Hahn polynomials and the representations of the rotation group SO(4) in four dimensional space, between the unitary irreducible representation of the Lorentz group SO(1,3) and the Hahn polynomials in an imaginary argument, and between the Racah polynomials and the representations of the group SU(3)[7]. The key facts about the representations of the groups SO(4), SO(1,3), and SU(3) will be assumed known and will be taken without proofs, in most cases, for which we refer the reader to [G13, N1, F2, P8].

5.5.1 The Hahn Polynomials and the Representation of the Rotation Group in Four-Dimensional Space

Consider a real four-dimensional Euclidean space with an orthonormal basis e_p, $p = 1, \ldots, 4$. All the possible rotations of the basis vectors by orthogonal matrices of unit determinant forms the *rotation group in four-dimensional space*, denoted by SO(4). Its representations may be evaluated in complete analogy with the problem of combining angular momenta [S25]. We establish the relation of the finite-dimensional irreducible representations of this group with the Hahn polynomials.

5.5.1.1. Let J_p be the infinitesimal operators corresponding to rotations in the three-dimensional space with the basis e_p, $p = 1, 2, 3$ and K_p be the infinitesimal operators of rotations in the planes (e_p, e_4), $p = 1, 2, 3$. The vectors J and K are known to commute according to the following rules (see, e.g. [G13, F2, S25])[8]:

$$[J_p, J_q] = i\varepsilon_{pqr}J_r \ ,$$

$$[J_p, K_q] = i\varepsilon_{pqr}K_r \ ,$$

$$[K_p, K_q] = i\varepsilon_{pqr}J_r \ ,$$

$$p, q, r = 1, 2, 3 \ . \tag{5.5.1}$$

---

[7] These groups frequently occur in physics. The group SO(4), for example, describes the symmetry of the quantum mechanical Coulomb problem in the case of a discrete spectrum. The Lorentz group SO(1,3) is of fundamental significance for relativistic physics. The representations of SU(3) are invoked for the classification of elementary particles.

[8] Here ε_{pqr} is the Levi-Civita symbol defined as follows: $\varepsilon_{pqr} = 0$ if at least two of its indices coincide, $\varepsilon_{pqr} = \varepsilon_{123} = 1$ if the indices p, q and r are obtained from the numbers 1, 2, and 3 as a result of an even permutation, and $\varepsilon_{pqr} = -\varepsilon_{123} = -1$ for an odd permutation of the indices. The summation in (5.5.1) and (5.5.2) is done over the repeated indices.

The transformation of the operators $A = (J + K)/2$, $B = (J - K)/2$ yields the following commutation rules for two independent angular momenta A and B:

$$[A_p, A_q] = i\varepsilon_{pqr} A_r ,$$
$$[B_p, B_q] = i\varepsilon_{pqr} B_r ,$$
$$[A_p, B_q] = 0 ,$$
$$p, q, r = 1, 2, 3 . \tag{5.5.2}$$

Therefore the construction of irreducible representation of the group SO(4) is closely connected with the problem of combining two angular momenta, $J = A + B$.

Because the operators A_3 and B_3 commute with each other, there exists a common system of their eigenvectors $\Phi_{m_1 m_2}$ in the representation space. Using the commutation rules (5.5.2) and following along the lines of argument used in proving Theorem 5.2 we may arrive at the form of the operator $A_\pm = A_1 \pm iA_2$, A_3, $B_\pm = B_1 \pm iB_2$, and B_3 in the basis $\Phi_{m_1 m_2}$, namely

$$A_\pm \Phi_{m_1 m_2} = \sqrt{(j_1 \mp m_1)(j_1 \pm m_1 + 1)} \Phi_{m_1 \pm 1, m_2} ,$$
$$A_3 \Phi_{m_1 m_2} = m_1 \Phi_{m_1 m_2} ,$$
$$B_\pm \Phi_{m_1 m_2} = \sqrt{(j_2 \mp m_2)(j_2 \pm m_2 + 1)} \Phi_{m_1, m_2 \pm 1} , \tag{5.5.3}$$
$$B_3 \Phi_{m_1 m_2} = m_2 \Phi_{m_1 m_2} .$$

where j_i are positive integers or half-integers, $-j_i \leq m_i \leq j_i$, $i = 1, 2$.

On the other hand, the infinitesimal operators J obey the commutation rules (5.1.4). Therefore the representation space has a canonical basis Ψ_{jm} on which the operators $J_\pm = J_1 \pm iJ_2$ and J_3 act by the Eqs. (5.1.11). This basis is referred to as the *Gelfand-Zetlin basis* for the group SO(4) and is denoted as follows:

$$\left| \begin{matrix} j_1 + j_2, & j_1 - j_2 \\ & j \\ & m \end{matrix} \right\rangle \equiv \Psi_{jm} . \tag{5.5.4}$$

(Gelfand-Zetlin bases have been introduced in [G9, G10] to describe the irreducible representations of the unitary matrix group SU(n) and the rotation group SO(n) in n-dimensional space. The basis used here at $n = 4$ differs from that introduced in [G10] by a phase multiplier. The infinitesimal operators J and K are elected to be Hermitian.)

Because $J = A + B$ the transformation between the bases Ψ_{jm} and $\Phi_{m_1 m_2}$ is effected by means of the Clebsch-Gordan coefficients, viz.

$$\Psi_{jm} = \sum_{m_1 + m_2 = m} \langle j_1 m_1 j_2 m_2 | j m \rangle \Phi_{m_1 m_2} . \tag{5.5.5}$$

The weight j takes on values in agreement with the triangle condition: $j_1 + j_2 \geq j \geq |j_1 - j_2|$.

5.5.1.2. From the infinitesimal standpoint the study of irreducible representations of the group SO(4) boils down to finding the form of J and K in the basis (5.5.4). Subject to the commutation relations (5.5.1) the action of these operators on the Gelfand-Zetlin basis Ψ_{jm} is completely defined by the action of the operators $J_\pm$, J_3, and K_3 [G10, G13]. For the operators $J_\pm$ and J_3 in the basis (5.5.4) the Eqs. (5.1.11) hold. In order to evaluate the matrix elements of $K_3 = A_3 - B_3$ we apply this operator to both sides of (5.5.5). Observing the recursion relation (5.2.17) for CGC we find

$$K_3\Psi_{jm} = a_{jm}\Psi_{j-1,m} + b_{jm}\Psi_{jm} + a_{j+1,m}\Psi_{j+1,m} \, , \tag{5.5.6}$$

where the values of a_{jm} and b_{jm} are also given in (5.2.17).

The recursion relation (5.2.17) is equivalent to that for the Hahn polynomials (see Sect. 5.2). Thus *the action of the infinitesimal operators J and K on the Gelfand-Zetlin basis (5.5.4) has been deduced with the help of the earlier evaluated properties of Hahn polynomials.* Specifically the Eq. (5.5.6) results as a consequence of the orthogonality of Hahn's polynomials. Conversely, proceeding from the formulae which define the action of J and K on the basis (5.5.4), one may obtain a *group-theoretical interpretation of the chief properties of Hahn polynomials.*

The aforementioned connections between the group representations and the Hahn polynomials have been discussed in [S16, N6]. A realization of Lie algebra representations of the group SO(4) by the Hahn polynomials introduced in [W2] has been constructed also in [M12] on the basis of the factorization technique.

5.5.3.1. Let $T(g)$ be an operator associated with an arbitrary finite rotation g in the space of the irreducible representation of SO(4). The matrix elements of this operator, when in the Gelfand-Zetlin basis, have the form

$$D^{[f]}_{jmj'm'}(g) = (\Psi_{jm}|T(g)\Psi_{j'm'})$$

(the symbol $[f]$ implies the first, upper line in (5.5.4)). Under a certain parametrization of SO(4) these matrix elements, i.e. the generalized spherical functions of this group, are known to be expressed through the Wigner functions $D^j_{mm'}(\alpha, \beta, \gamma)$ and the boost matrix elements[9]

$$d^{[f]}_{jj'm}(t) = (\Psi_{jm}|e^{-itK_3}\Psi_{j'm}) \, .$$

This function has been examined in [B8, F10, F9, L21, D10–13, B21, V7, S3]. By virtue of (5.5.5) and (5.5.3) it can be readily derived in the form

$$d^{[f]}_{jj'm}(t) = \sum_{m_1+m_2=m} \langle j_1 m_1 j_2 m_2 | jm \rangle e^{-it(m_1-m_2)} \langle j_1 m_1 j_2 m_2 | j'm \rangle \, .$$

It follows that the matrix elements of the boost $d^{[f]}_{jj'm}(t)$ are connected with the Hahn polynomials. In view of (5.2.13a) we have

---

[9] We call a rotation in the (x_3, x_4) plane a boost by analogy with the Lorentz group.

$$d^{[f]}_{jj'\,m}(t) = \sum_{\lambda} \tilde{h}_{j-\mu}(\lambda) e^{-it\lambda} \tilde{h}_{j'-\mu}(\lambda) \tilde{\varrho}(\lambda) \,, \tag{5.5.7}$$

where $\lambda = m_1 - m_2$, $\tilde{h}_n(\lambda) = d_n^{-1} h_n^{(\alpha,\beta)}(x, N)$, $\tilde{\varrho}(\lambda) = \varrho(x)$ at $x = (\alpha - \beta + 2N - 2)/4 + \lambda/2$, $\mu = (\alpha + \beta)/2$, and $\varrho(x)$ and d_n are the weight and norm of Hahn polynomials in (5.2.13).

Equation (5.5.7) enables us to examine the function $d^{[f]}_{jj'\,m}(t)$ by means of the Hahn polynomials. In particular, owing to the orthogonality of these polynomials, we have

$$d^{[f]}_{jj'\,m}(0) = \delta_{jj'} \,.$$

Linearizing the product of two Hahn polynomials in (5.5.7) yields

$$d^{[f]}_{jj'\,m}(t) = \sum_{j''} C^{jj'}_{j''}\, d^{[f]}_{j''mm}(t) \,. \tag{5.5.8}$$

The particular form of the function $d^{[f]}_{jmm}(t)$ arising here is expressed through a hypergeometric function in [B8]. The Eqs. (5.5.7) and (5.5.8) are generalized over the Lorentz group.

5.5.2 The Unitary Irreducible Representations of the Lorentz Group SO(1,3) and Hahn Polynomials in an Imaginary Argument

In this subsection we wish to examine the connections between the unitary irreducible representations of the Lorentz group and the Hahn polynomials in an imaginary argument. We shall elucidate the principal properties of the Lorentz group representations [G13, N1] by analogy with the familiar problem of combining two angular momenta.

5.5.2.1. Consider the Minkowski space, i.e., a four-dimensional, real-valued, pseudo-Euclidean space where the distance (interval) is defined by means of the quadratic form $s^2 = x_0^2 - x_1^2 - x_2^2 - x_3^2$. All rotations in the three-dimensional space (x_1, x_2, x_3) and boosts, i.e. the hyperbolic rotations in the planes (x_0, x_1), (x_0, x_2), and (x_0, x_3), form the proper Lorentz group SO(1,3).

Let J and K be the infinitesimal operator corresponding, respectively, to the spatial rotations and boosts of the Lorentz group SO(1,3). The operators J and K are known to obey the following commutation relations:

$$
\begin{aligned}
[J_p, J_q] &= i\varepsilon_{pqr} J_r \,, \\
[J_p, K_q] &= i\varepsilon_{pqr} K_r \,, \\
[K_p, K_q] &= -i\varepsilon_{pqr} J_r \,, \\
p, q, r &= 1, 2, 3 \,,
\end{aligned}
\tag{5.5.9}
$$

where ε_{pqr} is the Levi-Cevita symbol.

It is apparent that the transformation of operator $A = (J + iK)/2$ and $B = (J - iK)/2$ leads to the commutation relations of two independent angular momenta (5.5.2) [V4]. Therefore the construction of irreducible representations of the group SO(1,3) is intimately connected with the problem of combining two 'complex-conjugate momenta' A and B in a real-valued vector $J = A + B$.

For unitary representations[10] we have

$$J^+ = J , \quad K^+ = K , \quad A^+ = B .$$

According to the commutation rules (5.5.2), in the space of an irreducible representation of the group SO(1,3), we may construct a basis $\Phi_{m_1 m_2}$ which obeys the Eqs. (5.5.3) when subjected to the operators $A_\pm = A_1 \pm iA_2$, A_3, $B_\pm = B_1 \pm iB_2$, and B_3. (These formulae define the general form of operators A and B satisfying (5.5.2) in the basis of the eigenvectors of operators A_3 and B_3.) In the case of the Lorentz group, the constants j_1, j_2, m_1, and m_2 in (5.5.3) take on some complex values and $j_1^* = j_2$ and $m_1^* = m_2$ because $A^+ = B$.

Vectors $\Phi_{m_1 m_2}$ are the eigenvectors of two Hermitian operators $J_3 = A_3 + B_3$ and $K_3 = i^{-1}(A_3 - B_3)$, viz.

$$\begin{aligned}
J_3 \Phi_{m_1 m_2} &= m \Phi_{m_1 m_2} , \\
K_3 \Phi_{m_1 m_2} &= \lambda \Phi_{m_1 m_2} ,
\end{aligned} \tag{5.5.10}$$

and correspond to the real eigenvalues $m = m_1 + m_2$ and $\lambda = (m_1 - m_2)/i$; therefore

$$m_1 = (m + i\lambda)/2 , \quad m_2 = (m - i\lambda)/2 . \tag{5.5.11}$$

The basis $\chi_{\lambda m} \equiv \Phi_{m_1 m_2}$, where the quantum numbers are related by (5.5.11), is orthogonal and normalized, viz.

$$(\chi_{\lambda m} | \chi_{\lambda' m'}) = \delta_{mm'} \delta(\lambda - \lambda') , \tag{5.5.12}$$

where $\delta(\xi)$ is the Dirac delta-function.

On the other hand, the infinitesimal operators of rotation J obey the commutation rules (5.1.4). Therefore in the space of the irreducible representation there exists a basis Ψ_{jm} on which the operators $J_\pm = J_1 \pm iJ_2$ act by the Eqs. (5.1.11), where j is an integer or half-integer non-negative value, and $m = -j, -j + 1, \ldots, j - 1, j$.

5.5.2.2. As in the case of the SO(4) rotation group the study of the irreducible representations of the Lorentz group SO(1,3) from an infinitesimal standpoint reduces to evaluating the form of operators $K_\pm = K_1 \pm iK_2$ and K_3 in the basis Ψ_{jm} [G13, N1]. For this purpose we decompose Ψ_{jm} in the eigenfunctions of K_3, namely

$$\Psi_{jm} = \int_{-\infty}^{\infty} d\lambda \langle m_1 m_2 | jm \rangle \Phi_{m_1 m_2} , \tag{5.5.13}$$

---

[10] The Lorentz group SO(1,3) is not compact, therefore its irreducible representations are infinite-dimensional. A definition of the group representation, its unitary properties, etc., valid in this case may be found, e.g., in [N1, B7, G12].

subject to (5.5.11). The coefficients of this decomposition will be deduced as follows. Applying the operators $J_\pm = A_\pm + B_\pm$ to both sides of (5.5.13) leads by virtue of (5.1.11), (5.5.3) and (5.5.12) to the recursion relations (5.2.6) for these coefficients, the familiar equations of the theory of angular momentum. Incorporating the substitution

$$\langle m_2 m_2 | j m \rangle = \alpha_\pm C^\pm_{m_1 m_2 m} u^\pm_{jm}(m_1) \, , \tag{5.5.14}$$

where

$$C^\pm_{m_1 m_2 m} = \left[\frac{\Gamma(j_1 + m_1 + 1)\Gamma(j_2 + m_2 + 1)(j - m)!}{\Gamma(j_1 - m_1 + 1)\Gamma(j_2 - m_2 + 1)(j + m)!} \right]^{\pm 1/2} \, ,$$

$$\alpha_+^{-1} = \sin \pi(j_1 - m_1 + 1) \, , \quad \alpha_-^{-1} = \sin \pi(j_2 + m_2 + 1) \, ,$$

we obtain for $u^\pm_{jm}(m_1)$ the simple *formulae of difference differentiation*

$$u^+_{j,m+1}(m_1) = \nabla u^+_{jm}(m_1) \, , \quad u^-_{j,m-1}(m_1) = \Delta u^-_{jm}(m_1) \, . \tag{5.5.15}$$

Now, proceeding as in the derivation of Eq. (5.2.11) for CGC, we obtain

$$\langle m_1 m_2 | j m \rangle = \frac{(-1)^{j-m}}{\sin \pi(j_1 - m_1 + 1)} \frac{A}{(2j)!}$$

$$\times \sqrt{\frac{\Gamma(j_1 - m_1 + 1)\Gamma(j_2 - m_2 + 1)(j + m)!}{\Gamma(j_1 + m_1 + 1)\Gamma(j_2 + m_2 + 1)(j - m)!}}$$

$$\times \Delta^{j-m}_{m_1} \left[\frac{\Gamma(j_1 + m_1 + 1)\Gamma(j_2 + j - m_1 + 1)}{\Gamma(j_1 - m_1 + 1)\Gamma(j_2 - j + m_1 + 1)} \right] \, , \tag{5.5.16}$$

where A is a constant derived from the normalizing condition

$$\|\Psi_{jj}\|^2 = \int_{-\infty}^{\infty} d\lambda |\langle m_1 m_2 | j j \rangle|^2 = 1 \, . \tag{5.5.17}$$

The complex-conjugate values of the weights $j_1^* = j_2$ in (5.5.3) should be selected such that the condition (5.5.17) is satisfied.

Resorting to Barnes' lemma [B1, W5] it is an easy matter to verify that the normalizing condition (5.5.17) is satisfied for the unitary irreducible representations of the Lorentz group [G13, N1][11]

(i) for the principal series of $j_1 = j_2^* = (\mu - 1 + i\gamma)/2$, where $j = |j_1 + j_2 + 1|$, $|j_1 + j_2 + 1| + 1, \ldots = |\mu|, |\mu| + 1, \ldots$ (μ is an integer or half-integer, γ is real valued);

(ii) for a supplementary series of $j_1 = j_2 = (\delta - 1)/2$, where $-1 < \delta < 1$, $j = j_1 - j_2, j_1 - j_2 + 1, \ldots = 0, 1, 2, \ldots$.

In the circumstances we have

---

[11] We employ the notation related to that introduced in [G13] as follows: $l_0 = \mu$, $l_1 = i\gamma$, $l = j$, $\xi_{jm} = i^{j-m}\Psi_{jm}$.

$$\frac{2|A|}{\sqrt{\pi(2j)!}}$$

$$= \sqrt{\frac{2j+1}{\Gamma(j+j_1-j_2+1)\Gamma(j-j_1+j_2+1)\Gamma(j-j_1-j_2)\Gamma(j+j_1+j_2+2)}} \cdot \tag{5.5.18}$$

In the following it will be convenient to let $A = |A|\mathrm{i}^{j-m}$.

By analogy with the rotation group the coefficients of (5.5.13) are referred to as the *'complexified Clebsch-Gordan coefficients'* [S13, S15]. Using the Rodrigues formulae for the Hahn polynomials (1.2.8) these coefficients, by virtue of (3.10.24, 25), (5.5.16) and (5.5.18), can be expressed through the Hahn polynomials in an imaginary argument. For the principal series we have

$$\langle m_1 m_2 | j m \rangle = f \frac{\sqrt{\varrho(\lambda)}}{d_{j-m}} p_{j-m}^{(m-\mu, m+\mu)}(\lambda, \gamma) \,, \tag{5.5.19}$$

where $\varrho(\lambda)$ and d_n are the weight and norm of the polynomials $p_n^{(\alpha,\beta)}(\lambda, \gamma)$ (see Table 3.8).

For the supplementary series

$$\langle m_1 m_2 | j m \rangle = f \frac{\sqrt{\varrho(\lambda)}}{d_{j-m}} q_{j-m}^{(m)}(\lambda, \delta) \,, \tag{5.5.20}$$

where $\varrho(\lambda)$ and d_n are the weight and norm of the polynomials $q_n^{(\alpha)}(\lambda, \delta)$, and the factor f in the penultimate and last equations is as follows

$$f = \sqrt{\frac{\sin \pi(j_2 - m_2 + 1)}{\sin \pi(j_1 - m_1 + 1)}} \,, \quad ff^* = 1 \,.$$

The complexified Clebsch-Gordan coefficients are real-valued accurate to this factor.

5.5.2.3. The Eqs. (5.5.13, 19) and (5.5.20) enable one to derive the matrix elements of operators $K_\pm$ and K_3 on the basis Ψ_{jm}, drawing on the known properties of Hahn polynomials in an imaginary argument. From the recursion relation, say, we get

$$K_3\Psi_{jm} = a_{jm}\Psi_{j-1,m} + b_{jm}\Psi_{jm} + a_{j+1,m}\Psi_{j+1,m} \,, \tag{5.5.21}$$

where for the principal series

$$a_{jm} = \frac{1}{j}\sqrt{\frac{(j^2 - m^2)(j^2 - \mu^2)(j^2 + \gamma^2)}{4j^2 - 1}} \,,$$

$$b_{jm} = \frac{m\mu\gamma}{j(j+1)} \,.$$

For the supplementary series one is to put $\mu = 0$ and $\gamma = -\mathrm{i}\delta$.

The action of $K_\pm$ on the basis Ψ_{jm} is not hard to determine by means of (5.1.11), (5.5.21) and the comutation rules (5.5.9). This implies that the action of the infinitesimal operators J and K on the basis Ψ_{jm} of the irreducible representation of the Lorentz group SO(1,3) has been obtained with the help of the properties of the Hahn polynomials in an imaginary argument. Conversely the formulae specifying the action of J and K on the basis Ψ_{jm} [G13, N1] bring about a group-theoretical interpretation for the principal properties of these polynomials.

5.5.2.4. The preceding consideration paves the way for deriving the integral representation for the boost matrix elements

$$d^{(\mu,\gamma)}_{jj'm}(t) = (\Psi_{jm}|e^{-itK_3}\Psi_{j'm}) \ .$$

For the principal series (μ,γ) we have from (5.5.13) and (5.5.19)

$$d^{(\mu,\gamma)}_{jj'm}(t) = \int_{-\infty}^{\infty} \tilde{p}_{j-m}(\lambda)e^{-it\lambda}\tilde{p}_{j'-m}(\lambda)\varrho(\lambda)d\lambda \ , \tag{5.5.22}$$

where

$$\tilde{p}_{j-m}(\lambda) = d^{-1}_{j-m}p^{(m-\mu,m+\mu)}_{j-m}(\lambda,\gamma) \ .$$

For the supplementary series, a similar integral representation for the boost matrix can be readily obtained from (5.5.13) and (5.5.20).

The integral representation (5.5.22) is a tool to investigate the function $d^{(\mu,\gamma)}_{jj'm}(t)$ based on the known properties of the Hahn polynomials in an imaginary argument. Specifically from the property of orthogonality (3.10.20) we get $d^{(\mu,\gamma)}_{jj'm}(0) = \delta_{jj'}$. Linearization of the product of two Hahn polynomials in (4.5.22) leads to (5.5.8).

Integral representations for the boost matrix elements have been examined by a number of workers [V7, V6]. An integral representation of the type (5.5.22) has been obtained in [S15] and its relation with the Hahn polynomials has been established in [S26, A33]. Investigations of the function $d^{(\mu,\gamma)}_{jj'm}(t)$ from other standpoints have also been reported [S3, L17, K35, M2, L16, B9].

5.5.3 The Racah Polynomials and the Representations of the Group SU(3)

The unitary group SU(3) of unimodular matrices is a well known entity owing to its applications in the field of nuclear and particle physics (see, e.g. [G16, F2]). Investigation of unitary irreducible representations of this group is a way to give a group theoretical interpretation to the basic properties of Racah polynomials.

5.5.3.1. The infinitesimal operators A_{ik}, $i, k = 1, 2, 3$ of the group SU(3) satisfy the standard relations

$$[A_{ik}, A_{pq}] = \delta_{kp}A_{iq} - \delta_{iq}A_{pk} \ , \qquad A_{11} + A_{22} + A_{33} = 0 \ ,$$
$$A^{\dagger}_{ik} = A_{ki} \ .$$

The finite-dimensional irreducible representations of the group SU(3) are specified by the highest weight $\Lambda \equiv (\lambda, \mu)$, λ and μ being non-negative integers. For elementary particle applications, in the space of the irreducible representation of weight $\Lambda = (\lambda, \mu)$, three *canonical Gelfand-Zetlin* bases [G9] are selected which correspond to various reductions of SU(3) to the subgroup U(1) $\times$ SU(2) [P8], viz.

(1) A basis corresponding to the reduction

$$\mathrm{SU}(3) \supset \mathrm{U}_Y(1) \times \mathrm{SU}_T(2) \supset \mathrm{U}_{T_0}(1) \ , \tag{5.5.23}$$

where $\mathrm{SU}_T(2)$ is the T-spin subgroup with the intesimal operators $T_+ = A_{23}$, $T_- = A_{32}$, and $T_0 = (A_{22} - A_{33})/2$ which satisfy the ordinary commutation rules of angular momentum, namely $[T_0, T_\pm] = \pm T_\pm$ and $[T_+, T_-] = 2T_0$. In the theory of elementary particles the infinitesimal operator $Y = -(2A_{11} - A_{22} - A_{33})/3$ of the subgroup $\mathrm{U}_Y(1)$ is called the *hypercharge operator*. The operators of T-spin, $T_\pm$, T_0, and of hypercharge, Y, commute with each other.
We refer to the basis of the irreducible representation of SU(3) for the reduction of (5.5.23) as the *T-basis* and denote it as follows

$$\left| \begin{matrix} \lambda\mu \\ ytt_0 \end{matrix} \right\rangle \equiv |ytt_0\rangle \ .$$

In this basis the operators Y, T^2, and T_0 are diagonal,

$$Y|ytt_0\rangle = y|ytt_0\rangle \ ,$$
$$T^2|ytt_0\rangle = t(t+1)|ytt_0\rangle \ ,$$
$$T_0|ytt_0\rangle = t_0|ytt_0\rangle \ .$$

The operators $T_\pm$ act upon the T-basis by (5.1.11).

(2) In a very similar manner we introduce the basis for the reduction

$$\mathrm{SU}(3) \supset \mathrm{U}_Z(1) \times \mathrm{SU}_U(2) \supset \mathrm{U}_{U_0} \ , \tag{5.5.24}$$

where $\mathrm{SU}_U(2)$ is the U-spin subgroup with the infitinesimal operators $U_+ = A_{12}$, $U_- = A_{21}$, and $U_0 = (A_{11} - A_{22})/2$; in elementary particle physics the operator $Z = -(A_{11} + A_{22} - 2A_{33})/3$ is related to the charge of hadrons. The basis for the reduction (5.5.24) will be called the *U-basis* and denoted as

$$\left| \begin{matrix} \lambda\mu \\ zuu_0 \end{matrix} \right\rangle \equiv |zuu_0\rangle \ .$$

Here the operators Z, U^2, and U_0 are diagonal.

(3) The *V-basis* which is associated with the reduction

$$\mathrm{SU}(3) \supset U_X(1) \otimes \mathrm{SU}_V(2) \supset U_{V_0} \ , \tag{5.5.25}$$

where $\mathrm{SU}_V(2)$ is a V-spin subgroup formed by the operators $V_+ = A_{13}$, $V_- = A_{31}$, and $V_0 = (A_{11} - A_{33})/2$; the infinitesimal operator $X = -(A_{11} - 2A_{22} + A_{33})/3$ of the subgroup $U_X(1)$ is an analogue of hypercharge Y. To denote the V-basis we use the quantum numbers of the diagonal operators X, V^2, and V_0, viz.

$$\left| \begin{matrix} \lambda\mu \\ xvv_0 \end{matrix} \right\rangle \equiv |xvv_0\rangle .$$

The number of vectors in each of the aforementioned canonical bases equals the dimension of the irreducible representation (λ, μ), namely

$$\dim \varLambda = (\lambda + 1)(\mu + 1)(\lambda + \mu + 2)/2 .$$

For constraints imposed on the quantum numbers, see [P8].

5.5.3.2. The transformation between T-, U-, and V-bases is effected by the $6j$-symbols. The T- and U-bases, for example, are related by

$$|jtt_0\rangle = \sum_m W_{u,t}^{\varLambda}|\bar{f}u\bar{u}_0\rangle , \tag{5.5.26}$$

where

$$W_{u,t}^{\varLambda} = (-1)^{\lambda+\mu}\sqrt{(2t+1)(2u+1)} \left\{ \begin{matrix} j & \frac{\mu}{2} & t \\ \frac{\lambda}{2}+\frac{\mu}{2}-\bar{f}, & \frac{\lambda}{2}-j+\bar{f}, & u \end{matrix} \right\} ,$$

$$\bar{f} = \frac{1}{2}\left(\frac{\mu}{2} + j - t_0 \right) , \quad \bar{u}_0 = \frac{1}{2}\left(\lambda + \frac{\mu}{2} - 3j - t_0 \right) .$$

To denote the T- and U-bases Eq. (5.5.26) invokes new quantum numbers j and f in place of y and z, respectively, defined by the equations $y = -(2\lambda + \mu)/3 + 2j$ and $z = -(\lambda + 2\mu)/3 + 2f$. The T and V, and U and V-bases are related to each other in a similar manner (see [P8]).

The transformations of the type (5.5.26) are exploited in constructing generalized spherical functions of the group SU(3) [P8, C5], and find their use for derivation of mass formulae for hadrons [G16].

5.5.3.3. Equation (5.5.26) enables us to give a group-theoretical interpretation for the basic relations of the Racah polynomials. Recalling the formulae

$$U_+|jtt_0\rangle = a_{jtt_0}|j - \tfrac{1}{2}, t - \tfrac{1}{2}, t_0 - \tfrac{1}{2}\rangle$$
$$+ b_{jtt_0}|j - \tfrac{1}{2}, t + \tfrac{1}{2}, t_0 - \tfrac{1}{2}\rangle , \tag{5.5.27}$$

$$U_-|jtt_0\rangle = b_{j+\frac{1}{2},t-\frac{1}{2},t_0+\frac{1}{2}}|j + \tfrac{1}{2}, t - \tfrac{1}{2}, t_0 + \tfrac{1}{2}\rangle$$
$$+ a_{j+\frac{1}{2},t+\frac{1}{2},t_0+\frac{1}{2}}|j + \tfrac{1}{2}, t + \tfrac{1}{2}, t_0 + \tfrac{1}{2}\rangle , \tag{5.5.28}$$

where

$$a_{jtt_0} = \sqrt{\frac{(j + t - \mu/2)(\lambda + \mu/2 - j - t + 1)(t + j + \mu/2 + 1)(t + t_0)}{2t(2t + 1)}} ,$$

$$b_{jtt_0} = \sqrt{\frac{(j-t+\mu/2)(\lambda+\mu/2-j+t+2)(\mu/2-j+t+1)(t-t_0+1)}{(2t+1)(2t+2)}} \,,$$

we apply the operators $U_\pm$ to both sides of (5.5.26) to obtain the recursion relations (5.3.14) and (5.3.15) for the $6j$-symbols. In Sect. 5.3 we demonstrated that these recursion formulae are equivalent to those of difference differentiation for the Racah polynomials. In a similar manner, applying T^2 and U^2 to both sides of (5.5.26), we obtain a difference equation and a recursion relation for the Racah polynomials.

Conversely, using decompositions of the type (5.5.26) and the recurrence relations (5.3.14) and (5.3.15), one may derive matrix elements of the infinitesimal operators in the Gelfand-Zetlin basis, and specifically deduce the Eqs. (5.5.27) and (5.5.28). This implies that the basic properties of the Racah polynomials are closely connected with the representations of the group SU(3).

5.5.4 The Charlier Polynomials and Representations of the Heisenberg-Weyl Group

It is not hard to verify that the set of upper triangle matrices of the form

$$\begin{pmatrix} 1 & \alpha & \gamma \\ 0 & 1 & \beta \\ 0 & 0 & 1 \end{pmatrix} \equiv (\alpha, \beta, \gamma) \,, \tag{5.5.29}$$

where α, β and γ are arbitrary real numbers, forms a group under an ordinary operation of matrix multiplication. The relevant group composition law has the form

$$(\alpha, \beta, \gamma)(\alpha', \beta', \gamma') = (\alpha + \alpha', \beta + \beta', \gamma + \gamma' + \alpha\beta') \,,$$

the unit element, e, is $(0,0,0)$, and for an inverse element

$$(\alpha, \beta, \gamma)^{-1} = (-\alpha, -\beta, -\gamma + \alpha\beta) \,.$$

This group is known as the *three-dimensional Heisenberg-Weyl group*, denoted by N(3). As will be recalled its representations are closely connected with the canonical commutation algebra for the operators of coordinate and momentum in quantum mechanics (see, e.g., [B7]).

Let us establish the relation of representations of N(3) with the Charlier polynomials. Let $f(z)$ be an arbitrary element in a space of entire analytical functions of exponential growth. In this space the representations of N(3) will be defined as

$$T(g)f(z) = e^{-(\gamma+\beta z)}f(z+\alpha) \,. \tag{5.5.30}$$

Using the matrix composition law it is an easy matter to verify that such operators $T(g)$ do specify a representation of N(3), viz.

$$T(g)T(g') = T(gg') \,, \quad T(e) = 1 \,.$$

A natural basis in the space of entire analytical functions arises as a set of powers $\{z^n\}$, $n = 0, 1, \ldots$. Let us calculate matrix elements of representations (5.5.30) in this basis

$$T(g)z^n = \sum_{m=0}^{\infty} t_{mn}(g)z^m \ .$$ (5.5.31)

By virtue of (5.5.30) and (5.5.31) we have

$$e^{-(\gamma+\beta z)}(z + \alpha)^n = \sum_{p=0}^{\infty} t_{pn}(g)z^p \ .$$ (5.5.32)

(We assume that this series is convergent at sufficiently small $|z| < 1$.) To obtain the matrix elements $t_{mn}(g)$ we differentiate both sides of this equality m times with respect to z and let $z = 0$. Observing the Leibniz formula, we get

$$\begin{aligned}
t_{mn}(g) &= \frac{1}{m!} \left[(z + \alpha)^n e^{-(\gamma+\beta z)} \right]^{(m)} \Big|_{z=0} \\
&= \sum_{p=0}^{m} \frac{[(z+\alpha)^n]^{(p)}|_{z=0}[e^{-(\gamma+\beta z)}]^{(m-p)}|_{z=0}}{p!(m-p)!} \\
&= \frac{(-1)^m}{m!} e^{-\gamma} \alpha^n \beta^m {}_2F_0(-m, -n, -1/\alpha\beta) \ ,
\end{aligned}$$

whence by virtue of (2.7.9) we find the representation of the matrix elements (5.5.31) through the Charlier polynomials $c_n^\mu(x)$, namely

$$t_{mn}(g) = \frac{(-1)^m}{m!} e^{-\gamma} \alpha^n \beta^m c_n^{\alpha\beta}(m) \ .$$ (5.5.33)

By virtue of (2.7.10) and the symmetry property $c_n^\mu(m) = c_m^\mu(n)$ $(m, n = 0, 1, \ldots)$ these matrix elements also allow a representation through the Laguerre polynomials $L_n^\alpha(\mu)$ [V9], viz.

$$t_{mn}(g) = \begin{cases} (-1)^{m+n} \frac{n!}{m!} e^{-\gamma} \beta^{m-n} L_n^{m-n}(\alpha\beta) \ , & m \geq n \\ e^{-\gamma} \alpha^{n-m} L_n^{n-m}(\alpha\beta) \ , & m < n \ . \end{cases}$$ (5.5.34)

It follows that $t_{mn}(e) = \delta_{mn}$, and in view of the orthogonality of the Charlier polynomials we also have the relation

$$\sum_{n=0}^{\infty} t_{mn}(g) t_{nm'}(g^{-1}) = \delta_{mm'}$$

readily verifiable for $0 < \alpha\beta < 1$.

In turn, the Eqs. (5.5.32) and (5.5.33) lead to the generating function for the Charlier polynomials

$$e^t \left(1 - \frac{t}{\mu} \right)^x = \sum_{n=0}^{\infty} c_n^\mu(x) \frac{t^n}{n!} \ .$$

This expression has been derived for positive integers x. Its validity in a wider domain of x may be verified by expanding the left-hand side in a power series in t assuming $|t| < 1$.

Matrix elements similar to those discussed above arise in applications of representations of the Heisenberg-Weyl group to signal theory [S2]. Matrix elements of the type (5.5.34) enter also in the Schwinger formula for transition probabilities of a quantum oscillator subjected to external force [B15] where the said relations to the Charlier polynomials are also significant.

6. Hyperspherical Harmonics

This chapter is devoted to the construction of solutions to the n-dimensional Laplace equation

$$\Delta u = \frac{\partial^2 u}{\partial x_1^2} + \ldots + \frac{\partial^2 u}{\partial x_n^2} = 0$$

by separation of variables in spherical coordinates. An important class of special functions which naturally occur in this work is constituted by *hyperspherical harmonics*. In quantum mechanis these functions are used to construct basis functions in the K-harmonic method and in the translation-invariant model of shells thus enabling one to compute the fundamental physical characteristics of light nuclei; in representation theory these functions are exploited to study representations of the rotation group and motion group over an n-dimensional Euclidean space, to name just a few applications.

In an n-dimensional Euclidean space of $n \geq 3$, a large number of distinct systems of spherical coordinates can be set up, each of which has its own spherical harmonics. In applications it is important to know matrix elements of the transformation between the harmonics evolving in different coordinate systems. These quantities, called the T-coefficients, are closely related to the classical orthogonal polynomials of a discrete variable. Depending on the dimension of space they can be expressed through the Racah, Hahn, or Kravchuk polynomials. Hyperspherical harmonics are a useful tool in solving the Schrödinger equation for the n-dimensional harmonic oscillator in spherical coordinates. The structure of the oscillator wavefunctions yields a group-theoretical interpretation of the T-coefficients.

6.1 Spherical Coordinates in a Euclidean Space

Before we embark on the examination of the solutions to the equation

$$\Delta u = \frac{\partial^2 u}{\partial x_1^2} + \ldots + \frac{\partial^2 u}{\partial x_n^2} = 0 \tag{6.1.1}$$

we need to define the hyperspherical coordinates in the Euclidean space and set up the Laplace operator in these coordinates.

6.1.1 Setting up Spherical Coordinates

In a three-dimensional Euclidean space the position of a point can be readily defined by geometric considerations. In an n-dimensional space, where our geometric intuition is no longer applicable, the definition of hyperspherical coordinates is based on a more abstract construction which formalizes the familiar argument for low-dimension spaces as follows.

Consider an n-dimensional Euclidean space R which is decomposable into the *direct sum* of two orthogonal subspaces R_1 and R_2 of dimension n_1 and n_2, respectively,

$$R = R_1 \oplus R_2 , \quad n = n_1 + n_2 ;$$

at $n = 2$ we would elect any two perpendicular lines for R_1 and R_2, at $n = 3$ these would be a plane and a perpendicular to it. For an arbitrary vector x in R we have

$$\begin{aligned} x &= x_1 + x_2 , \\ x_1 \cdot x_2 &= 0 , \end{aligned} \qquad (6.1.2)$$

where x_i are the *orthogonal projections* of x onto the subspaces R_i; $x_i \in R_i$, $i = 1, 2$ (Fig. 6.1). It is worth recalling some notions of the theory of Euclidean spaces. A space R is said to be decomposed into the direct sum of its subspaces R_1 and R_2 if any vector x in R can be uniquely represented in the form

$$x = x_1 + x_2$$

of $x_1 \in R_1$ and $x_2 \in R_2$. By this definition the space R is decomposable into the direct sum of R_1 and R_2 provided that

(1) R_1 and R_2 have only one (zero) vector $x = 0$ in common, and

(2) the sum of dimensions of R_1 and R_2 equals the dimension of R.

(For a proof of this assertion, see, e.g., [G8, K27].) A vector $h \in R$ is said to be *orthogonal* to a subspace $R_1 \subset R$ if it is orthogonal to any vector $x_1 \in R_1$, i.e., $h \cdot x = 0$ (we use the designation $x \cdot y$ for the inner product of vectors x and y in this chapter). Subspaces R_1 and R_2 are orthogonal if every vector in one is orthogonal to the other subspace and vice versa. In the cirumstances $R_2(R_1)$ forms and *orthogonal complement* to subspace $R_1(R_2)$. In an n-dimensional Euclidean space the problem of droping a perpendicular of a point x in R to a subspace $R_1 \subset R$, i.e. the problem of constructing a vector x_1 in R_1 such that $h = x - x_1$ is orthogonal to R_1, has a unique solution. The vector $x_1 \in R_1$ derived in this manner is referred to as an *orthogonal projection* of x onto R_1 (see, e.g., [G8]).

In general all aforementioned concepts generalize familiar facts of elementary geometry over multidimensional cases.

Let s be a unit vector in the direction of x : $s = x/r$, $r = \sqrt{x^2}$. Then the decomposition (6.1.2) takes the form

$$s = s_1 \sin\theta + s_2 \cos\theta \, , \tag{6.1.3}$$

where θ is the angle between x and x_2 measured counterclockwise from x_2 to x in the plane of these vectors (see Fig. 6.1.) around the axis making a left-hand trihedral with x_1 and x_2 and s_i are unit vectors directed in x_i, $i = 1, 2$.[1]

Each subspace R_1 and R_2 is in turn represented as a direct sum of two orthogonal subspaces by establishing the respective angles θ_1 and θ_2. This process is continued until we arrive at unidimensional subspaces whose unit vectors $e_1, \ldots, e_n$ form by construction an orthogonal basis of R : $e_i \cdot e_k = \delta_{ik}$, $i, k = 1 \ldots, n$. As a result we get

$$x = x_1 e_1 + \ldots + x_n e_n \, ,$$

where x_i are the Cartesian coordinates of x in the basis e_i, $i = 1, \ldots, n$.

The collection of numbers r, Ω, where $\Omega = \{\theta, \theta_1, \theta_2, \ldots\}$ denotes the values of all angles that successfully arise in the process of construction as outlined above, uniquely defines a vector $x = x(r, \Omega)$. We shall refer to the numbers r and Ω as the *spherical coordinates* of the point x in the Euclidean space R. This name owes its existence to the fact that the values of angles Ω uniquely define a point s on the unit sphere S^{n-1} in R: $s = s(\Omega)$, $s \in S^{n-1}$.

It is quite obvious that a few distinct systems of spherical coordinates can be set up in R – every sequence of partitions of R into a direct sum of orthogonal subspaces would be associated with its own system of coordinates r, Ω.

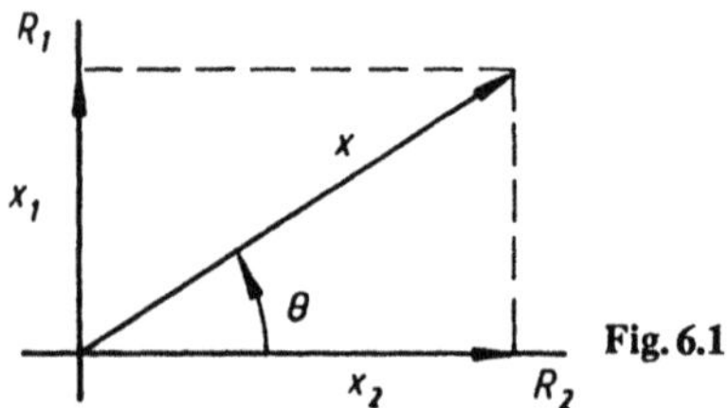

Fig. 6.1

6.1.2 A Metric and Elementary Volume

In order to write the Laplace operator

$$\Delta = \frac{\partial^2}{\partial x_1^2} + \ldots + \frac{\partial^2}{\partial x_n^2} \tag{6.1.4}$$

in spherical coordinates r, Ω we calculate the metric, that is, the squared differential of arc length, $(dx)^2$, in these coordinates. Since $x = rs$ and $s^2 = 1$, then $dx = r\,ds + s\,dr$ and $s \cdot ds = 0$, hence

---

[1] Thus we eliminate arbitrary choice of the axis and direction of rotation. For definiteness we shall introduce the angle subject to these conditions, although in some situations the opposite sense of rotation might be a more convenient choice.

$$dx^2 = dr^2 + r^2 ds^2 , \tag{6.1.5}$$

where $dx^2 = (dx)^2$, $dr^2 = (dr)^2$, and $ds^2 = (ds)^2$. Observing the orthogonality of R_1 and R_2, we find from (6.1.3) for the metric ds^2 on unit sphere S^{n-1} in R

$$ds^2 = d\theta^2 + \sin^2 \theta \, ds_1^2 + \cos^2 \theta \, ds_2^2 . \tag{6.1.6}$$

Here ds_i^2 is the metric on unit sphere $S^{n_i-1} \subset R_i$, $i = 1,2$ ($ds_i^2 = 0$ if R_i is a unidimensional subspace).

Applying (6.1.6) in succession to the spheres $S^{n_1-1} \subset R_1, S^{n_2-1} \subset R_2$, and so on up to unidimensional subspaces we get, upon collecting like terms,

$$ds^2 = \sum_\alpha g_\alpha^2 d\theta_\alpha^2 , \tag{6.1.7}$$

where the summation is over all angles $\Omega = \{\theta, \theta_1, \theta_2, \ldots\}$, and $g_\alpha = g_\alpha(\Omega)$ are certain coefficients. It follows from (6.1.5) and (6.1.7) that a spherical system of coordinates r, Ω is *orthogonal* in the Euclidean space R.

In agreement with (6.1.5) and (6.1.7) in the coordinates r, Ω the elementary volume $dv = dx_1 \ldots dx_n$ has the form

$$dv = r^{n-1} dr \, d\Omega , \quad d\Omega = \prod_\alpha g_\alpha d\theta_\alpha .$$

To recall the concepts used we suppose that in a Euclidean space R a vector x is given by curvilinear coordinates $x = x(\eta_1, \ldots, \eta_n)$. If the respective metric $dx^2 = (dx)^2$ has the form

$$(dx)^2 = \sum_{i=1}^n h_i^2 (d\eta_i)^2 , \tag{6.1.8}$$

where $h_i = h_i(\eta_1, \ldots, \eta_n)$ are certain coefficients, the coordinate system $\eta_1, \ldots, \eta_n$ is said to be orthogonal. The elementary volume in such a system is

$$dv = dx_1 \ldots dx_n = \prod_{i=1}^n h_i d\eta_i ,$$

and for the Laplace operator $\Delta = \mathrm{div\ grad}$ we have

$$\Delta = \frac{1}{h} \sum_{i=1}^n \frac{\partial}{\partial \eta_i} \left(\frac{h}{h_i^2} \frac{\partial}{\partial \eta_i} \right) . \tag{6.1.9}$$

where $h = \prod_{i=1}^n h_i$.

We give a simple derivation of (6.1.9) following along the line of [K28]. Consider a function u which possesses two derivatives in some area of R, vanishes on the boundary of this area, and is arbitrary elsewhere. Let us express the operator $(\mathrm{grad})^2$ in Cartesian x_i and curvilinear η_k coordinates. Noting that both coordinate systems are orthogonal we obtain that the squared gradient equals the sum of squared derivatives in coordinate arc lengths:

$$(\text{grad } u)^2 = \sum_i \left(\frac{\partial u}{\partial x_i}\right)^2 = \sum_k \frac{1}{h_k^2} \left(\frac{\partial u}{\partial \eta_k}\right)^2 .$$

Here we used the formula $ds_k = h_k d\eta_k$ for arc lenght along a coordinate, which corresponds to variation in only one variable η_k.

Integrating both parts of this equality over the considered domain with elementary volume $dv = \prod_i dx_i = \prod_k h_k d\eta_k$, we get

$$\int \sum_i \left(\frac{\partial u}{\partial x_i}\right)^2 dx_1 \ldots dx_n = \int \sum_k \frac{h}{h_k^2} \left(\frac{\partial u}{\partial \eta_k}\right)^2 d\eta_1 \ldots d\eta_n .$$

These integrals will be taken by parts with the use of the equalities

$$\frac{\partial}{\partial x_i}\left(u\frac{\partial u}{\partial x_i}\right) = \left(\frac{\partial u}{\partial x_i}\right)^2 + u\frac{\partial^2 u}{\partial x_i^2} ,$$

$$\frac{\partial}{\partial \eta_k}\left(u\frac{h}{h_k^2}\frac{\partial u}{\partial \eta_k}\right) = u\frac{h}{h_k^2}\frac{\partial^2 u}{\partial \eta_k^2} + u\frac{\partial}{\partial \eta_k}\left(\frac{h}{h_k^2}\frac{\partial u}{\partial \eta_k}\right) .$$

The surface integrals arising in this computation vanish because by assumption u becomes zero at the boundary of the domain. Finally we obtain

$$\int u \left[\sum_i \frac{\partial^2 u}{\partial x_i^2} - \frac{1}{h}\sum_k \frac{\partial}{\partial \eta_k}\left(\frac{h}{h_k^2}\frac{\partial u}{\partial \eta_k}\right)\right] dx_1 \ldots dx_n = 0 .$$

Observing the fact that u is arbitrary inside the domain yields (6.1.9).

6.1.3 The Laplace Operator

Now we set up the Laplace operator (6.1.4) in the spherical coordinates r, Ω. Because

$$dx^2 = dr^2 + r^2 ds^2 , \quad ds^2 = \sum_\alpha g_\alpha^2 d\theta_\alpha^2$$

we arrive by virtue of (6.1.8) and (6.1.9) at the equality

$$\Delta \equiv \sum_{i=1}^n \frac{\partial^2}{\partial x_i^2} = \Delta_r + \frac{1}{r^2}\Delta_\Omega , \tag{6.1.10}$$

where

$$\Delta_r = \frac{1}{r^{n-1}}\frac{\partial}{\partial r}\left(r^{n-1}\frac{\partial}{\partial r}\right) = \frac{\partial^2}{\partial r^2} + \frac{n-1}{r}\frac{\partial}{\partial r}$$

is the radial part of the Laplacian, and

$$\Delta_\Omega = \frac{1}{g}\sum_\alpha \frac{\partial}{\partial \theta_\alpha}\left(\frac{g}{g_\alpha^2}\frac{\partial}{\partial \theta_\alpha}\right) , \quad g = \prod_\alpha g_\alpha ,$$

is the Laplace operator on unit sphere S^{n-1}.

288

In agreement with the decomposition $R = R_1 \oplus R_2$ we evaluate the dependence of the operator Δ_Ω upon angle θ and derivatives with respect to this angle. By virtue of (6.1.6) we get

$$\Delta_\Omega = \Delta_\theta + \frac{\Delta_\Omega}{\sin^2 \theta} + \frac{\Delta_\Omega}{\cos^2 \theta} \, , \qquad\qquad (6.1.11)$$

where

$$\Delta_\theta = \frac{1}{\sin^{p_1} \theta \cos^{p_2} \theta} \frac{\partial}{\partial \theta} \left(\sin^{p_1} \theta \cos^{p_2} \theta \, \frac{\partial}{\partial \theta} \right)$$

$$= \frac{\partial^2}{\partial \theta^2} + (p_1 \cot \theta - p_2 \tan \theta) \frac{\partial}{\partial \theta}$$

and Δ_{Ω_i} are the Laplace operators on spheres $S^{p_i} \subset R_i$ ($p_i = \dim S^{p_i} = n_i - 1$), $i = 1, 2$. If a subspace R_i, $i = 1, 2$, is one-dimensional ($p_i = n_i - 1 = 0$), then Δ_{Ω_i} must be let to be identically zero in (6.1.11).

The relation (6.1.11) may be used further to derive the operators Δ_{Ω_1}, Δ_{Ω_2}, and so on until unidimensional subspaces are reached. As a result one could arrive at an explicit expression for the operator Δ_Ω in the spherical coordinates Ω.

6.1.4 A Graphical Approach

The preceding consideration suggests that in situations with arbitrary dimension of an Euclidean space R the metric dx^2 and Laplacian Δ assume a rather unwieldy form in spherical coordinates r, Ω. A considerable simplification results with a new form of writing based on the formalism of trees, the simplest entities of graph theory, as suggested for solving n-dimensional Laplace equations by Vilenkin, Kuznetzov and Smorodinskiĭ[V8, V1, V10].

6.1.4.1. The decomposition of a space R into the direct sum of orthogonal subspaces R_1 and R_2, $R = R_1 \oplus R_2$, will be represented graphically as the graph in Fig. 6.2.

The vertices of this graph are associated with the spaces R, R_1, and R_2. Vertex R has two adjacent vertices R_1 and R_2 with which is it connected by two edges RR_1 and RR_2, thus representing the formula $R = R_1 \oplus R_2$. This graphical work can be continued, representing the decomposition of subspaces R_1 and R_2 into direct sums, until we arrive at one-dimensional subspaces represented by edges incident on pendant vertices. As a result we obtain some tree T having n pendant vertices (Fig. 6.3a).

We digress for a moment to recall a few points of graph terminology. In the material that follows it will be sufficient to understand a graph as a plane pattern constituted by a set of *vertices* and a set of *edges* connecting these vertices pair-wise. An edge is said to be *incident on* its end vertices. Two vertices are *adjacent* if they are the end vertices of the same edge. The degree of a vertex

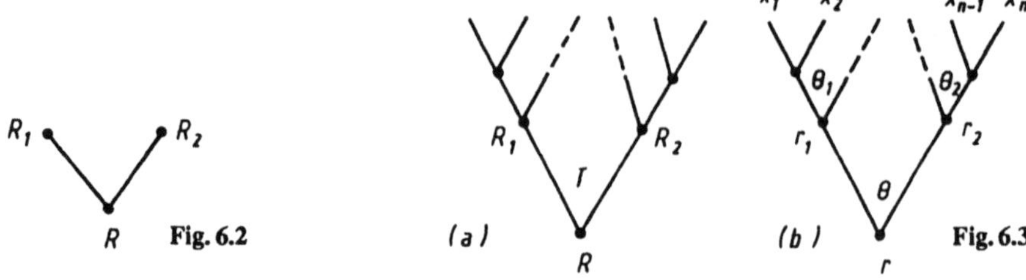

is the number of edges incident on this vertex. A vertex of degree 1 is called a
pendant vertex.

In this chapter we confine ourselves to considering only the case consistent
with the construction outlined above, where the out-degree of each vertex is
exactly two (it is usually assumed that any two vertices are concatenated by some
chain of edges). A family of directed graphs – with a direction assigned to each
edge – arising in this way is sometimes referred to as the one of 'dichotomous
root trees' [L20]. For brevity we shall refer to these binary trees as simply *trees.*
The crown structure of these trees is exceedingly simple–a root vertex of degree
two gives birth to two branches directed upwards, each of which devides in turn
and so on until pendant vertices (of degree 1) are attained (Fig. 6.3). For more
rigorous definitions of the basic concepts of graph theory we refer the reader to
[H8, A13, S36].

6.1.4.2. The tree T uniquely defines a system of spherical coordinates r, Ω and
enables us to formulate a simple *graphical rule* for writing the Cartesian coordinates of a vector $x \in R$ in terms of the spherical coordinates r, Ω,

$$x_i = x_i(r, \Omega) , \quad i = 1, \ldots, n .$$

For this purpose we shall attach a somewhat different meaning for the tree T. In
agreement with the decomposition $R = R_1 \oplus R_2$ for any arbitrary vector x in R
we have (6.1.2). Let r, r_1, and r_2 be the lengths of vectors x, x_1, and x_2, then

$$r_1 = r \sin \theta , \quad r_2 = \cos \theta , \tag{6.1.12}$$

where θ is the angle from x to x_2 (see Fig. 6.1). In graphical terms this rule is
represented by the tree in Fig. 6.4.

We shall refer to this type of tree consisting of vertices r, r_1 and r_2 with
incident edges rr_1 and rr_2 as a *node.* This node has an angle θ assigned to it;
the vertices r, r_1 and r_2 denote the lengths of x, x_1 and x_2. To arrive at r_1 or
r_2 we proceed from vertex r on the left, or on the right, and multiply r by $\sin \theta$,
or $\cos \theta$.

Applying this argument to R_1, R_2, and so on up to the one-dimensional
subspaces of T we arrive at a graphical representation of the Cartesian coordinates
$x_1, \ldots, x_n$ of x (Fig. 6.3b).[2] To express Cartesian coordinates x_i, $i = 1, \ldots, n$
through the spherical coordinates r, Ω, where $\Omega = \{\theta, \theta_1, \theta_2, \ldots\}$, the rule to

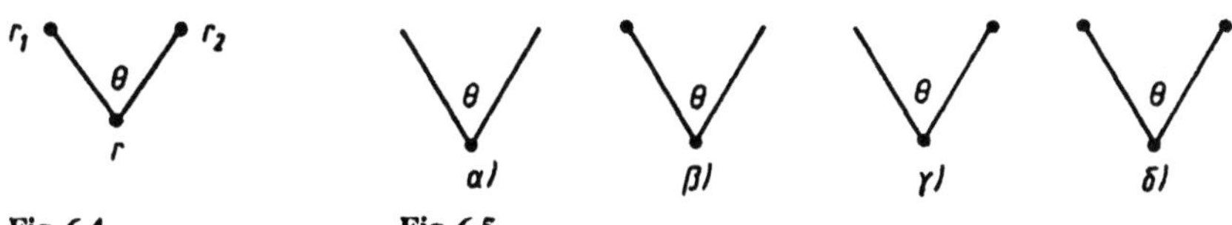

Fig. 6.4   Fig. 6.5

follow is this: *move from vertex r, the root of T, to the coordinate x_i (path $r \rightarrow x_i$) multiplying the length of the radius-vector traced to the adjacent vertex on the left (right) by the sine (cosine) of the respective angle.*

The node terminology will be as indicated in Fig. 6.5: a node-subgraph having two pendant vertices will be labelled α, those having one pendant vertex will be labelled β and γ, and the one without pendant vertices will be labelled δ. The nodes α, β and γ having pendant vertices will be termed *open*, while the node δ will be termed *closed*.

Now we require that the spherical coordinates r, Ω given by the tree T parametrize every point in R. The vector $\boldsymbol{x}$ in the decomposition (6.1.1) may belong to the subspace R_1 and R_2 (see Fig. 6.1), which defines the bounds of variaton of angle θ at the root of T. Depending on the type of node, the variation interval of this angle is defined as indicated in Fig. 6.6.

These bounds hold for all angles $\Omega = \{\theta, \theta_1, \theta_2, \dots\}$ of the tree T, and for almost every point on the sphere S^{n-1} (i.e. except the points constituting a set of lower dimension – the spherical coordinate system, as will be recalled, is local) one and only one set of angles Ω lying in these limits can be pointed out.

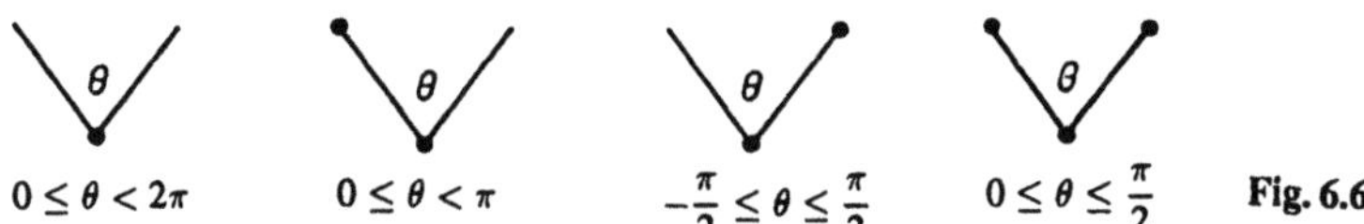

$$0 \le \theta < 2\pi \qquad 0 \le \theta < \pi \qquad -\frac{\pi}{2} \le \theta \le \frac{\pi}{2} \qquad 0 \le \theta \le \frac{\pi}{2} \qquad \text{Fig. 6.6}$$

6.1.4.3. Let us examine a few examples of spherical coordinates.

(1) On a plane only one system of spherical coordinates is possible (Fig. 6.7a):

$$x_1 = r \sin \varphi, \quad x_2 = r \cos \varphi . \tag{6.1.13}$$

(2) In a three-dimensional space there exist two systems of spherical coordinates (Fig. 6.7b, c), namely

$$x_1 = r \sin \theta \sin \varphi, \quad x_2 = r \sin \theta \cos \varphi, \quad x_3 = r \cos \theta ; \tag{6.1.14}$$

$$x_1 = r \sin \theta, \quad x_2 = r \cos \theta \sin \varphi, \quad x_3 = \cos \theta \cos \varphi . \tag{6.1.15}$$

---

[2] In general the Cartesian coordinates of $\boldsymbol{x}$ may be assigned to the pendant vertices of T in an arbitrary order. Changing the labels of the basis vectors e_i, $i = 1, \dots, n$, one can always manage the situation so that the coordinates $x_1, \dots, x_n$ follow in increasing order from left to right as indicated in Fig. 6.3b.

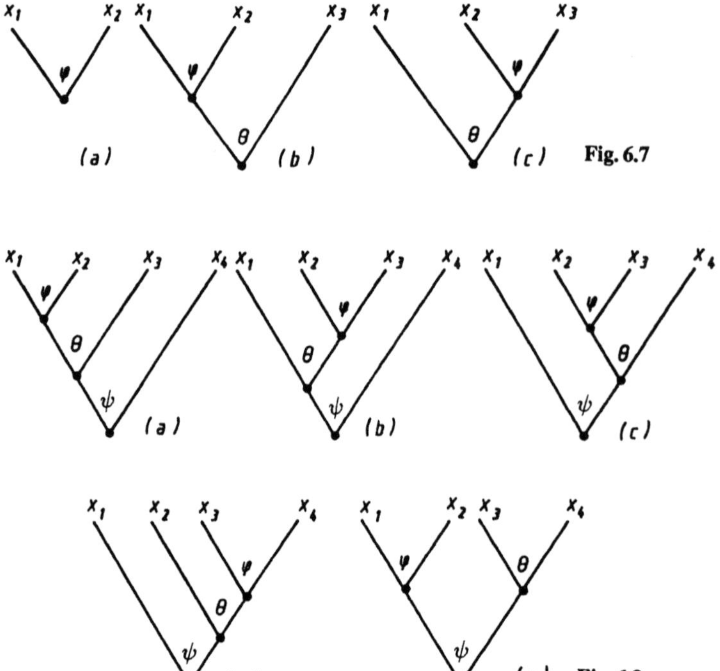

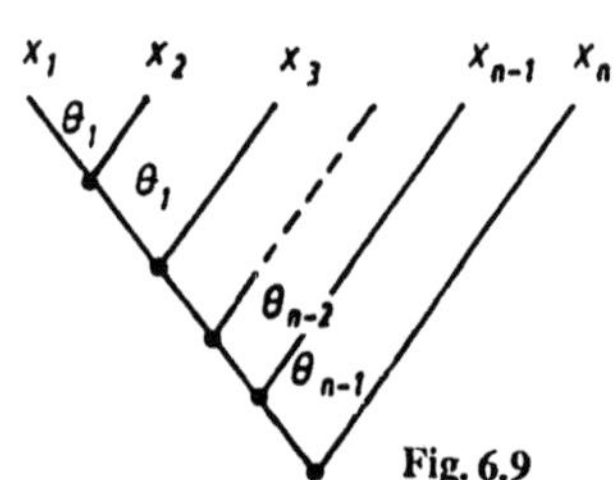

Fig. 6.9

(3) In a four-dimensional Euclidean space, five distinct coordinate systems may be set up (Fig. 6.8)

$$x_1 = r \sin\psi \sin\theta \sin\varphi , \qquad x_3 = r \sin\psi \cos\theta ,$$
$$x_2 = r \sin\psi \sin\theta \cos\varphi , \qquad x_4 = r \cos\psi , \tag{6.1.16}$$

$$x_1 = r \sin\psi \sin\theta , \qquad x_3 = r \sin\psi \cos\theta \cos\varphi ,$$
$$x_2 = r \sin\psi \cos\theta \sin\varphi , \qquad x_4 = r \cos\psi , \tag{6.1.17}$$

$$x_1 = r \sin\psi , \qquad x_3 = r \cos\psi \sin\theta \cos\varphi ,$$
$$x_2 = r \cos\psi \sin\theta \sin\varphi , \qquad x_4 = r \cos\psi \cos\theta , \tag{6.1.18}$$

$$x_1 = r \sin\psi , \qquad x_3 = r \cos\psi \cos\theta \sin\varphi ,$$
$$x_2 = r \cos\psi \sin\theta , \qquad x_4 = r \cos\psi \cos\theta \cos\varphi , \tag{6.1.19}$$

$$x_1 = r \sin \psi \sin \varphi \,, \qquad\qquad x_3 = r \cos \psi \sin \theta \,,$$
$$x_2 = r \sin \psi \cos \varphi \,, \qquad\qquad x_4 = r \cos \psi \cos \theta \,. \qquad (6.1.20)$$

(4) In an n-dimensional Euclidean space the following canonical system of spherical coordinates will be set up (Fig. 6.9):

$$x_1 = r \sin \theta_{n-1} \sin \theta_{n-2} \ldots \sin \theta_2 \sin \theta_1 \,,$$
$$x_2 = r \sin \theta_{n-1} \sin \theta_{n-2} \ldots \sin \theta_2 \cos \theta_1 \,,$$
$$x_3 = r \sin \theta_{n-1} \sin \theta_{n-2} \ldots \cos \theta_2 \,,$$
$$\ldots\ldots\ldots\ldots\ldots\ldots\ldots\ldots\ldots\ldots\ldots\ldots\ldots\ldots \qquad (6.1.21)$$
$$x_{n-1} = r \sin \theta_{n-1} \cos \theta_{n-2} \,,$$
$$x_n = r \cos \theta_{n-1} \,,$$

where $0 \le \theta_1 < 2\pi$, $0 \le \theta_\alpha \le \pi$, and $\alpha = 2, \ldots, n-1$.

Inspecting the trees in Figs. 6.7–9 in conjunction with the respective (6.1.13–21), we note that it will be not hard to perceive and adopt the aforementioned graphical rule of writing the Cartesian coordinates of a vector x in spherical form.

6.1.4.4. The metric ds^2 and Laplacian Δ_Ω on unit sphere S^{n-1} in an Euclidean space R may also be derived by invoking the tree T of spherical coordinates. Equation (6.1.6) suggests the following rule for constructing the coefficients g_α in (6.1.7): *if a path is to be traced out from the root of T to a vertex α, then assign unity to the root and moving on the left (right) multiply the previously derived term by the sine (cosine) function of the respective angle.*

By way of example we put down the metrix ds^2 on spheres S^1, S^2, S^3, and S^{n-1} using this rule for the coordinate system shown in Figs. 6.7a, b, 6.8a, e, and 6.9, viz.

$$ds^2 = d\theta^2 \,, \qquad (6.1.22)$$

$$ds^2 = d\theta^2 + \sin^2 \theta \, d\varphi^2 \,, \qquad (6.1.23)$$

$$ds^2 = d\psi^2 + \sin^2 \psi \, d\theta^2 + \sin^2 \psi \sin^2 \theta \, d\varphi^2 \,, \qquad (6.1.24)$$

$$ds^2 = d\psi^2 + \sin^2 \psi \, d\varphi^2 + \cos^2 \psi \, d\theta^2 \,, \qquad (6.1.25)$$

$$ds^2 = d\theta_{n-1}^2 + \sin^2 \theta_{n-1} \, d\theta_{n-2}^2 + \ldots$$
$$+ \sin^2 \theta_{n-1} \sin^2 \theta_{n-2} \ldots \sin^2 \theta_2 \, d\theta_1^2 \,. \qquad (6.1.26)$$

The respective Laplacians Δ_Ω obtained by (6.1.10) have the form

$$\Delta_\Omega = \partial^2 / \partial \varphi^2 \,, \qquad (6.1.27)$$

$$\Delta_\Omega = \frac{1}{\sin \theta} \frac{\partial}{\partial \theta} \left(\sin \theta \frac{\partial}{\partial \theta} \right) + \frac{1}{\sin^2 \theta} \frac{\partial^2}{\partial \varphi^2} \,, \qquad (6.1.28)$$

$$\Delta_\Omega = \frac{1}{\sin^2 \psi} \frac{\partial}{\partial \psi} \left(\sin^2 \psi \frac{\partial}{\partial \psi} \right) + \frac{1}{\sin^2 \psi \sin \theta} \frac{\partial}{\partial \theta} \left(\sin \theta \frac{\partial}{\partial \theta} \right)$$

$$+ \frac{1}{\sin^2 \psi \sin^2 \theta} \frac{\partial^2}{\partial \varphi^2} , \tag{6.1.29}$$

$$\Delta_\Omega = \frac{1}{\sin \psi \cos \psi} \frac{\partial}{\partial \psi} \left(\sin \psi \cos \psi \frac{\partial}{\partial \psi} \right) + \frac{1}{\sin^2 \psi} \frac{\partial^2}{\partial \varphi^2}$$

$$+ \frac{1}{\cos^2 \psi} \frac{\partial^2}{\partial \theta^2} , \tag{6.1.30}$$

$$\Delta_\Omega = \frac{1}{\sin^{n-2} \theta_{n-1}} \frac{\partial}{\partial \theta_{n-1}} \left(\sin^{n-2} \theta_{n-1} \frac{\partial}{\partial \theta_{n-1}} \right)$$

$$+ \frac{1}{\sin^2 \theta_{n-1} \sin^{n-3} \theta_{n-2}} \frac{\partial}{\partial \theta_{n-2}} \left(\sin^{n-3} \theta_{n-2} \frac{\partial}{\partial \theta_{n-2}} \right) + \ldots$$

$$+ \frac{1}{\sin^2 \theta_{n-1} \sin^2 \theta_{n-2} \ldots \sin^2 \theta_2} \frac{\partial^2}{\partial \theta_1^2} . \tag{6.1.31}$$

These expressions can be readily deduced also by successive application of (6.1.11). It is worth noting, however, that it is this equation (6.1.11) that will be valuable in our further search for the solution of (6.1.1) rather than the explicit expression for the angular part of the Laplacian.

For applications it will be convenient to represent the measure $d\Omega$ on sphere S^{n-1} in a Euclidean space R in the form

$$d\Omega = \prod_\alpha g_\alpha(\Omega) d\theta_\alpha = \prod_\alpha d\omega_\alpha(\theta_\alpha) ,$$

where each cofactor $d\omega_\alpha$ depends on one angle θ_α only. The quantities $d\omega_\alpha(\theta_\alpha)$ derived for all types of node of T by the Eq. (6.1.6) as indicated in Fig. 6.10. Here p_i is the dimension of unit sphere S^{p_i} in the subspace R_i, $i = 1, 2$.

Fig. 6.10

6.1.4.5. The tree T associated with some system of spherical coordinates in R allows a simple geometrical interpretation in this space. From an arbitrary point $\boldsymbol{x}$ in $R = R_1 \oplus R_2$ we drop perpendiculars $\boldsymbol{h}_1$ and $\boldsymbol{h}_2$ onto subspaces R_1 and R_2. This process will be continued in agreement with the tree T until we arrive at unidimensional subspaces. It is not hard to see that the collection of the perpendiculars $\boldsymbol{h}_1, \boldsymbol{h}_2$, etc., and their intersections forms in R a tree isomorphic to the original coordinate tree T pictured on the plane.

Examples of the aforementioned isomorphism (between the tree of coordinates and the "tree of perpendiculars") are depicted in Fig. 6.11 for $n = 2, 3, 4$. It will be instructive to trace the introduction of spherical coordinates (Eqs. (6.1.13–14) and (6.1.20) for all patterns shown by carrying out the respective construction work.

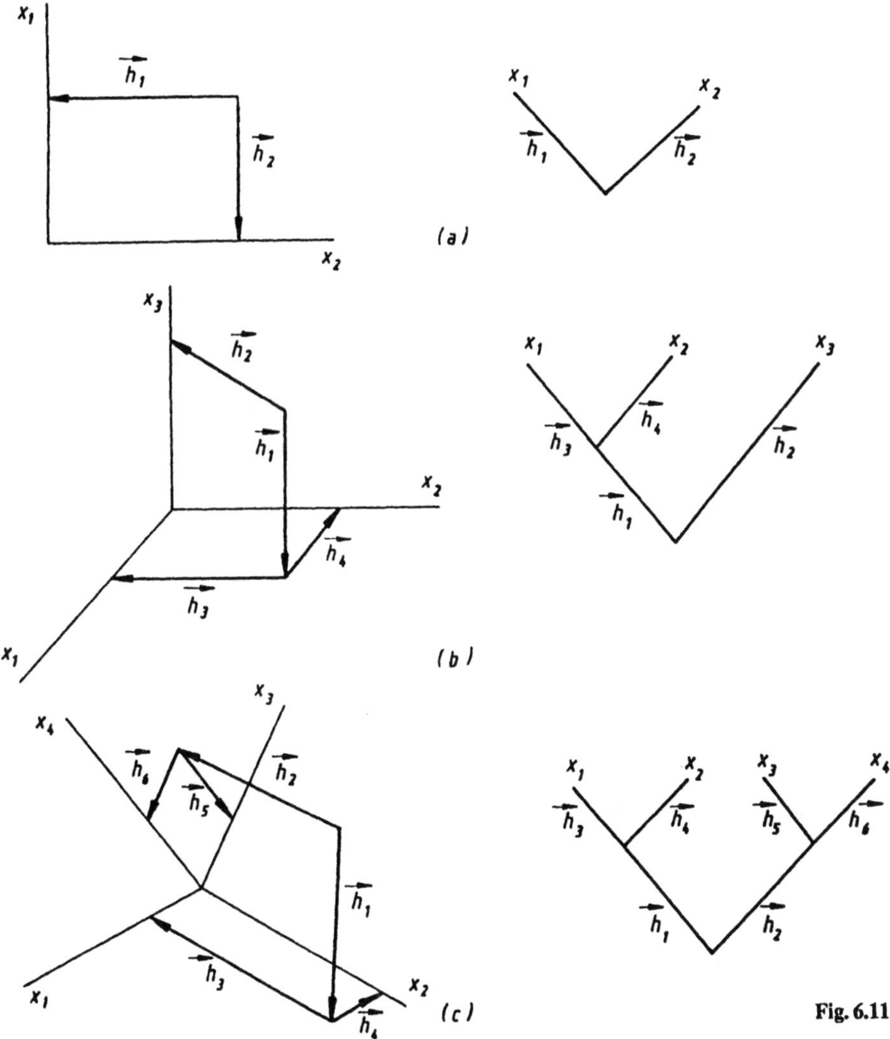

6.1.4.6. We have learned earlier that any n-dimensional Euclidean space R allows a few distinct systems of spherical coordinates each of which is uniquely defined by its own tree T. (Parallels and meridians on a hypersphere can be set up in a few different ways). It would be interesting to see how many such systems exist or, what is the same, how many respective trees may be drawn. First of all it will be convenient to establish the total number of trees having n pendant vertices with Cartesian coordinates x_i assigned to them. These Cartesian coordinates follow in a certain order, say that indicated in Fig. 6.3.b. It is quite obvious that from any given tree of this type one can obtain $n!$ trees of the same structure by permuting the labels of the coordinates assigned to the pendant vertices in all possible ways.

Denote by $N(n)$ the number of trees having n pendant vertices labelled in a certain order, then [K15, S9]

$$N(n) = \sum_{n_1+n_2=n} N(n_1)N(n_2) \, . \tag{6.1.32}$$

This relation implies that we are to carry out a consecutive enumeration of all co-ordinate systems for one-dimensional, two-dimensional, etc. up to n-dimensional subspaces and for their orthogonal complements.

To obtain an explicit expression for $N(n)$ from (6.1.32) it would pay if we first deduce the generation function $G(\xi)$ for these quantities,

$$G(\xi) = \sum_{k=0}^{\infty} N(k+1)\xi^k \, , \tag{6.1.33}$$

where ξ is an auxiliary variable. By virtue of (6.1.32) and (6.1.33) we have

$$\begin{aligned}
G^2(\xi) &= \left(\sum_{k=0}^{\infty} N(k+1)\xi^k \right)^2 \\
&= \sum_{m=0}^{\infty} \left[\sum_{i=0}^{m} N(i+1)N(m-i+1) \right] \xi^m \\
&= \frac{1}{\xi} \sum_{n=0}^{\infty} N(n+2)\xi^{n+1} \, .
\end{aligned}$$

Noting that $N(1) = 1$ we get the quadratic equation

$$G^2(\xi) = \frac{1}{\xi}[G(\xi) - 1]$$

with the roots

$$G(\xi) = \frac{1}{2\xi} \left[1 \pm \sqrt{1 - 4\xi} \right] \, . \tag{6.1.34}$$

Expanding the right-hand side of this expression as a power series in ξ and taking the lower sign, owing to the positivity contraint on $N(n)$, we arrive at the *general formula* for the number of trees having n pendant vertices enumerated to a certain order [K37]

$$N(n) = -\frac{2^{2n-1}}{n!} \left(-\frac{1}{2} \right)_n = \frac{(2n-2)!}{n!(n-1)!} \, . \tag{6.1.35}$$

Specifically $N(2) = 1$, $N(3) = 2$, and $N(4) = 5$; these trees are pictured in Figs. 6.6 and 6.7. The total number of trees and corresponding systems of spherical coordinates is

$$M(n) = n! N(n) = \frac{(2n-2)!}{(n-1)!} \, . \tag{6.1.36}$$

By way of example $M(2) = 2$, $M(3) = 12$, and $M(4) = 10$.

The penultimate and last formulae indicate that as n increases the functions $N(n)$ and $M(n)$ grow rapidly, which indicates that there exists considerable arbitrariness in choosing a system of spherical coordinates in an n-dimensional Euclidean space.

The expression (6.1.35) also solves the problem on binary tree enumeration in graph theory [L20].

6.2 Solution of the n-Dimensional Laplace Equation in Spherical Coordinates

This section deals with the construction of solutions to the Laplace equation (6.1.1) by separation of variables in spherical coordinates. The Eqs. (6.1.10) and (6.1.11) will be essential in this construction.

6.2.1 Separation of Variables

Let r, Ω be a system of spherical coordinates given by some tree T. Since by (6.1.10)

$$\Delta = \Delta_r + \frac{1}{r^2}\Delta_\Omega \,,$$

where the operators Δ_r and Δ_Ω act respectively on the radial, r, and angular, Ω, variables only, we shall seek a solution of (6.1.1) in the form

$$u = R(r)Y(\Omega) \,. \tag{6.2.1}$$

As a result we have

$$r^2\Delta_r R/R = -\Delta_\Omega Y/Y = \lambda = \text{const} \,,$$

that is

$$r^2\Delta_r R = \lambda R \,, \tag{6.2.2}$$

$$\Delta_\Omega Y + \lambda Y = 0 \,. \tag{6.2.3}$$

Thus the radial and angular variables are separated. The eigenvalue problem arising for the angular part of the Laplacian (6.2.3) allows further separation of variables. The node at the root of T may be one of the four types indicated in Fig. 6.5. We examine all the situations consecutively with reference to the Eq. (6.1.11) for the operator Δ_Ω.

(1) For the 'open' node α in Fig. 6.5 the function $Y = Y_0(\theta)$ satifies the equation

$$\frac{d^2 Y_0}{d\theta^2} + \lambda Y_0 = 0 \,. \tag{6.2.4}$$

(2) For a node β from (6.1.11) the operator Δ_Ω is

$$\Delta_\Omega = \Delta_\theta + \Delta_{\Omega_1}/\sin^2\theta ,$$

where

$$\Delta_\theta = \frac{1}{\sin^{p_1}\theta}\frac{\partial}{\partial\theta}\left(\sin^{p_1}\theta\,\frac{\partial}{\partial\theta}\right) ,$$

Δ_{Ω_1} is the Laplace operator on the sphere $S^{p_1} \subset R_1$ of smaller dimension, and dim $S^{p_1} = p_1 = n-2$. Therefore on substitution

$$Y = Y_0(\theta)Y_1(\Omega_1) ,$$

where Ω_1 are the coordinates on the sphere S^{p_1}, the variables in (6.2.3) again separate:

$$\frac{1}{\sin^{p_1}\theta}\frac{d}{d\theta}\left(\sin^{p_1}\theta\,\frac{dY_0}{d\theta}\right) + \left(\lambda - \frac{\lambda_1}{\sin^2\theta}\right)Y_0 = 0 , \tag{6.2.5}$$

$$\Delta_{\Omega_1}Y_1 + \lambda_1 Y_1 = 0 . \tag{6.2.6}$$

Here λ_1 is a constant.

(3) The separation of variables for the node γ in Fig. 6.5 $Y = Y_0(\theta)Y_2(\Omega_2)$ yields similar equations

$$\frac{1}{\cos^{p_2}\theta}\frac{d}{d\theta}\left(\cos^{p_2}\theta\,\frac{dY_0}{d\theta}\right) + \left(\lambda - \frac{\lambda_2}{\cos^2\theta}\right)Y_0 = 0 , \tag{6.2.7}$$

$$\Delta_{\Omega_2}Y_2 + \lambda_2 Y_2 = 0 , \quad \lambda_2 = \text{const} . \tag{6.2.8}$$

(We note also that (6.2.7) results from (6.2.5) by the substitution $\theta \to \pi/2$, $\lambda_1 \to \lambda_2$, and $p_1 \to p_2$.)

(4) Given a node δ, we have (6.1.11) for the Laplace operator Δ_Ω. We shall seek a solution in the form

$$Y = Y_0(\theta)Y_1(\Omega_1)Y_2(\Omega_2) ,$$

yielding

$$\frac{1}{\sin^{p_1}\cos^{p_2}\theta}\frac{d}{d\theta}\left(\sin^{p_1}\cos^{p_2}\theta\,\frac{dY_0}{d\theta}\right)$$
$$+ \left(\lambda - \frac{\lambda_1}{\sin^2\theta} - \frac{\lambda_2}{\cos^2\theta}\right)Y_0 = 0 , \tag{6.2.9}$$

$$\Delta_{\Omega_1}Y_1 + \lambda_1 Y_1 = 0 , \quad \Delta_{\Omega_2}Y_2 + \lambda_2 Y_2 = 0 , \tag{6.2.10}$$

where λ_1 and λ_2 are some constants.

Thus we have separated the variables for a node at the root of T. The same procedure may be employed to separate variables in Eqs. (6.2.6, 8, 10), etc. until we arrive at open nodes α for which (6.2.4) holds. Hence we have proved the assertion that *the Laplace equation* (6.1.1) *allows a complete separation of variables in an arbitrary system of spherical coordinates* r, Ω.

6.2.2 Hyperspherical Harmonics

The separation of variables by (6.2.1) in Eq. (6.1.1) results in an eigenvalue problem (6.2.3):

$$\Delta_\Omega Y + \lambda Y = 0 \, ,$$

where Δ_Ω is the Laplace operator on a sphere S^p, $p = \dim S^p = n - 1$. We now wish to demonstrate that for this problem *there exist continuous solutions bound over the entire sphere S^p for the eigenvalues*

$$\lambda = \lambda_{pl} = l(l + p - 1) \, , \tag{6.2.11}$$

where l is a positive integer for $p > 1$.

6.2.2.1. Eigenvalues. Let us prove (6.2.11). Recall that in (6.2.3) the variables are separable in spherical coordinates r, Ω given by some tree T. In what follows we solve the concomitant system of ordinary differential equations.

(1) Consider first of all those nodes of T having two pendant vertices (α in Fig. 6.5) for which (6.2.4) is valid. The solution to this equation, continuous and bound on the interval $0 \leq \theta \leq 2\pi$ (see Fig. 6.6), may be selected as two equivalent systems of functions

$$Y_0(\theta) = N \exp(im\theta) \, , \quad m = 0, \pm 1, \pm 2, \ldots, \tag{6.2.12}$$

$$Y_0^{\pm}(\theta) = \begin{cases} N_+ \cos m\theta \\ N_- \sin m\theta \end{cases} \quad m = 0, 1, 2, \ldots, \tag{6.2.13}$$

where N and $N_{\pm}$ are normalizing constants. The eigenvalues $\lambda = \lambda_{1m} = m^2$ corresponds to the Eq. (6.2.11).

The reasoning that follows will be on the lines of mathematical induction. We assume that (6.2.11) defines the separation constants λ_1, λ_2, etc., for all vertices lying above some vertex θ adjacent to at most one pendant vertex and examine the remaining types of nodes.

(2) For the case of a node β the substitution [3] $Y_0 = y(\xi) \sin^{l_1} \theta$, $\xi = \cos \theta$ carries (6.2.5) for $Y_0(\theta)$

---

[3] The substitution $\xi = \cos \theta$ makes (6.2.5) a particular case of the generalized hypergeometric equation examined in [N18], therefore this substitution can be found by the technique outlined in this book.

$$\frac{1}{\sin^{p_1}\theta}\frac{d}{d\theta}\left(\sin^{p_1}\theta\,\frac{dY_0}{d\theta}\right)+\left[\lambda-\frac{l_1(l_1+p_1-1)}{\sin^2\theta}\right]Y_0=0$$

to the form

$$\sigma(\xi)y''+\tau(\xi)y'+\mu y=0\,,\tag{6.2.14}$$

where $\sigma(\xi)=1-\xi^2$, $\tau(\xi)=-(2l_1+p_1+1)\xi$, $\mu=\lambda-l_1(l_1+p_1)$. By Theorem 1.1 continuous solutions to (6.2.14), bounded on the interval $-1\le\xi\le1$ ($\xi=\cos\theta$, $0\le\theta\le\pi$, see Fig. 6.6), exist subject to

$$\mu=-k\tau'-\frac{k(k-1)}{2}\sigma''\,,\qquad k=0,1,\ldots,\tag{6.2.15}$$

whence $\lambda=l_1(l_1+p_1)+k(2l_1+p_1+k)$. Letting $l-l_1=k=0,1$, etc., leads to (6.2.11).

For a node β the function $Y_0(\theta)$ can be expressed in terms of Jacobi's polynomials $P_k^{(\alpha,\beta)}(\xi)$, viz.

$$Y_0(\theta)=N\sin^{l_1}\theta\,P_{l-l_1}^{(2j_1+1,2j_1+1)}(\cos\theta)\,,\tag{6.2.16}$$

where $2j_1+1=l_1+(p_1-1)/2$, and N is a normalizing constant. At $p_1=1$, $|l_1|$ must substitute for l_1 in this expression.

(3) The validity of (6.2.11) for a node γ is proved in a similar manner. Because the substitution $\theta\to\pi/2-\theta$, $l_1\to l_2$, $p_1\to p_2$ sends (6.2.5) to (6.2.7), the function $Y_0(\theta)$ for this node can be obtained from (6.2.16) upon this substitution, viz.

$$Y_0(\theta)=N\cos^{l_2}\theta\,P_{l-l_2}^{(2j_2+1,2j_2+1)}(\sin\theta)\,,\tag{6.2.17}$$

where $l=l_2,l_2+1,\ldots$; $2j_2+1=l_2+(p_2-1)/2$, N is a constant, and $l_2\to|l_2|$ at $p_2=1$.

(4) It remains to verify the above assertion of induction for a node δ. Now for $Y_0(\theta)$ we have (6.2.9)

$$\frac{1}{\sin^{p_1}\theta\cos^{p_2}\theta}\frac{d}{d\theta}\left(\sin^{p_1}\theta\cos^{p_2}\theta\,\frac{dY_0}{d\theta}\right)$$
$$+\left[\lambda-\frac{l_1(l_1+p_1-1)}{\sin^2\theta}-\frac{l_2(l_2+p_2-1)}{\cos^2\theta}\right]Y_0=0\,.$$

Substituting $Y_0(\theta)=\sin^{l_1}\theta\cos^{l_2}\theta\,y(\xi)$, $\xi=\cos2\theta$, in this equation gives for $y(\xi)$ the hypergeometric Eq. (6.2.14), where

$$\sigma(\xi)=1-\xi^2\,,$$
$$\tau(\xi)=l_2-l_1+\frac{p_2-p_1}{2}-\left(l_1+l_2+\frac{p_1+p_2}{2}+1\right)\xi\,,$$
$$\mu=\tfrac{1}{4}[\lambda-(l_1+l_2)(l_1+l_2+p_1+p_2)]\,.$$

Continuous solutions to (6.2.14), bounded on the interval $-1 \leq \xi \leq 1$ ($\xi = \cos 2\theta$, $0 \leq \theta \leq \frac{\pi}{2}$ (see Fig. 6.6), are feasible subject to the condition (6.2.15), whence

$$\lambda = (l_1 + l_2)(l_1 + l_2 + p_1 + p_2) + 2k(2l_1 + 2l_2 + p_1 + p_2 + 2k)$$
$$k = 0, 1, \ldots \; .$$

Letting $l - l_1 - l_2 = 2k = 0, 2, \ldots$ yields the Eq. (6.2.11).

For a node δ

$$Y_0(\theta) = N \sin^{l_1} \theta \cos^{l_2} \theta P_{j-j_1-j_2-1}^{(2j_1+1, 2j_2+1)}(\cos 2\theta) \,, \tag{6.2.18}$$

where $2j_i + 1 = l_i + (p_i - 1)/2$, $i = 1, 2$; and N is a normalizing constant. (For $p_i = 1$, $l_i \rightarrow |l_i|$, $i = 1, 2$.) Thus the validity of (6.2.11) is proved completely. We have established the explicit forms of the solutions to (6.2.4, 5, 7), and (6.2.9) and the constraints imposed on the separation constants for all types of nodes in the tree of coordinates.

6.2.2.2. Eigenfunctions. Summarizing the above considerations we arrive at the following statements on the solutions to (6.2.3).

(i) *In a system of coordinates $\Omega = \{\theta_\alpha\}$ given on a sphere S^p by some tree T equation (6.2.3) has the continuous and bounded solutions $Y = Y(\Omega)$ corresponding to the eigenvalues (6.2.11), viz.*

$$Y(\Omega) = \prod_\alpha Y_{0\alpha}(\theta_\alpha) \,. \tag{6.2.19}$$

This product is taken over all vertices of T, the multiples $Y_{0\alpha}(\theta_\alpha)$ being defined by (6.2.12, 13), or (6.2.16–18) depending on the type of node at vertex θ_α.

(ii) *The solutions (6.2.19) form a complete orthogonal system of functions on the sphere S^p. (This property follows from the orthogonality and completeness of the trigonometric set $\{e^{\pm im\varphi}\}$, $m = 0, 1, \ldots$, and the system of Jacobi polynomials $P_k^{(\alpha,\beta)}(\xi)$.)*

We elect the normalization of the eigenfunctions (6.2.19) as follows:

$$\int_{S^p} |Y(\Omega)|^2 d\Omega = 1 \,.$$

The normalizing constants N computed for $Y_0(\theta)$ subject to this condition are listed in Table 6.1. In the said integration limits for θ we have also $\int |Y_0(\theta)|^2 \times d\omega(\theta) = 1$.

We shall refer to the functions (6.2.19) as *hyperspherical harmonics*.

Thus the tree T uniquely defines a coordinate system $\Omega = \{\theta_\alpha\}$ on the sphere S^p and the solutions of Eq. (6.2.3) in these coordinates, i.e. spherical harmonics. For more convenience of construction of the spherical harmonics (6.2.19) by the tree T we assign to each vertex of this tree, θ_α, the respective separation constant l_α. The graphical rules for writing the functions (6.2.19) are also listed in Table 6.1.

6.2.2.3. Solutions of the Laplace Equation (6.1.1). By virtue of (6.2.2) and (6.2.11) we obtain for the function $R(r)$ in (6.2.1) the Euler equation

$$r^2 R'' + (n-1)r R' - l(l+n-2)R = 0 \, ,$$

whose general solution has the form

$$R = C_1 r^l + C_2 r^{-l-n+2} \, ,$$

where C_1 and C_2 are some constants. Consequently the *particular solutions* to the Laplace equation (6.1.1) will be the functions

$$u_1 = r^l Y_{lm}(\Omega) \, ,$$

$$u_2 = r^{-l-n+2} Y_{lm}(\Omega) \, .$$

Here $Y_{lm}(\Omega)$ are the spherical harmonics associated with a certain tree T, l is the separation constant at the root of T, and the subscript $m = \{l_\alpha\}$ runs over the collection of separation constants l_α at all other vertices of T.

6.2.3 Illustrative Examples

6.2.3.1. For the Laplace equation on the plane $\partial^2 u/\partial x^2 + \partial^2 u/\partial y^2 = 0$ we have in the spherical coordinates r, φ (Fig. 6.12a) the particular solutions as follows

$$u(r,\varphi) = \frac{1}{\sqrt{2\pi}} r^m e^{\pm i m \varphi} = \frac{1}{\sqrt{2\pi}}(x \pm iy)^m \, , \quad m = 0, \pm 1, \pm 2, \ldots \, .$$

We note also that in the complex variables $z = x+iy$, $z^* = x-iy$ the Laplace equation is rewritten as

$$\frac{\partial^2 u}{\partial x^2} + \frac{\partial^2 u}{\partial y^2} = 4\frac{\partial^2 u}{\partial z \partial z^*} = 0$$

and the following familiar solutions arise:

$$u = \begin{cases} f(z) \, , \\ f(z^*) \, , \end{cases}$$

where

$$f(z) = \sum_m c_m z^m \, .$$

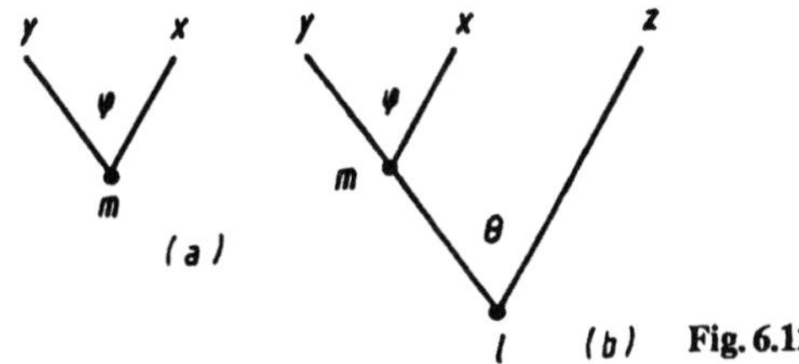

Fig. 6.12

6.2.3.2. In a three-dimensional Euclidean space R^3 the particular solutions to the Laplace equation in the spherical coordinates r, θ, φ (Fig. 6.12b) will be the functions

$$u(r, \theta, \varphi) = N r^{\pm(l+1/2)-1/2} \sin^{|m|} \theta P_{l-|m|}^{(|m|,|m|)}(\cos\theta) \frac{e^{im\varphi}}{\sqrt{2\pi}} ,$$

where

$$N = \frac{1}{2^{|m|}l!} \sqrt{\frac{2l+1}{2}(1+m)!(l-m)!} , \quad l \geq |m| .$$

A different choice of phases is commonly used. The *spherical harmonics* $Y_{lm}(\theta, \varphi)$, $-l \leq m \leq l$, are defined as follows:[4]

$$Y_{lm}(\theta, \varphi) = \begin{cases} (-1)^m N \sin^m \theta P_{l-m}^{(m,m)}(\cos\theta)\frac{e^{im\varphi}}{\sqrt{2\pi}} , & m \geq 0 . \\ (-1)^m Y_{l,-m}^*(\theta, \varphi) , & m < 0 . \end{cases} \qquad (6.2.20)$$

The functions $r^l Y_{lm}(\theta, \varphi)$ and $r^{-l-1} Y_{lm}(\theta, \varphi)$ are taken as solutions to the Laplace equation.

6.2.3.3. Let us construct the spherical harmonics for two different systems of coordinate on a sphere S^3 in a four-dimensional Euclidean space (Fig. 6.13):

(a) $\quad Y_{l'lm}(\psi, \theta, \varphi) = A \sin^l \psi P_{l'-l}^{(l+1/2, l+1/2)}(\cos\psi)$

$$\times \sin^{|m|} \theta P_{l-|m|}^{(|m|,|m|)}(\cos\theta)e^{im\varphi} ,$$

$$A = \frac{\sqrt{(2l+1)(2l'+2)(l-m)!(l+m)!(l'-l)!(l'+l+1)!}}{\sqrt{\pi}2^{l+|m|-2}l!\,\Gamma(l'+3/2)} , \qquad (6.2.21)$$

$$l' \geq l \geq |m| ;$$

(b) $\quad Y_{ll_1l_2}(\theta, \varphi_1, \varphi_2) = \frac{N}{2\pi} \sin^{|l_1|} \theta \cos^{|l_2|} \theta$

$$\times P_{\frac{l-|l_1|-|l_2|}{2}}^{(|l_1|,|l_2|)}(\cos 2\theta)$$

---

[4] In defining the spherical harmonics $Y_{lm}(\theta, \varphi)$ a choice of the phase multiplier is not unique. We adhere to the normalization adopted in the book by Varshalovich et al. [V5].

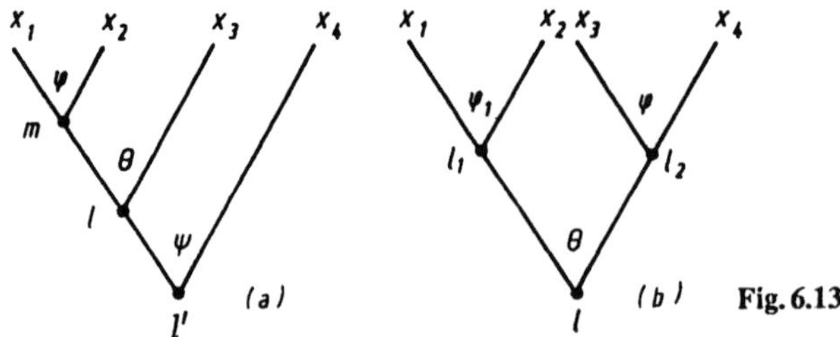

$$\times \exp i(l_1\varphi_1 + l_2\varphi_2) \,, \tag{6.2.22}$$

$$N = \sqrt{\frac{(2l+2)[(l-|l_1|-|l_2|)/2]![(l+|l_1|+|l_2|)/2]!}{[(l+|l_1|-|l_2|)/2]![(l-|l_1|+|l_2|)/2]!}} \,,$$

$$l = |l_1| + |l_2|,\ |l_1| + |l_2| + 2, \ldots \,.$$

Using arbitrary phase multipliers the spherical harmonic corresponding to the tree in Fig. 6.13 may be written in the form

$$Y_{ll_1 l_2}(\theta, \varphi_1, \varphi_2) = \sqrt{(2j+1)/2\pi^2} D^j_{mm'}(\alpha, \beta, \gamma) \,,$$

where $D^j_{mm'}(\alpha, \beta, \gamma)$ are the generalized spherical functions, $l_1 = m - m'$, $l_2 = m + m'$, $l = 2j$, $\varphi_1 = (\alpha - \gamma)/2$, $\varphi_2 = (\alpha + \gamma)/2$, $\theta = \beta/2$.

Therefore the matrix elements of the three-dimensional rotation group $D^j_{mm'}(\alpha, \beta, \gamma)$ also appear to be the eigenfunctions of the Laplace operator Δ_Ω on the sphere S^3 in a four-dimensional Euclidean space.

6.2.3.4. Now we turn to an n-dimensional Euclidean space R^n to examine the system of spherical coordinates (6.1.21) corresponding to the canonical binary tree depicted in Fig. 6.9. To each vertex θ_α of this tree we assign an integer separation constant l_α, $\alpha = 1, \ldots, n-1$. The spherical harmonics $Y = Y_{l_{n-1}\ldots l_1}(\theta_{n-1}\ldots\theta_1)$ have the form

$$Y = A \sin^{l_{n-2}} \theta_{n-1} P^{(2j_{n-2}+1, 2j_{n-2}+1)}_{l_{n-1}-l_{n-2}}(\cos\theta_{n-1})\ldots$$
$$\times \sin^{|l_1|} \theta_2 P^{(|2j_1+1|, |2j_1+1|)}_{l_2-l_1}(\cos\theta_2) \exp(il_1\theta_1) \,,$$

where $l_{n-1} \geq l_{n-2} \geq \ldots \geq l_2 \geq |l_1|$, $2j_{n-k+1} + 1 = l_{n-k+1} + (n-k)/2 - 1$, $k = 1, 2, \ldots, n-2$, and A is a normalizing constant.

The examples elucidated in this section are rather simple. The procedure outlined here is, however, fairly straightforward, and by invoking the graphical rules listed in Table 6.1 one may construct harmonics for all $M(n) = (2n-2)!/(n-1)!$ systems of spherical coordinates in any Euclidean space R^n.

Table 6.1. Principal characteristics for all types of nodes in a coordinate tree

Node type	Range of θ	Measure $d\omega$	$Y_0(\theta)$	Normalizing constant N
$(\alpha)\ m$	$[0, 2\pi]$	$d\theta$	(a) $\dfrac{1}{\sqrt{2\pi}}\,e^{im\theta}:(m = 0,\ \pm 1, \ldots)$ (b) $\dfrac{1}{\sqrt{2\pi}},\ \dfrac{1}{\sqrt{\pi}}\cos m\theta,$ $\dfrac{1}{\sqrt{\pi}}\sin m\theta$ $(m = 1, 2, \ldots)$	(a) $\dfrac{1}{\sqrt{2\pi}}$ (b) $\dfrac{1}{\sqrt{2\pi}},\ \dfrac{1}{\sqrt{\pi}}$
$(\beta)\ l$	$[0, \pi]$	$\sin^{p_1}\theta\,d\theta$ $(p_1 = n_1 - 1)$	$N\sin^{l_1}\theta\,P^{(2j_1 + 1, 2j_1 + 1)}_{l - l_1}(\cos\theta)$ $\left(2j_1 + 1 = l_1 + \dfrac{p_1 - 1}{2};\, l - l_1 = 0, 1, \ldots\right)$	$\dfrac{[(2l + p_1)(l - l_1)!(l + l_1 + p_1 - 1)!]^{1/2}}{2^{l_1 + p_1/2}\Gamma(l + (p_2 + 1)/2)}$
$(\gamma)\ l$	$\left[-\dfrac{\pi}{2}, \dfrac{\pi}{2}\right]$	$\cos^{p_2}\theta\,d\theta$ $(p_2 = n_2 - 1)$	$N\cos^{l_2}\theta\,P^{(2j_2 + 1, 2j_2 + 1)}_{l - l_2}(\sin\theta)$ $\left(2j_2 + 1 = l_2 + \dfrac{p_2 - 1}{2};\, l - l_2 = 0, 1, \ldots\right)$	$\dfrac{[(2l + p_2)(l - l_2)!(l + l_2 + p_2 - 1)!]^{1/2}}{2^{l_2 + p_2/2}\Gamma(l + (p_2 + 1)/2)}$
$(\delta)\ l$	$\left[0, \dfrac{\pi}{2}\right]$	$\sin^{p_1}\theta\cos^{p_2}\theta\,d\theta$ $(p_1 = n_1 - 1,$ $p_2 = n_2 - 1)$	$N\sin^{l_1}\theta\cos^{l_2}\theta\,P^{(2j_1 + 1, 2j_2 + 1)}_{j - j_1 - j_2 - 1}(\cos 2\theta)$ $\left(2j + 1 = l + \dfrac{p - 1}{2},\, j - j_1 - j_2 - 1 = 0, 1, \ldots\right)$	$\left[\dfrac{(4j + 2)(j - j_1 - j_2 - 1)!\,\Gamma(j + j_1 + j_2 + 2)}{\Gamma(j + j_1 - j_2 + 1)\Gamma(j - j_1 + j_2 + 1)}\right]^{1/2}$

6.3 Transformation of Harmonics Derived in Different Spherical Coordinates

In this section we turn to the major objective of this chapter – we wish to evaluate how harmonics constructed in various systems of hyperspherical coordinates are related to each other. In applications matrix elements of the transformation between different hyperspherical harmonics arise, for example, in the construction of the wave function of the K-harmonic method, which possess certain properties of symmetry with respect to particle permutations, in the solution of the Schrödinger equation for a Coulomb attraction field in momentum representation, and in the computing of generalized spherical functions for the rotation and motion groups of an n-dimensional Euclidean space. As has been learned these matrix elements can be expressed in terms of the Racah, Hahn, and Kravchuk polynomials, thus obtaining a simple form to study the properties of these quantities.

6.3.1 Transpositions and Transplants

To evaluate the structure of a transformation between two different harmonics, let Y and Y' be spherical harmonics given by trees T and T', respectively. Because Y and Y' form two complete orthogonal systems of functions on a sphere S^{n-1}, any harmonic Y may be expanded as a series in functions Y' and vice versa. Consequently there exists the *unitary transformation* $Y = UY'$, $Y' = U^+Y$.

A transition from T to T' can be made by sequentially performing two *elementary operations*:

(1) the *transposition* $P = P_{ab}$ of edges at the node ab of the tree T (Fig. 6.14a), which replaces vertices a and b leaving the crown structure unchanged;

(2) the *transplant* $C = C_{a(b)c}$ of one of the edges, as illustrated in Fig. 6.14b.

These operations result in a sequence of trees

$$T \to T_1 \to T_2 \to \ldots \to T_{m-1} \to T' \, ,$$

where each transformation is either a transposition P or a transplant C. Accordingly the matrix U factors into the ones corrseponding to elementary operations P and C, namely $U = U_1 U_2 \ldots U_m$. An example of a transformation $T \to T'$ is illustrated in Fig. 6.15.

Thus to dertermine a matrix U it is sufficient to evaluate the *matrix elements of transformations* P and C.

Now we wish to prove the aforementioned statement that *a transformation between two arbitrary trees T and T' can be effected by sequential application of two elementary operations P and C*. First of all we demonstrate that these operations reduce a tree T of arbitrary structure to the canonical tree T_0 depicted in Fig. 6.9, i.e., $T \to T_1 \to \ldots T_p \to T_0$. Such a transformation may be carried out, say, in two steps as follows.

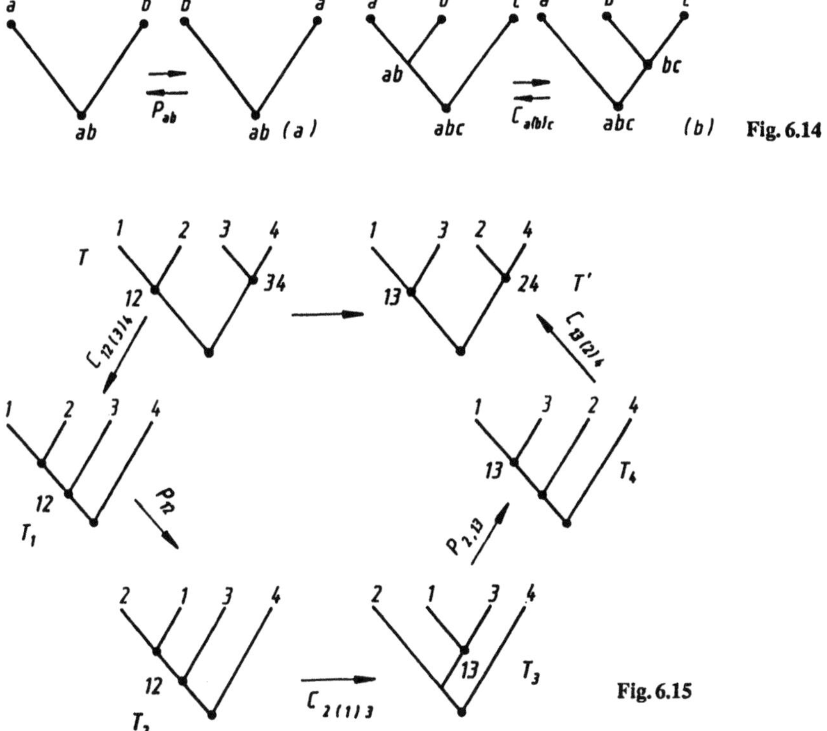

(i) First we reduce the tree T to the type of the canonical tree T_0 with the exception that the coordinates remain assigned to pendant vertices in the order they had in the original tree T. For this purpose we shall go from the root of T, to the right, and up to the terminal pendant vertex, transplanting a left edge at each vertex en route.

Having reached the terminal vertex we return to the root to start this time to the left of the adjacent vertex and from this vertex as a point of departure repeat the above procedure the necessary number of times. As a result we arrive at a tree T_0' isomorphic to the canonical tree T_0 but having its coordinates assigned to pendant vertices, generally speaking, arbitrarily.

(ii) To complete the transformation of an arbitrary tree to the canonical one T_0 we are left only to transpose the pendant vertices of T_0' to the order indicated in Fig. 6.9 so that their labelling increases from left to right as 1 to n. For this purpose we keep the structure of the tree unchanged and first transpose the pendant vertex associated with the coordinate x_1 to the leftmost position. Clearly this arrangement may always be achieved by operations of 'pendant vertex transposition' indicated in Fig. 6.16. Sequential applications of similar transformations to the pendant vertices with coordinates $x_2, x_3, \ldots, x_{n-1}$ carries T_0' to the canonical tree T_0.

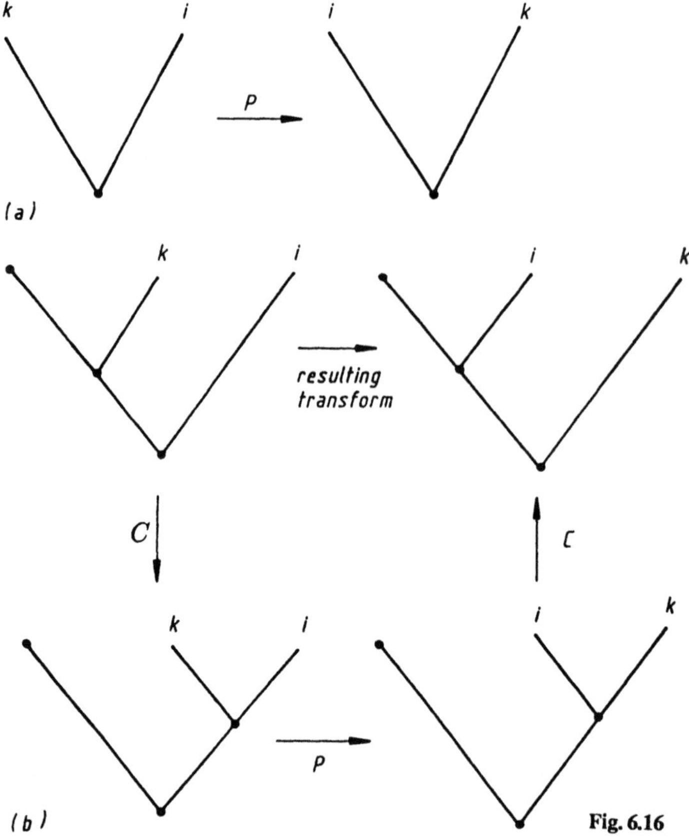

Fig. 6.16

Because all the operations performed above are reverse we have also proved the *converse statement* that an arbitrary tree is obtainable from the canonical one by consecutive application of elementary operations P and C, i.e. $T_0 \rightarrow T_p \rightarrow \ldots T_1 \rightarrow T$.

Now to demonstrate the feasibility of such transition between trees T and T' of arbitrary structure it suffices to carry out the transformation $T \rightarrow T_0 \rightarrow T'$. This proves the above assertion completely. It is quite clear also that the resultant transformation $T \rightarrow T'$ can be made in a few different ways embracing different numbers of steps.

A more detailed account of transformations between the trees can be found in [K15, S9, K37]. We note only that the collection of all such transformations – reduceable as has been demonstrated to consecutive performing the generic operations P and C – forms an algebraic structure called a *groupoid of transformations*, G [K15]. What distinguishes this set from a group is that the product (binary operation) is not defined for all pairs of its elements (C^2 for example requires special explanations since the transplant C is not defined with respect to a specific tree on which it acts), and that the multiplication is not, generally speaking, an associative operation.

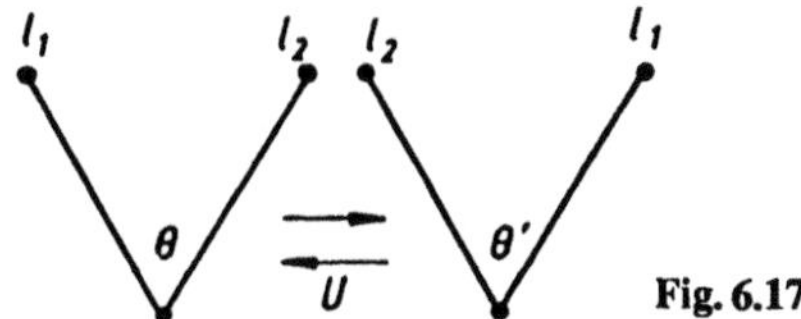

Fig. 6.17

It would be natural to deem also that the transformation matrices U to be derived further give representations of the groupoid G in the space of hyperspherical harmonics because a product of elements of the groupoid is associated with the product of the respective matrices. The number of elements in G sharply increases as n grows with the number of harmonics.

Let us return to the examination of the elementary operations. It is not hard to see that a transformation P leads to a simple phase factor for spherical harmonics. Comparing, for example, the functions $Y_0(\theta)$ for a transposition in a closed node (Fig. 6.17) we have

$$\sin^{l_1}\theta\cos^{l_2}\theta P_{j-j_2-j_1-1}^{(2j_1+1,2j_2+1)}(\cos 2\theta)$$
$$= (-1)^{j-j_1-j_2-1}\sin^{l_2}\theta'\cos^{l_1}\theta' P_{j-j_2-j_1-1}^{(2j_2+1,2j_1+1)}(\cos 2\theta')\ ,$$

where $\theta' = \pi/2 - \theta$, that is in this case $U = (-1)^{j-j_1-j_2-1}$. A similar approach is used to derive the matrix U for transpositions P in open nodes. It remains, consequently, to examine the transplant C. The elements of U are referred to as the T-coefficients. They have been calculated by Kildyushev [K10] (see also [K11, K30, K36, K38–40]).

6.3.2 The T-Coefficients for a Transplant Involving Closed Nodes

Matrix elements of a transplant C may be expressed through the classical orthogonal polynomials in a discrete variable.

6.3.2.1. We calculate the T-coefficients for a transplant involving *closed nodes* only (Fig. 6.18). The transformation between the harmonics $Y_{j_{12}m}(s)$ and $Y'_{j_{23}m}(s)$, where $s = x/r$ is an arbitrary point on the sphere S^{n-1}, and m is the collection of the remaining indices, has the form

$$Y'_{j_{23}m}(s) = \sum_{j_{12}} U(j_{12}, j_{23})Y_{j_{12}m}(s)\ . \tag{6.3.1}$$

Upon cancelling out we get

$$Y_j^{j_1 j_{23}}(\vartheta)Y_{j_{23}}^{j_2 j_3}(\psi) = \sum_{j_{12}} U(j_{12}, j_{23})Y_{j_{12}}^{j_1 j_2}(\varphi)Y_j^{j_{12} j_3}(\theta)\ , \tag{6.3.2}$$

where $Y_{j_{ik}}^{j_i j_k}(\theta)$ stand for the respective factors of the type (6.2.18) for the closed nodes shown in Fig. 6.18. Here in place of an integer separation constant l_α it

309

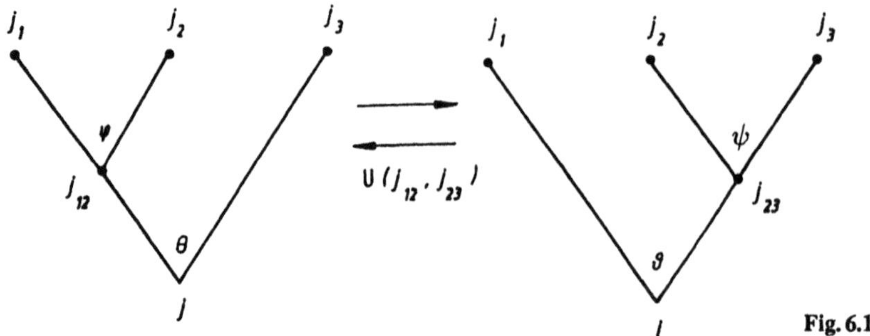

is convenient to assign to each vertex α a quantity j_α defined by the ordinary relation $2j_\alpha + 1 = l_\alpha + (p_\alpha - 1)/2$ (see Table 6.1).

Expressing $Y^{j_i j_k}_{j_{ik}}$ through the Jacobi polynomials, multiplying both sides of the decomposition (6.3.2) by $(\cos\theta)^{-l_3}$, observing that $\cos\theta = \cos\vartheta \cos\psi$, and letting $\theta = \psi = \pi/2$, $\varphi = \vartheta$, we get

$$2^{-j_{23}+j_2+j_3+1}\,\bar{P}^{(2j_2+1,2j_3+1)}_{j_{23}-j_2-j_3-1}(-1)(1+\xi)^{j_{23}-j_2-j_3-1}\,\bar{P}^{(2j_1+1,2j_{23}+1)}_{j-j_1-j_{23}-1}(\xi)$$

$$= \sum_{j_{12}} U(j_{12}, j_{23})\bar{P}^{(2j_{12}+1,2j_3+1)}_{j-j_{12}-j_3-1}(-1)\bar{P}^{(2j_1+1,j_2+1)}_{j_{12}-j_1-j_2-1}(\xi)\,, \tag{6.3.3}$$

where $\xi = \cos 2\vartheta$, $\bar{P}^{(\alpha,\beta)}_k(\xi) = N P^{(\alpha,\beta)}_k(\xi)$ are the Jacobi polynomials, and N is a normalizing constant for a closed node (see Table 6.1).

Noting the property of orthogonality of Jacobi polynomials $P^{(\alpha,\beta)}_k(\xi)$ in Eq. (6.3.3) we obtain the *integral representation* for the T-coefficients, viz.

$$U(j_{12}, j_{23}) = A \int_{-1}^{1} \bar{P}^{(2j_1+1,2j_2+1)}_{j_{12}-j_1-j_2-1}(\xi)\bar{P}^{(2j_1+1,2j_{23}+1)}_{j-j_1-j_{23}-1}(\xi)(1-\xi)^{2j_1+1}$$

$$\times (1+\xi)^{j_{23}+j_2-j_3}d\xi\,, \tag{6.3.4}$$

where

$$A = 2^{-2j_1-j_2+j_3-j_{23}-3}\,\bar{P}^{(2j_2+1,2j_3+1)}_{j_{23}-j_2-j_3-1}(-1)/\bar{P}^{(2j_{12}+1,2j_3+1)}_{j-j_{12}-j_3-1}(-1)\,.$$

6.3.2.2. To compute the integral (6.3.4) we resort to the equality (proved in the Addendum to Chap. 6)

$$\int_{-1}^{1} P^{(\alpha,\beta)}_k(\xi)P^{(\alpha,2s+1)}_{b-s-1}(\xi)(1-\xi)^\alpha(1+\xi)^{s-\alpha+\beta}d\xi$$

$$= B\frac{(-1)^k(\beta+1)_k(a+b+\alpha+1)_k(b-a-1)!}{k!(b-a-k-1)!}$$

$$\times {}_4F_3\left(\begin{array}{c}-k,\ \alpha+\beta+k+1,\ a-s,\ a+s+1\\ \beta+1,\ a+b+\alpha+1,\ -b+a+1\end{array}\Bigg|1\right)\,, \tag{6.3.5}$$

where

$$B = (-1)^{s-b+1} 2^{s-a+\alpha+\beta+1} \frac{\Gamma(s-a+\beta+1)\Gamma(b+\alpha-s)\Gamma(a+b-\beta-k)}{\Gamma(a-\beta+s+1)\Gamma(b-s)\Gamma(b-a+\alpha+\beta+k+1)},$$

${}_4F_3(1)$ is a value of the generalized hypergeometric function ${}_4F_3(z)$ (see (2.7.1)).

By virtue of (6.3.4) and (6.3.5) we have

$$U(j_{12},j_{23}) = \sqrt{(2j_{12}+1)(2j_{23}+1)C(j_{12})D(j_{23})}$$
$$\times \frac{\Gamma(1-b_2)}{\Gamma(b_1)\Gamma(b_3)} {}_4F_3 \left(\begin{matrix} a_1,a_2,a_3,a_4 \\ b_1,b_2,b_3 \end{matrix} \middle| 1 \right), \tag{6.3.6}$$

where $a_1 = -j_{12}+j_1+j_2+1$, $a_2 = j_1+j_2+j_{12}+2$, $a_3 = -j_{23}+j_2+j_3+1$, $a_4 = j_2+j_3+j_{23}+2$; $b_1 = 2j_2+2$, $b_2 = j_1+j_2+j_3-j+2$, $b_3 = j_1+j_2+j_3+j+3$;

$$C(j_{12})$$
$$= \frac{\Gamma(j-j_{12}+j_3+1)\Gamma(j+j_{12}+j_3+2)\Gamma(j_{12}-j_1+j_2+1)\Gamma(j_{12}+j_1+j_2+2)}{\Gamma(j+j_{12}-j_3+1)(j-j_{12}-j_3-1)!\Gamma(j_{12}+j_1-j_2+1)(j_{12}-j_1-j_2-1)!},$$

$$D(j_{23})$$
$$= \frac{\Gamma(j+j_1-j_{23}+1)\Gamma(j+j_1+j_{23}+2)\Gamma(j_{23}+j_2-j_3+1)\Gamma(j_{23}+j_2+j_3+2)}{\Gamma(j-j_1+j_{23}+1)(j-j_1-j_{23}-1)!\Gamma(j_{23}-j_2+j_3+1)(j_{23}-j_2-j_3-1)!}.$$

According to the representation (6.3.6) the T-coefficients of the transplant C in Fig. 6.18 are symmetric with respect to the substitution $j_1 \to j_3$ and $j_{12} \to j_{23}$.

6.3.2.3. The representation (6.3.6) enables one to express the T-coefficients of a transplant in terms of the Racah polynomials. It is not hard to see that the hypergeometric function ${}_4F_3(1)$ in (6.3.6) is a polynomial $y_k(x) = {}_4F_3(1)$ of degree $k = j_{12}-j_1-j_2-1$ in $x = j_{23}(j_{23}+1)$.

The orthogonality of the T-coefficients

$$\sum_{j_{23}=j_2+j_3+1}^{j-j_1-1} U(j_{12},j_{23})U(j'_{12},j_{23}) = \delta_{j_{12}j'_{12}}$$

leads to the orthogonality of the polynomials $y_k(x) = {}_4F_3(1)$ on the interval $[a, b-1]$ with the weight of the Racah polynomials $u_k(x) = u_k^{(\alpha,\beta)}(x,a,b)$ at $a = j_2+j_3+1$, $b = j-j_1$, $\alpha = 2j_1+1$, $\beta = 2j_2+1$. Consequently the polynomials $y_k(x)$ and $u_k(x)$ are identical accurate to the factor which can be determined by comparing the coefficients of the highest power. As a result we obtain the *representation of the T-coefficients of a transplant through the Racah polynomials* $u_k^{(\alpha,\beta)}(x,a,b)$

$$U(j_{12},j_{23}) = \frac{(-1)^{j_{12}-j_1-j_2-1}}{d_{j_{12}-j_1-j_2-1}} \sqrt{\varrho(j_{23})(2j_{23}+1)}$$
$$\times u_{j_{12}-j_1-j_2-1}^{(2j_1+1,2j_2+1)}[j_{23}(j_{23}+1), j_2+j_3+1, j-j_1], \tag{6.3.7}$$

where $\varrho(s)$ and d_k are the weight and norm of the polynomials $u_k^{(\alpha,\beta)}(x,a,b)$, $x = s(s+1)$. Suslov [S29] has established the relation of the T-coefficients with the classical orthogonal polynomials in a discrete variable.

We recall that in constructing the spherical harmonic $Y_{j_{12}m}(s)$ the node j_{12} indicated in Fig. 6.18 is associated with the factor (6.2.18). Since the Racah polynomials $u_k^{(\alpha,\beta)}(x,a,b)$ are difference analogues of Jacobi polynomials $P_k^{(\alpha,\beta)}(x)$ on a square mesh $x = s(s+1)$, the T-coefficient of a transplant (6.3.7) is, accurate to a normalizing factor, a difference analogue of the function (6.2.18).

Observing the property of symmetry of the matrix $U(j_{12}, j_{23})$ with respect to the transposition $j_1 \to j_3$, $j_{12} \to j_{23}$, we can write (6.3.7) in another form:

$$U(j_{12}, j_{23}) = \frac{(-1)^{j_{23}-j_2-j_3-1}}{d_{j_{23}-j_2-j_3-1}} \sqrt{\varrho(j_{12})(2j_{12}+1)}$$

$$\times u_{j_{23}-j_2-j_3-1}^{(2j_3+1,2j_2+1)}[j_{12}(j_{12}+1), j_1+j_2+1, j-j_3], \qquad (6.3.8)$$

This relation is consistent with the second relation of orthogonality of the T-coefficients

$$\sum_{j_{12}} U(j_{12}, j_{23})U(j_{12}, j'_{23}) = \delta_{j_{23}j'_{23}} .$$

6.3.2.4. The relations (6.3.7) and (6.3.8) enable us to obtain for the T-coefficients the difference equation, Rodrigues-type formula, recurrence relations, etc. with the use of the known properties of Racah polynomials. By way of example we give the *asymptotic formula of second order of accuracy* valid for the coefficients $U(j_{12}, j_{23})$ for $j \gg j_1 \sim j_2 \sim j_3 \sim j_{12}$, viz.

$$U(j_{12}, j_{23}) \approx (-1)^{j_{12}+j_1-j_2} \sqrt{\frac{(2j_{12}+1)(2j_{23}+1)}{(j-j_3)(j+j_3+1)}} d_{j_1+j_2+1, j_2-j_1}^{j_{12}}(\theta) ,$$

where the functions $d_{mm'}^{j_{12}}(\theta)$ are defined by (5.1.26),

$$\cos\theta = \frac{(2j_{23}+1)^2 - (j-j_3)^2 - (j+j_3-1)^2}{2(j-j_3)(j+j_3+1)} .$$

The attendant geometric interpretation of the angle θ is indicated in Fig. 6.19.

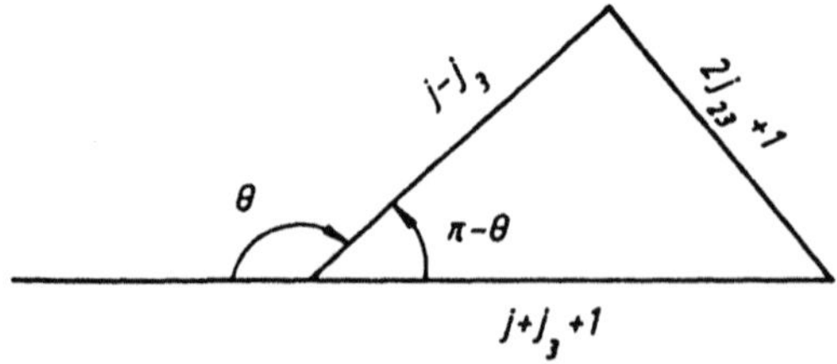

Fig. 6.19

312

6.3.3 Open Nodes

We have calculated the matrix elements of a transplant C for the case of closed nodes (see Fig. 6.18 depicting vertices of degree 3 or 2, which is possible for a root vertex only).

In a transformation between two arbitrary trees T and T' there occur also transplants involving nodes with pendant vertices, i.e. vertices of degree one. All such transplants can be readily derived from the general case in Fig. 6.18 by the consecutive inspection of the occurrence of one, two, and three pendant vertices. This results in 8 types of transplants graphed in Fig. 6.20 of which only the case at (a) has been elucidated thus far.

It emerged that the T-coefficients for all transplants with open nodes could be obtained from (6.3.7) and (6.3.8) as particular cases. Before we embark on this route we wish to examine how the functions $Y_0(\theta)$ listed in Table 6.1 for various types of nodes are related.

6.3.3.1. Let us derive $Y_0(\theta)$ for open nodes α, β and γ from Eq. (6.2.18) corresponding to node δ.

The first to examine is node β. To begin with we note that Eq. (6.2.9), where (6.2.11) is valid, reduces to (6.2.5) at $p_2 = 0$ and $l_2 = 0, 1$. Let us look at the eigenfunctions. The functions (6.2.18) will be denoted by $Y_j^{j_1 j_2}(\theta)$, where

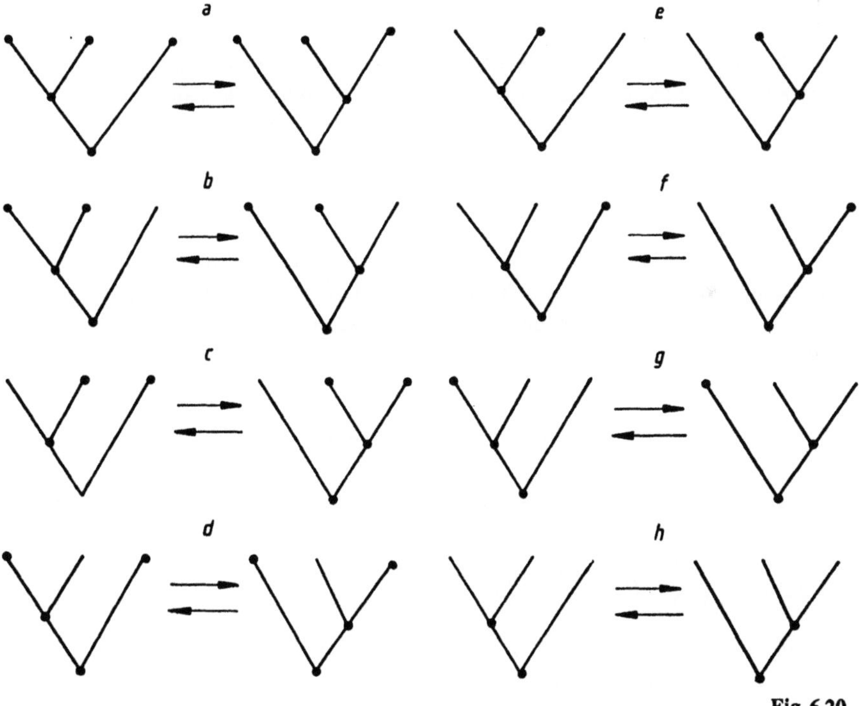

Fig. 6.20

313

as usual $2j_i + 1 = l_i + (p_i - 1)/2$, (6.2.16) by $Y_l^{l_1}(\theta)$. With reference to the relations between the Jacobi polynomials $P_k^{(\alpha,\pm 1/2)}(2s^2 - 1)$ and $P_k^{(\alpha,\alpha)}(s)$ (see the Eqs. (1.4.12) and (1.4.13) and to the duplication formulae for the Γ-function it is not hard to verify that

$$Y_l^{l_1}(\theta) = \frac{1}{\sqrt{2}} Y_j^{j_1 j_2}(\theta)\Big|_{j_2 = -(2\pm 1)/4} \, , \tag{6.3.9}$$

where the values of $j_2 = -3/4$ ($p_2 = 0, l_2 = 0$, upper sign) and $j_2 = -1/4$ ($p_2 = 0, l_2 = 1$, lower sign) correspond, respectively, to the even degrees ($l - l_1 = 0, 2, 4, \ldots$) and odd degrees ($l - l_1 = 1, 3, \ldots$) of the polynomials $P_{l-l_1}^{(2j_1+1, 2j_1+1)}(\cos\theta)$ in (6.2.16).

Thus the function $Y_l^{l_1}(\theta)$ for a β open node can indeed be obtained from the function $Y_j^{j_1 j_2}(\theta)$ for a δ node by assigning, in a formal manner, the constant $j_2 = -3/4$ or $j_2 = -1/4$ to the pendant vertex in agreement with the formula $2j_2 + 1 = l_2 - 1/2$. The rule $j \geq j_1 + j_2 + 1$ (see (6.3.9)) remains valid in this case as well.

A similar argument is applicable to the type γ of a node. Equation (6.2.9) becomes (6.2.7) at $p_1 = 0$ and $l_1 = 0$ or 1. If we denote the function (6.2.17) by $\tilde{Y}_l^{l_2}(\theta)$, then

$$\tilde{Y}_l^{l_2}(\theta) = \frac{(-1)^{\varphi_\pm}}{\sqrt{2}} Y_j^{j_1 j_2}(\theta)\Big|_{j_1 = -(2\pm 1)/4} \, , \tag{6.3.10}$$

where the values of $j_1 = -3/4$ ($p_1 = 0, l_1 = 0$) and the phase $\varphi_+ = (l - l_2)/2$ correspond to the even values of $l - l_2 = 0, 2, 4, \ldots$, while the values of $j_1 = -1/4$ ($p_1 = 0, l_1 = 1$) and $\varphi_- = (l - l_2 - 1)/2$ correspond to the odd values of $l - l_2 = 1, 3, 5, \ldots$. (The Eq. (6.3.10) follows also from (6.3.9) subject to the substitution $\theta \rightarrow \pi/2 - \theta$, $l_1 \rightarrow l_2$, and $p_1 \rightarrow p_2$; in the circumstances the symmetry property $Y_j^{j_2 j_1}(\pi/2 - \theta) = (-1)^{j - j_1 - j_2 - 1} Y_j^{j_1 j_2}(\theta)$ may also be handy.)

The remaining case to examine is that of node α. Equation (6.2.4) obviously results from (6.2.9), (6.2.5), and (6.2.7) if we formally let $p_1 = p_2 = 0$; $l_1 = 0, 1$; and $l_2 = 0, 1$. We take up the real-valued functions (6.2.13) denoted by $Y_m^\pm(\theta)$. Using the relations of the Jacobi polynomials $P_k^{(\pm 1/2, \pm 1/2)}(s)$ and $P_k^{(\pm 1/2, \mp 1/2)}(2s^2 - 1)$ with the Chebyshev polynomials of the first and second kind, $T_k(s)$ and $U_k(s)$ (see (1.4.2–4) and relations

$$T_{2n+1}(s) = \frac{n!}{(1/2)_n} s P_n^{(-1/2, 1/2)}(2s^2 - 1) \, ,$$

$$U_{2n}(s) = \frac{n!}{(1/2)_n} P_n^{(1/2, -1/2)}(2s^2 - 1)$$

(derived by comparing (1.4.13) with $\alpha = -1/2$ and (1.4.3), (1.4.12) with $\alpha = 1/2$ and (1.4.4), respectively) we consecutively obtain from (6.2.18)

$$Y_m^{\pm}(\theta) = \tfrac{1}{2}Y_j^{j_1 j_2}(\theta)\big|_{j_1=j_2=-(2\pm 1)/4} \quad \text{for } 2j+1 = m = 2,4,6,\dots$$

$$= \tfrac{1}{2}Y_j^{j_1 j_2}(\theta)\big|_{j_1=-j_2-1=-(2\pm 1)/4} \quad \text{for } 2j+1 = m = 1,3,5,\dots \,, \tag{6.3.11}$$

$$Y_0^+(\theta) = \frac{1}{\sqrt{2}}\lim_{m\to 0} Y_m^+$$

$$= \frac{1}{\sqrt{2}}Y_{-1/2}^{-3/4,-3/4} = 1/\sqrt{2\pi}\,.$$

By virtue of (6.3.9) and (6.3.10) these relations can be rewritten in the form

$$Y_m^{\pm}(\theta) = \frac{1}{\sqrt{2}}Y_m^{l_1}(\theta)\big|_{l_1=(1\mp 1)/2} \quad (m = 1,2,\dots)\,; \tag{6.3.12}$$

$$Y_m^{\pm}(\theta) = \frac{(-1)^{(m+1\mp 1)/2}}{\sqrt{2}}\tilde{Y}_m^{l_2}(\theta)\big|_{l_2=(1\mp 1)/2} \quad (m = m^+ = 2,4,\dots)$$

$$= \frac{(-1)^{(m-1)/2}}{\sqrt{2}}\tilde{Y}_m^{l_2}(\theta)\big|_{l_2=(1\pm 1)/2} \quad (m = m^- = 1,3\dots)\,. \tag{6.3.13}$$

These formulae follow also directly from (6.2.13, 16), and (6.2.17) if we again resort to the relation of the Jacobi polynomials with Chebychev's polynomials of the first and second kind.

As a result we arrive at the following property: in view of the Eqs. (6.3.9, 10) and (6.3.11) the *functions* $Y_m^{\pm}(\theta)$, $Y_l^{l_1}(\theta)$ and $\tilde{Y}_l^{l_2}(\theta)$ defined by the relations (6.2.13, 16) and (6.2.17) *for open nodes* α, β, and γ, respectively, *may be deemed particular cases of the function* $Y_j^{j_1 j_2}(\theta)$ *for closed node* δ (see (6.2.18)), if in using the graphical technique one formally assignes to the pendant vertices the constants $j_{1,2} = -3/4$ and $j_{1,2} = -1/4$. The relevant phase and normalizing factors must also be taken into account. It is worth emphasizing (see (6.3.9–11)) that in the said modification of the graphical rules, *for each of the open nodes* $\alpha, \beta,$*and* γ *the "momentum addition law"* $j \geq j_1 + j_2 + 1$ *holds* that was valid for node δ.

These considerations will be taken into account in order to avoid calculating anew the matrix elements for all remaining types of transplants in Fig. 6.20, but rather to exploit the above findings for the general case of closed nodes.

6.3.3.2. Now we turn to the main subject of this section and begin examination of *transplants with open nodes.* Here the decompositions of the type (6.3.1) and (6.3.2) take place too. The relations (6.3.2, 9–11) lead us to the rule of evaluating the matrix elements for the case of real-valued harmonics which reads as follows. *In the general formulae for the T-coefficients* (6.3.7) *and* (6.3.8) *one is to assign to pendant vertices* i *the constants* $j_i = -3/4$ *and* $j_i = -1/4$ *and take into account the respective phase and normalizing factors. For each node the rule* $j_{ik} \geq j_i + j_k + 1$ *remains valid.*

This property was established here by direct computations. Its group-theoretical implication, including the interpretation of 'momenta' j, will be elucidated in the ensuing section.

By way of example consider some transplants with open nodes for which the general Eqs. (6.3.7) and (6.3.8) simplify substantially.

(1) We calculate the matrix elements for the g-type transplant in Fig. 6.20. The decomposition (6.3.2) has the form

$$Y_j^{j_1 j_{23}}(\vartheta)Y_{l_{23}}^{\pm}(\psi) = \sum_{l_{12}} U'_{\pm}(l_{12}, l_{23})Y_{l_{12}}^{l_1}(\varphi)Y_l^{l_{12}}(\theta) \ . \tag{6.3.14}$$

Here the upper (lower) sign is associated with the transformation between harmonics which are even (odd) functions on substitution $\psi \to -\psi$, $\varphi \to \pi - \varphi$, $x_{n-1} \to -x_{n-1}$ (these angles were identified in Fig. 6.18, and (x_{n-1}, x_n) is a plane in which the angle ψ is specified for the open node α under consideration). These situations are illustrated in Fig. 6.21, at (a) and (b) respectively, where in turn the functions are singled out which are even and odd with respect to the substitution $\psi \to \pi - \psi$, $\theta \to \pi - \theta$, $x_n \to -x_n$. This results in four possible decompositions (6.3.14) associated with different properties of symmetry of spherical harmonics in reflection of the coordinate axes in the plane (x_{n-1}, x_n). The respective values of quantum numbers are determined according to the rule $j_{ik} \geq j_i + j_k + 1$ for all nodes in Fig. 6.21.

For matrix elements we have by virtue of (6.3.9) and (6.3.11)

$$U'_{\pm}(l_{12}, l_{23}) = \begin{cases} U(j_{12}, j_{23})\Big|_{\substack{j_2=-3/4 \\ j_3=-(2\pm1)/4}} \\ U(j_{12}, j_{23})\Big|_{\substack{j_2=-1/4 \\ j_3=-(2\pm1)/4}} \end{cases} , \tag{6.3.15}$$

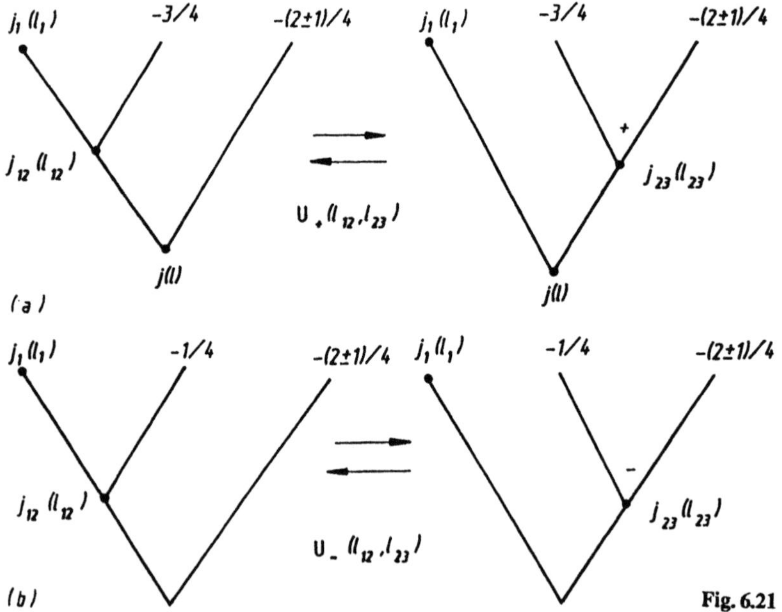

Fig. 6.21

$$U'_+(l_{12}, 0) = \frac{1}{\sqrt{2}} \lim_{l_{23} \to 0} U'_+(l_{12}, l_{23}) \ .$$

Hence with the use of the realtions between Racah's polynomials of the form $u_k^{(\alpha, \pm 1/2)}(x, \pm 1/2, b)$, $u_k^{(\alpha, \pm 1/2)}(x, 0, b)$ and the Hahn polynomials $h_k^{\alpha, \alpha}(x, N)$ — see the Eqs. (3.7.23, 24) — we find that

$$U'_\pm(l_{12}, l_{23}) = (-1)^{\varphi_\pm} \sqrt{2} \sqrt{\varrho \left(\frac{l - l_1 - l_{23}}{2} \right)} d^{-1}_{l_{12} - l_1}$$
$$\times h^{(2j_1+1, 2j_1+1)}_{l_{12} - l} \left(\frac{l - l_1 - l_{23}}{2}, l - l_1 + 1 \right) \ , \tag{6.3.16}$$

where

$$\varphi_\pm = (l_{12} - l_1)/2 \quad \text{for} \quad l_{12} - l_1 = 0, 2, 4, \ldots,$$
$$= (l_{12} - l_1 + 1)/2 \quad \text{for} \quad l_{12} - l_1 = 1, 3, 5, \ldots,$$

and $\varrho(s)$ and d_k are the weight and norm of the Hahn polynomials. In addition

$$U'_+(l_{12}, 0) = (-1)^{(l_{12} - l_1)/2} \sqrt{\varrho \left(\frac{l - l_1}{2} \right)} d^{-1}_{l_{12} - l_1}$$
$$\times h^{(2j_1+1, 2j_1+1)}_{l_{12} - l_1} \left(\frac{l - l_1}{2}, l - l_1 + 1 \right) \ . \tag{6.3.17}$$

These relations assume a more compact form if we employ the complex-valued function $\exp(im_{23}\psi)/\sqrt{2\pi}$, $l_{23} = |m|$, for the node j_{23} in Fig. 6.21. Denoting this function by $Y_{m_{23}}$ we have from (6.3.14) and (6.3.16)

$$Y_j^{j_1 j_{23}} Y_{m_{23}} = \sum_{l_{12}} U'(l_{12}, m_{23}) Y_{l_{12}}^{l_1} Y_l^{l_{12}} \ , \tag{6.3.18}$$

where

$$U(l_{12}, m_{23}) = (-i)^{l_{12} - l_1} \sqrt{\varrho \left(\frac{l - l_1 - m_{23}}{2} \right)} d^{-1}_{l_{12} - l_1}$$
$$\times h^{(2j_1+1, 2j_1+1)}_{l_{12} - l_1} \left(\frac{l - l_1 - m_{23}}{2}, l - l_1 + 1 \right) \ . \tag{6.3.19}$$

Noting the relation between the Hahn polynomials $h_k^{(\alpha, \beta)}(s, N)$ and the dual Hahn polynomials $w_n^{(c)}(x, a, b)$ — see (3.5.14) — the coefficients of the inverse transformation

$$Y_{l_{12}}^{l_1} Y_l^{l_{12}} = \sum_{m_{23}} U^{*\prime}(l_{12}, m_{23}) Y_j^{j_1 j_{23}} Y_{m_{23}} \tag{6.3.20}$$

(here the asterisk stands for the complex conjugate) may be obtained as follows:

$$U^{*\prime}(l_{12}, m_{23}) = (-i)^{l_{12}-l_1}(-1)^{(l-l_1-m_{23})/2} \tag{6.3.21}$$

$$\times \sqrt{\varrho\left(l_{12} + \frac{p_1 - 1}{2}\right)(2l_{12} + p_1)d^{-1}_{(l-l_1-m_{23})/2}}$$

$$\times w^{(0)}_{(l-l_1-m_{23})/2}\left[\left(l_{12} + \frac{p_1 - 1}{2}\right)\left(l_{12} + \frac{p_1 + 1}{2}\right), l_1 + \frac{p_1 - 1}{2}, l + \frac{p - 1}{2}\right] ,$$

where $\varrho(s)$ and d_n are the weight and norm of the dual Hahn polynomials $w_n(x)$, $x = s(s+1)$.

The Eqs. (6.3.18–21) completely solve the problem at hand. Similar relations also hold for the matrix elements of transplant f in Fig. 6.20.

To illustrate, at $p_1 = 1$ the Eqs. (6.3.19) and (6.3.21) define the matrix elements of the transformation between the harmonics (6.2.21) and (6.2.22) on the sphere S^3 in a four-dimensional space (see Fig. 6.13, where for the tree on the right it is convenient to substitue $l \to l'$, $\theta \to \vartheta$, $\varphi_1 \to \varphi$, $\varphi_2 \to \varphi'$, $l_1 \to m$, and $l_2 \to m'$). Representing the Clebsch-Gordan coefficients $C^{jm}_{j_1 m_1 j_2 m_2}$ in terms of the Hahn polynomials $h^{(\alpha,\beta)}_k(x, N)$ – see (5.2.13) – we may write in the aforementioned notation

$$Y_{l'lm}(\psi, \theta, \varphi) = i^{l-|m|} \sum_{m'} (-1)^{(l'+m'+|m|-2m)/2}$$

$$\times C^{lm}_{l'/2, m-(|m|+m')/2, l'/2, (|m|+m')/2} Y_{l'mm'}(\vartheta, \varphi, \varphi') .$$

This transformation arises, for example, in the Coulomb problem of nonrelativistic quantum mechanics (discrete spectrum, momentum representation, see [F5, F6, B15, E6]). Its group-theoretical meaning was examined in Chap. 5 (see (5.5.5)) while investigating irreducible representations of the rotation group in a four-dimensional space. It is worth noting that in this case, i.e. in the space of functions on the sphere S^3, only those representations of SO(4) are realized for which $j_1 = j_2 = l'/2$ [S25].

(2) Let us examine a *transplant with three pendant vertices* (case h in Fig. 6.20). It would pay to invoke complex harmonics here as well. The sought decomposition has the form

$$Y_m(\varphi)Y_l^{|m|}(\theta) = \sum_{m'} U^{*\prime\prime}(m, m')\tilde{Y}_l^{|m'|}(\vartheta)Y_{m'}(\psi) \tag{6.3.22}$$

(see Fig. 6.22). The respective T-coefficients $U''(m, m'$ can be deduced from

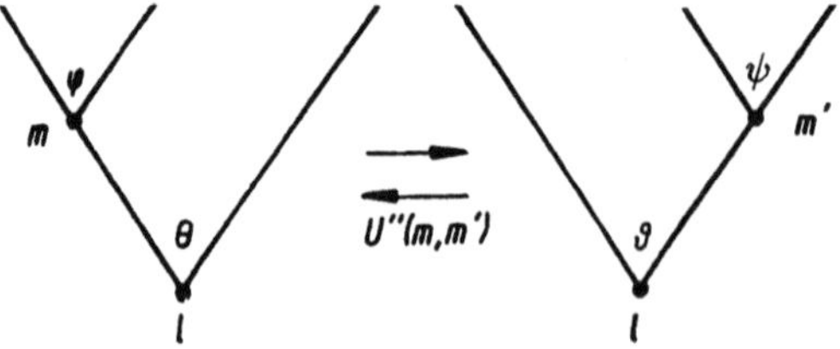

Fig. 6.22

318

(6.3.20) and (6.3.21). Letting $l_1 = 0, 1$ ($j_1 = -3/4, -1/4$), we find that it is convenient to obtain first the decompositions of the type (6.3.22) for the real-valued functions $Y_{|m|}^{\pm}(\varphi)$ with the use of the relations between the dual Hahn polynomials $w_n^{(0)}(x, \pm 1/2, b)$ and Kravchuk polynomials $k_n^{(1/2)}(s, N)$, namely (3.7.25, 26), and the Eqs. (6.3.10) and (6.3.12), and then to go over to the complex form with the use of the Euler formula. This approach results in

$$U''(m, m') = \mathrm{i}^{m' - |m'| + |m|} \sqrt{\varrho(l - m)} d_{l-m'}^{-1} k_{l-m'}^{(1/2)}(l - m, 2l) , \tag{6.3.23}$$

where $\varrho(s)$ and d_n are the weight and norm of Kravchuk's polynomials.

These matrix elements could also be derived from purely geometric considerations. For the case under examination we can use the rule for transformation of spherical harmonics $Y_{lm}(\theta, \varphi)$ in rotations of the coordinate system to write

$$Y_{lm}\left(\tfrac{\pi}{2} - \vartheta, \psi\right) = \sum_m D_{mm'}^l\left(\tfrac{\pi}{2}, \tfrac{\pi}{2}, \pi\right) Y_{lm}(\theta, \varphi) ; \tag{6.3.24}$$

where $D_{mm'}^l(\alpha, \beta, \gamma)$ are the generalized spherical functions (see Sect. 5.1).

Since in view of (6.2.20) we have

$$Y_{lm}(\theta, \varphi) = (-1)^{(m+|m|)/2} Y_m(\varphi) Y_l^{|m|}(\theta) ,$$

$$Y_{lm'}(\pi/2 - \vartheta, \psi) = (-1)^{(m'+|m'|)/2} \tilde{Y}_l^{|m'|}(\vartheta) Y_{m'}(\psi) ,$$

then we find by (6.3.24) for the T-coefficients

$$U''(m, m') = \mathrm{i}^{m' - |m'| + m} d_{mm'}^l(\pi/2) . \tag{6.3.25}$$

Noting the relation between the d-functions and Kravchuk's polynomials – see (5.2.27) – we again arrive at the representation (6.3.23).

We have discussed only those transplants with open nodes which allow substantial simplification of general Eqs. (6.3.7) and (6.3.8). One may readily derive the matrix elements for the remainig cases b through e in Fig. 6.20 by assigning the momenta $-3/4$ and $-1/4$ by the rules we have formulated above.

It will be useful for the reader to deduce matrix elements for the remaining cases in Fig. 6.20. Then derive integral representations of the type (6.3.4) for transplants with open nodes and compare the resultant expressions with the findings of [K10]. Thus far we have resorted to the tree technique to achieve tractable presentation of various systems of spherical coordinates and spherical harmonics in an n-dimensional Euclidean space. This application of the examined graphs is far from unique. They proved useful in solving other problems, specifically in the quantum theory of angular momentum where another interpretation of trees (we call it Wigner tree after [S9]) naturally arises [K37, K39].

6.4 Solution of the Schrödinger Equation for the n-Dimensional Harmonic Oscillator

The quantum-mechanical problem of the n-dimensional oscillator plays an important part in nuclear physics, molecular and crystalline vibration theory, and radiation theory to name just a few applications. Schrödinger's equation for an oscillator in n dimensions has an exact solution in Cartesian and spherical coordinates. The transformation coefficients between the respective wave functions are closely connected with the Hahn polynomials.

In view of the specific symmetry properties of an oscillator's Hamiltonian the structure of its wave functions reveals the group theoretical meaning of the T-coefficients by relating these quantities to the $3nj$-symbols of the noncompact group SU(1,1). Since the T-coefficients are expressable through Racah, Hahn, and Kravchuk polynomials, these can also arise in studies of the SU(1,1) representations owing to the said analogy.

6.4.1 Wave Functions of the Harmonic Oscillator in n Dimensions

Consider the Schrödinger equation for the n-dimensional harmonic oscillator

$$H\Psi = E\Psi ,\tag{6.4.1}$$

where

$$H = \sum_{s=1}^{n} \left(\frac{p_s^2}{2m} + \frac{m\omega^2}{2} q_s^2 \right) , \quad p_s = \frac{\hbar}{\mathrm{i}} \frac{\partial}{\partial q_s} ,$$

m is the mass, ω is the frequency of vibrations, and $\hbar$ is Planck's constant.

For the dimesionless coordinates $x_s = q_s / \sqrt{\hbar/m\omega}$, $s = 1,\ldots,n$, and energy $\varepsilon = E/\hbar\omega$ Eq. (6.4.1) takes the form

$$\frac{1}{2} \sum_{s=1}^{n} \left(-\frac{\partial^2}{\partial x_s^2} + x_s^2 \right) \Psi = \varepsilon\Psi .\tag{6.4.2}$$

We solve this equation by separation of variables.

6.4.1.1. In a Cartesian coordinate system, separation of variables in (6.4.1) yields n independent problems of a unidimensional oscillator. Therefore Eq. (6.4.2) has a solution

$$\Psi = \prod_{s=1}^{n} \Psi_{N_s}(x_s) ,\tag{6.4.3}$$

where

$$\Psi_{N_s}(x_s) = \frac{e^{-x_s^2/2}}{\sqrt{2^{N_s} N_s! \sqrt{\pi}}} H_{N_s}(x_s) \tag{6.4.4}$$

are the normalized wave functions of the unidimensional oscillator [5],

$$\int_{-\infty}^{\infty} \Psi_{N_s}^2(x_s)dx_s = 1 \ ,$$

and $H_k(\xi)$ are the Hermite polynomials ($s = 1, \ldots, n$).

The energy values ε corresponding to the solution (6.4.3) are as follows:

$$\varepsilon = N_1 + N_2 + \ldots + N_n + n/2 \ . \tag{6.4.5}$$

6.4.1.2. In spherical coordinates r, Ω given by a tree T we have by (6.1.10)

$$\Delta_r \Psi + \frac{1}{r^2} \Delta_\Omega \Psi + (2\varepsilon - r^2)\Psi = 0 \ , \tag{6.4.6}$$

where

$$\Delta_r = \frac{1}{r^{n-1}} \frac{\partial}{\partial r} \left(r^{n-1} \frac{\partial}{\partial r} \right)$$

is the radial part of the Laplace operator, and Δ_Ω its angular part.

For separation of variables we seek the solution of (6.4.6) in the form

$$\Psi = Y_{K\nu}(\Omega)R(r) \ ,$$

where $Y_{K\nu}$ is the spherical harmonic constructed by the tree T, K is the integer separation constant at the root of T (denoted by K according to the tradition of the method of K-harmonics and of the translation-invariant model of shells), and ν is the set of other subscripts. The equation for $R(r)$

$$R'' + \frac{n-1}{r}R' + \left[2\varepsilon - r^2 - \frac{K(K + n - 2)}{r^2} \right] R = 0$$

is carried by the substitution [6] $R(r) = \exp(-r^2/2)r^K y(\xi)$, $\xi = r^2$, for the function $y(\xi)$ to the hypergeometric equation (6.2.14) at $\sigma(\xi) = \xi$, $\tau(\xi) = K + n/2 - \xi$, and $\mu = (\varepsilon - K - n/2)/2$.

A normalized solution of (6.2.14) is enabled subject to the condition (6.2.15) yielding the energy levels as follws:

$$\varepsilon = N + n/2 \ , \quad (N - K)/2 = k = 0, 1, 2, \ldots \ . \tag{6.4.7}$$

---

[5] For the solution of the quantum mechanical problem of the unidimensional harmonic oscillator, see, e.g., [N18].

[6] This substitution may be found by the method elucidated in the book [N18].

The radial wave function $R(r)$ which meets the normalization condition

$$\int_0^\infty R_{NK}^2(r)r^{n-1}dr = 1 \, ,$$

is as follows:

$$R_{NK}(r) = \sqrt{\frac{2[(N-K)/2]!}{\Gamma[(N+K+n)/2]}}\,\exp{(-r^2/2)}r^K L_{(N-K)/2}^{K+n/2-1}(r^2) \, , \qquad (6.4.8)$$

where $L_K^\alpha(\xi)$ is a Laguerre polynomial.

The solution to (6.4.6) in the spherical coordinates r, Ω is given as

$$\Psi = \Psi_{NK\nu}(r, \Omega) = Y_{K\nu}(\Omega)R_{NK}(r) \, , \qquad (6.4.9)$$

where $Y_{K\nu}(\Omega)$ is the spherical harmonic associated with the coordinate tree T, and $R_{NK}(r)$ is defined by (6.4.8).

6.4.1.3. The wave function of the unidimensional oscillator (6.4.4) may be obtained as a particular case of (6.4.9). By letting $n = 1$ and $K = 0, 1$ and invoking the relaton between the Laguerre and Hermite polynomials $L_k^\alpha(\xi)$ and $H_k(\xi)$ (Eqs. (1.4.10) and (1.4.11)) we obtain for the wave function of the unidimensional oscillator

$$\Psi_N(x) = \begin{cases} \dfrac{(-1)^{N/2}}{\sqrt{2}}\Psi_{N0}(x) \, , \\ \dfrac{(-1)^{(N-1)/2}}{\sqrt{2}}\Psi_{N1}(x) \, , \end{cases} \qquad (6.4.10)$$

where $\Psi_{N0}(x) = R_{N0}(x)$ and $\Psi_{N1}(x) = R_{N1}(x)$ respectively for even and odd N, and $R_{N0}(x)$ and $R_{N1}(x)$ are found by (6.4.8) at $n = 1$.

6.4.2 Transformation between Wave Functions
of the Oscillator in Cartesian and Spherical Coordinates

The wave functions (6.4.3) and (6.4.9) form two complete orthonormal systems of functions. Accordingly between them there exists a unitary transformation whose matrix elements are closely connected with Hahn's polynomials. We demonstrate this connection below.

6.4.2.1. To obtain a decomposition of the wave function of the n-dimensional harmonic oscillator expressed in spherical coordinates (6.4.9) in wave functions expressed in Cartesian coordinates (6.4.3) consider the following problem. Suppose we decompose the initial space R into a direct sum of two mutually orthogonal subspaces R_1 and R_2, $R = R_1 \oplus R_2$, of dimension n_1 and $n_2 = n - n_1$, respectively. Using some trees T_1 and T_2, we introduce in R_1 and R_2 spherical coordinates r_1, Ω_1 and r_2, Ω_2 and build the oscillator functions $\Psi_{N_1 K_1 \nu_1}(r_1, \Omega_1)$ and $\Psi_{N_2 K_2 \nu_2}(r_2, \Omega_2)$ of the type (6.4.9). Their product

$$\Psi' = \Psi_{N_1 K_1 \nu_1}(r_1, \Omega_1)\Psi_{N_2 K_2 \nu_2}(r_2, \Omega_2) \tag{6.4.11}$$

forms a solution to the Schrödinger equation (6.4.2) corresponding to the energy level $\varepsilon = N_1 + N_2 + n/2$. To introduce spherical coordinates r, Ω over the entire space R we let $r_1 = r \sin \theta$ and $r_2 = r \cos \theta$. In the spherical coordinates r, Ω Eq. (6.4.2) has the solution (6.4.9) corresponding to the energy (6.4.7).

Consider now *the transformation between the solutions* (6.4.9) *and* (6.4.11) *associated with the same energy level*

$$\Psi_{N K \nu} = \sum_{N_1 + N_2 = N} \langle N_1 K_1 N_2 K_2 | N K \rangle \Psi_{N_1 K_1 \nu_1} \Psi_{N_2 K_2 \nu_2} \, , \tag{6.4.12}$$

and find *the coefficients of this transformation.*[7]

We shall use real-valued spherical harmonics, i.e. select a real-valued solution (6.2.13) for open nodes of the coordinate tree. Then the transformation inverse to (6.4.12) has the form

$$\Psi_{N_1 K_1 \nu_1} \Psi_{N_2 K_2 \nu_2} = \sum_{K} \langle N_1 K_1 N_2 K_2 | N K \rangle \Psi_{N K \nu} \, . \tag{6.4.13}$$

The angular parts of $\Psi_{N K \nu}, \Psi_{N_1 K_1 \nu_1}$ and $\Psi_{N_2 K_2 \nu_2}$ given by (6.4.9) are related by

$$Y_{K \nu}(\Omega) = Y_{K_1 \nu_1}(\Omega_1) Y_{K_2 \nu_2}(\Omega_2) Y_K(\theta) \, ,$$

where

$$Y_K(\theta) = N_K \sin^{K_1} \theta \cos^{K_2} \theta P_{j-j_1-j_2-1}^{(2j_1+1, 2j_2+1)}(\cos 2\theta) \, , \tag{6.4.14}$$

$2j_1 + 1 = K_i + n_i/2 - 1$, $i = 1, 2$; and N_K is the normalizer listed in Table 6.1. Therefore from (6.4.13) it follows that

$$R_{N_1 K_1}(r_1) R_{N_2 K_2}(r_2) = \sum_{K} \langle N_1 K_1 N_2 K_2 | N K \rangle Y_K(\theta) R_{N K}(r) \, ,$$

$$r_1 = r \sin \theta \, , \quad r_2 = r \cos \theta \, .$$

Multiplying both sides of this equality by $\exp(r^2/2) r^{-N}/\sqrt{2}$ and letting $r \to \infty$ by (6.4.8), we obtain

$$\frac{\sqrt{2}(\sin \theta)^{N_1}(\cos \theta)^{N_2}}{\sqrt{\left(\dfrac{N_1 - K_1}{2}\right)! \Gamma\left(\dfrac{N_1 + K_1 + n_1}{2}\right)\left(\dfrac{N_2 - K_2}{2}\right)! \Gamma\left(\dfrac{N_2 + K_2 + n_2}{2}\right)}}$$

---

[7] We denote the coefficients of the decomposition (6.4.12) by the same symbols as CGC without any risk of confusion as they never arise together.

$$= \sum_K \frac{\langle N_1 K_1 N_2 K_2 | N K \rangle}{\sqrt{\left(\frac{N-K}{2}\right)! \, \Gamma\left(\frac{N+K+n}{2}\right)}} (-1)^{(K-K_1-K_2)/2} Y_K(\theta) \; .$$

Using the Eq. (6.4.14) and the property of orthogonality of the Jacobi polynomials $P_k^{(\alpha,\beta)}(\xi)$ for $\xi = \cos 2\theta$ (or, what is the same, the fact that $Y_K(\theta)$ are orthonormal with respect to the measure $d\omega(\theta)$ listed in Table 6.1), we arrive at the *integral representation for the coefficients of* (6.4.12)

$$\langle N_1 K_1 N_2 K_2 | N K \rangle = \frac{(-1)^{j-j_1-j_2-1}}{2^{2j_1+2j_2+k_1+k_2+4}} A N_K$$

$$\times \int_{-1}^{1} (1-\xi)^{2j_1+k_1+1}(1+\xi)^{2j_2+k_2+1} P_{j-j_1-j_2-1}^{(2j_1+1,2j_2+1)}(\xi) d\xi \; , \qquad (6.4.15)$$

where

$$A = \sqrt{\frac{2k! \, \Gamma(k+2j+2)}{k_1! \, \Gamma(k_1+2j_1+2) k_2! \, \Gamma(k_2+2j_2+2)}} \; ,$$

$$k = (N-K)/2 \; , \quad k_i = (K_i - N_i)/2 \; , \quad i = 1,2 \; .$$

Because an analogous integral representation is valid for Hahn's polynomials (see (2.7.21)) we get

$$\langle N_1 K_1 N_2 K_2 | N K \rangle = \frac{(-1)^{j-j_2-j_1-1}}{d_{j-j_1-j_2-1}} \sqrt{\varrho(m_2 - j_2 - 1)}$$

$$\times h_{j-j_1-j_2-1}^{(2j_1+1,2j_2+1)}(m_2 - j_2 - 1, m - j_1 - j_2 - 1) \; , \qquad (6.4.16)$$

where $\varrho(x)$ and d_k are the weight and norm of Hahn polynomials $h_k^{(\alpha,\beta)}(x, N)$, $j_i = K_i/2 + n_i/4 - 1$, $m_i = N_i/2 + n_i/4$, $i = 1,2$; $m = m_1 + m_2$.

In view of (3.5.14) the coefficients of the decomposition (6.4.12) also have a representation in terms of the dual Hahn polynomials $w_k^{(c)}(x, a, b)$, namely

$$\langle N_1 K_1 N_2 K_2 | N K \rangle = \frac{(-1)^{m_2-j_2-1}}{d_{m_2-j_2-1}} \sqrt{\varrho(j)(2j+1)}$$

$$\times w_{m_2-j_2-1}^{(j_2-j_1)}[j(j+1), j_1 + j_2 + 1, m] \; , \qquad (6.4.17)$$

where $\varrho(s)$ and d_k are the weight and norm of the polynomials $w_k(x)$, $x = s(s+1)$.

Equations (6.4.16) and (6.4.17) completely solve the problem. The major properties of the decomposition coefficients of (6.4.12) are readily derivable with the help of the earlier evaluated properties of Hahn's polynomials. The relation (2.4.18) in particular leads to a symmetry property as follows:

$$\langle N_1 K_1 N_2 K_2 | N K \rangle = (-1)^{(K-K_1-K_2)/2} \langle N_2 K_2 N_1 K_1 | N K \rangle \; .$$

While in the derivations of the representations (6.4.15–17) only the more
general case of node δ corresponding to the function (6.4.14) was evaluated, the
expressions (6.4.15–17) remain valid for open nodes provided that real-valued
harmonics are used. By analogy with the reasoning of Sect. 6.3.3 and in view of
(6.4.10) it is recommended to re-derive the representations of the type (6.4.16)
and (6.4.17) for the decomposition coefficents of (6.4.12) for open nodes in order
to establish the correct phase and normalizing factors and, in all likelihood, to
simplify the results. These coefficients will be discussed under 6.4.2.3 below in
more detail with the use of the complex solution (6.2.12) for a node α.

6.4.2.2. The outlined considerations enable the decomposition of a wave function
of the n-dimensional harmonic oscillator in the spherical coordinates (6.4.9) to be
found in the wave functions in Cartesian coordinates (6.4.3). The transformation
(6.4.12) is applied one by one to the functions $\Psi_{NK\nu}, \Psi_{N_1 K_1 \nu_1}, \Psi_{N_2 K_2 \nu_2}$ and
so on in accordance with the tree T of the spherical coordinates, until the wave
functions of unidimensional oscillators (see (6.4.10)) are attained. The result is
the desired decomposition of (6.4.9) in the functions (6.4.3).

6.4.2.3. To conclude we take up some illustrative examples that allow further
simplification of the general results obtained above.

(i) The first example to come under our consideration is that of the decomposition of a wave function of the *two-dimensional isotropic oscillator* written in
the polar coordinates $\Psi_{NM}(r, \varphi)$ in the wave functions in Cartesian coordinates,
viz.

$$\Psi_{NM}(r, \varphi) = \sum_{N_1 + N_2 = N} \langle N_1 N_2 | NM \rangle \Psi_{N_1}(x_1) \Psi_{N_2}(x_2) \,. \tag{6.4.18}$$

Here

$$\Psi_{MN}(r, \varphi) = \frac{e^{iM\varphi}}{\sqrt{2\pi}} \sqrt{\frac{2[(N - \Lambda)/2]!}{[(N + \Lambda)/2]!}} \, e^{-r^2/2} r^\Lambda L^\Lambda_{(N-\Lambda)/2}(r^2) \,,$$

$M = -N, -N+2, \ldots, N-2, N$; $\lambda = |M|$ (we have chosen the complex form of
the solution for the angular part); the real-valued functons $\Psi_{N_i}(x_i)$, $i = 1, 2$, are
defined by virtue of (6.4.4); and $x_1 = r \sin \varphi$ and $x_2 = r \cos \varphi$ (see Fig. (6.7a)).
The coefficients of the decomposition (6.4.18) take on, generally speaking, some
complex values, unlike the general case discussed above.

The coefficients of the decomposition (6.4.18) are convenient to compute
without resorting to the general case. Taking the inverse transformation in (6.4.18)

$$\sum_M \langle N_1 N_2 | NM \rangle^* \sqrt{\frac{2[(N - \Lambda)/2]!}{[(N + \Lambda)/2]!}} r^\Lambda L^\Lambda_{(N-\Lambda)/2}(r^2) \frac{e^{iM\varphi}}{\sqrt{2\pi}}$$

$$= (2^N N_1! N_2! \pi)^{-1/2} H_{N_1}(x_1) H_{N_2}(x_2) \,,$$

letting r go to infinity and observing the property of orthogonality for the angular part, we arrive at the *integral representation*

$$\langle N_1 N_2 | NM \rangle = (-1)^{(N-|M|)/2} 2^{N/2} \sqrt{\frac{[(N+M)/2]![(N-M)/2]!}{N_1! N_2!}}$$

$$\times \frac{1}{2\pi} \int_0^{2\pi} (\sin \varphi)^{N_1} (\cos \varphi)^{N_2} e^{iM\varphi} d\varphi \,. \qquad (6.4.19)$$

To compute this integral it is sufficient to resort to the decompositions

$$(\sin \varphi)^{N_1} = \left(\frac{e^{i\varphi} - e^{-i\varphi}}{2i} \right)^{N_1} = \left(\frac{i}{2} \right)^{N_1} \sum_k (-1)^k \frac{N_1!}{k!(N_1 - k)!}$$

$$\times \exp\left[i(2k - N_1)\varphi\right] \,,$$

$$(\cos \varphi)^{N_2} = \left(\frac{e^{i\varphi} + e^{-i\varphi}}{2} \right)^{N_2}$$

$$= \left(\frac{1}{2} \right)^{N_2} \sum_l \frac{N_2!}{l!(N_2 - l)!} \exp\left[i(2l - N_2)\varphi\right] \,,$$

and the property of orthogonality of the system $\{\exp(iM\varphi)\}$. Using the representation of the Kravchuk polynomials $k_n^{(p)}(x, N)$ through the hypergeometric function (see (2.7.11)) we find

$$\langle N_1 N_2 | NM \rangle = (-1)^{(N-|M|)/2} (-i)^{N_1} 2^{(N_1 - N_2)/2}$$

$$\times \sqrt{\frac{N_1! N_2!}{[(N+M)/2]![(N-M)/2]!}} \, k_{N_1}^{(1/2)} \left(\frac{N-M}{2}, N \right) \,. \qquad (6.4.20)$$

This expression is real-valued accurate to the simple factor $(-i)^{N_1}$. In view of (5.1.27) the coefficients of the decomposition (6.4.18) also allow the representation through d-functions, namely

$$\langle N_1 N_2 | NM \rangle = (-1)^{(N-|M|)/2} i^{N_1} d_{(N_2 - N_1)/2, M/2}^{N/2} (\tfrac{\pi}{2}) \,. \qquad (6.4.21)$$

This relation is convenient to use in computations because tables of values of $d_{mm'}^j(\pi/2)$ for small $j = N/2$ are available (see, e.g. [V5]). A number of examples of the transformation (6.4.18) is considered in [F4].

If necessary, the expressions (6.4.20) and (6.4.21) may be rewritten in other equivalent forms using the properties of symmetry of the d-functions and Kravchuk polynomials.

To discuss the relationship with the case evaluated above we consider the transition in (6.4.18) to the wave functions of the two-dimensional oscillator,

$$\Psi_{N\Lambda}^{\pm} = \begin{cases} \sqrt{2}\,\mathrm{Re}\{\Psi_{N\Lambda}\} \\ \sqrt{2}\,\mathrm{Im}\{\Psi_{N\Lambda}\} \end{cases} \quad (\Lambda = |M| > 0) \,,$$

$$\Psi_{N0}^{+} = \frac{1}{\sqrt{2}} \lim_{\Lambda \to 0} \Psi_{N\Lambda}^{+} \,,$$

which correspond to the real-valued harmonics

$$Y_{\Lambda}^{+} = \tfrac{1}{\sqrt{\pi}} \cos \Lambda\varphi$$
$$Y_{\Lambda}^{-} = \tfrac{1}{\sqrt{\pi}} \sin \Lambda\varphi \quad (\Lambda = |M| > 0) \,,$$
$$Y_{0}^{+} = \frac{1}{\sqrt{2}} \lim_{\Lambda \to 0} Y_{\Lambda}^{+} = \frac{1}{\sqrt{2\pi}} \,.$$

The decomposition (6.4.18) now takes the form

$$\Psi_{N\Lambda}^{\pm}(r,\varphi) = \sum_{N_1 + N_2 = N} \langle N_1 N_2 | N\Lambda \rangle_{\pm} \Psi_{N_1}(x_1)\Psi_{N_2}(x_2) \,, \tag{6.4.22}$$

where

$$\langle N_1 N_2 | N\Lambda \rangle_{\pm} = \begin{cases} \sqrt{2}\,\mathrm{Re}\{\langle N_1 N_2 | N\Lambda \rangle\} \\ \sqrt{2}\,\mathrm{Im}\{\langle N_1 N_2 | N\Lambda \rangle\} \end{cases} \quad (\Lambda = |M| > 0) \,,$$
$$\langle N_1 N_2 | N0 \rangle_{+} = \frac{1}{\sqrt{2}} \lim_{\Lambda \to 0} \langle N_1 N_2 | N\Lambda \rangle_{+} \,.$$

Now to obtain the coefficients of the decomposition (6.4.22) it is sufficient to separate the real and imaginary parts in (6.4.20) and (6.4.21).

Because

$$\mathrm{Re}\{i^{N_1}\} = \begin{cases} (-1)^{N_1/2} & \text{for even } N_1 = N_1^{+} = 0, 2, 4, \ldots, \\ 0 & \text{for odd } N_1 = N_1^{-} = 1, 3, 5, \ldots, \end{cases}$$

and

$$\mathrm{Im}\{i^{N_1}\} = \begin{cases} 0 & \text{for } N_1 = N_1^{+} = 0, 2, 4, \ldots, \\ (-1)^{(N_1-1)/2} & \text{for } N_1 = N_1^{-} = 1, 3, 5, \ldots, \end{cases}$$

by virtue of (6.4.21) we find

$$\langle N_1^{\pm} N_2 | N\Lambda \rangle_{\pm} = (-1)^{\varphi_{\pm}} \sqrt{2}\, d_{(N_2 - N_1^{\pm})/2, \Lambda/2}^{N/2}\left(\tfrac{\pi}{2}\right) \,, \tag{6.4.23}$$

where $\varphi_{+} = N_1^{+}/2$ and $\varphi_{-} = (N_1^{-} - 1)/2$;

$$\langle N_1^{+} N_2 | N^{+}0 \rangle_{+} = (-1)^{N_1^{+}/2} d_{(N_2 - N_1^{+})/2, 0}^{N/2}\left(\tfrac{\pi}{2}\right) \,, \tag{6.4.24}$$

$$\langle N_1^{\mp} N_2 | N\Lambda \rangle_{\pm} = 0 \,. \tag{6.4.25}$$

As above, N_1^{+} and N_1^{-} designate respectively even and odd values of N_1.

The nonzero coefficients of the decomposition (6.4.22) also allow the representation through the Kravchuk polynomials

$$\langle N_1^{\pm} N_2 | N\Lambda \rangle_{\pm} = (-1)^{\varphi_{\pm}} \sqrt{2} \sqrt{\varrho[(N+\Lambda)/2]}\, d_{N_2}^{-1}$$
$$\times\, k_{N_2}^{(1/2)}((N+\Lambda)/2, N)\,, \tag{6.4.26}$$

where $\varphi_+ = N_1^+/2$, $\varphi_- = (N_1^- - 1)/2$, and $\Lambda = |M| > 0$;

$$\langle N_1^+ N_2 | N^+0 \rangle = (-1)^{N_1^+/2} \sqrt{\varrho(N^+/2)}\, d_{N_2}^{-1}$$
$$\times\, k_{N_2}^{(1/2)}(N^+/2, N^+)\,, \tag{6.4.27}$$

where $\varrho(x)$ and d_n are the weight and norm of the polynomials $k_n^{(p)}(x, N)$.

On the other hand, these relations may be obtained directly from the general representation (6.4.16) and (6.4.17) if all feasible cases corresponding to even and odd values of the quantum numbers N_1 and N_2 are consecutively evaluated with account of (6.4.10) and all the necessary simplifications are effected by means of (3.7.25, 26). It is not hard to see, however, that complex-valued harmonics are more convenient to use in the case of the two-dimensional oscillator.

(ii) The next to consider is the *three-dimensional harmonic oscillator*, for which it is common to use a choice of phases different from that of the general case. Let $\Psi_{NLM}(r, \theta, \varphi)$ be an oscillator wave function associated with the tree T in Fig. 6.7b, viz.

$$\Psi_{NLM}(r, \theta, \varphi) = Y_{LM}(\theta, \varphi) R_{NL}(r) \tag{6.4.28}$$

where $Y_{LM}(\theta, \varphi)$ is the spherical harmonic defined by (6.2.20), and the radial function R_{NL} is given by (6.4.8) with $K = L$ and $n = 3$; the quantum numbers N, L and M assume the values as follows: $L = N, N-2, N-4, \ldots 0$; $M = -L, -L+1, \ldots L-1, L$; and $N = 0, 1, 2, \ldots$.

It is obvious that to derive the decomposition of the function $\Psi_{NLM}(r, \theta, \varphi)$ and the functions $\Psi_{N_1}(x_1)\Psi_{N_2}(x_2)\Psi_{N_3}(x_3)$ it is sufficient to compute coefficents of the decomposition

$$\Psi_{NLM}(r, \theta, \varphi) = \sum_{N_{12}+N_3=N} \langle N_{12}MN_3 | NL \rangle \Psi_{N_{12}M}(r, \varphi)\Psi_{N_3}(x_3)\,, \tag{6.4.29}$$

and then make use of the decomposition (6.4.18) and (6.4.21) for the wave function of the two-dimensional oscillator $\Psi_{N_{12}M}(r, \varphi)$ written in the polar coordinates $r_1, \varphi(r_1^2 = x_1^2 + x_2^2, \tan\varphi = x_1/x_2)$ in the functions $\Psi_{N_1}(x_1)$ and $\Psi_{N_2}(x_2)$. It is quite clear that the coefficients of the decomposition (6.4.29) take on real values.

Similar to the reasoning used in the derivation of the integral representation (6.4.15) we find from (6.4.29) that

$$\langle N_{12}MN_3 | NL \rangle = (-1)^{(M+\Lambda)/2 - (N_3-L+\Lambda)/2} A N_L$$
$$\times \int_{-1}^{1} (1-\xi^2)^{(N_{12}+\Lambda)/2} \xi^{N_3} P_{L-\Lambda}^{(\Lambda,\Lambda)}(\xi) d\xi\,, \tag{6.4.30}$$

where

$$A = 2^{N_3/2} \sqrt{\frac{\left(\dfrac{N-L}{2}\right)! \, \Gamma\left(\dfrac{N+L+3}{2}\right)}{\left(\dfrac{N_{12}-M}{2}\right)! \left(\dfrac{N_{12}+M}{2}\right)! \, N_3! \sqrt{\pi}}} \, ,$$

N_L is the normalizing constant in (6.2.20), and $\Lambda = |M|$.

Clearly the coefficients of the decomposition (6.4.29) are other than zero only when the difference $L - \Lambda - N_3$ assumes even values (otherwise the integrand is an odd function). In this case by virtue of Eqs. (1.4.12, 13) and (2.7.21) we obtain

$$(-1)^{(L+M-N_3)/2}\langle N_{12}MN_3|NL\rangle$$
$$= \begin{cases} \sqrt{\varrho(N_3/2)}\,d^{-1}_{(L-\Lambda)/2} \\ \times h^{(\Lambda,-1/2)}_{(L-\Lambda)/2}(N_3/2,(N-\Lambda)/2+1)\,, \\ \sqrt{\varrho\left(\frac{N_3-1}{2}\right)}\,d^{-1}_{(L-\Lambda-1)/2} \\ \times h^{(\Lambda,1/2)}_{(L-\Lambda-1)/2}[(N_3-1)/2,(N-\Lambda+1)/2]\,, \end{cases} \tag{6.4.31}$$

provided that $L - \Lambda$ and N_3 simultaneously assume even or odd values. Here $\varrho(x)$ and d_k are the weight and norm of the Hahn polynomials $h_k^{(\alpha,\pm1/2)}(x)$, $\Lambda = |M|$, and we made use of the familiar relation

$$\sqrt{\pi}\,\Gamma(2z) = 2^{2z-1}\Gamma(z)\Gamma(z+1/2)\,.$$

It is worth noting that the Eqs. (6.4.31) follow from the general representation (6.4.16) at $j_2 = -3/4$ and $j_2 = -1/4$ (which corresponds to even and odd values of $L - \Lambda$, respectively) if all the necessary phase and normalizing factors are taken into account correctly by virtue of (6.2.20), (6.3.9) and (6.4.10).

This completes our consideration of the transformation between wave functions of the three-dimensional harmonic oscillator written in Cartesian and spherical coordinates for the tree T in Fig. 6.7b. This approach paves the way for straight forward evaluations of other cases of $n = 3$ (see, e.g., Fig. 6.7c). On the other hand, the Eqs. (6.4.31) may be readily generalized over the case of an open node for the n-dimensional oscillator.

6.4.3 The T-Coefficients as the $3nj$-Symbols of SU(1,1)

The study of the structure of wave functions of the harmonic oscillator in n dimensions reveals the group theoretical meaning of the T-coefficients.

6.4.3.1. First of all we note the group-theoretical properties of the oscillator wave function (6.4.9). Let the Hamiltonian of the harmonic oscillator (6.4.2) be

$$H = \frac{1}{2} \sum_{s=1}^{n} (a_s a_s^+ + a_s^+ a_s) , \tag{6.4.32}$$

where

$$a_s^+ = \frac{1}{i\sqrt{2}} \left(\frac{\partial}{\partial x_s} - x_s \right) \quad \text{and} \quad a_s = \frac{1}{i\sqrt{2}} \left(\frac{\partial}{\partial x_s} + x_s \right)$$

are the *creation* and *annihilation operators* having the commutation algebra [8]

$$[a_s, a_{s'}] = [a_s^+, a_{s'}^+] = 0 ,$$
$$[a_s, a_{s'}^+] = \delta_{ss'} , \quad s, s' = 1, \dots, n .$$

It is not hard to verify that the operators

$$J_+ = \frac{1}{2} \sum_{s=1}^{n} (a_s^+)^2 ,$$

$$J_- = \frac{1}{2} \sum_{s=1}^{n} (a_s)^2 , \tag{6.4.33}$$

$$J_0 = \frac{1}{4} \sum_{s=1}^{n} (a_s^+ a_s + a_s a_s^+) = \frac{1}{2} H$$

satisfy the commutation relations

$$[J_0, J_\pm] = \pm J_\pm , \quad [J_+, J_-] = -2J_0 . \tag{6.4.34}$$

The commutation rules (6.4.34) are valid for the infintesimal operators of the non-compact group SU(1,1) (see, e.g., [B5, F2]).

For the wave function of the n-dimensional harmonic oscillator (6.4.9) we introduce the following designation:

$$|jm\rangle \equiv \Psi_{NK\nu}(r, \Omega) = Y_{K\nu}(\Omega) R_{NK}(r) , \tag{6.4.35}$$

where $j = K/2 + n/4 - 1$, $m = N/2 + n/4$, and $m = j + 1, j + 2, \dots$. The inequality $m \geq j+1$ is satisfied because by the quantization rule (6.4.7) we have $N = K, K + 2, K + 4, \dots$.

Now we write the operators $J_\pm$ and J_0 in the spherical coordinates r, Ω. By the definition of a_s and a_s^+ we have

$$J_\pm = \frac{1}{2} \left(H - r^2 \pm \frac{n}{2} \pm r \frac{\partial}{\partial r} \right) \quad \text{and} \quad J_0 = \tfrac{1}{2} H .$$

In this derivation we observed the equality $x(\partial/\partial x) = r(\partial/\partial r)$ that is valid in view of the relations $x = x(r, \Omega) = rs(\Omega)$, $\Psi = \Psi(x(r, \Omega))$, and

---

[8] The creation and annihilation operators a_s and a_s^+ are defined accurate to the substitution $a_s \to ia_s$, $a_s^+ \to -ia_s^+$, which does not change the commutation algebra. For the following it is convenient to define these operators as in Problem 31 of S. Flügge's *Practical Quantum Mechanics* [F4].

$$r\frac{\partial \Psi}{\partial r} = r\frac{\partial \Psi}{\partial x}\frac{\partial x}{\partial r} = r\frac{\partial \Psi}{\partial x}s = x\frac{\partial \Psi}{\partial x} \ .$$

Now we apply the operators $J_\pm$ and J_0 to the wave function (6.4.35). Observing the differentiation formulae for Laguerre polynomials $L_k^\alpha(\xi)$

$$\xi\frac{d}{d\xi}L_k^\alpha(\xi) = kL_k^\alpha - (\alpha + k)L_{k-1}^\alpha(\xi)$$

$$= (k+1)L_{k+1}^\alpha(\xi) - (\alpha + k - \xi + 1)L_k^\alpha(\xi)$$

(following from Eqs. (1.2.13) and (1.4.14)) and Schrödinger's equation (6.4.2), we obtain[9]

$$J_\pm|jm\rangle = \sqrt{(m \mp j)(m \pm j \pm 1)}|j, m \pm 1\rangle \ ,$$
$$J_0|jm\rangle = m|jm\rangle \ , \tag{6.4.36}$$

whence $J^2|jm\rangle = j(j+1)|jm\rangle$, where $J^2 = J_0^2 + J_0 - J_-J_+ = J_0^2 - J_0 - J_+J_-$.

The relations (6.4.36) coincide with the formulae that define the action of the infinitesimal operators $J_\pm$ and J_0 of the group SU(1,1) on the basis $|jm\rangle$ of the irreducible representation D_+^j that belong to the *discrete positive series* (see [B5, F2, K17]). For the basis (6.4.35), depending on the number $n = \dim R$, the moment $j = K/2 + n/4 - 1$ of the group SU(1,1) and its projection $m = N/2 + n/4$ may assume integer, half-integer and quarter-integer values.

Consequently *the wave functions (6.4.9) of the n-dimensional harmonic oscillator form a basis of the two-valued irreducible representation D_+^j for the Lie algebra of* SU(1,1).

Let us look in particular at the group-theoretical properties of wave functions of the one-dimensional oscillator. By the definition of the creation and annihilation operators a^+ and a we have for the functions (6.4.4)

$$ia\Psi_N = \sqrt{N}\Psi_{N-1} \ , \quad -ia^+\Psi_N = \sqrt{N+1}\Psi_{N+1} \ ;$$

here we used the differentiation formulae for Hermite polynomials, namely $H_k' = 2kH_{k-1} = 2xH_k - H_{k+1}$. It follows that in this case the relations (6.4.36) will be met for the basis functions of the type

$$|jm\rangle = (-1)^{N/2}\Psi_N \ , \quad N = N^+ = 0, 2, 4, \ldots \quad \text{for } j = -3/4 \ ,$$
$$= (-1)^{(N-1)/2}\Psi_N \ , \quad N = N^- = 1, 3, 5, \ldots \quad \text{for } j = -1/4 \ , \tag{6.4.37}$$

where $m = N/2 + 1/4$.

Thus the even and odd wave functions of the unidimensional oscillator, Ψ_N, form, respectively, bases for two irreducible representations D_+^j of the algebra

---

[9] In the previous designations we have $J_\pm\Psi_{NK\nu} = \frac{1}{2}\sqrt{(N - K + 1 \pm 1)(N + K + n - 1 \pm 1)} \times \Psi_{N\pm2, K\nu}$.

SU(1,1), where the moments are $j = -3/4$ for the even values of $N = N^+$ and $j = -1/4$ for odd $N = N^-$.

In conclusion noting one more circumstance is in order. In Sects. 6.2 and 6.3 we learned that in graphical construction of hyperspherical harmonics and in subsequent studies of transplants it is convenient to assign to each vertex of the coordinate tree (including pendant vertices) the quantity $j = K/2+n/4-1$, where K is the respective integer in the Eq. (6.2.11) for the eigenvalue of the angular part of the Laplacian, Δ_Ω. As has been learned the quantity j that may assume quarter-integer values naturally occurs in the expressions describing transplants (explanation of this fact is deferred until later). On the other hand, the study into the group-theoretical properties of wave functions of the oscillator in n dimensions revealed that the quantity $j(j + 1)$ is an eigenvalue of the operator of squared momentum $J^2 = J_0^2 - J_0 - J_+J_-$ for the group SU(1,1) in the case of a discrete positive series. This implies that one constant $j = K/2 + n/4 - 1$ characterizes eigenvalues for two operators Δ_Ω and J^2 which have distinct group theoretical meanings. This connection is not accidental. Using the definition (6.4.33) for the operators J_+, J_-, and J_0 and the Laplace operator in spherical coordinates (6.1.10) for the case under examination we may obtain

$$J^2 = \frac{1}{4}\left(-\Delta_\Omega + \frac{n(n-4)}{4}\right) , \qquad (6.4.38)$$

which on converting to eigenvalues leads to the evaluated relation between j and K.

Thus the oscillator wave functions $\Psi_{NK\nu} = Y_{K\nu}R_{NK}$ are simultaneously the eigenfunctions of the operators Δ_Ω, J^2 and J_0 and by (6.4.38) there exists a one to one correspondence between the eigenvalues of the operators Δ_Ω and J^2, namely $j = K/2 + n/4 - 1$. A more detailed group-theoretical treatment of these interconnections based on the concept of complementness may be found in [M18, S9].

6.4.3.2. The Group-Theoretical Meaning of the T Coefficients.

Now that we are equipped with all the necessary concepts to discuss the group-theoretical properties of the T coefficients we return to the transformation (6.4.12). In the nomenclature outlined by (6.4.35) it takes the form

$$|jm\rangle = \sum_{m_1+m_2=m} \langle j_1 m_1 j_2 m_2 | jm\rangle |j_1 m_1\rangle |j_2 m_2\rangle , \qquad (6.4.39)$$

where

$$j = j_1 + j_2 + 1, j_1 + j_2 + 2, \ldots, m - 1 . \qquad (6.4.40)$$

We denote by $\boldsymbol{J}$ the set of operators J_+, J_-, J_0 for the group SU(1,1) defined by (6.4.33) and introduce such operators $\boldsymbol{J}_1$ and $\boldsymbol{J}_2$ for either subspace R_1 and R_2, respectively. Since the momenta of the group SU(1,1) combine in the transformation (6.4.12) as $\boldsymbol{J} = \boldsymbol{J}_1 + \boldsymbol{J}_2$, the coefficients of the decomposition

(6.4.39) defined by (6.4.16) are *the Clebsch-Gordan coefficients of the group* $SU(1,1)$ *for the multiplication of two discrete positive series* $D_+^{j_1} \otimes D_+^{j_2}$. By virtue of the Eqs. (6.4.16) and (6.4.17) these Clebsch-Gordan coefficients are expressable through the Hahn polynomials. The rule (6.4.40) yields the law for combining momenta of $SU(1,1)$ in the case of $D_+^{j_1} \otimes D_+^{j_2}$.

Let $\Psi_{NK\nu}$ be an oscillator wave function in the spherical coordinates r, Ω specified by some tree T. It has been demonstrated above that the function $\Psi_{NK\nu}$ can be constructed of wave functions of one-dimensional oscillators by consecutive application of the transformation (6.4.12) in accordance with the tree T. Since such a transformation is simply an addition of the respective momenta for the group $SU(1,1)$ (see Eqs. (6.4.39) and (6.4.40)) then the function (6.4.35) arises also as a result of consecutive addition of the momenta of the group $SU(1,1)$ by the coupling scheme defined by the tree T.

Thus a wave function of the n-dimensional harmonic oscillator (6.4.35) can be constructed in two ways:

(1) by solving the Schrödinger equation (6.4.2) by separation of variables in a certain system of spherical coordinates given by a tree T;

(2) from the wave functions of one-dimensional harmonic oscillators (6.4.4) as a result of consecutive combining of momenta of the group $SU(1,1)$ by the coupling scheme represented by the same tree T.

It follows that the transformation between the oscillator functions (6.4.35), obtained in different spherical coordinates T and T' and involving by (6.4.9) and (6.4.8) only the angular part of the wave function, is performed, on the one hand, by means of the T-coefficients, and on the other hand, by definition, with the help of the $3nj$-symbols of the group $SU(1,1)$. Therefore these quantities coincide. In particular, the T-coefficient of a transplant is proportional to the $6j$-symbol of $SU(1,1)$. If the transplant involves open nodes, one is to assign the $SU(1,1)$ momenta equal to $-3/4$ and $-1/4$, typical of one-dimensional oscillator wave functions to the pendant vertices.

The relationship of T-coefficients with the $3nj$-symbols for discrete positive series of the group $SU(1,1)$ has been established by Knyr et al. [K17] (see also the paper of Kuznetsov and Smorodinskiĭ [K36] and the review papers of Smirnov and Shitikova [S9] and Smorodinskiĭ [S12]).

6.4.3.3. A few illustrative examples are in order to clarify what we have said above.

(i) Consider a two-dimensional oscillator associated with the tree in Fig. 6.7a. From the view point of the representation of the $SU(1,1)$ algebra, the decomposition (6.4.22) implies subject to (6.4.37) the vectorial coupling of two $SU(1,1)$ momenta, j_1 and j_2, each of which equals either $-3/4$ or $-1/4$. The summation rule has the form $j_1 + j_2 + 1 \leq j_{12} \leq m_{12} - 1$, where $2j_2 + 1 = \Lambda$ and $2m_{12} = N_{12} + 1$. According to the formulae $m_i = N_i/2 + 1/4$, the equality $m_{12} = m_1 + m_2$ means that $N_{12} = N_1 + N_2$. As a result we obtain a real-valued wave function of two-dimensional oscillator in the polar coordinates $\Psi^{\pm}_{N_{12}\Lambda}$, the summation of the momenta $j_1 = -3/4$ and $j_2 = -3/4$ ($j_2 = -1/4$) leading to even (or odd) values of Λ and an angular function of the type $\cos\Lambda\varphi$, whereas the summation of $j_1 = -1/4$ and $j_2 = -1/4$ ($j_2 = -3/4$) yielding even (odd) values of Λ and an angular dependence of the type $\sin\Lambda\varphi$ (see (6.3.11)).

Because the coefficients of the decomposition (6.4.22) coincide accurate to simple phase factors with the respective CGC for the group SU(1,1), the latter reduce to Kravchuk's polynomials in view of (6.4.26) and (6.4.27).

(ii) Let us examine the oscillator functions for the tree T in Fig. 6.7b. We have recourse to the wave function constructed for a two-dimensional oscillator by combining momenta in the plane (x_1, x_2) to perform a further addition of the momentum j_{12} with the momentum j_3 equal to $-3/4$ or $-1/4$ and corresponding to the oscillator function in the coordinate x_3. By virtue of (6.4.29) we arrive at the real-valued function for the *three-dimensional oscillator*:

$$\Psi^{\pm}_{NL\Lambda} = \begin{cases} \sqrt{2}\,\mathrm{Re}\,\{\Psi_{NL\Lambda}\} \\ \sqrt{2}\,\mathrm{Im}\,\{\Psi_{NL\Lambda}\} \end{cases} (\Lambda = |M| > 0)\,,$$

$$\Psi^{+}_{NL0} = \frac{1}{\sqrt{2}}\lim_{\Lambda\to 0}\Psi^{+}_{NL\Lambda}\,,$$

where $\Psi_{NL\Lambda}$ is defined by (6.4.28), $N = 2m - 3/2$, $L = 2j + 1/2$, and all the constraints on the quantum numbers N, L and Λ are imposed by the rule of combining of the momenta of the group SU(1,1): $j_{12} + j_3 + 1 \le j \le m - 1$. In a very similar manner, combining the momenta of the group SU(1,1) leads us to the wave function of the three-dimensional oscillator for the tree T in Fig. 6.7c. The transformaton between the wave functions constructed by the coupling schemes b and c in Fig. 6.7 is carried out by definition with the help of the respective $6j$-symbol of the group SU(1,1). On the other hand, this transformation involves only the angular parts, spherical harmonics, owing to the structure of the oscillator wave functions and is effected by the T–coefficient of the transplant in Fig. 6.22, thus reducing to the Kravchuk polynomials. Therefore the $6j$-symbols for the discrete positive series of the group SU(1,1) allow a representation through the Kravchuk polynomials for a transplant with three pendant vertices.

(iii) The last example is devoted to the *general type of transplant* elucidated in Sect. 6.3. If the trees in Fig. 6.18 are treated as two distinct coupling schemes for the momenta of SU(1,1), then the transformation matrix $U(j_{12}, j_{23})$ (T-coefficient of the transplant) will be proportional to the *6j-symbol for* the case of multiplication of *three discrete* positive series of this group. For the arising $6j$-symbols all the basic properties of similar quantities for the three-dimensional rotation group remain valid (taking account of the momentum combining rules of SU(1,1)) that were derived in Sect. 5.3.1. In particular, there exists a representation through CGC which, as has been demonstrated above, must be determined by the Eqs. (6.4.16) and (6.4.17) in the case of the group SU(1,1). The relations (6.3.7) and (6.3.8) yield representations of the examined $6j$-symbols of SU(1,1) through the Racah polynomials. Possible simplifications of the general formulae associated with open nodes were elucidated earlier.

6.4.3.4. Multivaluedness of Trees.

Summing up, we should recognize that the simplest graphs, trees, proved useful in solving the Laplace equation in hyperspherical coordinates. The tree T is a convenient artifice to introduce a system of spherical coordinates (see Sect. 6.2). The study of transformations between trees of various structure yields a description of the T coefficients that relate harmonics constructed in various hyperspherical coordinates (see Sect. 6.3). In addition the tree T depicts a wave function of the n-dimensional harmonic oscillator and the equivalent scheme of coupling of the momenta of SU(1,1) (see Sect. 6.4).

The tree approach enables simple and tractable solutions to several multivariate problems that seem entirely different on first glance. The method reveals analogies between these problems that might remain unnoticed without this graphical technique.

6.4.4 Matrix Elements of SU(1,1)

As shown in Sect. 6.4.3 the oscillator wave functions (6.4.35) may be interpreted as the basis functions of the irreducible representation D_+^j (discrete positive series) of the group SU(1,1). The infinitesimal operators (6.4.33) act on this basis as defined by (6.4.36).

For this realization of D_+^j, one can readily evaluate the generalized spherical functions of SU(1,1), i.e. the matrix elements of the "finite rotation" operator, which are closely connected with the Jacobi and Meixner polynomials.

6.4.4.1. By virtue of (6.4.33) we introduce the Hermitian operators

$$J_x = \tfrac{1}{2}(J_+ + J_-) , \quad J_y = \frac{1}{2i}(J_+ - J_-) , \quad J_z = J_0 , \tag{6.4.41}$$

which in view of (6.4.33) satisfy the following commutation relations:

$$[J_x, J_y] = -iJ_z , \quad [J_y, J_z] = iJ_x , \quad [J_z, J_x] = iJ_y . \tag{6.4.42}$$

These commutation rules differ from (5.1.4) valid for the three-dimensional rotation group SO(3) only in the sign in the first equation. The relations (6.4.42) can be shown to take place for the generators of the Lorentz group SO(2,1) associated with the three-dimensional pseudo-Euclidean space in which a distance (interval) is defined by means of the quadratic form $r^2 = x^2 + y^2 - z^2$. The group SU(1,1) is the universal covering group of the Lorentz group SO(2,1) and obeys the same commutation rules.

The representations theory of SO(2,1) may be developed on the basis of the commutation relations (6.4.42) by analogy with the representation theory of the three-dimensional rotation group with appropriate modifications [B5]. In this section we discuss only the generalized spherical functions of SU(1,1) for the discrete positive series D_+^j. These functions may be defined by analogy with (5.1.14), viz.

$$T^j_{mm'}(\alpha, \tau, \gamma) = e^{-im\alpha} t^j_{mm'}(\tau) e^{-im'\alpha} \, ,$$

where

$$t^j_{mm'}(\tau) = \langle jm| e^{-i\tau J_y} |jm'\rangle \, . \tag{6.4.43}$$

They are referred to as the *Bargman functions*. Below we represent them in terms of the Jacobi and Meixner polynomials.

6.4.4.2. To obtain the explicit expression for (6.4.43) let us use the realization of D_+^j in terms of the oscillator wave functions (6.4.35). Writing the operators $J_\pm$ in the spherical coordinates r, Ω (see Sect. 6.4.3) we have

$$iJ_y = \tfrac{1}{2}(J_+ - J_-) = \frac{n}{4} + \tfrac{1}{2} r \frac{\partial}{\partial r} \, .$$

Thus

$$e^{-i\tau J_y} |jm'\rangle = e^{-i\tau J_y} \Psi_{N'K\nu}$$
$$= e^{-n\tau/4} Y_{K\nu}(\Omega) \left(e^{-(\tau/2) r \partial/\partial r} R_{N'K}(r) \right)$$
$$= e^{-n\tau/4} Y_{K\nu}(\Omega) R_{N'K} \left(e^{-\tau/2} r \right) \, .$$

Here we have used the equations $\exp(A+B) = \exp A \, \exp B$ subject to $[A, B] = 0$, and $\exp[\alpha r(\partial/\partial r)] f(r) = f(e^\alpha r)$.

Therefore for the matrix elements under consideration,

$$\langle jm| e^{-i\tau J_y} |jm'\rangle = \int \Psi^*_{NK\nu}(r, \Omega) \left(e^{-i\tau J_y} \Psi_{N'K\nu}(r, \Omega) \right) r^{n-1} dr \, d\Omega \, ,$$

the following *integral representation* is valid:

$$t^j_{mm'}(\tau) = e^{-n\tau/4} \int_0^\infty R_{NK}(r) R_{N'K}(e^{-\tau/2} r) r^{n-1} dr \, , \tag{6.4.44}$$

where the functions $R_{NK}(r)$ are defined by (6.4.8) with $j = K/2 + n/4 - 1$, $m = N/2 + n/4$, and $m' = N'/2 + n/4$.

By virtue of (6.4.8)

$$t^j_{mm'}(\tau) = \sqrt{\frac{(m - j - 1)!(m' - j - 1)!}{(m + j)!(m' + j)!}} \, \beta^{j+1}$$
$$\times \int_0^\infty t^{2j+1} e^{-(1+\beta)t/2} L^{2j+1}_{m-j-1}(t) L^{2j+1}_{m'-j-1}(\beta t) dt \, , \quad \beta = e^{-\tau} \, ,$$

where the integral can be evaluated by the relation

$$\int_0^\infty e^{-(1+\beta)t/2} t^\alpha L_p^\alpha(t) L_q^\alpha(\beta t)\,dt$$

$$= (-1)^p \frac{\Gamma(\alpha+p+1)\Gamma(\alpha+q+1)}{\Gamma(\alpha+1)p!q!} \left(\frac{1+\beta}{2}\right)^{-\alpha-1} \left(\frac{1-\beta}{1+\beta}\right)^{p+q}$$

$$\times F\left(-p,-q,\alpha+1,-\left(\frac{\beta^{-1/2}-\beta^{1/2}}{2}\right)^{-2}\right) \tag{6.4.45}$$

proved in the Addendum to Chap. 6. Here $L_p^\alpha(t)$ are the Laguerre polynomials, and $F(a,b,c,t)$ is the hypergeometric functions. Finally

$$t_{mm'}^j(\tau) = \frac{(-1)^{m-j-1}}{\Gamma(2j+2)} \sqrt{\frac{(m+j)!(m'+j)!}{(m-j-1)!(m'-j-1)!}}$$

$$\times \left(\sinh\frac{\tau}{2}\right)^{-2j-2} \left(\tanh\frac{\tau}{2}\right)^{m+m'}$$

$$\times F\left(-m+j+1,-m'+j+1,2j+2,-\frac{1}{\sinh^2(\tau/2)}\right) . \tag{6.4.46}$$

According to (6.4.46) we have

$$t_{mm'}^j(\tau) = (-1)^{m-m'} t_{m'm}^j(\tau) . \tag{6.4.47}$$

6.4.4.3. Comparing (6.4.46) and (2.7.12), one can find [K21, S11]

$$t_{mm'}^j(\tau) = (-1)^n \frac{\sqrt{\varrho(x)}}{d_n} m_n^{(\gamma,\mu)}(x) . \tag{6.4.48}$$

Here $\varrho(x)$ and d_n are the weight and the norm of the Meixner polynomials, respectively; $n = m - j - 1$, $x = m' - j - 1$, $\gamma = 2j + 2$, and $\mu = \tanh^2(\tau/2)$.

With the aid of (6.7.47) the function $t_{mm}^j(\tau)$ can be transformed to the form with $m' - m \geq 0$. In this case in view of (2.7.15) we finally obtain

$$t_{mm'}^j(\tau) = 2^m \sqrt{\frac{(m-j-1)!(m'+j)!}{(m+j)!(m'-j-1)!}} (s-1)^{(m'-m)/2}$$

$$\times (s+1)^{-(m'+m)/2} P_{m-j-1}^{(m'-m,-m'-m)}(s) , \quad s = \cosh\tau . \tag{6.4.49}$$

According to (6.4.48) and (6.4.49) the main properties of the functions $t_{mm'}^j(\tau)$ and $T_{mm'}^j(\alpha,\tau,\gamma)$ follow from the known properties for the Meixner and Jacobi polynomials (cf. Sect. 5.1.3). We note only the relations

$$\sum_{m''=j+1}^\infty t_{mm''}^j(\tau) t_{m'm''}^j(\tau) = \delta_{mm'} \tag{6.4.50}$$

and

$$\int_0^\infty t^j_{m\,m'}(\tau)t^{j'}_{m\,m'}(\tau)\sinh\tau\,d\tau = \frac{2}{2j+1}\delta_{jj'}\;. \tag{6.4.51}$$

The Eq. (6.4.50) follows from (6.4.48) and the orthogonality property of the Meixner polynomials. In view of (6.4.49) Eq. (6.4.51) corresponds to the orthogonality on the infinite interval $(1,+\infty)$ for the finite set of the Jacobi polynomials $\{P_n^{(\alpha,\beta)}(s)\}$ with $\alpha > 0$, $\beta < 0$ and $\alpha+\beta+2n < 0$. One can easily establish this property with the aid of (1.3.1) and evaluate the squared norm by using (1.3.8) and the relation

$$\int_0^\infty \frac{t^{x-1}}{(1+t)^{x+y}}dt = \frac{\Gamma(x)\Gamma(y)}{\Gamma(x+y)}\;.$$

Likewise for the functions $t^j_{m\,m'}(\tau)$ and $T^j_{m\,m'}(\alpha,\tau,\gamma)$ we can derive the differential and difference equations, differentiation formulae, asymptotic representations and so on. According to (5.1.26) and (6.4.49) the functions $t^j_{m\,m'}(\tau)$ are the analitic continuation of d-functions:

$$t^j_{m\,m'}(\tau) = d^{-j-1}_{-m,-m'}(\mathrm{i}\tau)\;.$$

6.4.4.4. In conclusion we note that the study of the wave functions of the n-dimensional harmonic oscillator evaluates all main quantities of the Wigner-Racah algebra for the discrete positive series of SU(1,1). Indeed the generalized spherical functions of SU(1,1) are expressed by (6.4.48) and (6.4.49) in terms of the Jacobi and Meixner polynomials; the Clebsch-Gordan coefficients (6.4.39) for coupling of two discrete positive series are represented by (6.4.16) and (6.4.17) through the Hahn polynomials; the $6j$-symbols proportional to T-coefficients are related with the Racah polynomials subject to (6.3.7) and (6.3.8). The simplification resulting for open nodes were discussed above. Thus there are close connections between the polynomials under consideration and the representations of SU(1,1).

6.4.5 Harmonic Oscillator and Matrix Elements of the Heisenberg-Weyl Group N(3)

Some applications require the integrals of the form

$$I = \int_{-\infty}^\infty \Psi_N(x)\mathrm{e}^{\mathrm{i}\lambda x}\Psi_{N'}(x+\alpha)dx\;,$$

where $\Psi_N(x)$ are the wave functions of the one-dimensional oscillator, and λ and α are real constants. This integral is closely connected with the matrix elements of the Heisenberg-Weyl group and with the Charlier polynomials.

6.4.5.1. Let N(3) be the three-dimensional group of real matrices of the form

$$\begin{pmatrix} 1 & \alpha & \gamma \\ 0 & 1 & \beta \\ 0 & 0 & 1 \end{pmatrix} = (\alpha, \beta, \gamma) .$$

The map

$$T(\alpha, \beta, \gamma)\Psi(x) = e^{i\lambda(\gamma + \beta x)}\Psi(x + \alpha) \tag{6.4.52}$$

(λ is a real constant) defines a unitary representation of N(3) in the space of all integrable squared functions $\Psi \in L^2(-\infty, \infty)$.

By definition the generators of one-parameter subgroups of N(3) corresponding to the parameters α, β and γ have the form

$$P = \frac{1}{i}\left.\frac{\partial T(\alpha, 0, 0)}{\partial\alpha}\right|_{\alpha=0} = \frac{1}{i}\frac{\partial}{\partial x} ,$$

$$Q = \frac{1}{i}\left.\frac{\partial T(0, \beta, 0)}{\partial\lambda\beta}\right|_{\beta=0} = x ,$$

$$I = \frac{1}{i}\left.\frac{\partial T(0, 0, \gamma)}{\partial\lambda\gamma}\right|_{\gamma=0} = 1 .$$

These operators coincide with the momentum and position operators in nonrelativistic quantum mechanics and obey the Heisenberg commutation relations

$$[P, Q] = -iI . \tag{6.4.53}$$

By using the formula

$$e^A e^B = e^{A + B + [A, B]/2} \tag{6.4.54}$$

valid under the assumption $[A, [A, B]] = [B, [A, B]] = 0$ (for a simple derivation of (6.4.54), see [D1]) one can also rewrite (6.4.53) in the Weyl form [B7]

$$e^{i\alpha P} e^{i\beta Q} = e^{i\alpha\beta} e^{i\beta Q} e^{i\alpha P} .$$

The creation and annihilation operators defined by (6.4.32) are related with P and Q by

$$a^+ = \frac{P + iQ}{\sqrt{2}} , \qquad a = \frac{P - iQ}{\sqrt{2}} . \tag{6.4.55}$$

The Hamiltonian of the one-dimensional harmonic oscillator has the form

$$H = \tfrac{1}{2}(P^2 + Q^2) = \tfrac{1}{2}(aa^+ + a^+a) . \tag{6.4.56}$$

6.4.5.2. The set $\{\Psi_N\}$, $N = 0, 1, 2, \ldots$, of the eigenfunctions (6.4.4) of the Hamiltonian (6.4.56) forms a complete orthogonal set in $L^2(-\infty, \infty)$. Let us evaluate the matrix elements of the representation (6.4.52) in this basis:

$$T(\alpha,\beta,\gamma)\Psi_N(x) = \sum_{M=0}^{\infty} T_{MN}(\alpha,\beta,\gamma)\Psi_M(x) ,$$

$$T_{MN}(\alpha,\beta,\gamma) = (\Psi_M|T(\alpha,\beta,\gamma)\Psi_N)$$

$$= \int_{-\infty}^{\infty} \Psi_M^*(x)e^{i\lambda(\gamma+\beta x)}(x+\alpha)dx . \tag{6.4.57}$$

By virtue of (6.4.52, 54) and (6.4.55) we can write operator $T(\alpha,\beta,\gamma)$ as

$$T(\alpha,\beta,\gamma) = e^{i\lambda\gamma I}e^{i\lambda\beta Q}e^{i\alpha P} = e^{i\lambda(\gamma-\alpha\beta/2)}e^{-(\alpha^2+\lambda^2\beta^2/4)}e^{ua^+}e^{va} ,$$

where

$$u = \frac{i\alpha+\lambda\beta}{\sqrt{2}} , \qquad v = \frac{i\alpha-\lambda\beta}{\sqrt{2}} .$$

Therefore

$$T_{MN}(\alpha,\beta,\gamma) = \exp\left[i\lambda(\gamma-\alpha\beta/2) - (\alpha^2+\lambda^2\beta^2)/4\right]\left(e^{u^*a}\Psi_M|e^{va}\Psi_N\right) .$$

Since

$$e^{va}\Psi_N = \sum_{k=0}^{\infty} \frac{(va)^k}{k!}\Psi_N = \sum_{k=0}^{N} \frac{v^k}{k!}(a^k\Psi_N) ,$$

$$a^k\Psi_N = (-i)^k \sqrt{\frac{N!}{(N-k)!}}\,\Psi_{N-k} ,$$

we have

$$(e^{u^*a}\Psi_M|e^{va}\Psi_N) = \begin{cases} \sqrt{\dfrac{M!}{N!}}\,\dfrac{(iu)^{M-N}}{(M-N)!}F(-N,M-N+1,-uv), \\[4pt] M-N \geq 0 , \\[10pt] \sqrt{\dfrac{N!}{M!}}\,\dfrac{(-iv)^{N-M}}{(N-M)!}F(-M,N-M+1,-uv), \\[4pt] M-N < 0 , \end{cases}$$

where $F(a,b,\mu)$ is the confluent hypergeometric function (2.7.2). These relations and (2.7.9) lead us to the representations of matrix elements (6.4.57) in terms of the Charlier polynomials

$$T_{MN}(\alpha,\beta,\gamma) = \frac{i^{M-N}}{\sqrt{M!N!}}e^{i\lambda(\gamma-\alpha\beta/2)}e^{-\mu/2}$$

$$\times \left(\frac{i\alpha+\lambda\beta}{\sqrt{2}}\right)^M \left(\frac{i\alpha-\lambda\beta}{\sqrt{2}}\right)^N c_M^{(\mu)}(N), \quad \mu = \tfrac{1}{2}(\alpha^2+\lambda^2\beta^2) . \tag{6.4.58}$$

By using (2.7.10) we find the relationship between the matrix elements of N(3) and Laguerre polynomials, viz.

$$(-i)^{M-N}e^{-i\lambda(\gamma-\alpha\beta/2)}e^{\mu/2}T_{MN}(\alpha,\beta,\gamma)$$
$$= \begin{cases} \sqrt{\dfrac{N!}{M!}}\left(\dfrac{i\alpha+\lambda\beta}{\sqrt{2}}\right)^{M-N}L_N^{M-N}(\mu)\,, & M-N\geq 0\,, \\[2ex] \sqrt{\dfrac{M!}{N!}}\left(\dfrac{i\alpha-\lambda\beta}{\sqrt{2}}\right)^{N-M}L_M^{N-M}(\mu)\,, & N-M\geq 0\,, \end{cases} \qquad (6.4.59)$$

where $\mu = \frac{1}{2}(\alpha^2 + \lambda^2\beta^2)$.

6.4.5.3. For the transformations $T(\alpha,0,0)$ the matrix elements have the form

$$\tilde{t}_{MN}(\alpha) = T_{MN}(\alpha,\beta,\gamma)\big|_{\beta=\gamma=0}$$
$$= \frac{(-1)^M}{\sqrt{M!N!}}\mu^{(M+N)/2}e^{-\mu/2}c_M^{(\mu)}(N)\,, \qquad \mu = \alpha^2/2\,. \qquad (6.4.60)$$

Similarly

$$T_{MN}(0,\beta,0) = \frac{i^{M+N}}{\sqrt{M!N!}}\mu^{(M+N)/2}e^{-\mu/2}c_M^{(\mu)}(N)\,, \qquad \mu = \lambda^2\beta^2/2\,. \qquad (6.4.61)$$

The matrix elements (6.4.60) and (6.4.61) are similar, thus reflecting the well-known fact that the Hamiltonian of an oscillator has the same form in the coordinate and momentum representations.

Addendum to Chapter 6

1. We wish to prove the equality (6.3.5) that was used in deriving the general expression for the T-coefficients. Denote the integral on the left-hand side of this equality by I. Since

$$P_k^{(\alpha,\beta)}(\xi) = \frac{(-1)^k(\beta+1)_k}{k!} F\left(-k, \alpha+\beta+k+1, \beta+1, \frac{1+\xi}{2}\right) \,,$$

then

$$I = 2^{\alpha+\beta+s-a}\frac{(-1)^k(\beta+1)_k}{k!} \sum_{p=0}^{k} \frac{(-k)_p(\alpha+\beta+k+1)_p}{(\beta+1)_p p!} \, I_p \,,$$

where

$$I_p = \int_{-1}^{1} P_{b-s-1}^{(\alpha,2s+1)}(\xi) \left(\frac{1-\xi}{2}\right)^{\alpha} \left(\frac{1+\xi}{2}\right)^{s-a+\beta+p} \, d\xi \,.$$

Developing the Jacobi polynomials $P_{b-s-1}^{(\alpha,2s+1)}(\xi)$ in a power series in $(1-\xi)/2$ and computing the integral with the known property of the beta function yields

$$I_p = 2\frac{\Gamma(b+\alpha-s)\Gamma(s-a+\beta+p+1)\Gamma(-a-s+\beta+p)}{(b-s-1)!\,\Gamma(b-a+\alpha+\beta+p+1)\Gamma(-a-b+\beta+p+1)} \,.$$

Therefore

$$I = 2^{\alpha+\beta+s-a+1}\frac{(-1)^{k+b-s-1}(\beta+1)_k\Gamma(b+\alpha-s)\Gamma(s-a+\beta+1)\Gamma(a+b-\beta)}{k!(b-s-1)!\,\Gamma(b-a+\alpha+\beta+1)\Gamma(a-\beta+s+1)}$$

$$\times \,_4F_3\left(\begin{matrix} -k, \alpha+\beta+k+1, s-a+\beta+1, -a-s+\beta \\ \beta+1, b-a+\alpha+\beta+1, -a-b+\beta+1 \end{matrix} \Big| 1\right) \,.$$

Using Eq. (1) in page 56 of Bailey's book of 1935 [B1] leads us to the relation (6.3.5). A similar integral is contained in the book of Baitmen and Erdelyi [E8] and computed in [S7].

2. Let us evaluate the integral (6.4.45)

$$I = \int_{0}^{\infty} e^{-(1+\beta)t/2}t^{\alpha} L_p^{\alpha}(t)L_q^{\alpha}(\beta t)dt \,.$$

Representing the polynomials $L_p^{\alpha}(t)$ and $L_q^{\alpha}(\beta t)$ in the form (2.7.4), we have

$$I = \frac{(\alpha+1)_p(\alpha+1)_q}{p!\,q!} \sum_{n=0}^{p} \frac{(-p)_n}{(\alpha+1)_n n!} \sum_{k=0}^{q} \frac{(-q)_k \beta^k}{(\alpha+1)_k k!}$$
$$\times \int_0^\infty e^{-(1+\beta)t/2} t^{\alpha+n+k}\,dt \ ,$$

and evaluating the integral with the aid of the gamma function

$$\int_0^\infty e^{-(1+\beta)t/2} t^{\alpha+n+k}\,dt = \left(\frac{2}{1+\beta}\right)^{\alpha+n+k+1} \Gamma(\alpha+n+k+1)$$

we find

$$I = \frac{(\alpha+1)_p(\alpha+1)_q}{p!\,q!} \Gamma(\alpha+1) \sum_{n=0}^{p} \frac{(-p)_n}{n!} \left(\frac{2}{1+\beta}\right)^{\alpha+n+1}$$
$$\times F\left(-q, \alpha+n+1, \alpha+1, \frac{2\beta}{1+\beta}\right) \ .$$

If we use the transformation [N18]

$$F(a,b,c,t) = (1-t)^{c-a-b} F(c-a, c-b, c, t) \ ,$$

then

$$I = \frac{(\alpha+1)_p(\alpha+1)_q}{p!\,q!} \Gamma(\alpha+1) \left(\frac{2}{1+\beta}\right)^{\alpha+1} \left(\frac{1-\beta}{1+\beta}\right)^{q}$$
$$\times \sum_k S_k \frac{(\alpha+q+1)_k}{(\alpha+1)_k} \left(\frac{2\beta}{1+\beta}\right)^{k} \ ,$$

where

$$S_k = \sum_n \frac{(-p)_n (-n)_k}{n!} \left(\frac{2}{1-\beta}\right)^{n}$$
$$= (-1)^k p! \sum_{n=k}^{p} \left(\frac{2}{\beta-1}\right)^{n} / (n-k)!(p-n)!$$
$$= (-1)^{p-k} \frac{p!}{(p-k)!} \left(\frac{2}{1-\beta}\right)^{k} \left(\frac{1+\beta}{1-\beta}\right)^{p-k} \ .$$

Therefore

$$I = \frac{(\alpha+1)_p(\alpha+1)_q}{p!\,q!} \Gamma(\alpha+1) \left(\frac{2}{1+\beta}\right)^{\alpha+1} \left(\frac{1+\beta}{1-\beta}\right)^{p-q}$$
$$\times (-1)^p F\left(-p, \alpha+q+1, \alpha+1, \frac{4\beta}{(1+\beta)^2}\right) \ ,$$

whence by virtue of

$$F(-p, b, c, t) = (1-t)^p F\left(-p, c-b, c, \frac{t}{t-1}\right)$$

we finally come to (6.4.45).

Appendix. Classification of Polynomial Solutions of Difference Equations of Hypergeometrical Type[1]

1. Let us consider the problem of classifying systems of polynomial solutions $\tilde{y}_n[x(s)] \equiv y_n(s)$ of a difference equation of hypergeometrical type. In Sect. 3.11.2 it was shown that in the most general case of nonuniform lattice

$$x(s) = \tilde{c}_1 q^s + \tilde{c}_2 q^{-s} + \tilde{c}_3 = \tilde{c}_1(q^s + q^{-s-\mu}) + \tilde{c}_3 (q^\mu = \tilde{c}_1/\tilde{c}_2) \quad ,$$

where $q \neq 1$, μ is a finite number and the function

$$\sigma(s) = A \prod_{i=1}^{4} \psi_q(s - s_i)$$

has four zeros, formula (3.11.25),

$$y_n(s) = \left[\frac{A}{\tilde{c}_1 q^{-\mu/2} \kappa^2} \right]^n B_n (s_1 + s_2 + \mu | q)_n (s_1 + s_3 + \mu | q)_n (s_1 + s_4 + \mu | q)_n$$

$$\times \; {}_4F_3 \left[\begin{array}{c} -n, \; \sum_{i=1}^{4} s_i + 2\mu + n - 1, \; s_1 - s, \; s_1 + s + \mu \\ s_1 + s_2 + \mu, \; s_1 + s_3 + \mu, \; s_1 + s_4 + \mu \end{array} \middle| q, 1 \right] \quad , \quad \text{(A.1)}$$

enables us to derive explicit expressions for polynomial solutions of a difference equation of hypergeometrical type at given values of q, μ and $s_i (i = 1, 2, 3, 4)$.

1.1. In the case where the function $\sigma(s)$ at a given μ has less than four zeros (or does not have zeros at all), the formulas can be derived from general formulas for $\sigma(s)$ and $y_n(s)$ using limit transitions at $q^{s_i} \to 0$ for some chosen values i (or $q^{-s_i} \to 0$) (see Sect. 3.11.5.1).

Given $\sigma(s)$, the functions

$$\frac{\sigma(-s - \mu) - \sigma(s)}{\Delta x(s - \frac{1}{2})} \qquad \text{and} \qquad \tfrac{1}{2}[\sigma(-s - \mu) + \sigma(s)]$$

are polynomials of at most the first and second degree, respectively, in $x(s)$, which coincide with polynomials $\tilde{\tau}[x(s)]$ and $\tilde{\sigma}[x(s)]$ in the initial difference

---

[1] This section was written by A.F. Nikiforov and V.B. Uvarov at the final stage of preparation of this book, in fact when the printing of the main text was over. If it were not for the help of Yu.A. Danilov, this work could not have been finished in such a short time. The authors wholeheartedly express their gratitude to Yu.A. Danilov for his cooperation and to Springer-Verlag for understanding the importance of having in our book a visual scheme of the interrelations between different systems of classical orthogonal polynomials.

344

equation of hypergeometrical type (3.1.5). Considering at different values of μ the cases where the function $\sigma(s)$ has a different number of zeros, we at the same time consider all possible forms of difference equations of hypergeometrical type and of the polynomial solutions corresponding to them.

According to the Rodrigues formula (3.2.19), the polynomial $\tilde{\tau}[x(s)] \equiv \tau(s)$ up to a constant factor coincides with the polynomial $y_1(s)$: $\tilde{y}_1[x(s)] = B_1\tau(s)$. This is why from now on we shall consider only such limit transitions, which leave the function $\tilde{\tau}[x(s)]$ as a polynomial of the first degree. In particular, when $\sigma(s)$ has less than four finite zeros, one need consider only such limit transitions under which $q^{s_i} \to 0$ at all chosen values of i. In the opposite case we get $\tilde{\tau}[x(s)] = \text{const}$ (see Sect. 3.11.2). As will be shown further, the cases of $\sigma(s)$ with different numbers of zeros at $q = 1$, when $x(s) = c_1 s^2 + c_2 s + c_3$ $(\mu = C_2/C_1)$, can be easily investigated using analogous considerations.

1.2. It is also necessary to consider the limit transitions corresponding to infinite values of μ. Because $x(s) = \tilde{c}_2(q^{s+\mu} + q^{-s}) + \tilde{c}_3$ $(q \neq 1)$, the case where $q^\mu \to 0$ $(\tilde{c}_2 = \text{const}, \tilde{c}_3 = \text{const})$ brings us, as can be easily shown, to the lattice function $x(s) = \tilde{c}_2 q^{-s} + \tilde{c}_3$ and the case $q^{-\mu} \to 0$ $(\tilde{c}_1 = \text{const}, \tilde{c}_3 = \text{const})$ brings us to the lattice function $x(s) = \tilde{c}_1 q^s + \tilde{c}_3$. At $q = 1$ the function $x(s)$ has the form $x(s) = C_2 s(1 + s/\mu) + C_3$. This is why at $\mu \to \pm\infty$ we arrive at the linear lattice function $x(s) = C_2 s + C_3$.

1.3. In the most general case, when $q \neq 1$, μ is a finite number and the function $\sigma(s)$ has four zeros, the difference equation of hypergeometric type allows polynomial solutions $y_n(s) \equiv \tilde{y}_n[x(s)]$ $(n = 0, 1, \ldots)$ only for the values

$$\lambda = \lambda_n = -\frac{A}{B^2}\psi_q(n)\psi_q\left(\sum_{i=1}^{4} s_i + 2\mu + n - 1\right) \quad . \tag{A.2}$$

Here A is a constant that enters in the formula for $\sigma(s)$, and $B = \tilde{c}_1 q^{-\mu/2}\kappa^2$ enters in the formula for $\Delta x(s)$: $\Delta x(s) = B\psi_q(2s + \mu + 1)$. Equation (A.2) for λ_n can be easily derived if we rewrite (3.1.5), using the formulas

$$\tilde{\sigma}[x(s)] = \tfrac{1}{2}[\sigma(-s - \mu) + \sigma(s)] \quad ,$$

$$\tilde{\tau}[x(s)] = \frac{\sigma(-s - \mu) - \sigma(s)}{\Delta x(s - \tfrac{1}{2})} \quad ,$$

as

$$\sigma(-s - \mu)\frac{\Delta y_n(s)}{\Delta x(s)} - \sigma(s)\frac{\nabla y_n(s)}{\nabla x(s)} + \lambda_n \Delta x(s - 1/2)\, y_n(s) = 0$$

and equate the coefficients of leading powers of q^s.

It is necessary to emphasize that explicit expressions for polynomial solutions of a difference equation of hypergeometrical type can be derived by the method mentioned above not only in the case where these polynomials do possess the property of orthogonality on a discrete point set or the property of continuous orthogonality on some contour C in the plane of complex variable s (the poly-

345

nomials possess these properties only with some additional restrictions on the coefficients of the difference equation). Carrying out the limit transitions mentioned above, we derive in explicit form all polynomial solutions of difference equations of hypergeometric type, including those which possess the property of orthogonality.

2. Let us construct the scheme of limit transitions for different families of polynomial solutions of difference equations of hypergeometric type. The most important and at the same time the simplest particular case of (A.1), when q, μ, s_i $(i = 1, 2, 3, 4)$ are finite numbers, is the classical case of a quadratic lattice corresponding to $q = 1$. At $q \to 1$ one can easily derive from (A.1) and (A.2) (see Sect. 3.11.3)

$$
y_n(s) = \left[\frac{A}{C_1}\right] B_n(s_1 + s_2 + \mu)_n (s_1 + s_3 + \mu)_n (s_1 + s_4 + \mu)_n
$$

$$
\times {}_4F_3 \left[\begin{matrix} -n, \; \sum_{i=1}^4 s_i + 2\mu + n - 1, \; s_1 - s, \; s_1 + s + \mu \\ s_1 + s_2 + \mu, \; s_1 + s_3 + \mu, \; s_1 + s_4 + \mu \end{matrix} \middle| 1\right] , \qquad \text{(A.3)}
$$

$$
\lambda_n = -\frac{A}{C_1^2} n \left(\sum_{i=1}^4 s_i + 2\mu + n - 1\right) , \qquad \text{(A.4)}
$$

where

$$
\sigma(s) = A \prod_{i=1}^4 (s - s_i), \quad x(s) = C_1 s^2 + C_2 s + C_3
$$

$$
(\mu = C_2/C_1 , \; C_1 = \tilde{c}_1 q^{-\mu/2} \kappa^2 = B) .
$$

A particular case of (A.3) is the formula for the Racah polynomials $u_n^{(\alpha,\beta)}[x(s)]$ (see Sect. 3.5) at $\mu = 1$, $C_1 = 1$, $C_3 = 0$ (i.e. when $x(s) = s(s + 1)$), $s_1 = a$, $s_2 = -b (b - a = N)$, $s_3 = \beta - a$, $s_4 = b + \alpha$. On the scheme given below, the polynomials, defined by equations (A.1,3), and also the Racah polynomials are situated on the upper horizontal line, which corresponds to four zeros of the function $\sigma(s)$.

2.1. In order to derive from (A.3) limit cases, corresponding to the case where the number of zeros of $\sigma(s)$ is less than four, one needs first of all to keep in mind that the case where $\sigma(s)$ is a polynomial of at most the second degree in s is not of interest, because at finite values of μ one finds

$$
\tau(s) = \frac{\sigma(-s - \mu) - \sigma(s)}{\Delta x(s - 1/2)} = \frac{\sigma(-s - \mu) - \sigma(s)}{c_1(2s + \mu)} = \text{const} \quad .
$$

Only the case where $\sigma(s)$ has three zeros, $\sigma(s) = \prod_{i=1}^3 (s - s_i)$, is of interest. This case follows from the general one (A.3), when $\sigma(s) = A \prod_{i=1}^4 (s - s_i)$, using the limit transition $s_4 \to \infty$, where $A = -1/s_4$. In the limit one derives from (A.3) and (A.4)

$$y_n(s) = (-1)^n \frac{B_n}{C_1^n} (s_1 + s_2 + \mu)_n (s_1 + s_3 + \mu)_n$$

$$\times \; {}_3F_2 \left[\begin{array}{c} -n, s_1 - s, s_1 + s + \mu \\ s_1 + s_2 + \mu, \; s_1 + s_3 + \mu \end{array} \Bigg| 1 \right] \; , \tag{A.5}$$

$$\lambda_n = \frac{n}{C_1^2} \; . \tag{A.6}$$

The particular case of (A.5) is the following formula for the dual Hahn polynomials $w_n^{(c)}[x(s)]$ when $\mu = 1$, $C_1 = 1$ (i.e. for $x(s) = s(s+1)$), $s_1 = a$, $s_2 = -b$, $s_3 = c$, $B_n = (-1)^n/n!$:

$$w_n^{(c)}[x(s)] = \frac{(a+1-b)_n (a+1+c)_n}{n!} \; {}_3F_2 \left[\begin{array}{c} -n, \; a - s, \; a + 1 + s \\ a + 1 - b, \; a + 1 + c \end{array} \Bigg| 1 \right] \; .$$

In the scheme enclosed, the considered polynomial families are situated on the second, third, fourth and fifth horizontal lines (counting from above) to the right from the vertical line drawn downwards from the rectangle, which corresponds to polynomials of general form at $q = 1$, where μ and $s_i \, (i = 1, 2, 3, 4)$ are given numbers.

2.2. When $q = 1$, it remains to consider the limit transitions at $\mu \to \infty$. If $C_1 = 1/\mu$, $C_2 = 1$, $C_3 = 0$, then the function $x(s)$ coincides in the limit with the function $x(s) = s$. When $x(s) = s$, the function $\sigma(s)$ is a polynomial of at most the second degree, and the function $\tau(s)$ has to be a polynomial of the first degree which does not have any reference to the polynomial $\sigma(s)$. This is why in this case, instead of giving four zeros of the function $\sigma(s)$ (in the most general case at finite values of μ) as parameters allowing one to derive expressions for $y_n(s)$ at $\mu = \infty$, it is convenient to choose two zeros s_1 and s_2 of the function $\sigma(s)$ and two zeros $\bar{s}_1$ and $\bar{s}_2$ of the polynomial $\sigma(s) + \tau(s)$ (these very functions enter (3.5.1) for the weight $\rho(s)$). Let us remark that polynomials $\sigma(s)$ and $\sigma(s) + \tau(s)$ are polynomials of the second degree with equal leading coefficient s:

$$\sigma(s) = \bar{A}(s - s_1)(s - s_2) \, ,$$
$$\sigma(s) + \tau(s) = \bar{A}(s - \bar{s}_1)(s - \bar{s}_2) \, ,$$
$$(\bar{A} = \text{const}) \quad . \tag{A.7}$$

Formula (A.7) can be derived at $\mu \to \infty$ from the general case of a quadratic lattice $x(s) = C_1 s^2 + C_2 s + C_3$, if one chooses constant $A = A(\mu)$ and $s_i = s_i(\mu)$ in

$$\sigma(s) = A \prod_{i=1}^{4} (s - s_i) \, ,$$

$$\sigma(s) + \tau(s)\Delta x(s - 1/2) = \sigma(-s - \mu) = A \prod_{i=1}^{4} (s + s_i + \mu) \quad ,$$

in such a way that the latter would go into (A.7). The imposed conditions are satisfied if one puts

$$s_1(\mu) = s_1 \,, \ s_2(\mu) = s_2 \,, \ s_3(\mu) = -\bar{s}_1 - \mu \,, \ s_4(\mu) = -\bar{s}_2 - \mu \,, \ A = \bar{A}/\mu^2 \quad .$$

As a result we derive from (A.3) and (A.4) the following expressions for $y_n(s)$ and λ_n:

$$y_n(s) = \bar{A}^n B_n (s_1 - \bar{s}_1)_n (s_1 - \bar{s}_2)_n$$
$$\times \, {}_3F_2 \left[\begin{matrix} -n, \ s_1 + s_2 - \bar{s}_1 - \bar{s}_2 + n + 1, \ s_1 - s \\ s_1 - \bar{s}_1, \ s_1 - \bar{s}_2 \end{matrix} \, \bigg| \, 1 \right] \quad , \tag{A.8}$$

$$\lambda_n = -\bar{A} n (s_1 + s_2 - \bar{s}_1 - \bar{s}_2 + n - 1) \quad . \tag{A.9}$$

2.2.1. In particular, for the Hahn polynomials $h_n^{(\alpha,\beta)}(s)$ (see Sect. 2.4.2) one gets

$$\sigma(s) = s(N + \alpha - s) \,, \ \sigma(s) + \tau(s) = (s + \beta + 1)(N - 1 - s) \,, \ B_n = (-1)^n/n! \,.$$

This corresponds to $s_1 = 0$, $s_2 = N + \alpha$, $\bar{s}_1 = -\beta - 1$, $\bar{s}_2 = N - 1$, $\bar{A} = -1$, from which

$$h_n^{(\alpha,\beta)}(s) = \frac{1}{n!}(\beta + 1)_n(-N + 1)_n \, {}_3F_2 \left[\begin{matrix} -n, \ \alpha + \beta + n + 1, \ -s \\ \beta + 1, \ -N + 1 \end{matrix} \, \bigg| \, 1 \right] \quad ,$$

$$\lambda_n = n(n + \alpha + \beta + 1) \quad .$$

2.2.2. The case where $\sigma(s)$ and $\sigma(s) + \tau(s)$ are polynomials of at most the first degree can be derived from (A.8) using limit transitions at $s_2 \to \infty$ and correspondingly chosen $s_1 = s_1(s_2)$, $\bar{s}_1 = \bar{s}_1(s_2)$, $\bar{s}_2 = \bar{s}_2(s_2)$, $\bar{A} = \bar{A}(s_2)$. For the case of the polynomials of Meixner $m_n^{(\gamma,\mu)}(s)$, Kravchuk $k_n^{(p)}(s)$ and Charlier $c_n^{(\mu)}(s)$, the expressions for $\sigma(s)$ and $\sigma(s) + \tau(s)$ can be derived at $s_2 \to \infty$, $s_1 = 0$, $\bar{A} = -1/s_2$ and with the following choice of $\bar{s}_1$, $\bar{s}_2$, B_n:

(1) $\bar{s}_1 = -\gamma$, $\bar{s}_2 = \mu s_2$, $B_n = 1/\mu^n$ for $m_n^{(\gamma,\mu)}(s)$;

(2) $\bar{s}_1 = N$, $\bar{s}_2 = -\dfrac{p}{1-p} s_2$, $B_n = \dfrac{(-1)^n (1-p)^n}{n!}$ for $k_n^{(p)}(s)$;

(3) $\bar{s}_1 = \sqrt{s_2}$, $\bar{s}_2 = -\mu\sqrt{s_2}$, $B_n = 1/\mu^n$ for $c_n^{(\mu)}(s)$.

As a result we arrive in these cases at the following expressions for $y_n(s)$ and λ_n:

$$m_n^{(\gamma,\mu)}(s) = (\gamma)_n \, {}_2F_1 \left[\begin{matrix} -n, -s \\ \gamma \end{matrix} \, \bigg| \, 1 - \frac{1}{\mu} \right], \quad \lambda_n = n(1 - \mu) \quad ;$$

$$k_n^{(p)}(s) = \frac{p^n}{n!}(-N)_n \, {}_2F_1 \left[\begin{matrix} -n, -s \\ -N \end{matrix} \, \bigg| \, \frac{1}{p} \right], \quad \lambda_n = \frac{n}{1 - p} \quad ;$$

$$c_n^{(\mu)}(s) = {}_2F_0 \left(-n, -s \, \bigg| \, -\frac{1}{\mu} \right), \quad \lambda_n = n \quad .$$

The case when $\sigma(s) = \text{const}$ is not of interest because it does not lead to any new polynomials.

In the scheme enclosed all the cases at $x(s) = s$ considered above are situated to the left of the vertical line corresponding to $q = 1$. First in the third line from above a general case is given which corresponds to $q = 1$, $\mu = \infty$ and given values of $s_1, s_2, \bar{s}_1, \bar{s}_2$. To the right of it a particular case of Hahn polynomials is situated. In the fourth line one finds the cases which are derived from the general one at $s_2 \to \infty$ and lead to the Meixner, Kravchuk and Charlier polynomials. In the fifth line the case of $\sigma(s) = \text{const}$ is given which is not of interest.

3. Let us consider now different cases at $q \neq 1$. Particular cases of formula (A.1) are for example, Askey-Wilson polynomials $p_n(x, a, b, c, d|q)$, corresponding to $x(s) = (q^s + q^{-s})/2$ at $q^s = e^{i\theta}$. For these polynomials we have $\mu = 0$, $a = q^{s_1}$, $b = q^{s_2}$, $c = q^{s_3}$, $d = q^{s_4}$. In the scheme enclosed, Askey-Wilson polynomials are situated in the first horizontal line to the right of the general case (A.1). The Askey-Wilson polynomials depend on five parameters, a, b, c, d, q, and polynomials (A.49) depend on six parameters: $y_n(s) = f(x; s_1, s_2, s_3, s_4, q, \mu)$. Nevertheless, we can use the fact that the form of difference equation of hypergeometric type is preserved under the linear transformation $s = s' - \mu/2$ (see Sect. 3.4). This transformation carries the nonuniform lattice

$$x(s) = \tilde{c}_1 q^s + \tilde{c}_2 q^{-s} = \tilde{c}_1(q^s + q^{-s-\mu}) \quad (q^\mu = \tilde{c}_1/\tilde{c}_2)$$

to the form

$$x(s') = \tilde{c}_1'(q^{s'} + q^{-s'}), \quad \text{where} \quad \tilde{c}_1' = \tilde{c}_1 q^{-\mu/2}.$$

Consequently, the general case of nonuniform lattice may be reduced to the case $\mu = 0$, and any polynomials of the form (A.1) after the change $s \to s - \mu/2$ (and $s_i \to s_i - 1/2$) may be represented in a simpler form with $\mu = 0$ (except the case $\mu = \infty$):

$$y_n(s) = \left(\frac{A}{\tilde{c}_1 \kappa^2}\right)^n B_n (s_1 + s_2|q)_n \, (s_1 + s_3|q)_n \, (s_1 + s_4|q)_n$$

$$\times \, {}_4F_3\left(\begin{array}{c} -n, \, \sum_{i=1}^4 s_1 + n - 1, \, s_1 - s, \, s_1 + s \\ s_1 + s_2, \, s_1 + s_3, \, s_1 + s_4 \end{array} \middle| q, 1\right).$$

In fact, these polynomials are Askey-Wilson polynomials. However, we decided to preserve the parameter μ in our exposition because for the construction of q-polynomials (in particular, q-analogs) it is convenient to work with the same value of μ (see Sect. 3.11.3, Tables 3.3–5).

For example, in order to construct q-Racah polynomials we use the characteristics of Racah polynomials $u_n^{(\alpha,\beta)}(x)$ with $x = x(s) = s(s+1)$ (i.e. $q = 1$, $C_1 = 1$, $\mu = 1$, $C_3 = 0$) and

$$s_1 = a, \, s_2 = -b \quad (b - a = N), \, s_3 = \beta - a, \, s_4 = b + \alpha, \, A = -1$$

(See Table 3.6). We have $\tilde{c}_1 = c_1 q^{\mu/2}/\kappa^2$, $\tilde{c}_2 = c_1 q^{-\mu/2}/\kappa^2$. From (A.1) and (3.11.28) we obtain the following expression for the q-Racah polynomials:

$$y_n(s) \equiv \tilde{y}_n[x(s)] = (-1)^n B_n (a - b + 1|q)_n (\beta + 1|q)_n (a + b + \alpha + 1|q)_n$$

$$\times \, {}_4F_3 \left(\begin{matrix} -n, \, \alpha + \beta + n + 1, \, a - s, \, a + s + 1 \\ a - b + 1, \, \beta + 1, \, a + b + \alpha + 1 \end{matrix} \, \middle| \, q, 1 \right) \; x(s) = \psi_q(s)\,\psi_q(s + 1)$$

3.1. The lines drawn downwards and to the left of the fundamental rectangle, which corresponds to $q \neq 1$ and given values of $\mu, s_i \, (i = 1, 2, 3, 4)$, bring us to the rectangles which correspond to the cases where the function $\sigma(s)$ has less than four zeros. These cases can be derived from the general one using the limit transition when for some values of i one considers the limit transition at $q^{s_i} \to 0$. In addition a certain dependence of the factor A on the corresponding values of q^{s_i} in the formula for $\sigma(s)$ is chosen. Let us consider the corresponding limit transitions.

1. At $q^{s_4} \to 0$, $A = -1/\psi_q(s_4)$ one gets (see (3.11.35)):

$$\sigma(s) = \lim_{q^{s_4} \to 0} \left[-\frac{\prod_{i=1}^4 \psi_q(s - s_i)}{\psi_q(s_4)} \right] = q^{3/2} \prod_{i=1}^3 \psi_q(s - s_i) \quad . \tag{A.10}$$

Using (A.1) and (A.2) we arrive at the following expression for $y_n(s)$ (see (3.11.36)) and λ_n:

$$y_n(s) = \frac{(-1)^n B_n}{(\tilde{c}_1 \kappa^2)^n} q^{-n(s_1 + (n-1)/2)/2} (s_1 + s_2 + \mu|q)_n (s_1 + s_3 + \mu|q)_n$$

$$\times \, {}_3F_2 \left[\begin{matrix} -n, \, s_1 - s, \, s_1 + s + \mu \\ s_1 + s_2 + \mu, \, s_1 + s_3 + \mu \end{matrix} \, \middle| \, q, \, q^{-(s_2 + s_3 + \mu + n - 1)/2} \right] \; ; \tag{A.11}$$

$$\lambda_n = \frac{\psi_q(n)}{(\tilde{c}_1 q^{-\mu/2} \kappa^2)^2} q^{-\left(\sum_{i=1}^3 2\mu + n - 1 \right)/2} \quad . \tag{A.12}$$

At $q \to 1$ (A.11) and (A.12) coincide, as can be easily seen with (A.5) and (A.6). Because at $\mu = 1$, $C_1 = 1$, $s_1 = a$, $s_2 = -b$, $s_3 = c$, $B_n = (-1)^n/n!$ (A.5) coincides with that for the dual Hahn polynomials, it is natural to call polynomials (A.11) at $\mu = 1$ the dual q-Hahn polynomials.

2. At $q^{s_3} \to 0$, $q^{s_4} \to 0$, $A = 1/(\psi_q(s_3)\psi_q(s_4))$ we have (see (3.11.44))

$$\sigma(s) = q^s \psi_q(s - s_1)\psi_q(s - s_2) \quad . \tag{A.13}$$

In this case to find $y_n(s)$ it is convenient to use (3.11.43), which is equivalent to (A.1), and to substitute s_2 for s_3. As a result one gets at $q^{s_3} \to 0$, $q^{s_4} \to 0$

$$y_n(s) = \frac{(-1)^n B_n}{(\tilde{c}_1 \kappa^2)^n} q^{n(s - s_1 - n + 1)/2} (s - s_2 - n + 1|q)_n$$

$$\times \, {}_2F_1 \left[\begin{matrix} -n, \, s + s_1 + \mu \\ s - s_2 - n + 1 \end{matrix} \, \middle| \, q, \, q^{(s_1 - s_2 - 2s - \mu)/2} \right] \quad . \tag{A.14}$$

From (A.2) one derives a formula for λ_n:

$$\lambda_n = \frac{\psi_q(n)}{\tilde{c}_1^2 \kappa^3} q^{-(s_1+s_2+n-1)/2} \quad . \tag{A.15}$$

3. At $q^{s_i} = 0 \, (i = 2, 3, 4)$, $A = -\dfrac{1}{\prod_{i=2}^4 \psi_q(s_i)}$ one gets (see (3.11.39))

$$\sigma(s) = q^{3s/2} \psi_q(s - s_1) \quad . \tag{A.16}$$

Using (A.1) and (A.2) one derives the following expressions for $y_n(s)$ (see (3.11.40)) and λ_n:

$$y_n(s) = \frac{(-1)^n B_n}{(\tilde{c}_1 \kappa^2)^n} q^{-3n[s_1+(n-1)/2]/2-\mu n}$$

$$\times \sum_{k=0}^{n} \frac{(-n|q)_k \, (s_1 - s|q)_k \, (s_1 + s + \mu|q)_k}{(1|q)_k} \kappa^{2k} q^{-k(n-k-2s_1-\mu)/2} , \tag{A.17}$$

$$\lambda_n = \frac{\psi_q(n)}{(\tilde{c}_1 \kappa)^2} q^{-(s_1+n-1)/2} \tag{A.18}$$

4) At $q^{s_i} \to 0 \, (i = 1, 2, 3, 4)$, $A = \dfrac{1}{\prod_{i=1}^4 \psi_q(s - s_i)}$ one gets (see (3.11.48))

$$\sigma(s) = q^{2s} \quad . \tag{A.19}$$

Using (3.11.43), which is equivalent to (A.1), one finds in this case the following formulas for $y_n(s)$ (see (3.11.49)) and λ_n:

$$y_n(s) = \frac{(-1)^n B_n}{(\tilde{c}_1 \kappa)^n} q^{-3n(n-1)/4} \sum_{k=0}^{n} (-1)^k \frac{(-n|q)_k}{(1|q)_k} q^{(n-2k)s+k(n-k-2\mu)/2} , \tag{A.20}$$

$$\lambda_n = \frac{\psi_q(n)}{\tilde{c}_1^2 \kappa} q^{-(n-1)/2} \quad . \tag{A.21}$$

The limit transitions at $q^{-s_i} \to 0$ for some i do not result in new forms of polynomials $y_n(s)$. This is due to the fact that the initial formulas for $\sigma(s)$, $y_n(s)$ and $x(s)$ at given $s_i \, (i = 1, 2, 3, 4)$ remain unchanged under substitution $1/q$ for q and $\tilde{c}_2$ for $\tilde{c}_1$, but under such substitution the limit transition $q^{s_i} \to 0$ must be substituted for $q^{-s_i} \to 0$. One can easily convince himself of this, for example, by substituting $1/q$ for q and $\tilde{c}_2$ for $\tilde{c}_1$ in (3.11.37,38) and comparing the resulting expressions with (3.11.35,36).

3.2 When $q \neq 1$ it remains to consider the limit transition at $\mu \to \infty$. As shown above, the case of $q^\mu \to 0$ leads to a lattice function of the form $\tilde{c}_2 q^{-s} + \tilde{c}_3$ and the case of $q^{-\mu} \to 0$ leads to $x(s) = \tilde{c}_1 q^s + \tilde{c}_3$. As in the case of $q = 1$, in the cases under consideration the functions $\sigma(s)$ and $\sigma(s) + \tau(s)\Delta x(s - 1/2)$ are polynomials of at most the second degree in $x(s)$.

Let us consider first the limit transition at $q^{-\mu} \to 0$. In the most general case the functions $\sigma(s)$ and $\sigma(s) + \tau(s)\Delta x(s - 1/2)$ are in the limit polynomials of the second degree with respect to q^s. Because $\Delta x(s - 1/2) = \tilde{c}_1 \kappa q^s$ when $x(s) = \tilde{c}_1 q^s + \tilde{c}_3$, in the limit both polynomials have the same free terms. It follows from this that they can be brought to the form

$$\sigma(s) = \bar{A}(q^{s-s_1} - 1)(q^{s-s_2} - 1) \quad (\bar{A} = \text{const}) \quad , \tag{A.22}$$

$$\sigma(s) + \tau(s)\Delta x(s - 1/2) = \bar{A}(q^{s-\bar{s}_1} - 1)(q^{s-\bar{s}_2} - 1) \quad . \tag{A.23}$$

These expressions can be derived from the general case of nonuniform lattice $x(s) = \tilde{c}_1 q^s + \tilde{c}_2 q^{-s} + \tilde{c}_3$ $(\tilde{c}_1/\tilde{c}_2 = q^\mu)$ at $q^{-\mu} \to 0$, if the constants $A = A(\mu)$ and $s_i = s_i(\mu)$ in the formulas

$$\sigma(s) = A \prod_{i=1}^{4} \psi_q(s - s_i) \quad ,$$

$$\sigma(s) + \tau(s)\Delta x(s - 1/2) = \sigma(-s - \mu) = A \prod_{i=1}^{4} \psi_q(s + s_i + \mu)$$

are chosen in such a way that at $q^{-\mu} \to 0$ these formulas become (A.22) and (A.23). For this case it is sufficient to put

$$s_1(\mu) = s_1 \, , \; s_2(\mu) = s_2 \, , \; s_3(\mu) = -\bar{s}_1 - \mu \, , \; s_4(\mu) = -\bar{s}_2 - \mu \, ,$$

$$A(\mu) = \kappa^4 q^{-\mu - (s_1 + s_2 + \bar{s}_1 + \bar{s}_2)/2} \bar{A} \quad .$$

As a result one derives from (A.1) and (A.2) the following expressions for $y_n(s)$ and λ_n:

$$y_n(s) = \left[\frac{\bar{A}\kappa}{\tilde{c}_1}\right]^n B_n \, q^{-n[\bar{s}_1 + \bar{s}_2 - (n-1)/2]/2} (s_1 - \bar{s}_1|q)_n \, (s_1 - \bar{s}_2|q)_n$$

$$\times \, {}_3F_2\left[\begin{array}{c} -n, \; s_1 + s_2 - \bar{s}_1 - \bar{s}_2 + n - 1, \; s_1 - s \\ s_1 - \bar{s}_1, \; s_1 - \bar{s}_2 \end{array} \middle| \, q, q^{(s-s_2)/2}\right], \tag{A.24}$$

$$\lambda_n = -\frac{\bar{A}}{\tilde{c}_1^2} q^{-(s_1 + s_2 + \bar{s}_1 + \bar{s}_2)/2} \psi_q(n) \, \psi_q(s_1 + s_2 - \bar{s}_1 - \bar{s}_2 + n - 1) \quad . \tag{A.25}$$

(Formula (A.24) coincides with (3.11.53) if one puts $d_1 = -\bar{s}_1$, $d_2 = -\bar{s}_2$).

When $s_1 = 0$, $s_2 = N + \alpha$, $\bar{s}_1 = -\beta - 1$, $\bar{s}_2 = N - 1$, $\bar{A} = -1/\kappa^2$, $\tilde{c}_1 = 1/\kappa$, $\tilde{c}_3 = -1/\kappa$, (A.24) coincides at $q \to 1$ with that for the Hahn polynomials. This is why it is natural to call the polynomials derived at the given values of parameters according to formula (A.24) q-Hahn polynomials.

Further, together with (A.24) we shall sometimes use another equivalent formula, which can be derived from (A.1) using the same limit transition after preliminary substitution in (A.1) s_3 for s_1, s_4 for s_2, s_1 for s_3, s_2 for s_4 (as can be easily seen, the initial formula for $\sigma(s)$ remains unchanged under such substitutions):

$$y_n(s) = (-1)^n \left[\frac{\bar{A}\kappa}{\bar{c}_1}\right]^n B_n q^{-n[s_1+s_2+(n-1)/2]/2} (s_1 - \bar{s}_1|q)_n (s_2 - \bar{s}_1|q)_n$$

$$\times \, {}_3F_2\left[\begin{matrix} -n, \; s_1 + s_2 - \bar{s}_1 - \bar{s}_2 + n - 1, \; s - \bar{s}_1 \\ s_1 - \bar{s}_1, \; s_2 - \bar{s}_1 \end{matrix} \middle| q, q^{(s-\bar{s}_2)/2}\right] . \quad (A.26)$$

We considered the fundamental case when functions $\sigma(s) + \tau(s)\Delta x(s - 1/2)$ and $\sigma(s)$ are polynomials of the second degree with respect to q^s. Each of these polynomials turns to zero at two finite values of s (they have two finite zeros). There are also possible other cases when these polynomials may have degree less than two, and cases when the number of finite zeros is less than the degree of the polynomial. All these cases can be derived from the fundamental one, using the limit transitions at $q^{s_i} \to 0$ or $q^{-s_i} \to 0$ under some values of i, and also at $q^{\bar{s}_i} \to 0$ or $q^{-\bar{s}_i} \to 0$. At $q^{-s_i} \to 0$ for a given value of i the degree of polynomial $\sigma(s)$ with respect to q^s becomes less by one, and at $q^{s_i} \to 0$ the degree of the polynomial remains unchanged, but the number of finite zeros diminishes by one (the same remark is also valid for $\sigma(s) + \tau(s)\Delta x(s - 1/2)$ at $q^{-\bar{s}_i} \to 0$ and $q^{\bar{s}_i} \to 0$).

Let us arrange all possible forms of $\sigma(s)$ and indicate the limit transitions which allow one to derive them from the fundamental case (A.22):

1) $\sigma(s) = A_1 q^s (q^{s-s_1} - 1)$ $(q^{s_2} \to 0, \; \bar{A} = A_1 q^{s_2}, \; A_1 = \text{const})$;

2) $\sigma(s) = A_1 q^{2s}$ $(q^{s_2} \to 0, \; q^{s_1} \to 0, \; \bar{A} = A_1 q^{s_1+s_2}, \; A_1 = \text{const})$;

3) $\sigma(s) = -\bar{A}(q^{s-s_1} - 1)$ $(q^{-s_2} \to 0, \; \bar{A} = \text{const})$;

4) $\sigma(s) = -A_1 q^s$ $(q^{-s_2} \to 0, \; q^{s_1} \to 0, \; \bar{A} = -A_1 q^{s_1}, \; A_1 = \text{const})$;

5) $\sigma(s) = \bar{A}$ $(q^{-s_2} \to 0, \; q^{-s_1} \to 0, \; \bar{A} = \text{const})$.

Similarly, changing q^{s_i} for $q^{\bar{s}_i}$ it is possible to derive from (A.23) all possible forms of the polynomial $\sigma(s) + \tau(s)\Delta x(s - 1/2)$:

1) $\sigma(s) + \tau(s)\Delta x(s - 1/2) = A_2 q^s (q^{s-\bar{s}_1} - 1)$
$(q^{\bar{s}_2} \to 0, \; \bar{A} = A_2 q^{\bar{s}_2}, \; A_2 = \text{const})$;

2) $\sigma(s) + \tau(s)\Delta x(s - 1/2) = A_2 q^{2s}$
$(q^{\bar{s}_2} \to 0, \; q^{\bar{s}_1} \to 0, \; \bar{A} = A_2 q^{\bar{s}_1+\bar{s}_2}, \; A_2 = \text{const})$;

3) $\sigma(s) + \tau(s)\Delta x(s - 1/2) = -\bar{A}(q^{s-\bar{s}_1} - 1)$ $(q^{-\bar{s}_2} \to 0, \; \bar{A} = \text{const})$;

4) $\sigma(s) + \tau(s)\Delta x(s - 1/2) = -A_2 q^s$
$(q^{-\bar{s}_2} \to 0, \; q^{\bar{s}_1} \to 0, \; \bar{A} = -A_2 q^{\bar{s}_1}, \; A_2 = \text{const})$;

5) $\sigma(s) + \tau(s)\Delta x(s - 1/2) = \bar{A}$ $(q^{-\bar{s}_2} \to 0, \; q^{-\bar{s}_1} \to 0, \; \bar{A} = \text{const})$.

Let us remark that instead of $\bar{A}$, which was some constant in (A.22) and (A.23), we achieve in the limit for $\sigma(s)$ and $\sigma(s) + \tau(s)\Delta x(s - 1/2)$ either the former constant $\bar{A}$ (in cases 3 and 5) or two new constants A_1 and A_2 (in cases $1, 2, 4$). In cases $1, 2, 4$, the constant $\bar{A}$ in the limit turns out to be zero.

Because free terms of the polynomials $\sigma(s)$ and $\sigma(s) + \tau(s)\Delta x(s - 1/2)$ must coincide, one can consider $\sigma(s)$ and $\sigma(s) + \tau(s)\Delta x(s - 1/2)$ simultaneously, for example, when $\bar{A} = $ const. If the free term of $\sigma(s)$ is zero (this corresponds to the limit transitions at $\bar{A} \to 0$), then the free term of $\sigma(s) + \tau(s)\Delta x(s - 1/2)$ has also to be zero (this also corresponds to the limit transitions at $\bar{A} \to 0$).

For us the cases when $\tau(s)$ is a polynomial of the first degree in respect to q^s are of interest. This is why in the case when $\sigma(s)$ is a polynomial of at most the first degree, the polynomial $\sigma(s) + \tau(s)\Delta x(s - 1/2)$ must be of the second degree. Considering the limit transitions, one has to keep in mind that at $q^{-s_i} \to 0$ under some values of i the limit transition at $q^{-\bar{s}_i} \to 0$ is impossible under any value of i.

In the case of limit transitions which correspond to $\bar{A} \to 0$ in the formulas relating $\bar{A}$ with A_1 and A_2 one can find factors q^{s_i} and $q^{\bar{s}_i}$ under values of i for which $q^{s_i} \to 0$, $q^{\bar{s}_i} \to 0$. Equating the initial expressions for $\bar{A}$ in $\sigma(s)$ and $\sigma(s) + \tau(s)\Delta x(s - 1/2)$, one can derive in a simple form the relations for corresponding values of s_i and $\bar{s}_i$, introducing the constant δ defined by $A_2/A_1 = q^\delta$.

We now list all possible forms of $\sigma(s)$ and $\sigma(s) + \tau(s)\Delta x(s - 1/2)$ and indicate the limit transitions which allow one to derive them from (A.22) and (A.23) for the fundamental case:

No.	$\sigma(s)$	$\sigma(s) + \tau(s)\Delta x(s - \frac{1}{2})$	Limit transition
1	$\bar{A}(q^{s-s_1} - 1)(q^{s-s_2} - 1)$	$-\bar{A}(q^{s-\bar{s}_1} - 1)$	$q^{-\bar{s}_2} \to 0$ $(\bar{A} = \text{const})$
2	$\bar{A}(q^{s-s_1} - 1)(q^{s-s_2} - 1)$	$\bar{A}$	$q^{-\bar{s}_2} \to 0$, $q^{-\bar{s}_1} \to 0$ $(\bar{A} = \text{const})$
3	$A_1 q^s(q^{s-s_1} - 1)$	$A_1 q^{s+\delta}(q^{s-\bar{s}_1} - 1)$	$q^{s_2} \to 0,\ q^{\bar{s}_2} \to 0,\ s_2 - \bar{s}_2 = \delta$ $\bar{A} = A_1 q^{s_2}$ $(A_1 = \text{const})$
4	$A_1 q^s(q^{s-s_1} - 1)$	$A_1 q^{2s+\delta}$	$q^{s_2} \to 0,\ q^{\bar{s}_1} \to 0,$ $q^{\bar{s}_2} \to 0\ \ s_2 - \bar{s}_1 - \bar{s}_2 = \delta$ $\bar{A} = A_1 q^{s_2}$ $(A_1 = \text{const})$
5	$A_1 q^s(q^{s-s_1} - 1)$	$A_1 q^{s+\delta}$	$q^{s_2} \to 0,\ q^{\bar{s}_1} \to 0,\ q^{-\bar{s}_2} \to 0,$ $s_2 - \bar{s}_1 = \delta + \frac{i\pi}{\ln q}$ $\bar{A} = A_1 q^{s_2}$ $(A_1 = \text{const})$
6	$-\bar{A}(q^{s-s_1} - 1)$	$\bar{A}(q^{s-\bar{s}_1} - 1)(q^{s-\bar{s}_2} - 1)$	$q^{-s_2} \to 0$ $(\bar{A} = \text{const})$
7	$A_1 q^{2s}$	$A_1 q^{s+\delta}(q^{s-\bar{s}_1} - 1)$	$q^{s_1} \to 0,\ q^{s_2} \to 0,$ $q^{\bar{s}_2} \to 0,\ s_1 + s_2 - \bar{s}_2 = \delta$ $\bar{A} = A_1 q^{s_1+s_2}$ $(A_1 = \text{const})$

8	$A_1 q^{2s}$	$A_1 q^{2s+\delta}$	$q^{s_1} \to 0,\ q^{s_2} \to 0,$ $q^{\bar{s}_1} \to 0,\ q^{\bar{s}_2} \to 0,$ $s_1 + s_2 - \bar{s}_1 - \bar{s}_2 = \delta$ $\bar{A} = A_1 q^{s_1 + s_2}\ (A_1 = \text{const})$
9	$A_1 q^{2s}$	$A_1 q^{s+\delta}$	$q^{s_1} \to 0,\ q^{s_2} \to 0,$ $q^{\bar{s}_1} \to 0,\ q^{-\bar{s}_2} \to 0,$ $s_1 + s_2 - \bar{s}_1 = \delta + \frac{i\pi}{\ln q}$ $\bar{A} = A_1 q^{s_1 + s_2}\ (A_1 = \text{const})$
10	$A_1 q^{s}$	$A_1 q^{s+\delta}(q^{s-\bar{s}_1} - 1)$	$q^{s_1} \to 0,\ q^{-s_2} \to 0,\ q^{\bar{s}_2} \to 0,$ $s_1 - \bar{s}_2 = \delta + \frac{i\pi}{\ln q}$ $\bar{A} = -A_1 q^{s_1}\ (A_1 = \text{const})$
11	$A_1 q^{s}$	$A_1 q^{2s+\delta}$	$q^{s_1} \to 0,\ q^{-s_2} \to 0,$ $q^{\bar{s}_1} \to 0,\ q^{\bar{s}_2} \to 0,$ $s_1 - \bar{s}_1 - \bar{s}_2 = \delta + \frac{i\pi}{\ln q}$ $\bar{A} = -A_1 q^{s_1}\ (A_1 = \text{const})$
12	$\bar{A}$	$\bar{A}(q^{s-\bar{s}_1} - 1)(q^{s-\bar{s}_2} - 1)$	$q^{-s_1} \to 0,$ $q^{-s_2} \to 0\ (\bar{A} = \text{const})$

Using the limit transitions indicated above and (3.11, 31–34), it is possible to derive expressions for $y_n(s)$ and λ_n from (A.24) and (A.25) (in cases 6, 10, 12 instead of (A.24) the equivalent formula (A.26) was used; in case 8 to simplify things the limit transition in (A.24) was carried out at $s_1 = s_2$, $\bar{s}_1 = \bar{s}_2$):

1)
$$y_n(s) = \left[-\frac{\bar{A}}{\tilde{c}_1} \right]^n B_n q^{-n(s_1+\bar{s}_1)/2} (s_1 - \bar{s}_1 | q)_n$$
$$\times\ {}_2F_1 \left[\begin{matrix} -n,\ s_1 - s \\ s_1 - \bar{s}_1 \end{matrix} \middle|\ q\,,\ q^{(s-2s_2+\bar{s}_1-n+1)/2} \right],$$
$$\lambda_n = \frac{\bar{A}}{\kappa \tilde{c}_1^2} q^{-[s_1+s_2+(n-1)/2]} \psi_q(n);$$

2)
$$y_n(s) = \left[\frac{\bar{A}}{\kappa \tilde{c}_1} \right]^n B_n q^{-n[s_1+(n-1)/4]}$$
$$\times \sum_{k=0}^{n} \frac{(-n|q)_k (s_1 - s|q)_k}{(1|q)_k} (-\kappa)^k q^{k[s+s_1-2s_2-n+1+(k-1)/2]/2}\,,$$
$$\lambda_n = \frac{\bar{A}}{\kappa \tilde{c}_1^2} q^{-[s_1+s_2+(n-1)/2]} \psi_q(n);$$

3)
$$y_n(s) = \left[\frac{A_1}{\tilde{c}_1} \right]^n B_n q^{n(s_1-\bar{s}_1+2\delta+n-1)/2} (s_1 - \bar{s}_1 | q)_n$$

$$\times \sum_{k=0}^{n} \frac{(-n|q)_k(s_1 - \bar{s}_1 + \delta + n - 1|q)_k(s_1 - s|q)_k}{(s_1 - \bar{s}_1|q)_k(1|q)_k} \kappa^k q^{k[s-s_1-\delta-(k-1)/2]/2},$$

$$\lambda_n = -\frac{A_1}{\tilde{c}_1^2} q^{-(s_1+\bar{s}_1-\delta)/2}\psi_q(n)\,\psi_q(s_1 - \bar{s}_1 + \delta + n - 1);$$

4) $\quad y_n(s) = \left[\dfrac{A_1}{\kappa \tilde{c}_1}\right]^n B_n q^{n[s_1+\delta+3(n-1)/4]}$

$$\times \sum_{k=0}^{n} \frac{(-n|q)_k(s_1 + \delta + n - 1|q)_k(s_1 - s|q)_k}{(1|q)_k} \kappa^{2k} q^{k(s-2s_1-\delta-k+1)/2},$$

$$\lambda_n = -\frac{A_1}{\tilde{c}_1^2} q^{(\delta-s_1)/2}\psi_q(n)\psi_q(s_1 + \delta + n - 1);$$

5) $\quad y_n(s) = \left[\dfrac{A_1}{\kappa \tilde{c}_1}\right]^n B_n q^{n[\delta+(n-1)/4]}$

$$\times \sum_{k=0}^{n} \frac{(-n|q)_k(s_1 - s|q)_k}{(1|q)_k} (-\kappa)^k q^{k[s-s_1-2\delta-n+1-(k-1)/2]/2},$$

$$\lambda_n = \frac{A_1}{\kappa \tilde{c}_1^2} q^{-[s_1+(n-1)/2]}\psi_q(n);$$

6) $\quad y_n(s) = (-1)^n \left[\dfrac{\bar{A}}{\tilde{c}_1}\right]^n B_n q^{-n(s_1+\bar{s}_1)/2}(s_1 - \bar{s}_1|q)_n$

$$\times {}_2F_1\left[\begin{matrix} -n,\ s - \bar{s}_1 \\ s_1 - \bar{s}_i \end{matrix} \,\Bigg|\, q\,,\ q^{(s+s_1-2\bar{s}_2+n-1)/2}\right],$$

$$\lambda_n = -\frac{\bar{A}}{\kappa \tilde{c}_1^2} q^{-[\bar{s}_1+\bar{s}_2-(n-1)/2]}\psi_q(n);$$

7) $\quad y_n(s) = \left[-\dfrac{A_1}{\kappa \tilde{c}_1}\right]^n B_n q^{n[\delta+(n-1)/4]}$

$$\times \sum_{k=0}^{n} \frac{(-n|q)_k(\delta - \bar{s}_1 + n - 1|q)_k}{(1|q)_k} \kappa^k q^{ks-k[\bar{s}_1+\delta+(k-1)/2]/2},$$

$$\lambda_n = \frac{-A_1}{\tilde{c}_1^2} q^{(\delta-\bar{s}_1)/2}\psi_q(n)\,\psi_q(\delta - \bar{s}_1 + n - 1);$$

8) $\quad y_n(s) = \left[\dfrac{A_1}{\tilde{c}_1}\right]^n B_n q^{n\delta/2}(\delta + n - 1|q)_n q^{ns},$

$$\lambda_n = -\frac{A_1}{\tilde{c}_1^2} q^{\delta/2}\psi_q(n)\,\psi_q(\delta + n - 1);$$

9) $\quad y_n(s) = \left[\dfrac{A_1}{\kappa \tilde{c}_1}\right]^n B_n q^{n[\delta+(n-1)/4]} \sum_{k=0}^{n} \frac{(-n|q)_k}{(1|q)_k} q^{k[s-\delta+1-(n+k)/2]},$

$$\lambda_n = \frac{A_1}{\kappa \tilde{c}_1^2} q^{-(n-1)/2} \psi_q(n);$$

10)
$$y_n(s) = (-1)^n \left[\frac{A_1}{\kappa \tilde{c}_1}\right]^n B_n q^{-n(n-1)/4}$$

$$\times \sum_{k=0}^{n} (-1)^k \frac{(-n|q)_k (s - \bar{s}_1|q)_k}{(1|q)_k} \kappa^k q^{k[s - \bar{s}_1 + 2\delta + n - 1 + (k-1)/2]/2},$$

$$\lambda_n = -\frac{A_1}{\kappa \tilde{c}_1^2} q^{\delta - \bar{s}_1 + (n-1)/2} \psi_q(n);$$

11)
$$y_n(s) = (-1)^n \left[\frac{A_1}{\kappa \tilde{c}_1}\right]^n B_n q^{-n(n-1)/4} \sum_{k=0}^{n} \frac{(-n|q)_k}{(1|q)_k} q^{ks + k[\delta - 1 + (n+k)/2]},$$

$$\lambda_n = -\frac{A_1}{\kappa \tilde{c}_1^2} q^{\delta + (n-1)/2} \psi_q(n);$$

12)
$$y_n(s) = (-1)^n \left[\frac{\bar{A}}{\kappa \tilde{c}_1}\right]^n B_n q^{n[(n-1)/4 - \bar{s}_1]}$$

$$\times \sum_{k=0}^{n} \frac{(-n|q)_k (s - \bar{s}_1|q)_k}{(1|q)_k} \kappa^k q^{k[s + \bar{s}_1 - 2\bar{s}_2 + n - 1 - (k-1)/2]/2},$$

$$\lambda_n = -\frac{\bar{A}}{\kappa \tilde{c}_1^2} q^{(n-1)/2 - \bar{s}_1 - \bar{s}_2} \psi_q(n);$$

One has to emphasize that some of the formulas given above for $y_n(s)$ were derived in Sect. 3.11.5.2 in another form. However, here, in this Conclusion, the derivation of the corresponding formulas is more consistent.

If for case 9 one puts $\tilde{c}_1 = 1$, $\tilde{c}_3 = 0$ (i.e. $x(s) = q^s$), $A_1 = 1$, $\delta = 3/2$, then we come to the Stieltjes–Wigert polynomials [S24, W6, C18], for which $\sigma(s) = q^{2s}$, $\rho(s) = q^{-s^2/2}$.

In case 5 at $q \to 1$, if one puts $B_n = \mu^{-n}$, $s_1 = 0$, $\tilde{c}_1 = A_1 = 1/\kappa$, $q^\delta = \mu \kappa$, one produces in the limit the Charlier polynomials $c_n^{(\mu)}(s)$. Similarly, in case 3 at $q \to 1$, by putting $\tilde{c}_1 = A_1 = 1/\kappa$, $s_1 = 0$ one produces at $B_n = \mu^{-n}$, $\bar{s}_1 = -\gamma$, $q^\delta = \mu$ the Meixner polynomials $m_n^{(\gamma, \mu)}(s)$, and at $B_n = (-1)^n (1-p)^n / \tilde{\Gamma}_q(n+1)$, $\bar{s}_1 = N$, $q^\delta = -p/(1-p)$ the Kravchuk polynomials $k_n^{(p)}(s)$.

We have considered the possible forms of polynomials $y_n(s)$ at $x(s) = \tilde{c}_1 q^s + \tilde{c}_3 (q^{-\mu} \to 0)$. The limit transition $q^\mu \to 0$, when $x(s) = \tilde{c}_2^1 q^{-s} + \tilde{c}_3$, does not lead to new forms of polynomials, because for the derivation of corresponding polynomials it is sufficient to substitute $1/q$ for q, $\tilde{c}_1$ for $\tilde{c}_2$ in the formulas derived above for $\sigma(s)$, $y_n(s)$ and λ_n.

The considered families of polynomials are situated in the lower right corner of the scheme, in the rectangles connected by lines with the rectangle corresponding to the fundamental case at $\mu = \infty$ when $q \neq 1$. The rectangles situated in the third line from above correspond to the case of two finite zeros of the function $\sigma(s)$. In the fourth line rectangles correspond to one finite zero, in the fifth line (the bottom one) rectangles correspond to the absence of finite zeros. In the upper part of each rectangle the degrees of polynomials of $\sigma(s)$ and $\sigma(s) + \tau(s)\Delta x(s - 1/2)$ (with respect to q^s) are given. In the lower part at the values of s_i and $\bar{s}_i$ the values of index i are given in brackets ($s_i(0)$ indicates the absence of finite zeros of the function $\sigma(s)$).

It is obvious that all formulas for polynomials $\tilde{y}_n[x(s)] \equiv y_n(s)$ given in this Conclusion can be brought into other forms using the property of symmetry $x(s) = x(-s - \mu)$, permutations of s_i and different initial formulas of $y_n(s)$ for limit transitions (for example, instead of (3.11.25) one can use (3.11.16,43)).

Classification of polynomial solutions of difference equations of hypergeometric type (A. Nikiforov and V. Uvarov, 1991)

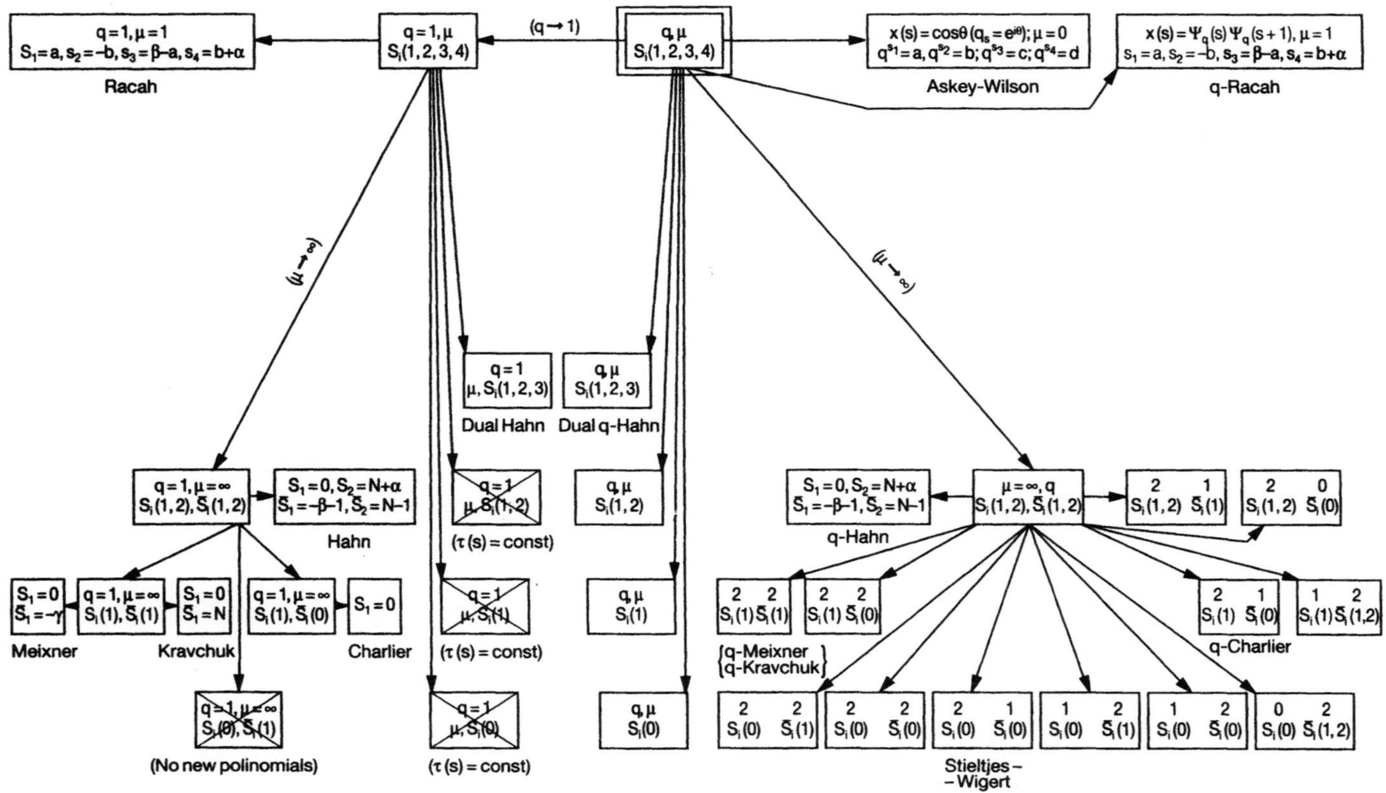
Racah
$q = 1,\ \mu = 1$
$S_1 = a,\ s_2 = -b,\ s_3 = \beta - a,\ s_4 = b + \alpha$
$q = 1,\ \mu$
$S_i(1, 2, 3, 4)$
$(q \to 1)$
$q,\ \mu$
$S_i(1, 2, 3, 4)$
Askey-Wilson
$x(s) = \cos\theta\,(q_s = e^{i\theta});\ \mu = 0$
$q^{s_1} = a,\ q^{s_2} = b;\ q^{s_3} = c;\ q^{s_4} = d$
q-Racah
$x(s) = \Psi_q(s)\,\Psi_q(s+1),\ \mu = 1$
$s_1 = a,\ s_2 = -b,\ s_3 = \beta - a,\ s_4 = b + \alpha$
$(\mu \to \infty)$
$(\mu \to \infty)$
Dual Hahn
$q = 1$
$\mu,\ S_i(1, 2, 3)$
Dual q-Hahn
$q,\ \mu$
$S_i(1, 2, 3)$
Hahn
$q = 1,\ \mu = \infty$
$S_i(1, 2),\ \bar{S}_i(1, 2)$
$S_1 = 0,\ S_2 = N + \alpha$
$S_1 = -\beta - 1,\ S_2 = N - 1$
$q = 1$
$\mu,\ S_i(1, 2)$
$(\tau(s) = \text{const})$
$q,\ \mu$
$S_i(1, 2)$
q-Hahn
$S_1 = 0,\ S_2 = N + \alpha$
$S_1 = -\beta - 1,\ S_2 = N - 1$
$\mu = \infty,\ q$
$S_i(1, 2),\ \bar{S}_i(1, 2)$
2   1
$S_i(1, 2)\ \bar{S}_i(1)$
2   0
$S_i(1, 2)\ \bar{S}_i(0)$
Meixner
$S_1 = 0$
$S_1 = -\gamma$
Kravchuk
$q = 1,\ \mu = \infty$
$S_i(1),\ \bar{S}_i(1)$
$S_1 = 0$
$S_1 = N$
$q = 1,\ \mu = \infty$
$S_i(1),\ \bar{S}_i(0)$
Charlier
$S_1 = 0$
$q = 1$
$\mu,\ S_i(1)$
$(\tau(s) = \text{const})$
$q,\ \mu$
$S_i(1)$
2   2
$S_i(1)\,\bar{S}_i(1)$
2   2
$S_i(1)\ \bar{S}_i(0)$
{q-Meixner
 q-Kravchuk}
2   1
$S_i(1)\ \bar{S}_i(0)$
q-Charlier
1   2
$S_i(1)\,\bar{S}_i(1,2)$
$q = 1,\ \mu = \infty$
$S_i(0),\ \bar{S}_i(1)$
(No new polinomials)
$q = 1$
$\mu,\ S_i(0)$
$(\tau(s) = \text{const})$
$q,\ \mu$
$S_i(0)$
2   2
$S_i(0)\ \bar{S}_i(1)$
2   2
$S_i(0)\ \bar{S}_i(0)$
2   1
$S_i(0)\ \bar{S}_i(0)$
1   2
$S_i(0)\ \bar{S}_i(0)$
1   2
$S_i(0)\ \bar{S}_i(1)$
1   2
$S_i(0)\ \bar{S}_i(0)$
0   2
$S_i(0)\ \bar{S}_i(1, 2)$
Stieltjes-
-Wigert

Bibliography

A1   M. Abramowitz, I.A. Stegun (eds.): *Handbook of Mathematical Functions* (National Bureau of Standards, Washington 1964)

A2   A.K. Agarwal, E.G. Kalnins, W. Miller: "Canonical Equations and Symmetry Techniques for q-Series", SIAM J. Math. Anal. (to appear)

A3   N.A. Al-Salam: Duke Math. J. **133**, 109 (1966)

A4   W. Al-Salam, W.R. Allaway, R. Askey: Canad. Math. Bull. **27**, 329 (1984)

A5   W. Al-Salam, W.R. Allaway, R. Askey: Trans. Amer. Math. Soc. **284**, 39 (1984)

A6   W.A. Al-Salam, L. Carlitz: Math. Nachr. **30**, 47 (1965)

A7   W.A. Al-Salam, T.S. Chihara: SIAM J. Math. Anal. **7**, 16 (1976)

A8   W.A. Al-Salam, T.S. Chihara: SIAM J. Math. Anal. **18**, 228 (1987)

A9   W.A. Al-Salam, M.E.H. Ismail: J. Math. Anal. Appl. **55**, 125 (1976)

A10 W.A. Al-Salam, M.E.H. Ismail: Proc. Amer. Math. Soc. **67**, 105 (1977)

A11 W.A. Al-Salam, M.E.H. Ismail: "q-Beta Integrals and q-Hermite Polynomials", Pacific J. Math. (to appear)

A12 W.A. Al-Salam, A. Verma: Proc. Amer. Math. Soc. **85**, 360 (1982)

A13 B. Andrásfai: *Introductory Graph Theory* (Akadémiai Kiadó, Budapest 1977)

A14 G.E. Andrews: SIAM Review **16**, 447 (1974)

A15 G.E. Andrews, R. Askey: "Enumeration of Partitions: The Role of Eulerian Series and q-Orthogonal Polynomials", in *Higher Combinatorics*, ed. by M. Aigner (D. Reidel Publishing Company, Dordrecht, Boston 1977) pp. 3–26

A16 G.E. Andrews, R. Askey: Proc. Amer. Math. Soc. **81**, 97 (1981)

A17 G.E. Andrews, R. Askey: "Classical Orthogonal Polynomials", in *Polynômes orthogonaux et applications*, Lecture Notes in Mathematics, Vol. 1171 (Springer-Verlag, Berlin 1985) pp. 36–62

A18 M.G. Arutynyan, G.S. Pogosyan, V.M. Ter-Antonyan: Izv. Acad. Nauk Arm. SSR, Fiz. **13**, 152 (1978) [in Russian]

A19 R. Askey: *Orthogonal Polynomials and Special Functions* (Society for Industrial and Applied Mathematics, Philadelphia 1975)

A20 R. Askey: Appl. Anal. **8**, 125 (1978)

A21 R. Askey: Amer. Math. Month. **87**, 346 (1980)

A22 R. Askey: Indian J. Pure Appl. Math. **14**, 892 (1983)

A23 R. Askey: J. Phys. A. Math. Gen. **18**, L1017 (1985)

A24 R. Askey, M.E.H. Ismail: Proc. Amer. Math. Soc. **77**, 218 (1979)

A25 R. Askey, M.E.H. Ismail: "The Rogers q-Ultraspherical Polynomials", in *Approximation Theory III*, ed. by E.W. Cheney (Academic Press, New York 1980) pp. 175-182

A26 R. Askey, M.E.H. Ismail: Mem. Amer. Math. Soc. **300** (1984)

A27 R. Askey, J.A. Wilson: SIAM J. Math. Anal. **10**, 1008 (1979)

A28 R. Askey, J.A. Wilson: SIAM J. Math. Anal. **13**, 651 (1982)

A29 R. Askey, J.A. Wilson: Memoirs Amer. Math. Soc. **319** (1985)

A30 N.M. Atakishiyev: Theor. i Math. Phys. **58**, 254 (1984)

A31 N.M. Atakishiyev: Ann. Phys. (Leipzig) **42**, 31 (1985)

A32   N.M. Atakishiyev, R.M. Mir-Kasimov, Sh.M. Nagiyev: Ann. Phys. (Leipzig) **42**, 25 (1985)

A33   N.M. Atakishiyev, S.K. Suslov: "The Hahn and Meixner polynomials of an imaginary argument and some of their applications"; Zeuten preprint PHE 84-12 (1984); J. Phys. A. Math. Gen **18**, 1583 (1985)

A34   N.M. Atakishiyev, S.K. Suslov: "About One Class of Special Functions"; Preprint No. 497, IIMAS UNAM, Mexico (1987)

A35   N.M. Atakishiyev, S.K. Suslov: "Continuous Orthogonality Property for Some Classical Polynomials of a Discrete Variable"; Preprint No. 498, IIMAS UNAM, Mexico (1987)

A36   N.M. Atakishiyev, S.K. Suslov: "On the Moment of Classical and Related Polynomials"; Preprint No. 499, IIMAS UNAM, Mexico (1987)

B1   W.N. Bailey: *Generalized Hypergeometric Series*, Cambridge Tracts in Math. No. 32 (Cambridge Univ. Press, London 1935)

B2   N.S. Bakhvalov, G.M. Kobel'kov, Yu.V. Noskov: "On Practical Evaluation of Values for Orthogonal Polynomials of Continuous and Discrete Argument", Preprint No. 158, Department of Comp. Math. Acad. Sci. SSR, Moscow (1987)

B3   M. Bander, C. Itzykson: Rev. Mod. Phys. **38**, 330 (1966)

B4   E. Bannai, T. Ito: *Algebraic Combinatorics I. Association Schemes* (Benjamin/Cummings, Menlo Part, California, London, Amsterdam, Don Mills, Ontario, Sydney, Tokyo 1984)

B5   V. Bargmann: Ann. Math. **48**, 568 (1947)

B6   J.J. Bartko: Amer. Math. Month. **70**, 978 (1963)

B7   A.O. Barut, R. Raczka: *Theory of Group Representations and Applications* (PWH-Polish Scientific Publishers, Warszawa 1977)

B8   A.O. Barut, R. Wilson: J. Math. Phys. **17**, 900 (1976)

B9   D. Basu, S. Srinivasan: Czechoslovat J. Phys. **B27**, 629 (1977)

B10   H. Bateman: Tôhoky Math. J. **37**, 23 (1933)

B11   H. Bateman: Ann. Math. **35**, 767 (1934)

B12   H. Bateman: Proc. Nat. Acad. Sci. USA **20**, 63 (1934)

B13   H. Bateman: Proc. Nat. Acad. Sci. USA **28**, 374 (1942)

B14   B.F. Bayman: *Some Lectures on Groups and their Applications to Spectroscopy* (Russian transl.: Fizmatgiz, Moscow 1961)

B15   A.I. Baz', Ya. B. Zel'dovich, A.M. Perelomov: *Scattering, Reactions, and Decay in Nonrelativistic Quantum Mechanics* (Nauka, Moscow 1971) [in Russian]

B16   C.M. Bender, L.R. Mead, S.S. Pinsty: Phys. Rev. Lett. **56**, 2445 (1986)

B17   C.M. Bender, L.R. Mead, S.S. Pinsty: J. Math. Phys. **28**, 509 (1987)

B18   H.A. Bethe, E.E. Salpeter: *Quantum Mechanics of One– and Two–Electron Systems* (Springer-Verlag, Berlin 1958)

B19   W.A. Beyer, J.D. Louck, P.R. Stein: J. Math. Phys. **28**, 497 (1987)

B20   L.C. Biedenharn: J. Math. and Phys. **31**, 287 (1953)

B21   L.C. Biedenharn: J. Math. Phys. **2**, 433 (1961)

B22   L.C. Biedenhardn, J.M. Blatt, M.E. Rose: Rev. Mod. Phys. **24**, 249 (1952)

B23   L.C. Biedenharn, P.J. Brussard: *Coulomb Excitation* (Clarendon Press, Oxford 1965) p. 107

B24   LC. Biedenharn, J.D. Louch: *Angular Momentum in Quantum Physics*, Encyclopedia of Mathematics and Its Applications, Vol. 8 (Addison–Wesley, Reading 1981)

B25   L.C. Biedenharn, J.D. Louck: *The Racah-Wigner Algebra in Quantum Theory*, Encyclopedia of Mathematics and Its Applications, Vol. 9 (Addison–Wesley, Reading 1981)

B26   L.C. Biedenharn, H. Van Dam (eds.): *Quantum Theory of Angular Momentum* (Academic Press, New York, London 1965)

B27   A. Bohr, B.R. Mottelson: *Nuclear Structure I. Single–Particle Motion* (Benjamin, New York, Amsterdam 1969)

B28   A. Bohr, B.R. Mottelson: *Nuclear Structure II. Nuclear Deformations* (Benjamin, New York, Amsterdam 1974)

B29 G. Boole: *A Treatise on the Calculus of Finite Differences*, 2nd. edn. (Macmillan, London 1872; reprinted Dover, New York 1960)

B30 F. Brahman: Canad. J. Math. **9**, 191 (1957)

B31 P.A. Braun, A.A. Kiselev: *Introduction to the Theory of Molecula Spectra* (Leningrad Univ. Press, Leningrad 1983) [in Russian]

B32 A. Bremner, S. Brudno: J. Math. Phys. **27**, 2613 (1986)

B33 D.M. Bressoud: Indiana Univ. Math. J. **29**, 577 (1980)

B34 D.M. Bressoud: Proc. London Math. Soc. (3). **42**, 478 (1981)

B35 D.M. Bressoud: SIAM J. Math. Anal. **12**, 161 (1981)

B36 S. Brudno: J. Math. Phys. **26**, 434 (1985)

B37 S. Brudno: J. Math. Phys. **28**, 124 (1987)

B38 S. Brudno, J.D. Louck: J. Math. Phys. **26**, 1125 (1985)

B39 S. Brudno, J.D. Louck: J. Math. Phys. **26**, 2092 (1985)

B40 P.J. Brussard, H.A. Tolhoek: Physica **10**, 955 (1957)

B41 P.R. Bunker: *Molecular Symmetry and Spectroscopy* (Academic Press, New York, San Francisco, London 1979)

C1   L. Carlitz: Ann. Math. Pure Appl. (4) **41**, 349 (1955)

C2   L. Carlitz: Duke Math. J. **24**, 521 (1957)

C3   L. Carlitz: Canad. J. Math. **9**, 188 (1957)

C4   L. Carlitz: Duke Math. J. **26**, 1 (1959)

C5   E. Chacon, M. Moshinsky: Phys. Lett. **23**, 567 (1966)

C6   C.V.L. Charlier: Artiv för matematik, astronomi och fysik **2**, 20 (1905–1906)

C7   P.L. Chebyshev: "Note sur une formule d'analyse" (1854), in *Complete Collection of Papers*, Vol. 2 (USSR Academy of Sciences, Moscow 1947) pp. 99–100 [in Russian]

C8   P.L. Chebyshev: "Extruction des memoire sur les fractions continues" (1855), ibid, pp. 101–102

C9   P.L. Chebyshev: "Sur les fractions continues" (1855), *ibid*, pp. 103–126

C10 P.L. Chebyshev: "Sur une novelle série" (1858), *ibid*, pp. 236–238

C11 P.L. Chebyshev: "Sur l'interpolation par la méthode des moindres carres" (1859), *ibid*, pp. 314–334

C12 P.L. Chebyshev: "Sur l'interpolation" (1864), *ibid*, pp. 357–374

C13 P.L. Chebyshev: "Sur l'interpolation des valeurs équidistantes (1875), in *Complete Collection of Papers*, Vol. 3 (USSR Academy of Sciences, Moscow 1948) pp. 66-87 [in Russian]

C14 L. Chihara, T.S. Chihara: J. Math. Anal. Appl. **126**, 275 (1987)

C15 L. Chihara, D. Stanton: Graphs and Comb. **2**, 101 (1986)

C16 T.S. Chihara: Duke Math. J. **35**, 505 (1968)

C17 T.S. Chihara: Duke Math. J. **38**, 599 (1971)

C18 T.S. Chihara: *An Introduction to Orthogonal Polynomials* (Gordon and Breach, New York, London, Paris 1978)

C19 T.S. Chihara: J. Comp. Appl. Math. **5**, 291 (1979)

C20 T.S. Chihara: Rocky Mount. J. Math. **15**, 705 (1985)

C21 T.S. Chihara, M.E.H. Ismail: Adv. Appl. Math. **3**, 441 (1982)

C22 E.U. Candon, G.H. Shortley: *The Theory of Atomic Spectra* (Cambridge Univ. Press, London 1935)

D1   A.S. Davydov: *Quantum Mechanics* (Nauka, Moscow 1973) [in Russian]

D2   A.S. Davydov: *Theory of Solid State* (Nauka, Moscow 1976) [in Russian]

D3   P. Delsarte: J. Comb. Theor. A**20**, 230 (1976)

D4   P. Delsarte: SIAM J. Appl. Math. **31**, 262 (1976)

D5   P. Delsarte: SIAM J. Appl. Math. **34**, 157 (1978)

D6   P. Delsarte: J. Comb. Theor. A**25**, 226 (1978)

D7   P. Delsarte, J.M. Goenthals: J. Comb. Theor. A**19**, 26 (1975)

D8   H. DeMeyer, G. Vanden Berghe, J. Van der Jeugt: J. Math. Phys. **25**, 751 (1984)

D9   G. Doetsch: Math. Ann. **109**, 257 (1933)

D10   A.Z. Dolginov: Zh. Eksp. Teor. Fiz. **30**, 746 (1956)

D11   A.Z. Dolginov, A.N. Moskalev: Zh. Eksp. Teor. Fiz. **37**, 1697 (1959) [English transl.: Sov. Phys.-JETP **10**, 1202 (1960)]

D12   A.Z. Dolginov, I.N. Toptygin: Zh. Eksp. Teor. Fiz. **35**, 794 (1958)

D13   A.Z. Dolginov, I.N. Toptygin: Zh. Eksp. Teor. Fiz. **37**, 1441 (1959) [English transl.: Sov. Phys.-JETP **10**, 1022 (1960)]

D14   C.F. Dunkl: Indiana Univ. Math. J. **25**, 335 (1976)

D15   C.F. Dunkl: SIAM J. Math. Anal. **9**, 627 (1978)

D16   C.F. Dunkl: Monat. Math. **85**, 5 (1978)

D17   C.F. Dunkl: SIAM J. Alg. Dis. Meth. **1**, 137 (1980)

D18   C.F. Dunkl: Pacific J. Math. **29**, 57 (1981)

D19   C.F. Dunkl, D.E. Ramirez: SIAM J. Math. Anal. **5**, 351 (1974)

E1   G.K. Eagleson: Studia Scient. Math. Hungurica **3**, 127 (1968)

E2   G.K. Eagleson: Austral. J. Statist. **11**, 29 (1969)

E3   P.J. Eberlein: Math. Comp. **18**, 296 (1964)

E4   A.R. Edmonds: *Angular Momentum in Quantum Mechamics* (Princeton Univ. Press, Princeton 1957)

E5   J.P. Elliott: Proc. Roy Soc. London A, **218**, 345 (1953)

E6   M.L. Englefield: *Group Theory and the Coulomb Problem* (Whiley Interscience, New York, London, Sydney, Toronto 1972)

E7   A. Erdelyi (ed.): *Higher Transcendental Functions*, Vol. 2 Bateman Manuscript Project (McGraw–Hill, New York 1953; reprinted Krieger, Malabar, Florida 1981)

E8   A. Erdelyi (ed.): *Tables of Integral Transforms*, Vol. 2 Bateman Manuscript Project (McGraw–Hill, New York, 1954)

F1   E. Feldheim: Dokl. Acad. Nauk SSSR **31**, 528 (1941) [English transl.: Soviet Math. Dokl.

F1a   W. Feller: *An Introduction to Probability Theory and Its Applications*, Vol. 1 (John Wiley & Sons, New York, Chichester, Brisbane, Toronto 1979)

F2   G.F. Filippov, V.I. Ovcharenko, Yu.F. Smirnov: *Microscopic Theory of Collective Excitation of Atomic Nuclei* (Naukova Dumka, Kiev 1981) [in Russian]

F3   J. Fischer: Ann. Phys. **8**, 821 (1931)

F4   S. Flügge: *Practical Quantum Mechanics I* (Springer-Verlag, Berlin, Heidelberg, New York 1971)

F5   V.A. Fock: Zs. Physik, **98**, 145 (1935)

F6   V.A. Fock: *The Principles of Quantum Mechanics* (Nauka, Moscow 1976) [in Russian]

F7   B.A. Fomin, V.D. Efros: Yad. Fiz. **34**, 587 (1981) [English transl.: Sov. J. Nucl. Phys.

F8   M.D. Frank–Kamenetskii, A.V. Lukashin: Usp. Fiz. Nauk **116**, 193 [English transl.: Sov. Phys. Usp. (1975)

F9   W.R. Frazer et al. Phys. Rev. **175**, 2047 (1968)

F10   D.Z. Freedman, J.M. Wang: Phys. Rev. **160**, 1560 (1967)

G1   D.H. Galvan, S.K. Saha, W.C. Henneberger: Amer. J. Phys. **51**, 953 (1983)

G2   G. Gasper: J. Math. Anal. Appl. **42**, 438 (1973)

G3   G. Gasper, J. Math. Anal. Appl. **45**, 176 (1974)

G4   G. Gasper, M. Rahman: SIAM J. Math. Anal. **14**, 409 (1983)

G5   G. Gasper, M. Rahman: J. Math. Anal. Appl. **95**, 304 (1983)

G6   G. Gasper, M. Rahman: SIAM J. Math. Anal. **15**, 768 (1984)

G7   G. Gasper, M. Rahman: SIAM J. Math. Anal. **17**, 970 (1986)

G7a   G. Gasper, M. Rahman: Basic Hypergeometric Series (Cambridge Univ. Press, 1990)

G8   I.M. Gel'fand: *The Lectures on Linear Algebra* (Nauka, Moscow 1971) [in Russian]

G9   I.M. Gel'fand, M.L. Cetlin: Dokl. Acad. Nauk SSSR **71**, 825 [English transl.: Soviet Math. Dokl.

G10   I.M. Gel'fand, M.L. Celtin: Dokl. Acad. Nauk SSSR **71**, 1017 [English transl.: Soviet Math. Dokl.

G11   I.M. Gel'fand, M.I. Graev: Izv. Acad. Nauk SSSR, Math. **29**, 1329 (1965)

G12   I.M. Gel'fand, M.I. Graev, N.Ya. Vilenkin: *Integral Geometry and Representation Theory*, Generalized Functions, Vol. 5 (Acad. Press, New York 1966)

G13   I.M. Gel'fand, R.A. Minlos, Z.Ya. Sapiro: *Representations of the Rotation Group and of the Lorentz Group and Their Applications* (MacMillan, New York, 1963)

G14   J. Geronimus: Ann. Math. **31**, 681 (1930)

G15   J. Geronimus: *Orthogonal Polynomials*, Appendix to Russian Translation of [S38] (Fizmatgiz, Moscow 1961) [English transl.: Amer. Math. Soc. Trans. (2) **108**, 37–130 (1977)]

G16   W.M. Gibson, B.R. Pallard: *Symmetry Principles in Elementary Particle Physics* (Cambridge Univ. Press, Cambridge, London, New York, Melbourne 1976)

G17   S.K. Godunov, V.S. Ryaben'kiĭ: *Difference Schemes* (Nauka, Moscow 1973) [in Russian]

G18   M.J. Gottlieb: Amer. J. Math. **60**, 453 (1938)

G19   K.D. Granzow: J. Math. Phys. **4**, 897 (1963)

G20   K.D. Granzow: J. Math. Phys. **5**, 1474 (1964)

G21   D.P. Grechukhin: *Collective Quadrupole Excitations of Nuclei* (Experimental data and phenomenological models, MIPhI Lecture Notes, Moscow 1971) [in Russian]

G22   D.P. Grechukhin: "Critical Review of Phenomenological Collective Models of Nucleus"; Preprint No. 2080, Kurchatov Atom. Energy Inst., Moscow (1971) [in Russian]

G23   D.P. Grechukhin: Yad. Fiz. **19**, 1226 (1974) [English transl.: Sov. J. Nucl. Phys.

G24   H.E.H. Greenleaf: Ann. Math. Statist: **3**, 204 (1932)

H1   W. Hahn: Math. Nachr. **2**, 4 (1949)

H2   W. Hahn: Math. Nachr. **2**, 263 (1949)

H3   W. Hahn: Math. Nachr. **2** 340 (1949)

H4   W. Hahn: Math. Nachr. **3**, 257 (1950)

H5   W. Hahn: J. Math. Anal. Appl. **96**, 52 (1983)

H6   W. Hahn: "Über Orthogonalpolynome, die Linearen Functionalgleichungen Genügen", in *Polynômes orthogonaux et applications*, Lecture notes in Mathematics, Vol. 1171 (Springer, Berlin, Heidelberg 1985)

H7   M. Hamermesh: *Group Theory and Its Applications to Physical Problems* (Addison–Wesley, Reading, Mass. 1962)

H8   F. Harary: *Graph Theory* (Addison–Wesley, Reading, Mass. 1969)

H9   E. Heine: J. Reine Angew. Math. **32**, 210 (1846); **34**, 285 (1847)

H10   E. Heine: *Handbuch der Kugelfunctionen*, Bd. I (Druck und Verlag von G. Reimer, Berlin 1878) pp. 97-125, 273-285

I1   E. Inönü, E.P. Wigner: Proc. Nat. Acad. Sci. USA **39**, 510 (1953)

I2   M.E.H. Ismail: J. Math. Anal. Appl. **57**, 487 (1977)

I3   M.E.H. Ismail, J.A. Wilson: J. Approx. Theor. **36**, 43 (1982)

I4   M.E.H. Ismail, J. Letessier, D.R. Masson, G. Valent: Orthogonal Polynomials: Theory and Practice, ed. by Pol Nevai, 229–255, Nato ASI Series C, Vol. 294 (Kluwer Academic Publishers 1990)

J1   F.H. Jackson: Quart. J. Pure Appl. Math. **41**, 193 (1910)

J2   C.G.J. Jacobi: J. Reine Angew. Math. **32**, 197 (1846)

J3   S. Jang: J. Math. Phys. **9**, 397 (1968)

J4   C. Jordan: Proc. London Math. Soc. (2) **20**, 297 (1922)

J5   B.R. Judd: *Topics in Atomic Theory* (University of Canterbury, New Zealand

K1   V.G. Kadyshevsky, R.M. Mir-Kasimov, N.B. Skachkov: Yad. Fiz. **9**, 212 (1969) English transl.: Sov. J. Nucl. Phys.

K2   V.G. Kadyshevsky, R.M. Mir-Kasimov, M. Freeman: Yad. Fiz. **9**, 646 (1969 English transl.: Sov. J. Nucl. Phys.

K3   E.G. Kalnins, W. Miller: "Symmetry Technique for q–Series: Askey–Wilson Polynomials"; Report 86–113, University of Minnesota, Minneapolis (1986)

K4   V.P. Karasev, L.A. Shelepin: Teor. i Math. Fiz. **17**, 67 (1973)

K5   V.P. Karasev, L.A. Shelepin: "Difference Methods and their Role in the Theory of Coherent Phenomena", in *Coherent Cooperative Phenomena*, (Proc. PhIAN, Vol. 87), ed. by N.G. Basov (Nauka, Moscow 1976) pp. 55–91 [in Russian]

K5a   S. Karlin: *A First Cource in Stochastic Processes* (Academic Press, New York and London 1968)

K6   S. Karlin, J. McGregor: Trans. Amer. Math. Soc. **85**, 489 (1957)

K7   S. Karlin, J. McGregor: J. Math. Mech. **7**, 643 (1958)

K8   S. Karlin, J. McGregor: J. Math. Anal. Appl. **1**, 163 (1960)

K9   S. Karlin, J. McGregor: Scriptam Math. **26**, 33 (1961)

K10   M.S. Kildyushov: Yad. Fiz. **15**, 197; **16**, 217 (1972) English transl.: Sov. J. Nucl. Phys.

K11   M.S. Kildyushov, G.I. Kuznetsov: Yad. Fiz. **17**, 1330 (1973) English transl.: Sov. J. Nucl. Phys.

K12   I.K. Kirichenko, Yu.P. Stepanovskii: Yad. Fiz. **20**, 554 (1974)

K13   A.A. Kirillov: *Elements of the Theory of Representations*, 2nd ed. (Nauka, Moscow 1978) English transl.: Springer (1976)

K14   F. Klein: *Vorlesungen über die Hypergeometrische Funktion.* Die Grundlehren der Math. Wissenschaften, Bd. 34 (Springer–Verlag, Berlin 1933)

K15   N.P. Klepikov: Yad. Fiz. **19**, 462 (1974) [English transl. Sov. J. Nucl. Phys.

K16   D. Kleppner, M. Littman, M. Zimmermann: "Rydberg Atoms in High Fields", in *Rydberg States of Atoms and Molecules*, ed. by R. Stebbings, F. Dunning (Cambridge Univ. Press, Cambridge, London, New York, New Rochell, Melbourne, Sydney)

K17   V.A. Knyr, P.P. Pipiraite, Yu.F. Smirnov: Yad. Fiz. **22**, 1063 (1975) [English transl.: Sov. J. Nucl. Phys. **22**, 554 (1975)]

K18   G.A. Kobatjanskii, V.I. Levenstein: Prob. Per. Inf. **14**, 3 (1978) [in Russian]

K19   I.V. Komarov, T.P. Grozdanov, R.K. Janev: J. Phys. **B13**, L573 (1980)

K20   T.H. Koornwinder: Nieuw arch. wisk. **29**, 140 (1981)

K21   T.H. Koornwinder: SIAM J. Math. Anal. **13**, 1011 (1982)

K22   T.H. Koornwinder: "Special Orthogonal Polynomial Systems Mapped onto Each Other by the Fourier-Jacobi Transform"; Report PM–R8501, Centre for Math. and Comp. Science, Amsterdam (1985)

K23   T.H. Koornwinder: "A Group Theoretic Interpretation of Wilson Polynomials"; Report PM–R8504, Centre for Math. and Comp. Science, Amsterdam (1985)

K24   T.H. Koornwinder: "Group Theoretic Interpretations of Askey's Scheme of Hypergeometric Orthogonal Polynomials"; Report PM–R8703, Centre for Math. and Comp. Science, Amsterdam (1987)

K25   G.Ya. Korenman: Vestnik Moscow State Univ., Fiz. Astr. **17**, 375 (1976) [in Russian]

K26   V.S. Koroluk, N.I. Portenko, A.V. Skorohod, A.F. Turbin: *Handbook on Probability Theory and Mathematical Statistics* (Nauka, Moscow 1985) [in Russian]

K27   A.I. Kostrikin, Yu.A. Manin: *Linear Algebra and Geometry* (Moscow State Univ. Publ., Moscow 1980) [in Russian]

K28   A. Kratzer, W. Franz: *Transzendente Funktionen* (Akademische Verlagsgesellschaft, Leipzig 1960)

K29   M. Krawtchouk: Compt. Rend. Acad. Sci. Paris. **189**, 620 (1929)

K30   G.I. Kuznetsov: Teor. i Math. Fix. **18**, 367 (1974)

K31   G.I. Kuznetsov: Yad. Fiz. **24**, 235 (1976) [English transl.: Sov. J. Nucl. Phys.

K32   G.I. Kuznetsov: Intern. J. Theor. Phys. **16**, 345 (1977)

K33   G.I. Kuznetsov: Yad. Fiz. **30**, 1158 (1979) [English transl.: Sov. J. Nucl. Phys.

K34   G.I. Kuznetsov: Yad. Fiz. **32**, 554 (1980) [English transl.: Sov. J. Nucl. Phys.

K35   G.I. Kuznetsov, M.A. Liberman, A.A. Makarov, Ya.A. Smorodinskiĭ: Yad. Fiz. **10**, 641 (1969)
[English transl.: Sov. J. Nucl. Phys.

K36   G.I. Kuznetsov, Ya.A. Smorodinskiĭ: Pis'ma v ZETF **22**, 378 (1975) [English transl.: JETP
Lett. **22**, 179 (1975)]

K37   G.I. Kuznetsov, Ya.A. Smorodinskiĭ: Yad. Fiz. **21**, 1135 (1975) [English transl.: Sov. J. Nucl.
Phys.

K38   G.I. Kuznetsov, Ya.A. Smorodinskiĭ: "The T-Coefficients and $6j$ Symbols"; Preprint No. 2659,
Kurchatov Ins. Atom. Energy, Moscow (1976)

K39   G.I. Kuznetsov, Ya.A. Smorodinskiĭ: Yad. Fiz. **25**, 447 (1977)

K40   G.I. Kuznetsov, Ya.A. Smorodinskiĭ: Yad. Fiz. **29**, 286 (1979)

L1   G. Labelle: Ann. sc. math. Quebec **6**, 163 (1982)

L2   C.D. Lai: Mathematical Chronicle **6**, 6 (1977)

L3   E.N. Lambina: Dokl. Akad. Nauk Bel. SSR **14**, 786 (1970)

L4   L.D. Landau, E.M. Lifshitz: *Quantum Mechanics* (Pergamon Press, Oxford 1977)

L5   I.L. Lanzewizky: Dokl. Acad. Nauk SSR **31**, 199 (1941) [English transl.: Soviet Math. Dokl.

L6   V.I. Levenshtein: Prob. Kib. **40**, 43 (1983) [in Russian]

L7   P.A. Lee: SIAM J. Appl. Math. **19**, 266 (1970)

L8   M. Leon, H. Bethe: Phys. Rev. **127**, 636 (1962)

L9   D. Leonard: SIAM J. Math. Anal. **13**, 656 (1982)

L10   P. Lesky: Monat. Math. **65**, 1 (1961); **66**, 203, 431 (1962)

L11   P. Lesky: Math. Leitschr. **78**, 29 (1962)

L12   P. Lesky: Monat. Math. **98**, 277 (1984)

L13   R.J. Levit: Bull. Amer. Math. Soc. **57**, 130 (1951)

L14   R.J. Levit: SIAM Review **9**, 191 (1967)

L15   K.L. Lezuo: J. Math. Phys. **8**, 1163 (1967)

L16   M.A. Liberman, G.I. Kuznetsov, Ya.A. Smorodinskiĭ: Fiz. Elem. Chastits At. Yadra 2, 105
(1971) English transl.: Sov. J. Part. Nucl. **2**, 70 (1971)

L17   M.A. Liberman, A.A. Makarov: Yad. Fiz. **9**, 1314 (1969)

L18   A. Lindner: J. Phys. A.: Math. Gen. **18**, 3071 (1985)

L19   A. Lindner: *Drehimpulse in der Quantenmechanik* (B.G. Teubner, Stuttgart 1984)

L20   E.P. Lipatov: *Graph Theory and Its Applications* (Znanie, Ser.: Mat. Kib. No. 2, Moscow
1986) [in Russian]

L21   H.M. Lipinski, D.R. Snider: Phys. Rev. **176**, 2054 (1968)

L22   G. Ljubarskii: *The Theory of Groups and Its Application to Physics* (Pergamon Press, Oxford
1960)

M1   S.D. Majumdar, D. Basu: J. Phys. A.: Math. Nucl. Gen. **7**, 787 (1974)

M2   A.A. Makarov, G.I. Shepelev: Yad. Fiz. **12**, 1092 (1970) English transl.: Sov. J. Nucl. Phys.

M3   I.A. Malkin, V.I. Man'ko: *Dynamical Symmetries and Coherent States of Quantum Systems*
(Nauka, Moscow 1979) [in Russian]

M4   O.I. Marichev, V.K. Tuan: Dokl. Acad. Nauk Bel. SSR **26**, 488 (1982) [in Russian]

M5   A.A. Markov: *On Some Applications of Algebraic Continued Fractions* (Thesis, St. Petersburg
1884) [in Russian]

M6   E.P. McCoy, J.G. Ross: Aust. J. Chem. **15**, 457 (1962)

M7   A. Meckler: Nuovo Cim. Suppl. **12**, 1 (1959)

M8   J. Meixner: J. London Math. Soc. **9**, 6 (1934)

M9   J. Meixner: Math. Zeit. **44**, 531 (1938)

M9a   J. Meixner: Deutsche Math. **6**, 341 (1942)

M10   A. Messiah: *Quantum Mechanics I, II* (North–Holland, Amsterdam 1961, 1962)

M11   A.B. Migdal, V.P. Krainov: *Approximate Methods of Quantum Mechanics* (Nauka, Moscow
1966) [in Russian]

M12   W. Miller: J. Math. Anal. Appl. **28**, 383 (1969)

M13 W. Miller: *Symmetry and Separation of Variables*, Encyclopedia of Mathematics and Its Application, Vol. 4 (Addison–Wesley, Reading, London etc. 1977)

M14 W. Miller: SIAM J. Math. Anal. **18**, 1221 (1987)

M15 D.S. Moak: Aequat. Math. **20**, 278 (1980)

M16 P.A.P. Moran: Proc. Cambr. Phil. Soc. **54**, 60 (1958)

M17 M. Moshinsky: *The Harmonic Oscillator in Modern Physics: from Atoms to Quarks* (Gordon and Breach, New–York, London, Paris, 1969)

M18 M. Moshinsky, C. Quesne: J. Math. Phys. **11**, 1631 (1970)

N1 M.A. Naimark: *Linear Representations of the Lorentz Group* MacMillan, New York, 1964)

N2 B. Nassrallah, M. Rahman: SIAM J. Math. Anal. **16**, 186 (1985)

N2a P. Nevai (ed.): Orthogonal Polynomials: Theory and Practice, 257–292, Nato ASI Series C, Vol. 294 (Kluwer Academic Publishers 1990)

N3 A.F. Nikiforov: *Computational Methods in Some Applied Quantum Mechanics Problems* (Moscow State Univ. Publ., Moscow 1981) [in Russian]

N4 A.F. Nikiforov, V.G. Novikov, V.B. Uvarov: Voprosy Atomnoi Nauki i Tekhniki **4** (6), 16 (1979) [in Russian]

N5 A. Nikiforov, V. Ouvarov: *Fonctions spéciales de la physique mathématique*(Mir, Moscow 1983)

N5a A. Nikiforov, V. Uvarov: Elements of the theory of special functions (Nauka, Moscow, 1974)

N6 A.F. Nikiforov, S.K. Suslov: "The Hahn Polynomials and Their Connection with Clebsh-Gordan Coefficients for SU(2) Group"; Preprint No. 83, Keldysh Inst. Appl. Math., Moscow (1982) [in Russian]

N7 A.F. Nikiforov, S.K. Suslov: "Systems of Classical Orthogonal Polynomials of a Discrete Variable on Nonuniform Lattices"; Preprint No. 8, Keldysh Inst. Appl. Math., Moscow (1985) [in Russian]

N8 A.F. Nikiforov, S.K. Suslov: Lett. Math. Phys. **11**, 27 (1986)

N9 A.F. Nikiforov, S.K. Suslov, V.B. Uvarov: "The Racah Polynomials and the Dual Hahn Polynomials as a Generalization of Classical Orthogonal Polynomials of a Discrete Variable"; Preprint No. 165, Keldysh Inst. Appl. Math., Moscow (1982) [in Russian]

N10 A.F. Nikiforov, S.K. Suslov, V.B. Uvarov: "Asymptotic Formulas of the Second Order of Accuracy for Clebsh-Gordan Coefficients and Wigner $6j$ Symbols"; Preprint No. 64, Keldysh Inst. Appl. Math., Moscow (1983) [in Russian]

N11 A.F. Nikiforov, S.K. Suslov, V.B. Uvarov: "Construction of Partial Solutions for the Difference Equation of Hypergeometric Type"; Preprint No. 142, Keldysh Inst. Appl. Math., Moscow (1984) [in Russian]

N12 A.F. Nikiforov, S.K. Suslov, V.B. Uvarov: Funktsional Anal. i Prilozhen. **19**, 22 (1985) [English transl.: Functional Anal. Appl. **19**, 182 (1985)]

N13 A.F. Nikiforov, S.K. Suslov, V.B. Uvarov: "Classical Orthogonal Polynomials of a Discrete Variable in the Theory of Group Representation", in *Group Theoretical Methods in Physics*", Vol. 1, ed. by M.A. Markov et al. (Harwood Academic Publishers, Chur, London, Paris, New York 1985) pp. 87–96

N14 A.F. Nikiforov, S.K. Suslov, V.B. Uvarov: *Classical Orthogonal Polynomials of a Discrete Variable* (Nauka, Moscow 1985) [in Russian]

N15 A.F. Nikiforov, S.K. Suslov, V.B. Uvarov: Dokl. Acad. Nauk SSSR **291**, 1056 (1986) [English transl.: Soviet Math. Dokl. **34**, 576 (1987)]

N16 A.F. Nikiforov, V.B. Uvarov: "Theory of Special Functions"; Preprint No. 51, Keldysh Inst. Appl. Math., Moscow (1974) [in Russian]

N17 A.F. Nikiforov, V.B. Uvarov: "Classical Orthogonal Polynomials of a Discrete Variable on Nonuniform Lattices"; Preprint No. 17, Keldysh Institute Appl. Math., Moscow (1983) [in Russian]

N18 A.F. Nikiforov, V.B. Uvarov: *Special Functions of Mathematical Physics* (Nauka, Moscow, 1984; Birkhauser Verlag, Basel, Boston 1988)

N19 A.F. Nikiforov, V.B. Uvarov: "Fundamentals of the Theory of Classical Orthogonal Polynomials of a Discrete Variable"; Preprint No. 56, Keldysh Inst. Appl. Math., Moscow (1987) [in Russian]

N20 A.F. Nikiforov, V.B. Uvarov: "Spherical Harmonics Orthogonal on a Discrete Set of Points"; Preprint No. 174, Keldysh Inst. Appl. Math., Moscow (1987) [in Russian]

N21 A.F. Nikiforov, V.B. Uvarov: "Construstion of q-Analogues of Classical Orthogonal Polynomials of a Discrete Variable on Nonuniform Lattices"; Preprint No. 179, Keldysh Inst. Appl. Math., Moscow (1987) [in Russian]

N22 A.F. Nikiforov, V.B. Uvarov: "Classical orthogonal polynomials of a discrete variable and their representation in the form of generalized q-hypergeometric series; Preprint No. 98, Keldysk. Inst. Appl. Math, Moscow (1989) [in Russian]

P1 D. Park: Zs. Physik **159**, 155 (1960)

P2 S. Pasternack: Proc. Nat. Acad. Sci. USA **23**, 71 (1937)

P3 S. Pasternack: Phil. Mag. (7) **28**, 209

P4 A.M. Perelomov, V.S. Popov: Zh. Eksp. Teor. Fiz. **50**, 179 (1966) [English transl.: Sov. Phys. JETP **23**, 118 (1966)]

P5 A.M. Perelomov, V.S. Popov: Zh. Eksp. Teor. Fiz. **54**, 1799 (1968) [English transl.: Sov. Phys. JETP **27**, 967 (1968)]

P6 M. Perlstadt: SIAM J. Alg. Disc. Meth. **4**, 94 (1983)

P7 M. Perlstadt: SIAM J. Math. Anal. **15**, 1043 (1984)

P8 Z. Pluhar, Yu.F. Smirnov, V.N. Tolstoy: "A Novel Approach to the SU(3) Symmetry Calculus"; Preprint Karlov Univ., Prague (1981)

P9 G.S. Pogosyan, V.M. Ter-Antonyan:: "Connection between Spherical and Parabolic Coulomb Wave Functions in Continuous Spectrum"; Preprint No. P2–80–318, Joint Institute for Nuclear Research, Dubna (1980) [in Russian]

P10 F. Pollaczek: Compt. Rend. de l'Acad. Sc. Paris **230**, 1563 (1950)

P11 V.E. Prudnikov: *P.L. Chebyshev* (Gosud. Uchebno Pedag. Izdat. Minist. Prosv. RSFSR, Moscow 1950) [in Russian]

R1 G. Racah: Phys. Rev. **61**, 186 (1941); **62**, 438 (1942); **63**, 367 (1943)

R2 M. Rahman: SIAM J. Math. Anal. **7**, 414 (1976); **9**, 891 (1978)

R3 M. Rahman: Canad. J. Math. **30**, 133 (1978)

R4 M. Rahman: SIAM J. Math. Anal. **10**, 438 (1979)

R5 M. Rahman: Canad. J. Math. **32**, 1501 (1980)

R6 M. Rahman: Canad. J. Math. **33**, 961 (1981)

R7 M. Rahman: Utilitas Mathematica **20**, 261 (1981)

R8 M. Rahman: SIAM J. Math. Anal. **12**, 145 (1981)

R9 M. Rahman: Trans. Amer. Math. Soc. **273**, 483 (1982)

R10 M. Rahman: Proc. Amer. Math. Soc. **92**, 413 (1984)

R11 M. Rahman: J. Math. Anal. Appl. **118**, 309 (1986)

R12 M. Rahman: SIAM J. Math. Anal. **17**, 1267, 1280 (1986)

R13 M. Rahman: A. Verma: SIAM J. Math. Anal. **17**, 1461 (1961)

R14 M. Rahman: A. Verma: Rocky Mount. J. Math. **17**, 371 (1987)

R15 E.D. Rainville: *Special Functions* (MacMillan, New York 1960)

R16 K.S. Rao: J. Math. Phys. **26**, 2260 (1985)

R17 K.K. Rebane: *Elementary Theory of Vibrational Spectra of Impurity Centers of Crystals* (Nauka, Moscow 1968) [in Russian]

R18 T. Regge: Nuovo Cim. **10**, 544 (1959)

R19 T. Regge: Nuovo Cim. **11**, 116 (1959)

R20 V.S. Ryaben'kii, I.L. Sofronov: "Difference Spherical Functions"; Preprint No. 75, Keldysh Inst. Appl. Math., Moscow 1983

R21 L.J. Rogers: Proc. London Math. Soc. **24**, 171, 337 (1893)

R22 L.J. Rogers: Proc. London Math. Soc. **25**, 318 (1894)

R23 L.J. Rogers: Proc. London Math. Soc. **26**, 15 (1895)

R24 V. Rojansky: Phys. Rev. **33**, 1 (1929)

R25 A.V. Rozenblum: Dokl. Acad. Nauk Bel. SSR **19**, 297 (1975) [in Russian]

R26 V.B. Ryvkin: Dokl. Acad. Nauk Bel. SSR **3**, 183 (1959) [in Russian]

S1 A.A. Samarskii: *Theory of Difference Schemes* (Nauka, Moscow 1977) [in Russian]

S2 W. Schempp: *Harmonic Analysis on the Heisenberg Nilpotent Lie Group with Applications to Signal Theory* (Longman, New York 1986)

S3 C.K.E. Schneider, R. Wilson: J. Math. Phys. **20**, 2380 (1979)

S4 D.B. Sears: Proc. London Math. Soc. (2) **52**, 158, 467 (1951)

S5 E.F. Sheka: Usp. Fiz. Nauk **104**, 593 (1971)

S6 L.A. Shelepin: "Calculation of Clebsch-Gordon Coefficients and Its Physical Applications", in: *Group Theoretical Methods in Physics* (Proc. PIAN, Vol. 70) ed. by D.V. Skobeltsyn (Nauka, Moscow 1973) pp. 3–119 [in Russian]

S7 K. Shyaml: Tamkang J. Math. **15**, 149 (1984)

S8 L.J. Slater: *Generalized Hypergeometric Functions* (at the University Press, Cambridge 1966)

S9 Yu.F. Smirnov, K.V. Shitikova: Fiz. Elem. Chastits At. Yadra **8**, 847 (1977) [English transl.: Sov. J. Part. Nuclei **8**, 344 (1977)]

S10 Yu.F. Smirnov, A.P. Shustov: Yad. Fiz. **34**, 626 (1981) [English transl.: Sov. J. Nucl. Phys.

S11 Yu.F. Smirnov, S.K. Suslov, A.M. Shirokov: J. Phys. A. Math. Gen. **17**, 2157 (1984)

S12 Ya.A. Smorodinskiĭ: Izv. Vuzov, Radiofizika **19**, 932 (1976) [in Russian]

S13 Ya.A. Smorodinskiĭ, M. Huzar: Teor. i Math. Fiz. **4**, 328 (1970)

S14 Ya.A. Smorodinskiĭ, L.A. Shelepin: Usp. Fiz. Nauk **106**, 3 (1972)

S15 Ya.A. Smorodinskiĭ, G.I. Shepelev: Yad. Fiz. **13**, 441 (1971) [English transl.: Sov. J. Nucl. Phys.

S16 Ya.A. Smorodinskiĭ, S.K. Suslov: Yad. Fiz. **35**, 192 (1982) [Sov. J. Nucl. Phys. **35**, 108 (1982)]

S17 Ya.A. Smorodinskiĭ, S.K. Suslov: Yad. Fiz. **36**, 1066 (1982) [Sov. J. Nucl. Phys. **36**, 623 (1982)]

S18 I.I. Sobel'man: *Introduction to the Theory of Atomic Spectra* (Nauka, Moscow 1977) [in Russian]

S19 D. Stanton: Not. Amer. Math. Soc. **24**, A86 (1977)

S20 D. Stanton: SIAM J. Math. Anal. **11**, 100 (1980)

S21 D. Stanton: Amer. J. Math. **102**, 625 (1980)

S22 D. Stanton: J. Comb. Theor. A**30**, 276 (1981)

S23 D. Stanton: "Orthogonal Polynomials and Chevalley Groups", in *Special Functions: Group Theoretical Aspects and Applications*, ed. by R. Askey, T. Koornwinder, W. Schempp (D. Reidel Publishing Company, Dordrecht 1984) pp. 87–128

S24 T.J. Stieltjes: "Recherches sur les fractions continues" (1894–1895), in Oeuvres, T.2 (Noordhoff, Groningen 1918) pp. 398–566

S25 A.P. Stone: Proc. Cambr. Phil. Soc. **52**, 424 (1956)

S26 S.K. Suslov: Yad. Fiz. **36** 1063 (1982) [English transl.: Sov. J. Nucl. Phys. **36**, 621 (1982)]

S27 S.K. Suslov: Yad. Fiz. **37**, 795 (1983) [English transl.: Sov. J. Nucl. Phys. **37**, 472 (1983)]

S28 S.K. Suslov: Yad. Fiz. **38**, 1102 (1983) [English transl. Sov. J. Nucl. Phys. **38**, 662 (1983)]

S29 S.K. Suslov: Yad. Fiz. **38**, 1367 (1983) [English transl.: Sov. J. Nucl. Phys. **38**, 829 (1983)]

S30 S.K. Suslov: "The Hahn and Meixner Polynomials of an Imaginary Argument"; Preprint No. 3915/1, Kurchatov Institute of Atomic Energy, Moscow (1984) [in Russian]

S31 S.K. Suslov: Yad. Fiz. **40**, 126 (1984) [English transl.: Sov. J. Nucl. Phys. **40**, 79 (1984)]

S32 S.K. Suslov: Lett. Math. Phys. **14**, 77 (1987)

S33 S.K. Suslov: "Construction of Partial Solutions for the Difference Equation of Hypergeometric Type on Nonuniform Lattices"; Preprint No. 4384/1, Kurchatov Inst. Atom. Energy, Moscow (1987) [in Russian]

S34   S.K. Suslov: "To the Theory of Difference Analogs of Hypergeometric Functions on Uniform Nets"; Preprint No. 4445/1, Kurchatov Inst. Atom Energy, Moscow (1987) [in Russian]

S35   D.T. Sviridov, Yu.F. Smirnov: *Theory of Optical Spectra of Ions of Transition Metals* (Nauka, Moscow 1977) [in Russian]

S36   M.N.S. Swamy, K. Thulasiraman: *Graphs, Networks, and Algorithms* (A Wiley Interscience Publication, New York, Chichester, Brisbane, Toronto 1981)

S37   G. Szegö: Preuss. Akad. Wiss. Phys. Math. **K129**, 242 (1926)

S38   G. Szegö: *Orthogonal Polynomials*, Amer. Math. Soc. Colloq. Pubb. **23**, fourth edition (Amer. Math. Soc., Providence, Phode Island 1975)

T1   C.B. Tarter: J. Math. Phys. **11**, 3192 (1970)

T2   P.L. Tchebychef: "Sur une nouvelle série" (1958), in Oeuvres, T.I. (Chelsea, New York 1961) pp. 381–384

T3   P.L. Tchebychef: "Sur l'interpolation des valeurs équidistantes" (1875), in *Œuvres*, T.II (Chelsea, New York 1961) pp. 219–242

T4   J. Thomae: Zeit. Math. Phys. **14**, 349 (1869); **16**, 146, 428 (1871)

T5   J. Thomae: J. Reine Angew. Math. **87**, 26 (1879)

T6   J. Thomae: Math. Ann. **2**, 427 (1870)

T7   J. Touchard: Canad. J. Math. **8**, 305 (1956)

V1   L.L. Vaksman, Ja.S. Soibelman: Funktsional Anal i Prilosken. 22, 3 (1988)

V2   G. Vander Berghe, H. De Meyer, J. Van der Jeugt: J. Math. Phys. **25**, 2585 (1984)

V3   J. Van der Jeugt, G. Vander Berghe, H. De Meyer: J. Phys. A. Math. Gen. **16**, 1377 (1983)

V4   B.L. Van der Waerden: *Die Gruppentheoretische Methode in der Quantenmechanik* (J.W. Edwards, Ann. Arbor 1944)

V5   D.A. Varshalovich, A.I. Moskalev, V.K. Khersonskii: *Quantum Theory of Angular Momentum* (Nauka , Leningrad 1975) [in Russian]

V6   Yi.A. Verdiyev: Yad. Fiz. **9**, 1310 (1969) [English transl. Sov. J. Nucl. Phys.

V7   N.Ya. Vilenkin: Dokl. Acad. Nauk SSSR **113**, 16 (1957)

V8   N.Ya. Vilenkin: Mat. Sbornik **68(110)**, 432 (1965)

V9   N.Ya. Vilenkin: *Special Functions and Theory of Group Representations*, Translations of Math. Monogr. **22** (Amer. Math. Soc. Providence, Rhode Island 1968)

V10   N.Ya. Vilenkin, G.I. Kuznetsov, Ya.A. Smorodinskiǐ: Yad. Fiz. **2**, 906 (1965) [English transl.: Sov. J. Nucl. Phys. **2**, 645 (1965)]

V11   N.Ya. Vilenkin, Ya.A. Smorodinskiǐ: Zh. Eksp. Teor. Fiz. **46**, 1793 (1964) [English transl.: Sov. Phys. JETP **19**, 1209 (1964)

W1   G.N. Watson: *A Treatise of the Theory of Bessel Functions* (1945)

W2   M. Weber, A. Erdelyi: Amer. Math. Month. **59**, 163 (1952)

W3   G. Wentzel: Zs. Physik **58**, 348 (1929)

W4   H. Weyl: *The Theory of Groups and Quantum Mechanics* (Dover, New York 1950)

W5   E.T. Whittaker, G.N. Watson: *A Course of Modern Analysis*, 4th edn. (Univ. Press, Cambridge 1927)

W6   S. Wigert: Arkiv för Matem., Astron. och Fysik **17**, 18 (1922–1923)

W7   E.P. Wigner: *Group Theory and Its Applications to the Quantum Mechanics of Atomic Spectra* (Academic Press, New York 1959)

W8   J.A. Wilson (1978): Hypergeometric series, recurrence relations and some new orthogonal polynomials, Thesis, Univ. of Wisconsin, Madison

W8a   J.A. Wilson: SIAM J. Math. Anal. **11**, 690 (1980)

W9   M.W. Wilson: SIAM J. Math. Anal. **1**, 131 (1970)

W10   M. Wyman, L. Moser: Canad. J. Math. **8**, 321 (1956)

Y1   A.P. Yutsis, A.A. Banzaitis: *Theory of Angular Momentum in Quantum Mechanics* (Mokslas, Vilnyus 1977) [in Russian]

Y2   A.P. Yutsis, I.B. Levinson, V.V. Vanagas: *The Theory of Angular Momentum* (Office of Technical Services, U.S. Detp. Commerce, Washington, D.C. 1962)

Z1   D. Zwanziger: J. Math. Phys. **8**, 1858 (1967)

Z2   D. Zwanziger: Phys. Rev. **176**, 1480 (1968)

Subject Index